Quantitative MRI in Cancer

IMAGING IN MEDICAL DIAGNOSIS AND THERAPY

William R. Hendee, Series Editor

Quality and Safety in Radiotherapy
Todd Pawlicki, Peter B. Dunscombe, Arno J. Mundt, and
Pierre Scalliet, Editors
ISBN: 978-1-4398-0436-0

Adaptive Radiation Therapy
X. Allen Li, Editor
ISBN: 978-1-4398-1634-9

Quantitative MRI in Cancer
Thomas E. Yankeelov, David R. Pickens, and
Ronald R. Price, Editors
ISBN: 978-1-4398-2057-5

Forthcoming titles in the series

Image-Guided Radiation Therapy
Daniel J. Bourland, Editor
ISBN: 978-1-4398-0273-1

Informatics in Radiation Oncology
Bruce H. Curran and George Starkschall, Editors
ISBN: 978-1-4398-2582-2

Adaptive Motion Compensation in Radiotherapy
Martin Murphy, Editor
ISBN: 978-1-4398-2193-0

Image Processing in Radiation Therapy
Kristy Kay Brock, Editor
ISBN: 978-1-4398-3017-8

Proton and Carbon Ion Therapy
Charlie C.-M. Ma and Tony Lomax, Editors
ISBN: 978-1-4398-1607-3

Monte Carlo Techniques in Radiation Therapy
Jeffrey V. Siebers, Iwan Kawrakow, and
David W. O. Rogers, Editors
ISBN: 978-1-4398-1875-6

Informatics in Medical Imaging
George C. Kagadis and Steve G. Langer, Editors
ISBN: 978-1-4398-3124-3

Informatics in Medical Imaging
George C. Kagadis and Steve G. Langer, Editors
ISBN: 978-1-4398-3124-3

Stereotactic Radiosurgery and Radiotherapy
Stanley H. Benedict, Brian D. Kavanagh, and
David J. Schlesinger, Editors
ISBN: 978-1-4398-4197-6

Cone Beam Computed Tomography
Chris C. Shaw, Editor
ISBN: 978-1-4398-4626-1

Handbook of Brachytherapy
Jack Venselaar, Dimos Baltas, Peter J. Hoskin, and
Ali Soleimani-Meigooni, Editors
ISBN: 978-1-4398-4498-4

Targeted Molecular Imaging
Michael J. Welch and William C. Eckelman, Editors
ISBN: 978-1-4398-4195-0

IMAGING IN MEDICAL DIAGNOSIS AND THERAPY

William R. Hendee, Series Editor

Quantitative MRI in Cancer

Edited by
Thomas E. Yankeelov
David R. Pickens
Ronald R. Price

CRC Press
Taylor & Francis Group
Boca Raton   London   New York

CRC Press is an imprint of the
Taylor & Francis Group, an **informa** business

A TAYLOR & FRANCIS BOOK

CRC Press
Taylor & Francis Group
6000 Broken Sound Parkway NW, Suite 300
Boca Raton, FL 33487-2742

First issued in paperback 2020

© 2012 by Taylor & Francis Group, LLC
CRC Press is an imprint of Taylor & Francis Group, an Informa business

No claim to original U.S. Government works

ISBN-13: 978-0-367-57687-5 (pbk)
ISBN-13: 978-1-4398-2057-5 (hbk)

Library of Congress Cataloging-in-Publication Data

Quantitative MRI in cancer / editors, Thomas Yankeelov, Ronald R. Price, and David R. Pickens.
    p. ; cm. -- (Imaging in medical diagnosis and therapy)
   Includes bibliographical references and index.
   ISBN 978-1-4398-2057-5 (hardback : alk. paper)
   1. Cancer--Magnetic resonance imaging. 2. Magnetic resonance imaging. I. Yankeelov, Thomas. II. Price, Ronald R. III. Pickens, David R. IV. Series: Imaging in medical diagnosis and therapy.
   [DNLM: 1. Magnetic Resonance Imaging--methods. 2. Neoplasms--diagnosis. 3. Magnetic Resonance Imaging--trends. QZ 241]

RC270.3.M33Q83 2011
616.99'40757--dc23

2011018372

Visit the Taylor & Francis Web site at
http://www.taylorandfrancis.com

and the CRC Press Web site at
http://www.crcpress.com

To Margie: Thank you for your love and devotion, your wisdom and humor, your beauty and elegance, and for our Lexie girl. You make me the luckiest and happiest guy on the planet.

—T.E.Y.

To Allison and our boys, Christopher and Richard: Thank you for your support and love, thank you for making life interesting and fun, and thank you for your patience with my many projects.

—D.R.P.

To the most important people in my life, Donna and Amanda, and thanks to all the students who have helped me challenge my thoughts.

—R.R.P.

Contents

PART I Physical Basis of Magnetic Resonance Imaging

PART II Characterizing Tissue Properties with Endogenous Contrast Mechanisms

PART III   Characterizing Tissue Properties with Exogenous Contrast Agents

PART IV   Image Processing in Cancer

PART V   Emerging Trends

Series Preface

Since their inception over a century ago advances in the science and technology of medical imaging and radiation therapy are more profound and rapid than ever before. Further, the disciplines are increasingly cross-linked as imaging methods become more widely used to plan, guide, monitor, and assess treatments in radiation therapy. Today, the technologies of medical imaging and radiation therapy are so complex and computer-driven that it is difficult for the people (physicians and technologists) responsible for their clinical use to know exactly what is happening at the point of care, when a patient is being examined or treated. The people best equipped to understand the technologies and their applications are medical physicists, and these individuals are assuming greater responsibilities in the clinical arena to ensure that what is intended for the patient is actually delivered in a safe and effective manner.

The growing responsibilities of medical physicists in the clinical arenas of medical imaging and radiation therapy are not without their challenges, however. Most medical physicists are knowledgeable in either radiation therapy or medical imaging, and expert in one or a small number of areas within their disciplines. They sustain their expertise in these areas by reading scientific articles and attending scientific talks at meetings. In contrast, their responsibilities increasingly extend beyond their specific areas of expertise. To meet these responsibilities, medical physicists periodically must refresh their knowledge of advances in medical imaging or radiation therapy, and they must be prepared to function at the intersection of these two fields. How to accomplish these objectives is a challenge.

At the 2007 annual meeting of the American Association of Physicists in Medicine in Minneapolis, this challenge was the topic of conversation during a lunch hosted by Taylor & Francis Publishers and involving a group of senior medical physicists (Arthur L. Boyer, Joseph O. Deasy, C.-M. Charlie Ma, Todd A. Pawlicki, Ervin B. Podgorsak, Elke Reitzel, Anthony B. Wolbarst, and Ellen D. Yorke). The conclusion of this discussion was that a book series should be launched under the Taylor & Francis banner, with each volume in the series addressing a rapidly advancing area of medical imaging or radiation therapy of importance to medical physicists. The aim would be for each volume to provide medical physicists with the information needed to understand technologies driving a rapid advance and their applications to safe and effective delivery of patient care.

Each volume in the series is edited by one or more individuals with recognized expertise in the technological area encompassed by the book. The editors are responsible for selecting the authors of individual chapters and ensuring that the chapters are comprehensive and intelligible to someone without such expertise. The enthusiasm of volume editors and chapter authors has been gratifying and reinforces the conclusion of the Minneapolis luncheon that this series of books addresses a major need of medical physicists.

The series Imaging in Medical Diagnosis and Therapy would not have been possible without the encouragement and support of the series manager, Luna Han, of Taylor & Francis Publishers. The editors and authors, and most of all I, are indebted to her steady guidance of the entire project.

William Hendee
Series Editor
Rochester, MN

Preface

Recent years have seen a tremendous explosion in both the number and quantity of imaging techniques that can be applied in the quantitative characterization of cancer. These techniques have come from all fields of noninvasive, *in vivo* medical imaging, including magnetic resonance imaging (MRI), optical imaging, computed tomography (CT), positron emission tomography (PET), single photon emission computed tomography (SPECT), and ultrasound. Relevant techniques that have been developed report on, for example, tumor cellularity, vessel perfusion and permeability, hypoxic fractions, as well as cellular and molecular signatures. It is a reasonable hypothesis that characterization of tissue status can offer increased sensitivity and specificity when diagnosing and grading tumors. Furthermore, as many current anticancer drugs are designed to alter these specific tumor characteristics, imaging metrics designed to report on those phenomena promise to offer improved methods of planning treatment as well as assessing the response of tumors to treatment.

Yet, for all the progress that has taken place in biomedical imaging science, very few of these techniques are used clinically. For example, the assessment of treatment response in the clinical setting is still dominated by the RECIST criteria. RECIST (i.e., Response Evaluation Criteria in Solid Tumors) is based entirely on changes in the sum of the longest dimension of target lesions. In other words, the current standard of care in assessing treatment response is based on one-dimensional morphological and anatomical changes. Again, it is reasonable to expect that some subset of the emerging imaging metrics described in this text will provide more specific information on treatment response and the physiological and cellular status of the tumor. Furthermore, since anatomical and morphological changes often occur temporally downstream from the underlying physiological, cellular, and molecular changes, the emerging imaging metrics may be applied early in the course of treatment to determine if a given treatment is efficacious. Indeed, there are efforts to improve RECIST criteria by incorporating some of the techniques described in this text.

MRI is currently one of the more powerful techniques available in medical imaging; in the same imaging session, MRI can acquire high-resolution anatomical data with excellent soft tissue contrast, as well as data reporting on physiological and even cellular information, all of which is accomplished without ionizing radiation. As there is currently no single resource bringing together the available and emerging quantitative MRI techniques

for assessing cancer, we felt it was appropriate to assemble such a volume. There are, for example, many excellent texts on imaging particular cancers and assessing treatment response with routine clinical methods. Additionally, there are several recent volumes on particular imaging techniques as applied to cancer. However, there is no current text that offers both a biophysical and theoretical explanation of the most relevant MRI techniques as well as example applications of these techniques in cancer. Our hope is that this textbook will help bring knowledge of these methods to a larger audience that includes clinicians and translational scientists so that quantitative MRI techniques can be moved from bench to bedside. It is anticipated that some subset of the techniques presented in this text will provide significant impact on personalized medicine in the 21st century.

One of our goals in assembling this textbook is that it could function as a self-contained resource. After the opening chapter on basic cancer biology, the text is divided into five major parts: Physical Basis of Magnetic Resonance Imaging, Characterizing Tissue Properties with Endogenous Contrast Mechanisms, Characterizing Tissue Properties with Exogenous Contrast Agents, Image Processing in Cancer, and Emerging Trends. Part I, Physical Basis of Magnetic Resonance Imaging, presents the basics of the theoretical aspects of NMR/MRI physics (Chapter 2), and background on the hardware (Chapter 3) required to form MR images (Chapter 4). The next three parts of the text (Chapters 5–18) comprise the bulk of the book and are generally arranged as follows. Each chapter begins with a qualitative description of the technique (with multiple figures) in order to provide the reader with an intuitive understanding of the material. This introduction is followed by a more rigorous treatment of the material. It should be noted that this part within each chapter can be skipped without significantly disrupting the flow of the material. Within these 14 chapters are parts on Characterizing Tissue Properties with Endogenous (Part II) and Exogenous (Part III) Contrast Mechanisms, and common Image Processing (Part IV) techniques relevant for cancer.

The second part, Characterizing Tissue Properties with Endogenous Contrast Mechanisms, begins with chapters on quantitative relaxometry (Chapter 5) and arterial spin labeling (Chapter 6), before moving to the more recent developments seen in diffusion-weighted MRI (Chapter 7), magnetization transfer (Chapter 8), and MR spectroscopy (Chapter 9). The third part, Characterizing Tissue Properties with Exogenous

Contrast Mechanisms, begins with a chapter on the relevant contrast agents themselves (Chapters 10 and 11) before moving to three of the most common uses of them: dynamic contrast-enhanced MRI (Chapter 12), dynamic susceptibility contrast MRI (Chapter 13), and contrast-enhanced angiography (Chapter 14). The final chapters of Part III cover measuring tissue oxygen status in tumors (Chapter 15).

As the two previous parts considered imaging acquisition techniques, it is appropriate that Part IV covers elements of image processing relevant for cancer MRI. Chapter 16 covers morphologic and volumetric assessment of tumors, while Chapter 17 covers common image segmentation schemes. Chapter 18 introduces the reader to methods for image registration, including the very difficult problem of longitudinal registration, which can assist in the comparison of imaging data obtained on separate days throughout the course of treatment. Chapter 19 concludes Part IV with a discussion on methods to synthesize multiparameter data sets that can frequently arise not only in MRI studies but also in multimodality studies.

The book concludes with Part V, which presents introductory material on four emerging areas of MR characterization of cancer: therapy planning (Chapter 20), cellular and molecular imaging (Chapter 20), pH imaging (Chapter 21), and methods of hyperpolarized MRI (Chapter 22).

The goal of the textbook is to provide a one-volume overview of the theoretical and experimental essentials of MRI, and the theoretical and experimental techniques of state-of-the-art cancer MR. Thus, the primary audience of this textbook is any scientist or clinician interested in learning the use of MRI for the diagnosis and therapy monitoring of cancer. Individuals may use the text to learn the theoretical and experimental aspects of applying existing methods in the preclinical or clinical setting. They may also be motivated by the desire to refine the existing methods or develop entirely new techniques and therefore require knowledge of what is currently available in cancer MR. This potential audience includes both students and advanced practitioners of MRI and extends from basic scientists to physician scientists. Each of the post-introductory chapters includes self-contained parts on qualitative and quantitative techniques, before proceeding to preclinical and clinical applications, thereby providing relevant material for those looking to quickly apply the technique, as well as those looking to advance and refine the technique. Throughout the book, chapters include sufficient references to form a starting point for in-depth study of each topic, should the reader wish additional information. The secondary readership would be instructors who are designing a one-semester course in MRI of cancer. In this case, knowledge of Chapters 1–4 could be assumed and the bulk of the course time could be spent on Chapters 5–22. For a typical one-semester course consisting of approximately 14 weeks with three lectures per week (for a total of 42 lectures minus days for examinations, presentations, etc.), we envision two lectures for each chapter. Of course, the emphasis and timing could be changed to reflect local needs and interests.

MATLAB® is a registered trademark of The MathWorks, Inc. For product information, please contact:

The MathWorks, Inc.
3 Apple Hill Drive
Natick, MA 01760-2098 USA
Tel: 508-647-7000
Fax: 508-647-7001
E-mail: info@mathworks.com
Web: www.mathworks.com

Acknowledgments

I would like to thank John C. Gore, PhD, and Jeremy J. Kaye, MD, for providing an absolutely wonderful environment in which to work. The graduate students (Richard Baheza, MS, Mary E. Loveless, PhD, and Jennifer Whisenant, MS) and post-doctoral fellows (Lori Arlinghaus, PhD, Nkiruka Atuegwu, PhD, Jacob Fluckiger, PhD, Xia Li, PhD, and David Smith, PhD) with whom I have the privilege to work are truly the brains behind the operation. I would also like to thank the excellent staff of the Vanderbilt University Institute of Imaging Science, especially Nancy Hagans, BA, Jarrod True, MS, and Danny Colvin, PhD. Our editor at Taylor & Francis, Luna Han, was a tremendous help in assembling this text.—**T.E.Y.**

Over the years, the Department of Radiology and Radiological Sciences at Vanderbilt University Medical Center has been tremendously supportive. My previous chairmen have all had the vision that an academic radiology department needs not only physicians, but also medical physicists and others who advance the field in many different ways. My current chair, Jeremy J. Kaye, MD, continues to foster an environment where the missions of clinical service, research, and education can thrive, for which I am very grateful. My colleagues in the Vanderbilt University Institute of Imaging Science provide those of us who do mostly clinical medical physics with continuous stimulation by developing new imaging approaches that will someday be used in the clinic. Our editor at Taylor & Francis, Luna Han, has been exceptionally helpful in structuring and developing this book. Finally, the work of my coeditors, Tom Yankeelov, PhD, and Ron Price, PhD, has been essential in making this book a useful and detailed text for those interested in understanding the quantitative power of MRI applied to cancer.—**D.R.P.**

I would like to acknowledge Dr. Ronald Arildsen and Dr. Jeffry Creasy for their valuable comments and clinical reviews, Ric Andal for his assistance in identifying appropriate illustrations, and the Department of Radiology for supporting the work of this project.—**R.R.P.**

About the Editors

 **Thomas E. Yankeelov** is director of Cancer Imaging Research at the Vanderbilt University Institute of Imaging Science and is an associate professor of radiology and radiological sciences, physics and astronomy, biomedical engineering, and cancer biology at Vanderbilt University. He received an MA in applied mathematics and an MS in physics from Indiana University in 1998 and 2000, respectively. His doctorate is in biomedical engineering from SUNY at Stony Brook, where he completed his dissertation at Brookhaven National Laboratory in 2003.

 **David R. Pickens** is associate professor of radiology and radiological sciences at Vanderbilt University School of Medicine. He is a fellow of the American Association of Physicists in Medicine and holds an MS and PhD in mechanical engineering from Vanderbilt University.

 **Ronald R. Price** is professor of radiology and radiological sciences and physics and astronomy at Vanderbilt University. His PhD is in physics from Vanderbilt University, and he is a fellow of the Society for Magnetic Resonance Imaging and a fellow of the American Association of Physicists in Medicine.

Contributors

Ronald C. Arildsen
Vanderbilt University
Nashville, Tennessee

Lori R. Arlinghaus
Vanderbilt University
Nashville, Tennessee

Dmitri Artemov
The Johns Hopkins University
Baltimore, Maryland

Deniz Aykac
Oak Ridge National Laboratory
Oak Ridge, Tennessee

Zaver M. Bhujwalla
The Johns Hopkins University
Baltimore, Maryland

Jason R. Buck
Vanderbilt University
Nashville, Tennessee

Eduard Y. Chekmenev
Vanderbilt University
Nashville, Tennessee

Anthony Cmelak
Vanderbilt University
Nashville, Tennessee

Jeffrey L. Creasy
Vanderbilt University
Nashville, Tennessee

Bruce M. Damon
Vanderbilt University
Nashville, Tennessee

George X. Ding
Vanderbilt University
Nashville, Tennessee

Mark D. Does
Vanderbilt University
Nashville, Tennessee

Richard D. Dortch
Vanderbilt University
Nashville, Tennessee

Christopher P. Elder
Vanderbilt University
Nashville, Tennessee

Shaun S. Gleason
Oak Ridge National Laboratory
Oak Ridge, Tennessee

Daniel F. Gochberg
Vanderbilt University
Nashville, Tennessee

Matthew R. Hight
Vanderbilt University
Nashville, Tennessee

Martin Lepage
Université de Sherbrooke
Sherbrooke, Quebec, Canada

Mia Levy
Vanderbilt University
Nashville, Tennessee

Xia Li
Vanderbilt University
Nashville, Tennessee

Mary E. Loveless
Vanderbilt University
Nashville, Tennessee

H. Charles Manning
Vanderbilt University
Nashville, Tennessee

Michelle D. Martin
Susan G. Kommen Foundation
Houston, Texas

J. Oliver McIntyre
Vanderbilt University
Nashville, Tennessee

Kathryn M. McMillan
G.E. Healthcare
Waukesha, Wisconsin

Nilesh Mistry
University of Maryland
College Park, Maryland

Noriko Mori
The Johns Hopkins University
Baltimore, Maryland

Michael Nickels
Vanderbilt University
Nashville, Tennessee

Ken J. Niermann
Vanderbilt University
Nashville, Tennessee

Vincent C. Paquit
Oak Ridge National Laboratory
Oak Ridge, Tennessee

Marie-France Penet
The Johns Hopkins University
Baltimore, Maryland

Wellington Pham
Vanderbilt University
Nashville, Tennessee

David R. Pickens
Vanderbilt University
Nashville, Tennessee

Ronald R. Price
Vanderbilt University
Nashville, Tennessee

C. Chad Quarles
Vanderbilt University
Nashville, Tennessee

Seth A. Smith
Vanderbilt University
Nashville, Tennessee

Dewei Tang
Vanderbilt University
Nashville, Tennessee

Kevin W. Waddell
Vanderbilt University
Nashville, Tennessee

E. Brian Welch
Vanderbilt University
Nashville, Tennessee

Fen Xia
Vanderbilt University
Nashville, Tennessee

Eddy S. Yang
Vanderbilt University
Nashville, Tennessee

Thomas E. Yankeelov
Vanderbilt University
Nashville, Tennessee

I

Physical Basis of Magnetic Resonance Imaging

The Biology and Imaging of Cancer

Michelle D. Martin
Vanderbilt University

J. Oliver McIntyre
Vanderbilt University

1.1 What Is Cancer?

Cancer can be broadly described as the result of genetic and other changes that lead to alterations in normal cellular function. The effect of these alterations is unlimited growth and expansion that occurs both locally and at distant sites within the body. It is estimated that in 2010, 1,529,560 new cases of cancer will be diagnosed in the United States, with 562,340 deaths [1]. Any individual can develop cancer, but the risk increases with age. The majority of all cancers occur sporadically, and only 5% are hereditary, although enhanced risk for some cancers may be familial. A variety of causes can contribute to malignant transformation and the development of a tumor. These can include genetic changes that either actively contribute to the development of cancer when altered (oncogenes) or lead to cancer when their protective effect is removed (tumor suppressor genes). In addition, nongenetic causes can also accelerate tumor progression, including exposure to carcinogens and other mutagenic agents, as well as chronic inflammation and some infectious agents, including hepatitis B virus, human papilloma virus, and human immunodeficiency virus (HIV). The development of cancer is also influenced by the normal cells surrounding the tumor cells, as the tumor cells interact with such "stromal" cells to create conditions that are favorable for their continued growth and survival. Once tumors reach a critical size, this interaction with the "tumor microenvironment" enables the tumors to co-opt the process of normal blood vessel development resulting in the formation of new blood vessels to further feed the tumor in a process known as angiogenesis.

The first findings that linked a genetic cause to cancer came with the discovery of cellular oncogenes. These genes actively contribute to cancer when one copy of the gene is altered. The first of these genes to be discovered was the cellular sarcoma (*c-src*) gene [2]. This gene was closely related to the previously described viral sarcoma (*v-src*) gene discovered by Francis Rous in 1911 [3]. However, while the *v-src* oncogene was a virus, the *c-src* gene was a normal cellular gene that became misregulated during cancer formation. This discovery highlighted the understanding of cancer as a disease in which normal cellular function goes wrong. Two other oncogenes that were subsequently discovered were the Ras sarcoma (*ras*) and myelocytomatosis (*myc*) oncogenes [4–6]. Subsequent experiments that placed either the myc or the ras oncogene into the germ lines of mice under the control of the mouse mammary tumor virus (MMTV) promoter determined that cancer was a multistep process, since even though one copy of either the *myc* or *ras* oncogene was expressed in most if not all of the mammary epithelial cells of these mice, it was months before the tumors in the mice actually formed [7]. These findings provided evidence that the presence of a single oncogene within a normal cell was not sufficient to cause tumor formation. In additional studies, double-transgenic mice were developed that carried both *MMTV-ras* and *MMTV-myc* transgenes; these mice contracted tumors at a greatly accelerated rate and not only provided additional support to the hypothesis that oncogenes cannot act alone but also that oncogenes can act together *in vivo* to generate tumors [8]. These results indicated that at least one additional "hit" or alteration of the genome was necessary for the tumors to form.

A group of genes that influence the development of cancer through a different mechanism of action was also discovered. This class of genes appeared to protect cells from the development of cancer and as a result was named tumor-suppressor genes. If these genes are inactivated, either through mutation or by other means, the risk of cancer development is increased. The first study on tumor-suppressor genes was conducted by Henry

Harris in Oxford [9]. This experiment involved the formation of cellular hybrids of tumor cells with normal cells. To his surprise, the resulting cell fusions were nontumorigenic, meaning that the genes of the normal cells that were important to normal cell functioning and proliferation were dominant to those genes of the tumor cells. Studies that involved the retinoblastoma (*RB*) tumor suppressor gene also were critical to understanding the mechanism of action of this class of genes. In 1971, Knudson [10] used statistical analysis to show that retinoblastoma is a cancer caused by two mutational events. Additional studies in 1983 revealed that an additional mutation must also occur on the second copy of the *RB* gene in order for abnormal cellular behavior and tumorigenesis to occur [11].

Yet another well-studied tumor-suppressor gene is *p53*. This gene is a transcription factor that normally functions as a checkpoint to stop the progress of the cell cycle once it is activated by stress signals within the cell. However, when this gene is mutated, its function is altered, predisposing cells to erroneous division cycles and higher mutation rates [12]. As a consequence, the *p53* gene has been shown to be mutated in over 50% of all human cancers, reflecting its important function in regulating the normal growth of cells. Individuals who harbor inherited mutations in the *p53* gene have a much higher propensity for developing cancer during their lifetimes [13].

There are nongenetic causes of cancer including exposure to high or chronic levels of a chemical or physical carcinogen and chronic inflammation that can contribute to tumor formation. One of the most well-known of these agents is tobacco smoke. Additionally, heterocyclic amine mutagens, created by the cooking of red meat at high temperatures, have also been identified as causative agents in colon cancers. Another important nongenetic influence that can lead to the formation of tumors is chronic inflammation, caused either by exposure to something in the environment or by an infectious agent. Other mechanisms include alteration of the host's genome. For example, in HIV, the infection results in immunosuppression of the host, leading to the decreased ability of the immune system to identify transformed or abnormal cells that would normally be attacked by healthy immune cells.

1.1.1 Types of Cancers

Over 200 types of human cancers have been identified and classified. Although cancers are primarily classified according to their site of origin, they can also be roughly divided into three types depending on origin: epithelial, mesenchymal, and hematopoetic. In addition, there are cancers that do not fall into any of the above categories. For example, melanomas, while occurring in the skin, are not classified as an epithelial tumor since the cells that melanomas are derived from originate in the neural crest.

Cancers of the epithelial cells, which line the body and cover the internal organs, are termed *carcinomas*. These make up the majority of cancers and include, among others, cancers of breast, prostate, lung, and skin. Mesenchymal cells develop into connective tissues, blood vessels, and lymphatic tissue. Cancers arising from these cells are known as sarcomas and include cancers of

the bone, muscle, and cartilage. Cancers of hematopoetic origin include cancers of the blood cells and are known as leukemias, while the cancers of the lymphatic system are known as lymphomas. There are also cancers that occur in the tissues of the central and peripheral nervous system. The majority of these types of tumors occurs in the brain, and since they arise in the glial cells, they are known as gliomas. Lastly, melanoma, while not falling into any of the major groups noted above, is the most common serious form of skin cancer, contributing to an estimated 8650 deaths in 2009 [14]. It is important to note that tumors forming in any of the cell types mentioned above can be diagnosed as benign or malignant. Benign tumors are not cancerous, and although they may grow quite large, they do not have the ability to spread to other parts of the body. Malignant tumors, by contrast, are cancerous in that such tumors can invade and destroy nearby tissue, as well as metastasize to different parts of the body.

Tumor cells, regardless of origin, have a number of characteristics that enable evasion of the normal checks on the cell that limit the cellular life span. These characteristics, originally specified by Hanahan et al. in 2000 [15] are termed the "hallmarks of cancer." These hallmarks of cancer cells include the ability of the cell to become immortalized, evade apoptosis, have abnormal growth regulation, have the ability to invade local tissue and then metastasize to distant sites in the body, and once the cells form a tumor, have the ability to recruit new blood vessels (angiogenesis) from which to receive nutrients and oxygen to sustain the growth of the tumor.

1.1.1.1 Epithelial Cancers

The majority of surfaces of the body consists of epithelial tissue, which can be further classified into a number of subtypes, including glandular epithelium. The tumors that form from these tissues are known as carcinomas and are responsible for more than 80% of the cancer-related deaths in the Western world [14]. One type of epithelium covers the body on both its external and internal surfaces. Tumors arising from this type of epithelium are known as squamous cell carcinomas. The other type of epithelial tissue is glandular epithelium, which originates from invaginated epithelia. Tumors arising from this tissue are known as adenocarcinomas. Normal epithelial tissues have a regular, ordered appearance, whereas cancerous tissue has a disorderly appearance, with cells containing irregular nuclei and oftentimes an increased nucleus/cytoplasm ratio. Cancers that form in epithelial tissue can either be benign, as in adenomas, papillomas, and lipomas, or malignant, as in carcinomas. Outlined below are a few examples of the most common cancers of epithelial origin.

1.1.1.1.1 Breast Cancer

Breast cancer is the second most frequently diagnosed cancer in women, following cancers of the skin [14]. Death rates for breast cancer have steadily decreased in women since 1990, and incidence rates have also decreased since 1999. It is speculated that the decrease in incidence rates is related to both the reduction of use of hormone replacement therapy, known to impact breast cancer risk in postmenopausal women, and delayed diagnosis resulting from a decrease in utilization of mammograms, which is currently the most often used method of imaging to diagnose breast cancer [14].

Both endogenous levels of estrogen and the length of a woman's lifetime exposure to estrogen positively contribute to their risk for developing breast cancer [16]. This finding is pertinent to the potential for enhanced breast cancer rates that may result from the current high rates of obesity in the United States, i.e., obese and overweight girls typically go through puberty at an earlier age than previously, leading to a lengthening of the lifetime exposure to estrogen and consequent increased risk for breast cancer. The involvement of estrogen in some breast cancers has also provided a target for therapy. Estrogen receptor–positive breast tumors can be treated with hormone therapy that either disrupts the binding of the hormone to the receptor, e.g., with selective estrogen receptor modulators (SERMs), such as tamoxifen (AstraZeneca, London, UK) and raloxifene (Eli Lilly, Indianapolis, IN), or disrupts the synthesis of estrogen in the body, e.g., with aromatase inhibitors, including anastrozole (Arimidex; AstraZeneca) and letrozole (Femara; Novartis, Basel, Switzerland). Additionally, for women whose cancer tests positive for the HER2/neu oncogene, FDA-approved targeted therapies include trastuzumab (Herceptin; Genentech, San Francisco, CA) and lapatinib (Tykerb; GlaxoSmithKline, London, UK) [17]. These drugs interfere with the HER2/neu signaling pathway, resulting in reduced proliferation. The disease also has an inherited predisposition that is involved in approximately 5%–10% of all breast cancer cases. This disposition involves inherited genetic mutations in the breast cancer susceptibility genes *BRCA1* and *BRCA2* [18]. Mutations in these genes have also been shown to predispose women to ovarian cancer. Breast cancers arising from *BRCA1* mutations are typically estrogen receptor–negative, while those arising from *BRCA2* mutations are usually estrogen receptor–positive [18].

1.1.1.1.2 Prostate Cancer

Prostate cancer is the most frequently diagnosed cancer in men and the second leading cause of death due to cancer [14]. As in breast cancer, prostate cancer is linked to a steroid hormone, in this case, testosterone. Also, similarly to breast cancer, drugs have been developed to target testosterone for cases in which the prostate cancer is testosterone responsive. Testosterone's effects are mediated through the androgen receptor. Prostate cancer typically starts out being androgen dependent but eventually becomes independent following androgen-deprivation therapy. Early screening began for prostate cancer in the late 1980s, which led to an increase in early diagnosis of the disease in addition to a decrease in mortality [19]. Screening consists of both physical examination and testing for prostate-specific antigen (PSA) levels, with increased levels of PSA indicating an increased likelihood for developing prostate cancer. Five-year survival rates are very high (approaching 100%) for the majority of prostate cancers that are diagnosed in the local and regional stages before the disease has acquired the ability to spread beyond the prostate, but markedly reduced (~30%) 5-year survival for patients with metastatic prostate cancer [14].

1.1.1.1.3 Lung Cancer

The majority of cancer-related deaths in men and women are due to lung cancer [14]. Exposure to tobacco smoke remains the most important risk factor, and risk increases with the amount and duration of exposure to this carcinogen, with smokers having a >10-fold increased risk for lung cancer (17% of male smokers) than that for never-smokers (<1.5%) [20]. Exposure to environmental carcinogens such as asbestos, radon, and certain metals, including cadmium and arsenic, have also been shown to affect the development of lung cancer [14], and lung cancer in never-smokers is an important public health issue [21]. Imaging techniques to diagnose lung cancer have progressed from the use of chest X-rays, which did not significantly impact the ability to diagnose, to the current use of low-dose spiral computed tomography (CT) scans, which have shown promise for the detection of tumors at earlier stages. The two main classifications of lung cancer are non–small cell and small cell lung cancer. The majority of lung cancer is non–small cell, with its subtypes including squamous cell lung carcinoma, adenocarcinoma, and large cell lung carcinoma. Small cell lung cancer tends to arise in the larger airways of the lungs and carries with it a worse prognosis. It is particularly important in the case of non–small cell cancer that the disease be accurately staged, as the stage of the disease process determines both treatment options and prognosis for lung cancer patients [22]. Treatment options include surgery, radiation, chemotherapy, and some biologically targeted therapies, including erlotinid (Tarceva; Genentech) and bevacizumab (Avastin; Genentech). Avastin works by cutting off the ability of the tumor to form new blood vessel networks, i.e., is an antiangiogenic agent, while Tarceva is a tyrosine kinase inhibitor that targets the epidermal growth factor receptor that is often highly expressed in cancer. Survival rates for lung cancer still remain relatively low (five-year survival rate for all stages combined is 15%), but are improved if the tumors are diagnosed and treatment is initiated when the tumor is still localized and has not spread (metastasized) to other organs [14]. For this reason, improvements in imaging and diagnostic techniques that can detect lung cancer at very early stages are crucial for improving the ability to survive this aggressive epithelial cancer.

1.1.1.1.4 Colon Cancer

Colorectal cancer is the third most common cancer in both men and women [14]. The incidence of colorectal cancer has been steadily decreasing in the United States since 1998 due in part to increased screening and the subsequent removal of polyps from the colon before progression to cancer [23]. Obesity, a diet high in red or processed meats, and heavy alcohol consumption have all been linked to an increased risk of colorectal cancer. In addition, some hereditary colorectal cancer syndromes exist, including familial adenomatous polyposis (FAP) and hereditary nonpolyposis colorectal cancer (HNPCC). FAP is due to a mutation in the *APC* (adenomatous polyposis coli) gene and in FAP patients, many (100–1000) adenomas form in the large bowel and rectum, with a portion progressing to colon cancer. The mutation in *APC* is autosomal dominant, affecting about 1 in 8000 individuals in the United States, but accounts for less than 0.5% of colorectal cancers [24]. Chronic inflammation is also involved in the development of colorectal cancer, as chronic inflammatory bowel disease and ulcerative colitis can lead to precursor lesions that can progress to cancer.

1.1.1.2 Mesenchymal Cancers

Mesenchymal cancers are derived from the various connective tissues of the body, which are formed from the mesoderm of the developing embryo. Known as sarcomas, mesenchymal cancers account for only about 1% of diagnosed tumors [25]. As the name suggests, these tumors are derived from mesenchymal cells, including fibroblasts and other related connective tissue cells; osteoblasts, which form bone; adipocytes, which store fat; and myocytes, which form muscle. Some of the more common sarcomas include osteosarcomas, liposarcomas, and fibrosarcomas. As with other tumors, treatment options for sarcomas depend on the type, size, grade, and stage of the tumor, as well as the location in the body and whether all of the tumor, can be removed by surgery.

1.1.1.2.1 Osteosarcomas

Osteosarcomas are derived from osteocytes, which are the cells that are responsible for the formation of bone. The characteristic feature of an osteosarcoma is the presence of osteoid (bone formation) within the tumor. These tumors are a broad group, and 80% of osteosarcomas occur in the extremities, and patients usually present with a dull or aching pain [26]. Radiographs are usually the first diagnostic study done to determine the cause of the pain, followed by a bone biopsy for a definitive diagnosis. Standard therapy includes limb-salvaging orthopedic surgery combined with high-dose chemotherapy.

1.1.1.2.2 Liposarcomas

Liposarcomas develop from cells closely related to the adipocytes, which are responsible for the storage of lipids. These rare tumors occur almost exclusively in adults and arise in fat cells located deep within soft tissue, often of the proximal extremities and the retroperitoneum. Well-differentiated liposarcoma is the most common subtype [27], and prognosis differs depending on a variety of factors, including site of origin, tumor size, and tumor type. Treatment of liposarcomas often involves surgery with a wide local excision. Recurrence rates vary depending upon the location and subtype of liposarcoma [28].

1.1.1.2.3 Fibrosarcomas

Fibrosarcomas are uncommon, malignant tumors arising from fibrous connective tissue. Such tumors can originate in the fibrous connective tissues of the bone but can also involve the periosteum and overlying muscle. Fibrosarcomas are characterized by immature proliferating fibroblasts or undifferentiated anaplastic spindle cells that metastasize to the lungs and bone in up to 65% of patients but rarely to the lymph nodes [28]. Five-year survival from these lesions is approximately 50%.

1.1.1.3 Hematopoetic Cancers

Hematopoetic cancers are derived from the various cell types that comprise the blood, including the cells of the immune system. The two major types of these malignancies are the leukemias and the lymphomas. Leukemias are malignant tumors of the blood cells, including erythrocytes and plasma cells. The lymphomas include tumors of the lymphoid lineage (B and T lymphocytes) and tend to form solid tumor masses located in the lymph nodes, unlike the leukemias that are typically single tumor cell populations circulating throughout the body.

1.1.1.3.1 Leukemia

Leukemias exist in both acute and chronic forms, as in acute lymphocytic leukemia (ALL), which accounts for approximately 70% of the leukemia cases among children 0–19 years old, and in chronic forms, such as chronic myelogenous leukemia (CML), which can progress slowly and with few symptoms [14]. Leukemia is diagnosed 10 times more often in adults than in children. The most common forms of leukemia in adults are acute myeloid leukemia (AML) and chronic lymphocytic leukemia (CLL). CML was the first cancer associated with a specific chromosomal abnormality. This abnormality is known as the Philadelphia chromosome and results from a balanced reciprocal translocation between chromosomes 9 and 22. The translocation causes the generation of the oncogene BCR-ABL1 that encodes a protein that is constitutively activated in CML. Leukemias are difficult to diagnose using any imaging technique, as they tend to be diffuse, single cell populations within the circulation, and therefore do not provide a target for a contrast agent as can be used for solid tumors. Survival from leukemia is highly dependent upon type, with chemotherapy being the most effective method of treatment.

1.1.1.3.2 Lymphoma

Lymphomas can be further divided into three subtypes that include Hodgkin lymphomas, non-Hodgkin lymphomas, and multiple myelomas. Multiple myeloma is characterized by the clonal proliferation of plasma cells of the B-cell lineage in the bone marrow. Hodgkin and non-Hodgkin lymphomas differ in the specific type of lymphocyte that is affected. Non-Hodgkin lymphoma is much more common and can originate in either B or T cells. Hodgkin lymphomas are characterized by the presence of a specific type of B lymphocyte known as a Reed–Sternberg cell. These cells were named after Dorothy Reed Mendenhall and Carl Sternberg, who provided the first microscopic description of Hodgkin disease [29]. Swelling of the involved lymph nodes typically occurs in both types of lymphoma. Survival varies depending on the cell type and stage of disease, as does the standard method of treatment.

1.1.1.4 Gliomas

Gliomas, the most common primary brain tumor, are formed from the glial cells of the brain that have their origin in the neuroectodermal cell layer of the developing embryo. These types of tumors account for 50%–60% of all primary brain tumors, and most are malignant [30]. Gliomas are classified in terms of their histology: gliomas resembling astrocytes are referred to as astrocytomas, those resembling oligodendrocytes are called oligodendrogliomas, and those resembling ependymal cells are known as ependymomas [31]. Patients presenting with a glioma have very different symptoms that depend upon the anatomical site of the brain that is affected. Treatment includes surgical resection, radiation therapy, and, for some gliomas, chemotherapy. Survival rate depends upon the specific type and stage of glioma at diagnosis.

1.1.1.5 Melanomas

Melanomas are derived from melanocytes, which are the cells responsible for the pigment found in the skin and the retina. Although the melanocytes are descended from the cells of the neural crest of the embryo, mature melanocytes have no direct contact with any part of the nervous system. The incidence of melanoma has been increasing for 30 years, and rates are 10 times higher in Caucasians than in African Americans [14]. The majority of melanomas arise in the skin, with a minority forming in either the eye or the internal mucosal membranes. Melanomas can be thought of as among the most malignant human cancers, as they can give rise to metastases in virtually every tissue [32]. Death from melanoma is due to distant metastasis to other vital organs. The most important prognostic variable in determining the metastatic potential of a lesion is the depth of invasion of the lesion into the skin [33]. The most significant environmental risk for developing melanoma is exposure to sunlight. However, patients who have genetic defects in their ability to correct mutations in their DNA caused by ultraviolet exposure have greater than a 100-fold higher risk for the development of melanoma [34]. Prevention is the best way to protect oneself from developing melanoma by minimizing sun exposure and wearing sun protection factor while in the sun. Melanoma is highly curable if detected in its earliest stages, with the 10-year survival rate being 90% [14].

1.2 Cancer Progression

The development of a benign cancer cell to a tumor and the subsequent spread of tumor cells to distant sites in the body are defined by different steps in cancer progression. These steps include growth at the primary site, the growth of new blood vessels to feed the tumor and, for malignant cancers, the dissociation of the tumor cells from the primary tumor, invasion of the tumor cells through the basement membrane into the surrounding tissue, and metastasis to and establishment of secondary lesions at other sites throughout the body. Each step in this process is necessary for a tumor cell to develop into a detectable tumor with metastatic potential, i.e., that has the ability to spread to distant sites within the body. Each step in this progression involves both regulated and dysregulated interactions among many genes.

1.2.1 Uncontrolled Cellular Growth

One of the earliest steps of cancer progression is uncontrolled cellular growth at the primary site. This unchecked growth results in the formation of a primary tumor mass that then can acquire the ability to invade surrounding tissues. Some tumors are not invasive and are said to be benign rather than malignant. However, other tumors do have the ability to invade surrounding tissues and spread to distant sites within the body.

1.2.2 Angiogenesis

After tumors reach a certain size, either at the primary site or at a resulting metastatic site, new blood vessels must be recruited via a process known as angiogenesis in order to continue growth. The new blood vessels have multiple functions, including supplying oxygen and needed nutrients to the tumor cells [35]. In addition, new vessels can serve as conduits for the spread of the tumor cells to other sites in the body. Work by Folkman et al. showed that tumors only have the capability to grow up to a size of approximately 1 mm in diameter, beyond which their own vasculature must be developed [36]. Angiogenesis is a rare process and normally only occurs during the processes of wound healing and the female reproductive cycle. However, tumors undergo an "angiogenic switch" in which processes that enable formation of new blood vessels are activated. One factor that has been shown to be crucial for this phenomenon is matrix metalloproteinase 9 (MMP-9). Work by Bergers et al. showed in a model of pancreatic carcinogenesis that (MMP-9) released vascular endothelial growth factor (VEGF) from pancreatic islets, enabling this factor to localize with its receptor, VEGF receptor 2 (VEGFR2), to induce angiogenesis in tumors [37]. Importantly, the MMP-9 involved was not expressed by the tumor cells but instead by the surrounding microenvironment, highlighting the importance of tumor/stromal interactions during tumor progression.

1.2.3 Invasion and Metastasis

In malignant tumors, some tumor cells separate from the primary tumor mass and invade into surrounding tissue. In addition, those cells that are destined to travel to a distant site must first escape from the primary site, then gain access to the bloodstream or the lymphatics. During this progression, tumor cells that are of epithelial origin are said to undergo a process known as epithelial to mesenchymal transition (EMT). This process aids in the ability of the cells to transverse through the basement membrane into the surrounding tissue. Tumor cells have been shown to migrate either as single cells or collectively as files or clusters of cells. Both methods are an efficient technique in the invasion of the tumor cells into surrounding tissue and vessels.

Tumor cells that have moved into the bloodstream (or lymphatics) must somehow evade the host's immune system and the shear forces present in the bloodstream, either of which could result in cell death. Next, cells must acquire the ability to leave the bloodstream and lodge in a site that is able to support their growth. There is a propensity for tumor cells to travel to and become established as metastases in particular organs, an observation that forms the basis for the "seed-and-soil" hypothesis, first postulated by Stephen Paget in 1889 (see [38]). Tumor cells sometimes spread to lymph nodes near the primary tumor. This is termed having "positive nodes" or is otherwise known as nodal involvement. Because of this, it is now common practice for physicians to check the lymph nodes nearest the primary tumor for evidence of metastatic spread. Some of the earliest research on metastasis was conducted by Issiah Fidler, who proposed that only a small subset of cells within the primary tumor had the ability to spread to other areas [39]. Research then began to focus on genes that were turned on or off in cancer cells that had metastasized. This led to the finding of the gene *nm23* by Pat Steeg in 1988 [40]. This gene, and others like it, are now known as metastasis-suppressor genes. When these genes are turned on in cells of the primary tumor, they seem to prevent spread of the cells. However, once they are turned off through inactivation or mutation, the cell is more likely to metastasize. Later work by Steeg et al. determined that metastasis-suppressor genes are important for a cancer cell's ability to colonize a distant site [41].

1.3 Classification of Tumors

1.3.1 Pathology

The technique that remains the most definitive method for classifying whether a tumor is benign or cancerous is histological examination of a biopsy specimen. This method is often used to correlate results from *in vivo* imaging in patients. Using this approach, a small section of the tumor is removed from the patient and processed, and smaller sections are taken and placed onto slides for staining and evaluation by a pathologist. Depending upon the type of cancer, different histological stains are employed to assist in the diagnosis and staging of disease. One common stain used on virtually all specimens is hematoxylin and eosin, commonly referred to as H&E stain. This method results in the staining of the nuclei of cells (and some other objects, including keratin granules) blue. The nuclear staining is followed by eosin counterstaining, which stains eosinophilic structures (including the cytoplasm of cells) in shades of pink to red. Pathologists utilize this staining procedure to quickly stain tumor sections and gain information on the type of cells

comprising the tumor and how they appear visually, including the regularity of the nuclei, the uniformity of cell size, and if the tumor cells have begun to invade into the surrounding tissue [42]. The pathological analysis from such biopsy samples and a variety of imaging techniques (see below) depending upon the cancer type and location form the basic tools for the complete pathological and clinical assessment of cancer.

1.4 Cancer Imaging Modalities and Strategies

A wide variety of imaging platforms are currently used routinely for both clinical assessment and preclinical studies. With the exception of ultrasound (US), the various imaging modalities exploit various frequencies throughout the electromagnetic spectrum, from high-energy X-rays to lower energy radiofrequencies (for MRI) and have a broad range of both resolution and sensitivity (see Table 1.1). The advent of X-ray imaging more than 100 years ago provided a new tool for the clinician that complemented clinical assessment and histological analyses, both based primarily on trained observation, the simplest form of optical imaging. The toolbox for *in vivo*, clinical imaging now includes a variety of modalities such as optical imaging, US, CT, nuclear imaging [scintigraphy, positron emission tomography (PET), and single photon emission computed tomography (SPECT)], and magnetic resonance imaging (MRI), each of which have application for clinical assessment, particularly in the diagnosis and staging of various types of cancer as well as in monitoring response of such diseases to therapy. The characteristics of the various modalities and examples of applications in cancer are summarized in Table 1.1 [43–45], although a comprehensive discussion of specific methods and applications is beyond the scope of this chapter. It should be noted than many, although by no means all, of the developments in the various imaging modalities

TABLE 1.1 Imaging Modalities for Detection, Diagnosis and Staging of Cancer: Characteristics and Examples of Applications

Imaging Modality	Energy Used	Spatial Resolution (mm)	Temporal Resolution (sec)	Molecular Sensitivity (mol/l)	Tissue Penetration (mm)	Examples of Cancer Applications
Optical	Visible to NIR light (350–1000 nm)	2–10	>10	10^{-9}–10^{-12}	<20	Intestinal tract/ oral cavity/lung
X-ray/CT	X-rays (17–70 keV)	0.05–0.2	<1	Not well characterized	No limit	Breast/lung
PET	Gamma rays from positron (511 keV)	6–10	>10	10^{-11}–10^{-12}	No limit	Lung/prostate/cervical cancers and metastatic disease; molecular imaging
SPECT	Gamma rays (100–200 keV)	7–15	>60	10^{-10}–10^{-11}	No limit	Bone/brain; molecular imaging
MRI	Radiofrequency (1–100 MHz)	0.2	>60	10^{-3}–10^{-5}	No limit	Most solid tumors and metastatic disease
Ultrasound	Sound waves (2–18 MHz)	0.05–1	0.1–100	Not well characterized	<200	Breast/liver/gastric pancreas/prostate

Source: Massoud T. F., Gambhir S. S., *Genes Dev.,* 17, 545–580, 2003; Scherer R. L. et al., in *Cancer Metastasis: Biologic Basis and Therapeutics,* Welch D. R., Lyden D., and Psila K., editors, in press.

were first explored and tested in small animal models of cancer such as mouse and rat models, the latter particularly for MRI studies. Such models will likely continue to be profoundly useful for future developments in the field despite the limitations of such rodent models due to differences in size and physiology as compared with humans [46]. The remainder of this chapter provides examples of applications of these various imaging tools, especially for cancer assessment and the potential for future developments in the field, particularly in the modern era of targeted drug therapies and molecular imaging (MI) modalities.

1.4.1 Optical Imaging

Optical imaging is routinely used in the clinical setting via a variety of fairly standard endoscopic tools, such as those designed for the intestinal tract [23] or for the lungs [47–48]. For example, optical imaging is used routinely for anatomical detection of colonic polyps by colonoscopy to screen for colorectal cancer [23] and clinical assessment of lung pathologies such as carcinoid tumors [47] that might first be suspected from X-ray images of the chest (see below). In addition, there are a number of developing optical imaging techniques for clinical assessment of cancer as are documented in a recent monograph by Eben Rosenthal and Kurt Zinn [49]. In particular, a number of novel optical cancer imaging protocols are being developed in preclinical models that make use of both endogenous and exogenous contrast agents, including fluorescent probes [50–51]. Furthermore, some of these techniques, such as those based for example on diffuse optical tomography, are beginning to find clinical application such as for breast cancer imaging notably in the near infrared (NIR) regime and in combination with MRI [52].

1.4.2 Ultrasound

The past decade has seen major developments in US technology, both in terms of instrumentation and contrast agents. In addition to routine use for image-guided biopsies, particularly in breast cancer [53], endoscopic, transrectal, and endobronchial US is becoming standard practice for the staging of gastric [54], pancreatic, and prostate cancers [55–56] and for mediastinal staging of non–small cell lung cancers (NSCLC) [22]. In addition, US provides an alternative modality for clinical assessment of cancers in patients with contraindications for MRI [57].

1.4.3 Radiography and CT

Classical 2D radiographs generated by X-rays are useful for screening for breast cancer (mammography) [58–59] and for lung abnormalities including cancer, although such standard chest X-ray images used in screening trials tend to identify "indolent" slow-growing tumors rather than the more aggressive fast-growing carcinomas [60]. However, the advent of high-resolution CT to generate 3D X-ray images is beginning to provide new insights into the pathology of lung adenocarcinomas and, despite relatively poor soft-tissue contrast, low-dose CT is being

explored for screening of subjects at high risk for lung cancer [60]. In addition, CT may find application as an alternative to the standard colonoscopy that is poorly accepted for colorectal cancer screening [61]. The power of CT is its high spatial and temporal resolution and facile delineation of the skeletal structure. In this context, it is often used to provide reference for other imaging modalities, particularly PET (see below). A caveat in the use of CT is that the radiation exposure from CT appears to be associated, at least in CT imaging of children, with an increased risk for fatal cancer [62], prompting evaluation of the risks from CT [63] and PET/CT [64]. Such risk assessment may receive more attention in all imaging modalities as the field progresses, particularly with the increasing use of multimodality imaging.

1.4.4 Nuclear Imaging (Scintigraphy, PET, and SPECT)

The most common means for detecting bone metastasis is skeletal scintigraphy, a 2D nuclear imaging technique that developed following the invention of the gamma camera in 1956 by Hal Anger (see [65]). In current practice, scintigraphy makes use of radionuclides, such as technetium-99m bound to methylene diphosphonate (^{99m}Tc-MDP), a compound that targets bone with enhanced metabolism [66–67]. PET has become a standard tool throughout the developed world, particularly for cancer imaging. This modality most often makes use of an ^{18}F-labeled analog of glucose, ^{18}F-FDG (fluorodeoxyglucose), and, in this context, is usually referred to as FDG-PET. PET has intrinsic high molecular sensitivity and is particularly useful for the detection of metastatic disease and the staging of a variety of cancers, although it has relatively poor spatial resolution (Table 1.1). Nuclear medicine imaging also includes SPECT, a technique that images radioisotope decay and is used predominantly for bone imaging [68], but also has applications for MI (see below) in a variety of cancers, particularly brain and neuroendocrine tumors [69]. As noted above, the poor spatial resolution of PET and SPECT can, at least in part, be compensated by coregistration with CT images using a PET/CT or SPECT/CT scanner; clinically, both PET and SPECT are now generally used in combination with CT, i.e., PET/CT and SPECT/CT [70]. For example, PET/CT is used clinically for imaging and staging a variety of cancers, including lung [71–73], prostate [74], and cervical [75] cancers. Recent studies, particularly of breast-to-bone metastasis, have begun to combine PET/CT imaging with either SPECT, which is reported to improve the diagnostic accuracy of bone scintigraphy [76], and/or with MRI [77].

1.4.5 Magnetic Resonance Imaging

MRI is used extensively for clinical diagnosis, yielding images that can reveal anatomic detail at resolutions approaching 0.2 mm. Specifics of both established and developing MRI methods are detailed in the subsequent chapters of this monograph. In conventional MRI, the images represent the properties of mobile protons and are dominated by the properties of water in tissues. Intrinsic contrast in the images is due to variations in the water content of different tissues and how the water interacts with macromolecules (see Chapters 2, 4, and 5) [78]. In addition, extrinsic contrast agents, such as gadolinium chelates or iron oxide particles, which are designed to alter the relaxation properties of the mobile hydrogen nuclei of water, can be used to probe specific tissues, explore blood flow, etc., and thereby provide additional contrast and with temporal resolution (see Chapters 9 and 10). The different types of imaging contrast (e.g., T_1- and T_2-weighted images) in MRI are obtained using exquisitely designed radiofrequency pulse sequences to drive spin transitions in atomic nuclei (protons) oriented within a uniform strong magnetic field, either 1.5 or 3 T for current clinical MRI and up to 9 or 15 T for clinical or preclinical research, respectively. Sophisticated algorithms are used to define the temporal and spatial location of the mobile protons (water) within the magnetic field and to generate 3D images. Selected applications for cancer imaging are noted here with more detailed information on current and developing MRI paradigms being described in subsequent chapters of this monograph. MRI can be extended to yield spatial information about biochemical compounds by methods referred to as magnetic resonance spectroscopy (MRS) or as magnetic resonance spectroscopic imaging (MRSI) as described, for example, in Chapter 9.

MRI provides high resolution images with good contrast, although a shortcoming of the modality is the relatively low molecular sensitivity (Table 1.1). However, MRI is in routine use in the clinical assessment of patients with a number of different types of cancer and is particularly useful for assessment of blood flow and perfusion [44], identifying soft-tissue tumors and tumor-like lesions [79], abdominal oncology [80], prostate cancer [81], and in neuro-oncology [82–83]. The clinical assessment of the use of MRI for breast cancer imaging showed early promise, but results of more extensive studies of the comparative effectiveness of MRI in breast cancer (COMICE) [84] remain equivocal. Thus, although preoperative MRI reveals the extent of the disease much better than other common imaging methods (mammography and US), this information does not yield better outcomes, at least as assessed in the COMICE trial that focused on surgical care and prognosis [85–86]. However, it would appear that the additional information provided by multiparametric MRI and functional imaging of breast cancers, noted for a broad etiology and a spectrum of pathologies (see above), may yet be useful clinically, particularly for accurate diagnosis and monitoring response to therapy for breast cancer patients [87]. Such combinatorial multimodal MRI/MRSI techniques may have broad application in the assessment of tumor status in a variety of cancer settings. Whole-body MRI for staging and follow-up of patients with metastatic disease [88] or lung cancer [89] reflect innovations in radiofrequency pulse design and data acquisition, which in combination with the advent of high-field scanners (already at 9 and 15 T for clinical and preclinical research, respectively), are likely to continue to expand the opportunities for the use of MRI in the management of cancer patients.

1.4.6 Molecular Imaging

In the past decade, a number of imaging modalities have been used to begin to image particular biological processes, developments that led to the establishment of an accepted definition of MI as being "the visualization, characterization, and measurement of biological processes at the molecular and cellular levels in humans and other living organisms" [90]. As elaborated in that definition, MI typically includes 2D or 3D imaging as well as quantification (a key element of MI) over time using a variety of techniques [nuclear (scintigraphy, PET, and SPECT), MRI, MRS, optical, US, and others] that usually require sophisticated imaging agents and instrumentation [43,90]. For example, MI is being used to assess tumor angiogenesis [91], neuroendocrine tumors [92], tumor metabolism (particularly by PET/CT and/or SPECT) [93]. The burgeoning field of MI shows promise as a robust method for cancer imaging, including the detection of tumors, staging tumors, locating metastatic disease, evaluating therapeutic targets, and monitoring treatment efficacy [94]. MRSI is beginning to be explored as an MI tool in cancers as, for example, in studies to evaluate the metabolic status of brain tumors [95] and in multiparametric imaging of breast cancer [87]. MI in MRI is beginning to be explored with recent innovations in MR "contrast" agents such as tumor-targeting molecular beacons [96], PARACEST reagents [97] (and Gochberg, Chapter 8, this monograph) and hyperpolarized reagents [98–99] (and Chekmenov and Waddell, Chapter 22, this monograph). Such developments may fundamentally change the sensitivity of MRI for metabolic and cancer imaging and portend a possible paradigm shift in the application of MRI in clinical practice particularly in oncology.

1.4.7 Multimodal Imaging

A current trend in the clinical practice of oncology is to use a combination of modalities to assess tumors, particularly for staging and detection of metastatic disease. Clinical PET imagers are now marketed in tandem with CT, and preclinical developments are exploring the combination of MRI with PET and/or CT. The basic strategy for such bimodal or multimodal clinical imaging systems is to couple a highly sensitive imaging modality (such as PET or SPECT) that has somewhat limited resolution with higher resolution imaging modalities (MRI and CT) (see Table 1.1). For example, the detection and staging of prostate cancer based on long-established transrectal US imaging is improved not only by 3D US and contrast-enhanced US techniques but also by combination with PET and MRI [100]. Likewise, multimodal imaging is standard for a variety of cancers such as hepatocellular carcinoma [101], head and neck tumors in children [102], gliomas [103], and in the clinical staging of lung cancer patients that currently includes physical and radiological (X-ray and CT) examinations together with endoscopic US, bronchoscopy, mediastinoscopy, and thoracoscopy [104]. These established imaging methods for lung cancer assessment may be supplemented, at least in some settings, by MRI, despite the intrinsic challenges in obtaining high-resolution MR images from the lung due to the magnetic susceptibility mismatch at the air–tissue interface, as well as cardiac and respiratory motion. Such pulmonary MRI is facilitated by the use of hyperpolarized gases [98] now being explored for clinical imaging [105]. The use of metabolic probes and combinations of imaging modalities are particularly useful for the detection and staging of metastatic disease, for example, imaging bone metastases in breast cancer [77] or for nodal staging of cancer patients [106]. Future directions for development of multimodal imaging include the use of not only different platforms, but also multicolor contrast agents, e.g., simultaneous imaging of two isotopes of different energy [107]. The particular combinations of modalities that prove useful in the clinical setting are expected to differ for different types of cancer as well as for cancer metastases, as is already evident from current radiological practice [44].

1.4.8 Imaging and Therapy

In selected tumor settings, imaging is being used to guide therapy, and both preclinical and clinical studies are in progress to examine imaging metrics to assess the response to therapy. For example, tumor-targeted focused US treatments, designed to induce "thermal surgery," are being developed based on MRI to monitor response at the target [108]. Likewise, multimodal imaging of brain tumors before and after treatment of patients provides insights into the biological responses to the therapy of both the tumor and its local environment [109]. In preclinical studies using a mouse model of breast cancer, Virotsko et al. [110] have shown that MI of the expression of VEGFR2 is a potential biomarker for assessing angiogenesis and the efficacy of antiangiogenic therapies. As noted by Bhujwalla et al. [87], these kinds of MI probes have potential for monitoring responses to therapies, particularly targeted therapeutics, using a variety of imaging modalities, notably nuclear PET or SPECT imaging as well as by MRI and MRSI. On another front, advances in 4D imaging are facilitating the refinement of targeted radiation therapies, and such strategies, combined with MI either with or without combination therapy, form the basis for the developing field of "theragnostic imaging" [111].

1.4.8.1 Assessing Response to Therapy

Currently, response assessment metrics for solid tumors are those designed first by the World Health Organization (WHO) (with bidimensional metrics) and subsequently modified by the Response Evaluation Criteria in Solid Tumors (RECIST) Working Group, resulting in a unidimensional metric [112]. Response to antitumor therapy, as measured according to the WHO and RECIST criteria, has been considered the primary endpoint to determine success in phase 2 clinical trials before proceeding to phase 3 studies [112–113]. However, the RECIST metric that uses only the measurement of maximal tumor diameter, has found mixed acceptance in the oncology community, and its limitations have become more apparent with the advent of tumor-targeted therapies where the RECIST criteria often fail to capture a positive effect from treatment such as the arrest of tumor growth [114]. These limitations have led to the development

of alternate metrics for assessing response in different tumor settings. For example, a modified RECIST paradigm (mRECIST) has been used to assess tumor response to therapy in hepatocellular carcinoma patients [115]. In those studies, it was noted that direct volumetric measurements to identify partial response versus progression may provide more robust assessment and should be examined in future clinical trials. Such volume measurements were recently reported as part of a clinical trial in renal cell carcinoma patients [116]. Likewise, an alternative tumor-response metric, referred to as "MDA" (having been developed at the University of Texas M. D. Anderson Cancer Center, Houston, TX) and based on multimodal imaging including metrics from CT, MRI, radiography, and skeletal scintigraphy, appears to be useful for the assessment of tumor response in patients with bone-only metastatic breast cancer [117]. Alternate metrics for assessing tumor response by ^{18}F-FDG-PET imaging include the use of internal standardization to control for variables in administration of the ^{18}F-FDG contrast agent [118]. From a number of clinical trials with targeted therapies, such as with kinase inhibitors like sorofenib, it is also evident that biomarkers and MI may be more pertinent for assessing response to therapy than the classic unidimensional tumor metric as defined by the RECIST criteria [113]. Thus, recent developments in the field suggest that more refined metrics [119], such as time-to-event endpoint metrics, are likely to be established for the assessment of tumor response in patients, particularly those with breast-to-bone metastases where RECIST is not useful [66] and those with solid tumors that are being treated with targeted molecular therapies. In this regard, recent work highlights the potential for using multimodal imaging, including MRI, MRSI, and PET as an early surrogate biomarker of tumor response to therapy [120].

1.4.9 Future Prospects for Developments in Cancer Imaging

There have been rapid developments, notably over the past decade, that have improved the tools in each of the major imaging modalities listed in Table 1.1 that are now available to the radiologist for clinical imaging and assessment, particularly for the cancer patient. Even so, there is evidence for potentially significant developments in the near future in a number of the established and developing imaging modalities. For example, new approaches for optical imaging, particularly in the NIR, are being pioneered in mouse models [121–122] and have potential for clinical application in a number of settings such as in the lymphatic system [123] and also for screening and early detection in the context of global cancer management [124]. New techniques for US, such as the development of US–CT for breast imaging [125] and modifications to CT that permit dynamic perfusion imaging [126], illustrate the kinds of improvements that are already being explored for clinical assessment. The development of new imaging modalities such as photoacoustic imaging, that uniquely combines the absorption contrast of light or radio-frequency waves with US resolution [127], and terahertz pulsed imaging with potential for applications for MI [128], may open new avenues for clinical imaging. In the nearer term, the developments of new MI agents for a variety of modalities, including multimodal nanoparticle reagent, and particularly targeted PET and SPECT probes, are likely to markedly broaden the available methodology for clinical assessment of cancer, metastatic disease, and the response to therapy over the next decade. In MRI, a host of potential applications in cancer imaging are described in detail throughout this monograph that documents current developments in this exciting field.

References

1. Jemal A, Siegel R, Xu J, Ward E. Cancer statistics, 2010. *CA Cancer J Clin* 2010.

2. Stehelin D, Fujita DJ, Padgett T, Varmus HE, Bishop JM. Detection and enumeration of transformation-defective strains of avian sarcoma virus with molecular hybridization. *Virology* 1977;76:675–684.

3. Rous P. A Sarcoma of the fowl transmissible by an agent separable from the tumor cells. *J Exp Med* 1911;13:397–411.

4. Santos E, Tronick SR, Aaronson SA, Pulciani S, Barbacid M. T24 human bladder carcinoma oncogene is an activated form of the normal human homologue of BALB- and Harvey-MSV transforming genes. *Nature* 1982;298:343–347.

5. Parada LF, Tabin CJ, Shih C, Weinberg RA. Human EJ bladder carcinoma oncogene is homologue of Harvey sarcoma virus ras gene. *Nature* 1982;297:474–478.

6. Dalla-Favera R, Gelmann EP, Martinotti S, Franchini G, Papas TS, Gallo RC, Wong-Staal F. Cloning and characterization of different human sequences related to the onc gene (v-myc) of avian myelocytomatosis virus (MC29). *Proc Natl Acad Sci U S A* 1982;79:6497–6501.

7. Land H, Parada LF, Weinberg RA. Tumorigenic conversion of primary embryo fibroblasts requires at least two cooperating oncogenes. *Nature* 1983;304:596–602.

8. Sinn E, Muller W, Pattengale P, Tepler I, Wallace R, Leder P. Coexpression of MMTV/v-Ha-ras and MMTV/c-myc genes in transgenic mice: Synergistic action of oncogenes *in vivo. Cell* 1987;49:465–475.

9. Harris H, Miller OJ, Klein G, Worst P, Tachibana T. Suppression of malignancy by cell fusion. *Nature* 1969;223: 363–368.

10. Knudson AG, Jr. Mutation and cancer: Statistical study of retinoblastoma. *Proc Natl Acad Sci U S A* 1971;68:820–823.

11. Cavenee WK, Dryja TP, Phillips RA, Benedict WF, Godbout R, Gallie BL, Murphree AL, Strong LC, White RL. Expression of recessive alleles by chromosomal mechanisms in retinoblastoma. *Nature* 1983;305:779–784.

12. Huang TT, D'Andrea AD. Regulation of DNA repair by ubiquitylation. *Nat Rev Mol Cell Biol* 2006;7:323–334.

13. Levine AJ HW, Fang Z. Tumor suppressor genes. In: Mendelsohn J, Howley PM, Israel, MA, Gray JW, Thompson CB, editor. *The Molecular Basis of Cancer.* 3rd ed. Philadelphia, PA: Saunders/Elsevier; 2008. pp. 31–38.

14. Jemal A, Siegel R, Ward E, Hao Y, Xu J, Thun MJ. Cancer statistics, 2009. *CA Cancer J Clin* 2009;59:225–249.

15. Hanahan D, Weinberg RA. The hallmarks of cancer. *Cell* 2000;100:57–70.

16. Veronesi U, Boyle P, Goldhirsch A, Orecchia R, Viale G. Breast cancer. *Lancet* 2005;365:1727–1741.

17. Macrinici V, Romond E. Clinical updates on EGFR/HER targeted agents in early-stage breast cancer. *Clin Breast Cancer* 2010;10 Suppl 1:E38–46.

18. Narod SA, Foulkes WD. BRCA1 and BRCA2: 1994 and beyond. *Nat Rev Cancer* 2004;4:665–676.

19. Baade PD, Coory MD, Aitken JF. International trends in prostate-cancer mortality: The decrease is continuing and spreading. *Cancer Causes Control* 2004;15:237–241.

20. Villeneuve PJ, Mao Y. Lifetime probability of developing lung cancer, by smoking status, Canada. *Can J Public Health* 1994;85:385–388.

21. Wakelee HA, Chang ET, Gomez SL, Keegan TH, Feskanich D, Clarke CA, Holmberg L, Yong LC, Kolonel LN, Gould MK, West DW. Lung cancer incidence in never smokers. *J Clin Oncol* 2007;25:472–478.

22. Navani N, Spiro SG, Janes SM. Mediastinal staging of NSCLC with endoscopic and endobronchial ultrasound. *Nat Rev Clin Oncol* 2009;6:278–286.

23. Cunningham D, Atkin W, Lenz HJ, Lynch HT, Minsky B, Nordlinger B, Starling N. Colorectal cancer. *Lancet* 2010; 375:1030–1047.

24. Dietrich WF, Lander ES, Smith JS, Moser AR, Gould KA, Luongo C, Borenstein N, Dove W. Genetic identification of Mom-1, a major modifier locus affecting Min-induced intestinal neoplasia in the mouse. *Cell* 1993;75:631–639.

25. Altekruse SF KC, Krapcho M, Neyman N, Aminou R, Waldron W, Ruhl J, Howlader N, Tatelovich Z, Cho H, Mariotto A, Eisner MP, Lewis DR, Cronin K, Chen HS, Fever EJ, Stincomb DG, Edwards BK. *SEER Cancer Statistics Review 1975–2007.* Bethesda, MD: National Cancer Institute; 2010.

26. Heare T, Hensley MA, Dell'Orfano S. Bone tumors: Osteosarcoma and Ewing's sarcoma. *Curr Opin Pediatr* 2009;21: 365–372.

27. Lucas DR, Nascimento AG, Sanjay BK, Rock MG. Well-differentiated liposarcoma. The Mayo Clinic experience with 58 cases. *Am J Clin Pathol* 1994;102:677–683.

28. Wu JM, Montgomery E. Classification and pathology. *Surg Clin North Am* 2008;88:483–520, v–vi.

29. Reed W. Recent Researches concerning the etiology, propagation, and prevention of yellow fever, by the United States Army Commission. *J Hyg (Lond)* 1902;2:101–119.

30. Havrda M IM. Primary brain tumors. In: Mendelsohn J, Howley PM, Israel, MA, Gray JW, Thompson CB, editor. *The Molecular Basis of Cancer.* 3rd ed. Philadelphia, PA: Saunders/Elsevier; 2008. pp. 487–493.

31. Gladson CL, Prayson RA, Liu WM. The pathobiology of glioma tumors. *Annu Rev Pathol* 2010;5:33–50.

32. Merghoub T, Polsky D, Houghton AN. Molecular biology of melanoma. In: Mendelsohn J, Howley PM, Israel, MA, Gray JW, Thompson CB, editor. *The Molecular Basis of*

Cancer. 3rd ed. Philadelphia, PA: Saunders/Elsevier; 2008. pp. 463–470.

33. Miller AJ, Mihm MC, Jr. Melanoma. *N Engl J Med* 2006; 355:51–65.

34. Kraemer KH, Lee MM, Andrews AD, Lambert WC. The role of sunlight and DNA repair in melanoma and nonmelanoma skin cancer. The xeroderma pigmentosum paradigm. *Arch Dermatol* 1994;130:1018–1021.

35. Adams RH, Alitalo K. Molecular regulation of angiogenesis and lymphangiogenesis. *Nat Rev Mol Cell Biol* 2007;8:464–478.

36. Gimbrone MA, Jr., Leapman SB, Cotran RS, Folkman J. Tumor dormancy in vivo by prevention of neovascularization. *J Exp Med* 1972;136:261–276.

37. Bergers G, Brekken R, McMahon G, Vu TH, Itoh T, Tamaki K, Tanzawa K, Thorpe P, Itohara S, Werb Z, Hanahan D. Matrix metalloproteinase-9 triggers the angiogenic switch during carcinogenesis. *Nat Cell Biol* 2000;2:737–744.

38. Fidler IJ. The pathogenesis of cancer metastasis: The 'seed and soil' hypothesis revisited. *Nat Rev Cancer* 2003;3:453–458.

39. Fidler IJ. Tumor heterogeneity and the biology of cancer invasion and metastasis. *Cancer Res* 1978;38:2651–2660.

40. Steeg PS, Bevilacqua G, Kopper L, Thorgeirsson UP, Talmadge JE, Liotta LA, Sobel ME. Evidence for a novel gene associated with low tumor metastatic potential. *J Natl Cancer Inst* 1988;80:200–204.

41. Leone A, Flatow U, King CR, Sandeen MA, Margulies IM, Liotta LA, Steeg PS. Reduced tumor incidence, metastatic potential, and cytokine responsiveness of nm23-transfected melanoma cells. *Cell* 1991;65:25–35.

42. Robbins SL, Kumar V, Cotran RS. *Robbins and Cotran Pathologic Basis of Disease.* Philadelphia, PA: Saunders/Elsevier; 2010. xiv, 1450 pp.

43. Massoud TF, Gambhir SS. Molecular imaging in living subjects: Seeing fundamental biological processes in a new light. *Genes Dev* 2003;17:545–580.

44. Wong FC, Kim EE. A review of molecular imaging studies reaching the clinical stage. *Eur J Radiol* 2009;70:205–211.

45. Scherer RL, Kobayashi, H, Lin CP. Cancer nanotechnology offers great promise for cancer research and therapy. In: Welch DR, Lyden D, and Psila K, editors. *Cancer Metastasis: Biologic Basis and Therapeutics.* Cambridge University; in press.

46. de Jong M, Maina T. Of mice and humans: Are they the same?—Implications in cancer translational research. *J Nucl Med* 2010;51:501–504.

47. Detterbeck FC. Management of carcinoid tumors. *Ann Thorac Surg* 2010;89:998–1005.

48. Herth FJ. Bronchoscopy/endobronchial ultrasound. *Front Radiat Ther Oncol* 2010;42:55–62.

49. Eben Rosenthal KZ, editor. *Optical Imaging of Cancer—Clinical Applications.* New York: Springer; 2009.

50. Scherer RL, VanSaun MN, McIntyre JO, Matrisian LM. Optical imaging of matrix metalloproteinase-7 activity in vivo using a proteolytic nanobeacon. *Mol Imaging* 2008; 7:118–131.

51. Blum G, Weimer RM, Edgington LE, Adams W, Bogyo M. Comparative assessment of substrates and activity based probes as tools for non-invasive optical imaging of cysteine protease activity. *PLoS One* 2009;4:e6374.

52. Karellas A, Vedantham S. Breast cancer imaging: A perspective for the next decade. *Med Phys* 2008;35:4878–4897.

53. O'Flynn EA, Wilson AR, Michell MJ. Image-guided breast biopsy: State-of-the-art. *Clin Radiol* 2010;65(4):259–270.

54. Hartgrink HH, Jansen EP, van Grieken NC, van de Velde CJ. Gastric cancer. *Lancet* 2009;374:477–490.

55. De Visschere P, Oosterlinck W, De Meerleer G, Villeirs G. Clinical and imaging tools in the early diagnosis of prostate cancer, a review. *JBR-BTR* 2010;93:62–70.

56. Galasso D, Carnuccio A, Larghi A. Pancreatic cancer: Diagnosis and endoscopic staging. *Eur Rev Med Pharmacol Sci* 2010;14:375–385.

57. Godfrey EM, Rushbrook SM, Carrol NR. Endoscopic ultrasound: A review of current diagnostic and therapeutic applications. *Postgrad Med J* 2010;86:346–353.

58. Schulz-Wendtland R, Fuchsjager M, Wacker T, Hermann KP. Digital mammography: An update. *Eur J Radiol* 2009; 72:258–265.

59. Smith RA, Cokkinides V, Brooks D, Saslow D, Brawley OW. Cancer screening in the United States, 2010: A review of current American Cancer Society guidelines and issues in cancer screening. *CA Cancer J Clin* 2010;60:99–119.

60. Chirieac LR, Flieder DB. High-resolution computed tomography screening for lung cancer: Unexpected findings and new controversies regarding adenocarcinogenesis. *Arch Pathol Lab Med* 2010;134:41–48.

61. Veerappan GR, Cash BD. Should computed tomographic colonography replace optical colonoscopy in screening for colorectal cancer? *Pol Arch Med Wewn* 2009;119:236–241.

62. Shah NB, Platt SL. ALARA: Is there a cause for alarm? Reducing radiation risks from computed tomography scanning in children. *Curr Opin Pediatr* 2008;20:243–247.

63. Huppmann MV, Johnson WB, Javitt MC. Radiation risks from exposure to chest computed tomography. *Semin Ultrasound CT MR* 2010;31:14–28.

64. Devine CE, Mawlawi O. Radiation safety with positron emission tomography and computed tomography. *Semin Ultrasound CT MR* 2010;31:39–45.

65. Tapscott E. Nuclear medicine pioneer: Hal O. Anger. First scintillation camera is foundation for modern imaging systems. *J Nucl Med* 1998;39:15N, 19N, 26N–27N.

66. Hamaoka T, Madewell JE, Podoloff DA, Hortobagyi GN, Ueno NT. Bone imaging in metastatic breast cancer. *J Clin Oncol* 2004;22:2942–2953.

67. Blake GM, Moore AE, Fogelman I. Quantitative studies of bone using (99m)Tc-methylene diphosphonate skeletal plasma clearance. *Semin Nucl Med* 2009;39:369–379.

68. Gnanasegaran G, Cook G, Adamson K, Fogelman I. Patterns, variants, artifacts, and pitfalls in conventional radionuclide bone imaging and SPECT/CT. *Semin Nucl Med* 2009;39:380–395.

69. Eary JF. Nuclear medicine in cancer diagnosis. *Lancet* 1999;354:853–857.

70. Cherry SR. Multimodality imaging: Beyond PET/CT and SPECT/CT. *Semin Nucl Med* 2009;39:348–353.

71. Divgi CR. Molecular imaging of pulmonary cancer and inflammation. *Proc Am Thorac Soc* 2009;6:464–468.

72. Erasmus JJ, Rohren E, Swisher SG. Prognosis and reevaluation of lung cancer by positron emission tomography imaging. *Proc Am Thorac Soc* 2009;6:171–179.

73. Groth SS, Whitson BA, Maddaus MA. Radiographic staging of mediastinal lymph nodes in non-small cell lung cancer patients. *Thorac Surg Clin* 2008;18:349–361.

74. Bouchelouche K, Capala J, Oehr P. Positron emission tomography/computed tomography and radioimmunotherapy of prostate cancer. *Curr Opin Oncol* 2009;21:469–474.

75. Grigsby PW. Nuclear imaging and cervical cancer. *Minerva Ginecol* 2009;61:45–51.

76. Ben-Haim S, Israel O. Breast cancer: Role of SPECT and PET in imaging bone metastases. *Semin Nucl Med* 2009;39:408–415.

77. Costelloe CM, Rohren EM, Madewell JE, Hamaoka T, Theriault RL, Yu TK, Lewis VO, Ma J, Stafford RJ, Tari AM, Hortobagyi GN, Ueno NT. Imaging bone metastases in breast cancer: Techniques and recommendations for diagnosis. *Lancet Oncol* 2009;10:606–614.

78. Gore JC. Principles and practice of functional MRI of the human brain. *J Clin Invest* 2003;112:4–9.

79. Wu JS, Hochman MG. Soft-tissue tumors and tumor-like lesions: A systematic imaging approach. *Radiology* 2009;253:297–316.

80. Sugita R, Ito K, Fujita N, Takahashi S. Diffusion-weighted MRI in abdominal oncology: Clinical applications. *World J Gastroenterol* 2010;16:832–836.

81. Futterer JJ, Barentsz J, Heijmijnk ST. Imaging modalities for prostate cancer. *Expert Rev Anticancer Ther* 2009; 9:923–937.

82. Waldman AD, Jackson A, Price SJ, Clark CA, Booth TC, Auer DP, Tofts PS, Collins DJ, Leach MO, Rees JH. Quantitative imaging biomarkers in neuro-oncology. *Nat Rev Clin Oncol* 2009;6:445–454.

83. Price SJ. The role of advanced MR imaging in understanding brain tumour pathology. *Br J Neurosurg* 2007;21:562–575.

84. Turnbull L, Brown S, Harvey I, Olivier C, Drew P, Napp V, Hanby A, Brown J. Comparative effectiveness of MRI in breast cancer (COMICE) trial: A randomised controlled trial. *Lancet* 2010;375:563–571.

85. Morris EA. Should we dispense with preoperative breast MRI? *Lancet* 2010;375:528–530.

86. Houssami N, Hayes DF. Review of preoperative magnetic resonance imaging (MRI) in breast cancer: Should MRI be performed on all women with newly diagnosed, early stage breast cancer? *CA Cancer J Clin* 2009;59:290–302.

87. Glunde K, Jacobs MA, Pathak AP, Artemov D, Bhujwalla ZM. Molecular and functional imaging of breast cancer. *NMR Biomed* 2009;22:92–103.

88. Schmidt GP, Reiser MF, Baur-Melnyk A. Whole-body MRI for the staging and follow-up of patients with metastasis. *Eur J Radiol* 2009;70:393–400.

89. Puls R, Kuhn JP, Ewert R, Hosten N. Whole-body magnetic resonance imaging for staging of lung cancer. *Front Radiat Ther Oncol* 2010;42:46–54.

90. Mankoff DA. A definition of molecular imaging. *J Nucl Med* 2007;48(6):18N, 21N.

91. Cai W, Chen X. Multimodality molecular imaging of tumor angiogenesis. *J Nucl Med* 2008;49 Suppl 2:113S–128S.

92. Oberg K. Molecular imaging in diagnosis of neuroendocrine tumours. *Lancet Oncol* 2006;7:790–792.

93. van der Meel R, Gallagher WM, Oliveira S, O'Connor AE, Schiffelers RM, Byrne AT. Recent advances in molecular imaging biomarkers in cancer: Application of bench to bedside technologies. *Drug Discov Today* 2010;15:102–114.

94. Elias DR, Thorek DL, Chen AK, Czupryna J, Tsourkas A. In vivo imaging of cancer biomarkers using activatable molecular probes. *Cancer Biomark* 2008;4:287–305.

95. Martinez-Bisbal MC, Celda B. Proton magnetic resonance spectroscopy imaging in the study of human brain cancer. *Q J Nucl Med Mol Imaging* 2009;53:618–630.

96. Jastrzebska B, Lebel R, McIntyre OJ, Paquette B, Neugebauer W, Escher E, Lepage M. Monitoring of MMPs activity in vivo, non-invasively, using solubility switchable MRI contrast agent. *Adv Exp Med Biol* 2009;611:453–454.

97. Jones CK, Li AX, Suchy M, Hudson RH, Menon RS, Bartha R. In vivo detection of PARACEST agents with relaxation correction. *Magn Reson Med* 2010;63:1184–1192.

98. Albert MS, Cates GD, Driehuys B, Happer W, Saam B, Springer CS, Jr, Wishnia A. Biological magnetic resonance imaging using laser-polarized 129Xe. *Nature* 1994;370: 199–201.

99. Viale A, Aime S. Current concepts on hyperpolarized molecules in MRI. *Curr Opin Chem Biol* 2010;14:90–96.

100. Afnan J, Tempany CM. Update on prostate imaging. *Urol Clin North Am* 2010;37:23–25, Table of Contents.

101. Sherman M. Hepatocellular carcinoma: Epidemiology, surveillance, and diagnosis. Semin Liver Dis 2010;30:3–16.

102. Lloyd C, McHugh K. The role of radiology in head and neck tumours in children. *Cancer Imaging* 2010;10:49–61.

103. Price SJ. Advances in imaging low-grade gliomas. *Adv Tech Stand Neurosurg* 2010;35:1–34.

104. Kligerman S, Abbott G. A radiologic review of the new TNM classification for lung cancer. *AJR Am J Roentgenol* 2010;194:562–573.

105. Terreno E, Castelli DD, Viale A, Aime S. Challenges for molecular magnetic resonance imaging. *Chem Rev* 2010; 110:3019–3042.

106. Ganeshalingam S, Koh DM. Nodal staging. *Cancer Imaging* 2009;9:104–111.

107. Kobayashi H, Longmire MR, Ogawa M, Choyke PL, Kawamoto S. Multiplexed imaging in cancer diagnosis: Applications and future advances. *Lancet Oncol* 2010;11: 589–595.

108. Hynynen K. MRI-guided focused ultrasound treatments. *Ultrasonics* 2010;50:221–229.

109. O'Connor JP, Jackson A, Asselin MC, Buckley DL, Parker GJ, Jayson GC. Quantitative imaging biomarkers in the clinical development of targeted therapeutics: Current and future perspectives. *Lancet Oncol* 2008;9:766–776.

110. Virostko J, Xie J, Hallahan DE, Arteaga CL, Gore JC, Manning HC. A molecular imaging paradigm to rapidly profile response to angiogenesis-directed therapy in small animals. *Mol Imaging Biol* 2009;11:204–212.

111. Bentzen SM. Theragnostic imaging for radiation oncology: Dose-painting by numbers. *Lancet Oncol* 2005;6:112–117.

112. Buckler AJ, Mulshine JL, Gottlieb R, Zhao B, Mozley PD, Schwartz L. The use of volumetric CT as an imaging biomarker in lung cancer. *Acad Radiol* 2010;17:100–106.

113. Llovet JM, Di Bisceglie AM, Bruix J, Kramer BS, Lencioni R, Zhu AX, Sherman M, Schwartz M, Lotze M, Talwalkar J, Gores GJ. Design and endpoints of clinical trials in hepatocellular carcinoma. *J Natl Cancer Inst* 2008;100:698–711.

114. Bruix J, Llovet JM. Major achievements in hepatocellular carcinoma. *Lancet* 2009;373:614–616.

115. Lencioni R, Llovet JM. Modified RECIST (mRECIST) assessment for hepatocellular carcinoma. *Semin Liver Dis* 2010;30:52–60.

116. Stein WD, Huang H, Menefee M, Edgerly M, Kotz H, Dwyer A, Yang J, Bates SE. Other paradigms: Growth rate constants and tumor burden determined using computed tomography data correlate strongly with the overall survival of patients with renal cell carcinoma. *Cancer J* 2009;15:441–447.

117. Hamaoka T, Costelloe CM, Madewell JE, Liu P, Berry DA, Islam R, Theriault RL, Hortobagyi GN, Ueno NT. Tumour response interpretation with new tumour response criteria vs the World Health Organisation criteria in patients with bone-only metastatic breast cancer. *Br J Cancer* 2010;102:651–657.

118. Wahl RL, Jacene H, Kasamon Y, Lodge MA. From RECIST to PERCIST: Evolving considerations for PET response criteria in solid tumors. *J Nucl Med* 2009;50 Suppl 1:122S–150S.

119. Desar IM, van Herpen CM, van Laarhoven HW, Barentsz JO, Oyen WJ, van der Graaf WT. Beyond RECIST: Molecular and functional imaging techniques for evaluation of response to targeted therapy. *Cancer Treat Rev* 2009;35:309–321.

120. Harry VN, Semple SI, Parkin DE, Gilbert FJ. Use of new imaging techniques to predict tumour response to therapy. *Lancet Oncol* 2010;11:92–102.

121. Kosaka N, Ogawa M, Choyke PL, Kobayashi H. Clinical implications of near-infrared fluorescence imaging in cancer. *Future Oncol* 2009;5:1501–1511.

122. Kaijzel EL, Snoeks TJ, Buijs JT, van der Pluijm G, Lowik CW. Multimodal imaging and treatment of bone metastasis. *Clin Exp Metastasis* 2009;26:371–379.

123. Rasmussen JC, Tan IC, Marshall MV, Fife CE, Sevick-Muraca EM. Lymphatic imaging in humans with near-infrared fluorescence. *Curr Opin Biotechnol* 2009;20:74–82.

124. Bedard N, Pierce M, El-Nagger A, Anandasabapathy S, Gillenwater A, Richards-Kortum R. Emerging roles for multimodal optical imaging in early cancer detection: A global challenge. *Technol Cancer Res Treat* 2010;9:211–217.

125. Hollenhorst M, Hansen C, Huttebrauker N, Schasse A, Heuser L, Ermert H, Schulte-Altedorneburg G. Ultrasound computed tomography in breast imaging: First clinical results of a custom-made scanner. *Ultraschall Med* 2010.

126. Petralia G, Bonello L, Viotti S, Preda L, d'Andrea G, Bellomi M. CT perfusion in oncology: How to do it. *Cancer Imaging* 2010;10:8–19.

127. Li C, Wang LV. Photoacoustic tomography and sensing in biomedicine. *Phys Med Biol* 2009;54:R59–97.

128. Pickwell-MacPherson E, Wallace VP. Terahertz pulsed imaging—A potential medical imaging modality? *Photodiagnosis Photodyn Ther* 2009;6:128–134.

2

Physics of MRI

Seth A. Smith
Vanderbilt University

2.1 Background

Since its inception in 1946, nuclear magnetic resonance (NMR) imaging has afforded an opportunity to noninvasively study the structure and, ultimately, function of tissues from the biochemical or cellular to the macroscopic or organ level. In 1946, Felix Bloch [1] and Edward Purcell [2] independently described that they could manipulate the residual magnetization of spins that are placed in a magnetic field by irradiating at a frequency corresponding to the energy difference (Larmor frequency) between the high- and low-energy proton spin states. In conjunction with an earlier discovery by Sir Joseph Larmor, they saw that small electromagnetic currents were induced in a nearby radiofrequency (RF) coil, which corresponded to the angular frequency of precession of the nuclei about the field and was subsequently named *nuclear magnetic resonance*.

While NMR was avidly pursued to noninvasively study molecular properties, magnetic resonance imaging (MRI) was not technically born until almost 30 years later when Paul Lauterbur published an article in *Nature* [3] entitled, "Image Formation by Induced Local Interaction; Examples Employing Magnetic Resonance." The monumental finding was that application of magnetic field gradients to two test tubes separated in space could be used to localize the NMR signal, which could be back-projected to form an image of both test tubes. He called this method zeugmatography (from the Greek, ζευγμω, meaning to yoke or bind together) indicating its roots in the yoking together of weak field gradients with the strong main magnetic field. This manuscript, in effect, made the transition from single and multidimensional NMR to multispatial dimension MRI.

A short time later, Sir Peter Mansfield, demonstrated a method to quickly sample, encode, and transform acquired resonant data into an image, coined echo planar imaging, which became the method to move functional MRI into the pragmatic regime [4]. All four of these seminal scientists were awarded the Nobel Prize (Bloch and Purcell, Physics, 1952; Lauterbur and Mansfield, Physiology or Medicine, 2003). Since then MRI has profoundly impacted the diagnosis and treatment of diseases of the human body and 30 years later is a quintessential diagnostic imaging technique.

2.2 Description of Magnetization

All substances can be parsed into their constituents. In the human body, for example, a system is composed of various organs performing a specific function, and each of those organs are made up of tissues that serve a specific purpose. Tissue can be further dissected into cells, and each cell is composed of organelles that are an efficient combination of molecules. Finally, molecules are collections of atoms, which are composed of charged (protons and electrons) and noncharged (neutrons) particles. The punch line is that nuclei composed of protons will exhibit a magnetic field derived from these nuclear, charged particles. Quantum mechanics tells us that a charged particle will exhibit spin, and an associated spin angular momentum. Additionally, particles bearing will also exhibit orbital angular momentum.

Recall that linear momentum, $\vec{p}$, is the product of the mass and the velocity of the object ($\vec{p} = m\vec{v}$). Angular momentum is the angular correlate in that the angular momentum of a particle, $\vec{L} = I\vec{\omega}$, where I is the moment of inertia of the particle and $\vec{\omega}$ is the angular velocity. For example, consider the gyroscope. When the string is pulled, the gyroscope experiences an angular momentum such that it does not simply fall over. Similarly,

TABLE 2.1 Details of Nuclei Active to an MRI Experiment

Nucleus	Spin (I)	Magnetic Moment (μ)	γ(MHz/T)	Concentration
^{1}H	1/2	2.79	42.58	100
^{17}O	5/2	1.89	5.77	50
^{19}F	1/2	2.63	40.08	4×10^{-6}
^{23}Na	3/2	2.22	11.27	8×10^{-2}
^{31}P	1/2	1.13	17.25	7.5×10^{-2}

Note: The most abundant and NMR reactive element is the hydrogen nuclei; thus, it is the focus of many MRI experiments.

protons existing in a nucleus bear a spin angular momentum, I (not to be confused with the moment of inertia), which can assume 0, whole integer, or half integer values. It should be pointed out that nuclei consisting of protons that bear a nuclear spin angular momentum $I = 0$ will be unaffected by magnetic fields, and nuclei with integer or half integer values for I not only can be affected by magnetic fields but also detected using magnetic resonance. Additionally, it should be pointed out that those nuclei with half integer spin numbers are of most interest when considering an MRI experiment.

As a brief aside, the question arises, why half integer spin numbers and what are these nuclei that are most often explored in magnetic resonance? The most abundant nucleus in the human body bearing a half integer spin number is hydrogen (one proton and one electron), especially given its place as part of the water molecule (the most abundant molecule in the human body). Therefore, the bulk of magnetic resonance experiments are performed seeking to exploit the signal that can be obtained from hydrogen molecules. Since protons by themselves exhibit a spin 1/2 number, any nuclei with an odd number of protons will similarly have a half integer spin number. However, there are many nuclei that contain an odd number of protons or neutrons that could be used in MRI, but those listed in Table 2.1 [5] are important biologically. It is important to note at this point that the focus of this book and this chapter is the description and exploitation of magnetization residing in tissues in the context of MRI; therefore, we will confine ourselves to looking

only at nuclei that are important to the biological framework of the human or animal. One thing to point out in the table is the biological concentration of these nuclei. It is clear that the most likely candidate for imaging will be the single proton of the hydrogen nucleus due to its high concentration, especially as found in water, and its strong magnetic moment. For more specific details on spin, the reader is referred to Liboff's *Introductory Quantum Mechanics* [6].

Without loss of generality, let us study the nature of the hydrogen molecule and its relationship to magnetic resonance and signal generation. As was previously noted, charged nuclei will bear spin. Next, a charged, spinning particle will exhibit an orbital angular momentum $\vec{L}$. The spinning nature of a charge gives rise to a magnetic moment, $\vec{\mu}$, defined as

$$\vec{\mu} = \gamma \cdot (\vec{L} + \vec{I}) = \gamma \cdot \vec{J} \qquad (2.1)$$

where $\vec{J}$ is the total angular momentum and γ is the so-called gyromagnetic ratio (which will be discussed shortly).

Quantum mechanics reveals that nuclei placed in an external magnetic field (for the sake of this chapter, defined as B_0 and given in units of tesla) with spin quantum number I has $2I + 1$ nondegenerate (discrete) energy levels (orientations) [6]. As was previously pointed out, it is the nuclei with half integer spins that are of most interest for MRI experiments, and thus for the proton with spin 1/2, only two possible energy states exist: α and β, parallel (low energy) and antiparallel (high energy) to B_0, respectively (Figure 2.1a). Note that in Figure 2.1a, the higher the B_0, the greater the separation between the high- and low-energy states becomes, which will be described briefly. Classically, a magnetic dipole in an external magnetic field will attempt to align itself with the field and thus experience a torque, which is defined as the rate of change of the magnetic moment:

$$\frac{d\vec{J}}{dt} = \vec{\mu} \times \vec{B} \qquad (2.2)$$

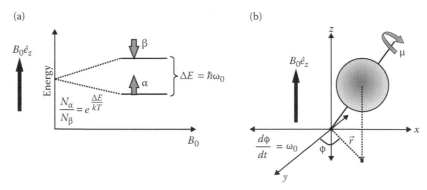

FIGURE 2.1 (a) Energy level diagram for spins existing in a high- (antiparallel to B_0, β) or low-energy (parallel to B_0, α) state. Note that as the B_0 increases, the energy separation (ΔE) increases and the disparity between the numbers of antiparallel spins (N_β) to parallel spins (N_α) will also increase as the energy difference increases. (b) Cartoon of a nuclear spin when tipped off-axis (i.e., the axis of B_0). It possesses magnetic moment, μ, and will precess about the axis of B_0 with angular frequency ω_0.

Recall that the total angular momentum, $\vec{J}$, is related to the magnetic moment of the nucleus, $\vec{\mu}$, and unless perfectly aligned with B_0 the magnetic moment, $\vec{\mu}$, will trace a circular path about the axis of B_0, which in this case is along the z-axis and is referred to as *precession* (Figures 2.1b and 2.2a). The rate of precession is defined as

$$\frac{d\varphi}{dt} = \omega_o = \gamma B_0 \text{ rad/s},\qquad(2.3)$$

and is denoted, ω_o, the Larmor frequency with γ the gyromagnetic ratio. The gyromagnetic ratio is defined as the ratio of the magnetic moment, $\vec{\mu}$, to the angular momentum, $\vec{L}$, of the nucleus of interest and is given in units of MHz/T. Again, we draw the reader to recall the gyroscope. Where once the internal wheel of the gyroscope is spinning, when it is tipped off its axis, it will precess about a fixed point with a rotational frequency related to the spin of the internal wheel. Nuclei with a net magnetic moment exhibit a similar behavior.

We have discovered from quantum mechanics that spins placed in a magnetic field will exhibit $2I + 1$ nondegenerate orientations, spin 1/2 nuclei in an external magnetic field will exist in one of two orientations, α and β. These nuclei, furthermore, are described as existing in the ground state energy level of a harmonic oscillator. The harmonic oscillator is a quantum mechanical analog to the classical harmonic oscillator, which can be thought of as any system that when its mass is displaced from equilibrium experiences a restoring force. In classical physics, the term *harmonic oscillator* can be thought of as a spring with an attached mass, which, when stretched, will experience a force to restore the spring to its original state. This "restorative force" follows Hooke's law, $F = -kx$, where the position, x, is a sinusoidal oscillation and k is the spring constant. Similarly in quantum mechanics, a particle exists in an "equilibrium well" or "ground state" when unperturbed. However, when perturbed, these particles can be promoted to higher and/or lower energy levels and will seek to regain equilibrium back to the ground state, similar to the spring being stretched. Therefore, to draw the reader back to the situation with nuclei, nuclei that are unperturbed are described as existing in the ground state energy level of a harmonic oscillator whose energy levels are described by

$$E_n = \left(n + \frac{1}{2}\right)\hbar\omega_o,\qquad(2.4)$$

where ω_o is the Larmor frequency of the nuclei. Note that the energy levels are quantized and the energy is discretized by $n = 1, 2, 3$, etc. Within the framework of nuclei placed in a magnetic field, nuclei can have either a high- or low-energy state (corresponding to β and α, respectively); i.e., $n = 1$ and $n = 2$. Then the energy difference between the α and β states is

$$\Delta E = \hbar\omega_o.\qquad(2.5)$$

For the longitudinal magnetization, NMR occurs when at least some of the magnetic moments change their spin state (from $\alpha \rightarrow \beta$, or vice versa); i.e., a perturbation is applied to promote the nuclei to a higher energy state. The most simplistic NMR or MRI experiment applies a transient, perpendicular magnetic field, B_1, to the system residing in a static magnetic field, B_0. The resonance condition, or the situation in which an applied external magnetic field, B_1, can influence the transition of a nuclear spin from an α to β energy level, must satisfy the following mathematical formalism:

$$\hbar\omega_1 = |E_\alpha - E_\beta|,\qquad(2.6)$$

where $\omega_1 = \gamma B_1$. Therefore, B_1 has rotational frequency on the order of the energy spacing between the two spin states. For typical MRI scanners, the frequency of B_1 will reside in the RF range and is termed applied RF field. The effect of B_1 on $\vec{\mu}$ is (1) a tipping away from the axis of B_0 and (2) subsequent precession. The amount of tip (flip angle, α) is related to the magnitude of the applied, time-varying field (B_1) over a specified amount of time (t) [7]:

$$\alpha = \gamma \int_{t=0}^{t_f} B_1(t)\,dt\qquad(2.7)$$

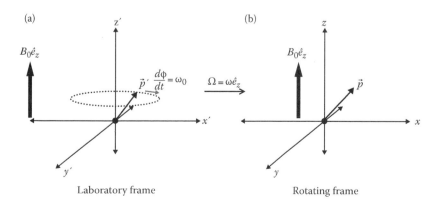

Laboratory frame Rotating frame

FIGURE 2.2 Diagram of a vector, $\vec{p}'$, in the laboratory (stationary frame), which will precess with rotational frequency, ω_0. When a transformation to a rotating reference frame, which rotates with Larmor frequency ω_0, is performed, the transformed vector $\vec{p}$ appears static in time.

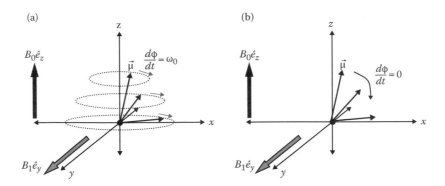

FIGURE 2.3   (a) Diagram of the effect of a B_1 RF field on a magnetic moment, μ as seen from the laboratory frame. Note that the rotation of μ is complex and is described as a precession (about B_0) and a nutation (about B_1). The amount of nutation is determined by the duration and power of the B_1 RF field. (b) However, when transformed to the rotating reference frame, the magnetic moment, μ is seen to only nutate (about B_1).

One can imagine that, since the effect of B_1 is not instantaneous, the magnetic moment will undergo two rotations (Figure 2.3a). The first is precession about B_0 once tipped away from parallel, and the second, azimuthal tip defined as nutation [7]. After application of B_1, the magnetic moment will corkscrew down to the transverse plane at an angle α, with respect to B_0. It should be clear, that further involvement of B_1 on the spin will only be effective if B_1 is perpendicular to B_0 and rotating at the same frequency as $\vec{\mu}$. The reader is further encouraged to note the distinction between the flip angle, α, and the low-energy state of a magnetic moment, α. Some manuscripts will refer to the flip angle by a Greek symbol θ, but in order to be consistent with the bulk of literature, we will refer to the flip angle by α when discussing a pulse sequence and reserve the energy level, α, for discussion of the quantum mechanics of MRI.

2.3 Rotating Reference Frame

A useful concept in understanding NMR is the idea of a rotating reference frame. By definition, a reference frame is any coordinate system employed to measure the position, orientation, and other properties of objects in it, or it may even refer to an observational reference frame tied to the state of motion of the observer. It is well known that the frame in which the measurement is made dictates the absolute magnitude of the measurement. For example, an apple will weigh less on the moon than it would if it were weighed on earth. Thus, while the mass stays the same (invariant under transformation), the observed weight (weight = $m \cdot g$, where m is the mass and g is the gravitational acceleration) is different (not invariant under transformation). While this principle describes a static measurement, the theory of relativity says that two observers looking at the same causal event will not see the same event if they are moving relative to each other [6]. An example of this is that a train moving past a station appears to be moving according to the observer standing on the platform. However, if the observer were to run alongside the train (at the same rate the train is moving) for a certain distance, then the quantities within the train will appear not to

move. In MRI, we employ a very similar set of principles to simplify the description of the time evolution of the magnetization after a perturbation. At this point, we have described a reference frame in terms of static events (i.e., the location of the measurement is changed) and temporal events (the linear movement of the observer with respect to the measurement); however, we have not described the rotational event (the rotation movement of the observer with respect to the measurement). Similar to the linearly moving observer, the rotational observer will proceed "around" the observation while the measurement is being made.

As we have just described in the previous section, the magnetization vector in an NMR or MRI experiment when influenced by an external magnetic field (B_1) will undergo two rotations: (1) precession and (2) nutation, when observed in the laboratory frame. That is, the observation is made by a stationary detector (or observer). However, a very simple transformation reduces the time evolution of the magnetization vector to only one rotation: nutation (Figure 2.2b). In the context of MRI, we will consider two reference frames: (1) the laboratory or stationary reference frame and (2) the rotating reference frame.

The laboratory frame (Figure 2.3a) was just described as a coordinate system (x, y, z) as viewed from a stationary observer. In this frame, a magnetic moment, $\vec{\mu}$, when tipped off axis with respect to B_0 will precess. Alternatively, the rotating reference frame (Figure 2.3b) can be described as a coordinate system (x', y', z') rotating with frequency

$$\Omega = \omega \hat{e}_z \qquad (2.8)$$

about the z-axis (where $\hat{e}_z$ denotes the unit vector along the axis parallel to B_0, usually taken to be along z) and the laboratory frame is related to the rotating reference frame by the spatial transformation:

$$x = x' \cos(\omega t) - y' \sin(\omega t)$$
$$y = x' \sin(\omega t) + y' \cos(\omega t) \qquad (2.9)$$

Mathematically, any vector, $\vec{p}$, in the rotating reference frame rotating with frequency ω, can be related to the laboratory frame by the following relationship:

$$\vec{p} = p_x + p_y + p_z$$

$$\begin{bmatrix} p_x \\ p_y \\ p_z \end{bmatrix} = \begin{bmatrix} \cos(\omega t) & -\sin(\omega t) & 0 \\ \sin(\omega t) & \cos(\omega t) & 0 \\ 0 & 0 & 1 \end{bmatrix} \begin{bmatrix} p'_x \\ p'_y \\ p'_z \end{bmatrix} \quad (2.10)$$

$$\vec{p} = \underline{R}\vec{p}'$$

where $\underline{R}$ is the so-called rotation matrix. Additionally, a vector $\vec{p}'$ in the laboratory frame can be related to a vector $\vec{p}$ in the rotating reference frame rotating with frequency ω, and can be described as

$$\vec{p} = p'_x + p'_y + p'_z$$

$$\begin{bmatrix} p'_x \\ p'_y \\ p'_z \end{bmatrix} = \begin{bmatrix} \cos(\omega t) & \sin(\omega t) & 0 \\ -\sin(\omega t) & \cos(\omega t) & 0 \\ 0 & 0 & 1 \end{bmatrix} \begin{bmatrix} p_x \\ p_y \\ p_z \end{bmatrix} \quad (2.11)$$

$$\vec{p}' = \underline{R}^T \vec{p}'$$

where $\underline{R}^T$ is the matrix transpose of the rotation matrix, $\underline{R}$. Note that in the rotating reference frame, the originally precessing vector, $\vec{p}'$, now appears as a static vector, $\vec{p}$.

So, the question becomes, "Why go through all of this work?" The result of transforming all fields and trajectories of the magnetic moment into the rotating frame is astounding. We have seen that the effect of an applied field, B_1, on the net magnetic moment, $\vec{\mu}$, results in a compound rotation: (1) precession at frequency $\omega = \omega_0$ and (2) nutation α. Thus, in the laboratory frame, the effect of B_1 on the magnetic moment results in corkscrewing paths of $\vec{\mu}$ after a B_1 pulse. However, in the rotating reference frame, when the rate of rotation of the rotating reference frame is equal to the Larmor frequency (i.e., $\omega = \omega_0$), this complex motion now appears as a simple azimuthal tip, or nutation, and is shown in Figure 2.2b. For ease, all terminology will be further defined with respect to the rotating reference frame. Description of the rotating reference frame was adapted from reference [8].

2.4 Definition of Magnetization

In reality, however, our system is composed of many protons, and thus for a system of N identical magnetic moments placed in an external magnetic field, not all spins completely align to B_0. At this point, it is instructive to define the magnetization, M, of a system, in the context of N dipoles. Note that in the zero-temperature limit, all spins align with B_0 and complete magnetization is formed. However, due to thermal agitation, when a system of magnetic moments is placed in a magnetic field and

thermally equilibrated, a slight disparity in population forms between spins parallel (N_α) and antiparallel (N_β) to the field and is related by the Boltzmann factor [9]:

$$\frac{N_\alpha}{N_\beta} = e^{\frac{\Delta E}{kT}} \quad (2.12)$$

where k is the Boltzmann constant (1.381×10^{-34} J/K) and T is the temperature (in kelvin) and $N = N_\alpha + N_\beta$. At standard clinical magnetic field strength at body temperature (37°C), the slight excess is on the order of 1 part in 10^6. The observable magnetization, then, is the vector sum of the individual magnetic moments, μ, and their angular orientation with respect to a particular axis, θ, of the system:

$$M_z = N\langle \mu\cos(\theta)\rangle = \frac{N}{kT}\frac{\partial}{\partial B}\ln(Q_1(B_0,T)) \quad (2.13)$$

where Q_1 is the single spin partition function [9], B_0 is the externally applied field, and θ is the angle between μ and B_0. Integrating over the elemental solid angle ($\sin\theta d\theta d\varphi$) representing a small range of orientations of the dipoles, and with N dipoles per unit volume, the total z-magnetization is

$$M_z = N\mu L(x) \quad (2.14)$$

where L(x) is the Langevin function [9] and $x = \frac{\mu B_0}{kT}$. In the weak field limit, the net equilibrium z-magnetization is approximately

$$M_0^z \approx \frac{N\mu^2}{kT}B_0 = N\frac{\gamma^2 h^2}{4}\frac{B_0}{kT} \quad (2.15)$$

where M_0^z is parallel to B_0 and is denoted longitudinal magnetization.

At this point, we have described the notion that spins placed in a magnetic field exhibit a polarization, the degree of polarization is determined by the Boltzman distribution, and that transitions between energy levels (magnetic resonance) can occur when an external magnetic field is applied on the order of the energy spacing between the nuclear energies. Additionally, when this perturbation is applied, we have described the subsequent rotations and nutation of the nuclear magnetic moment and, when many spins are considered, the formalism for the magnetization of a sample.

2.5 Relaxation

One point that was not mentioned in the previous sections is that when a system of protons exists in a substance, they exert and are influenced by interactions with other magnetic moments in the tissue. Thus, after a perturbation, the signal obtained is time varying due to the nature of the tissue, the interactions within the tissue, and applied, external forces. The observable ramification is that the signal obtained in an MRI experiment is time varying and the extent to which the signal varies is related to the tissue type. The umbrella term that describes the time evolution of the observable signal is called relaxation. The punch line and one of the most important principles in MRI is that the magnetization in disparate tissues gives rise to different relaxation and thus different observed signals [10,11]. This is simply described as contrast. The ability for the observed signal to report on the tissue in which the protons subsist is critical to the usage of MRI as a tool for measuring the health and welfare of tissues in the human body. Two main relaxation types are described here: longitudinal (T_1) and transverse (T_2), and a more thorough examination of quantitative relaxometry will be developed in Chapter 5.

2.5.1 Longitudinal Relaxation: T_1

As was described earlier in this chapter, when nuclei with a net magnetic moment are placed in an external magnetic field, the magnetic moment will seek to align in a low-energy state that is parallel (or antiparallel) to the external magnetic field. Heuristically, when the net z-magnetization is perturbed such that the magnetic moment is tipped away from parallel to B_0, it will seek to regain the equilibrium state by dissipation of the applied energy (since energy is put into the system to tip the magnetization away from parallel) and will eventually relax back to the low-energy state of being parallel to B_0. This is described as longitudinal relaxation, spin-lattice relaxation, or simply, T_1 relaxation [7]. The definition of the T_1 time is the time that it

takes for approximately 63% of the z-magnetization to return to equilibrium and is governed by a solution to the z-component of the Bloch equations (the mathematical formalism for the Bloch equations will be Section 2.6) [12]:

$$M_z = M_o(1 - e^{-t/T_1}) \qquad (2.16)$$

where M_z is the longitudinal magnetization, M_o is the equilibrium magnetization, t is the time, and T_1 is the relaxation time.

Different tissues exhibit different T_1 relaxation times [10,13]. An example of Equation 2.16 for different types of tissue, each with different T_1 values, is given in Figure 2.4a. In this example, breast adipose tissue has the shortest T_1 ($T_1 = 423$ ms [14]), that is, M_z recovers back to equilibrium ($M_z = 1$) in a shorter amount of time than blood ($T_1 = 1923$ ms [15,16]). Additionally, the longer the T_1, the less steep the relaxation curve is. The black dotted line indicates the 63% criterion for determining the T_1 value. For each of these curves, a corresponding point on the x-axis for which the M_z recovery curve passes the black dotted line is the T_1 value. An additional point to note is that tissues of the central nervous system (white matter, gray matter, spinal cord) demonstrate T_1 values very similar in magnitude. It is for this reason that much attention has been paid to devising MRI methods to detect contrast in the central nervous system. T_1 values (at 3 T) for each of the tissues plotted in Figure 2.4a are given in Table 2.2 [14,16,17].

Interestingly, T_1 is known to be field-dependent. What this means is that the T_1 measured in a particular tissue at one field strength (e.g., 1.5 T) will not be the same value measured at

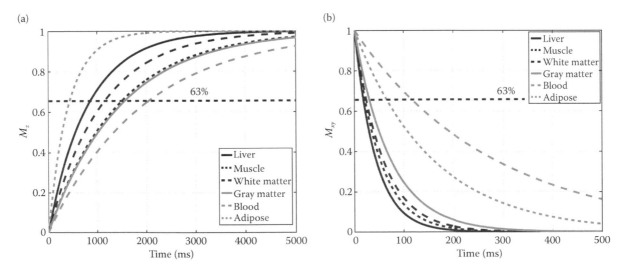

FIGURE 2.4 (a) T_1 recovery curves for sample tissues. T_1 relaxation time can be pictorially described at the time (x-axis) at which approximately 63% (black dotted line) of the z-magnetization (M_z) has recovered. Note that the tissues with the shortest T_1 (breast adipose tissue) recover most rapidly, while those with longer T_1. Also, note that T_1 values are field strength dependent and the values used in this figure were taken from the literature at 3 T [14–17]. (b) T_2 decay curves for sample tissues. T_2 relaxation time can be described as the time (x-axis) at which approximately 63% (black dotted line) of the xy-components of the magnetization (M_{xy}) have decayed. Note that tissues with the shortest T_2 times decay rapidly (liver), while those tissues that more similarly reflect water (blood) have much longer T_2 times. It is also important to note that while T_2 values are expected not to change as a function of field strength, the values reported here are taken from the literature at 3 T [14–17].

TABLE 2.2 T_1 and T_2 Values for Tissues in the Human Body

	T_1 (ms)	T_2 (ms)
Liver	812	42
Muscle	1412	50
Kidney	1194	56
White matter	1084	56
Gray matter	1470	71
Spinal cord	993	78
Blood	1932	275
Adipose	423	154

Note: T_1 and T_2 values were taken from the literature at 3 T [14–17].

other field strengths (e.g., 3 T). This field dependence has been described by Bottomley et al. [10] and is given as a power-law dependency:

$$T_1 = A\omega_0^B \qquad (2.17)$$

where A and B are constants related to the tissue of interest and ω_0 is the Larmor frequency given by $\omega_0 = \gamma B_0$ and B_0 is the magnetic field strength. It should be noted that this empirically determined relationship has not been tested for ultra-high field strengths (> 128 MHz or >3 T), so caution should be taken when exploring this in detail. However, as it influences the thought process of what to expect when performing T_1 measurements at higher field, this equation gives the simple interpretation that the higher the field, the longer the T_1.

2.5.2 Transverse Relaxation: T_2

Earlier in this chapter, we discussed that if the magnetization is tipped away from its equilibrium orientation (i.e., parallel to B_0), then it will precess with Larmor frequency ω_0. Thus, the transverse (x, y) components of the magnetization vector are seen to rotate about B_0 (oriented along the z-axis) with a particular frequency. However, again, protons exists in environments that are not completely homogeneous; thus, the energetic interactions in each tissue type give rise to a change in the precession of the transverse magnetization vector. Therefore, local field inhomogeneities that are experienced by the protons result in a dispersion of precession. Simply speaking, initially after a tip away from parallel to B_0, all of the spins are in a coherent "pack." As time evolves, this coherent collection of spins experience local field fluctuations and ultimately lose their coherence exponentially and are said to relax. The terminology for the loss of coherence of the transverse components is called transverse relaxation, spin–spin relaxation, or simply, T_2 relaxation. The definition of T_2 is the time at which approximately 63% of the original spin packet has lost its coherence and is governed by a solution to the transverse components of the Bloch equations:

$$M_\perp = M_o e^{-t/T_2} \qquad (2.18)$$

Just as the observed T_1 is different for different tissues, T_2 exhibits the same behavior; however, contrary to T_1, T_2-relaxation governed by Equation 2.18 exponentially "decays" rather than "recovers." That is, as a function of time, the M_{xy} component exponentially becomes more incoherent and thus the observed signal (explored later) will decay. For each tissue type, the T_2 relaxation time can be quite different. An example set of decay curves for different tissues in the human body is shown in Figure 2.4b. For each of these curves, a corresponding point on the x-axis for which the M_{xy} decay curve passes the black dotted line is the T_2 value. Tissues that are more densely packed, such as muscle and tissue in the central nervous system (white matter, gray matter, spinal cord) exhibit shorter T_2 relaxation times, whereas less densely packed tissues such as adipose tissue or blood show a much longer T_2-relaxation.

It is interesting to note that liver has the shortest T_2 time of the tissues examined. Recall that the decay of transverse magnetization is driven by spins experiencing inhomogeneous fields. The liver is dense in iron (in the form of hemosiderin), and this excess iron concentration causes the spins in the liver to experience a dramatically different field than they would in normal tissues and results in a reduced T_2 relaxation time. The absolute T_2 values (taken at 3 T) for each of the tissues plotted in Figure 2.4b are given in Table 2.2 [14,16,17].

One point to note is that unlike T_1, T_2 has been shown not to vary as significantly with field strength [18]. One way to rationalize this is that T_2-relaxation is driven by the interaction between spins, rather than the interaction with the magnetic field to reattain equilibrium. Since the interaction between spins is largely due to their proximity, orientation, and dipole–dipole interaction, the field strength plays a minor role in determining the time it takes for coherence to be lost.

2.5.3 More on T_2

In the previous section, an explanation of why the liver has a significantly lower T_2 when compared to other tissues was given. While correct, it is only heuristic; therefore, at this point, it is necessary to further develop the idea of T_2 as it pertains to susceptibility related experiments described in future chapters. For simplicity, introductory NMR texts often use T_2 to universally describe the decay of transverse magnetization. However, this is not the complete picture. Two factors actually contribute to the transverse decay: molecular interactions (T_2) and static local field inhomogeneities (T_2'). When performing an NMR experiment, it is often found that the NMR signal decays faster than transverse relaxation time, T_2, alone would predict, and thus the observed transverse relaxation is defined as T_2^*. Pure T_2 decay is a function of completely random interactions between spins. The assumption is that the main external B_0 field and the local, experienced field are absolutely homogeneous. However, every tissue has a different magnetic susceptibility, or local field, which distorts the field at tissue borders, particularly at air–tissue interfaces. This can radically reduce the observed transverse relaxation time. The relationship between the observed T_2^* resulting from both molecular interactions (T_2) and static field inhomogeneity (T_2') is

$$\frac{1}{T_2^*} = \frac{1}{T_2} + \frac{1}{T_2'} \qquad (2.19)$$

It should be noted that there are NMR and MRI experiments that are designed to mitigate the T_2' component (e.g., spin echo) via refocusing the static field inhomogeneities by employing a 180° pulse, but these techniques will not be discussed in detail here. If the reader is interested in a greater knowledge of this phenomenon, then refer to reference [7].

2.6 Mathematical Formalism for Relaxation: Bloch Equations

While the previous section was designed to give the reader an intuition into the mechanisms and the realization of relaxation of spins placed in a magnetic field, this section demonstrates the mathematical formalism for these same intuitions [12]. Felix Bloch (1905–1983) wrote down the so-called Bloch equations, which relate the time evolution of the magnetization to (1) the static and applied magnetic fields, (2) the relaxation times, and (3) the diffusion process (i.e., the thermal, random motions of spins within an environment). As the diffusion process is beyond the scope of this chapter (but will be considered in Chapter 7), we will focus on the relaxation process, and give a special case for the interaction with the magnetic fields. However, for completeness, the Bloch equations for the evolution of the magnetization over time are as follows:

$$\frac{d\vec{M}}{dt} = \gamma \vec{M} \times \vec{B} - \frac{M_x + M_y}{T_2} + \frac{(M_0 - M_z)}{T_1} + D\nabla^2 \vec{M} \qquad (2.20)$$

where $\vec{M}$ is the magnetization vector (with x-, y-, and z-components), $M_{x,y,z}$ are the x, y (transverse), and z (longitudinal) components of the magnetization vector, T_1 and T_2 are the longitudinal and transverse relaxation times, respectively, and D is the diffusion coefficient of the nuclear species under investigation. If we ignore the diffusion component, we can write the Bloch equations as

$$\frac{d\vec{M}}{dt} = \gamma \vec{M} \times \vec{B} - \frac{M_{x,y}}{T_2} + \frac{(M_0 - M_z)}{T_1} \qquad (2.21)$$

which can be further broken down into their components:

$$\frac{dM_x}{dt} = -\frac{M_x}{T_2} + \gamma M_y B_0$$

$$\frac{dM_y}{dt} = -\frac{M_y}{T_2} - \gamma M_x B_0 \qquad (2.22)$$

$$\frac{dM_z}{dt} = \frac{(M_0 - M_z)}{T_1}$$

The solutions to the differential equations in Equation 2.22 are given by

$$M_x = e^{-t/T_2}\left(M_x(0)\cos(\omega_0 t) + M_y(0)\sin(\omega_0 t)\right)$$

$$M_y = e^{-t/T_2}\left(-M_x(0)\sin(\omega_0 t) + M_y(0)\cos(\omega_0 t)\right) \qquad (2.23)$$

$$M_z = M_0 + \left(M_z(0) - M_0\right)e^{-t/T_1}$$

where $M_{x,y,z}(0)$ is the initial value for the x, y, and z components of the magnetization, and M_0 is the equilibrium magnetization.

Now consider two principles. First, use the complex representation for the transverse magnetization, and let $M_{xy} = M_x + iM_y$ where i is defined as $i = \sqrt{-1}$. Second, considering that immediately after 90° excitation, $M_z(0) = 0$, then we can write in a compact form the solution to the Bloch equations:

$$M_{xy} = M_{xy}(0)e^{(i\omega_0 t - t/T_2)}$$

$$M_z = M_0\left(1 - e^{-t/T_1}\right) \qquad (2.24)$$

Finally, consider transforming to the rotating reference frame as seen in the previous section, which rotates with Larmor frequency, ω_0. Then the solution to the Bloch equations is as follows:

$$M_{xy} = M_{xy}(0)e^{(-t/T_2)}$$

$$M_z = M_0\left(1 - e^{-t/T_1}\right) \qquad (2.25)$$

2.7 Formation of an Observable Quantity: Signal

At this point, we have described the existence of nuclear magnetization, how that magnetization is perturbed and eventually evolves, both heuristically and mathematically. However, magnetization in its equilibrium form (i.e., aligned parallel to B_0) is not directly detectable. Nevertheless, if it were detectable, how would one detect it? What is the relationship between the perturbation and subsequent evolution of magnetization and the observed signal, which will ultimately give rise to images? We have learned at this point that magnetic moments and the magnetization when tipped away from equilibrium will precess with a particular frequency. Additionally, from elementary physics, we know that moving charge is defined as current. Finally, we know from Faraday's principle of electromagnetic induction that a moving charge in a magnetic field will induce a current in a nearby circuit, or wire. At this point, it becomes clear how to measure the magnetization in a tissue. The simplest experiment is to tip the longitudinal magnetization away from its equilibrium stance, place a loop of wire perpendicular near to the path of precession, and record the induced current. While a basic description, this is precisely how an NMR or MRI experiment is performed. While many sophisticated experiments have been devised to study the properties of tissues through the association of the observed signal and the underlying tissue, the basics are the same as those just given. In a formal manner, the generation of an NMR or MRI signal follows.

When magnetization is tipped off its axis, it will precess with Larmor frequency, ω_0. To measure this, we follow Faraday's principle of electromagnetic induction, which says that a current will be induced in a nearby closed loop when the magnetic flux through a surface changes:

$$EMF = -\frac{d\Phi_B}{dt} \qquad (2.26)$$

where *EMF* is the induced electromotive force (in volts) and Φ_B is the magnetic flux. Here the change in magnetic flux is a result of a magnetization precessing about B_0 and thus creates a time varying magnetic perturbation (i.e., flux). Equation 2.11 can now be rewritten in terms of the observed signal in an MRI experiment by the following:

$$S \approx C\frac{\Delta\Phi_B}{\Delta t} = C\omega_o M_\perp \qquad (2.27)$$

where *C* is a constant relating to the reception coil properties (sensitivity, geometry) and $M_\perp$ is the magnetization of that sample that has been tipped away from the axis of B_0. Recall in the first section, we derived the formalism for the equilibrium magnetization as

$$M_o^z \approx \frac{N\mu^2}{kT}B_0 = N\frac{\gamma^2\hbar^2}{4}\frac{B_o}{kT} \qquad (2.28)$$

In this experiment, we assume that all of the equilibrium magnetization has been nutated into the transverse plane such that

$$M_\perp = M_0^z \qquad (2.29)$$

However, this can easily be modified for any flip angle (α) as $M_\perp = M_o \sin(\alpha)$. With these tools in our hands, we can derive the signal in an MRI experiment (i.e., the induced current in a nearby close loop) as

$$S \approx C\frac{\Delta\Phi_B}{\Delta t} = C\omega_o M_\perp = C\gamma B_o N\frac{\gamma^2\hbar^2}{4}\frac{B_o}{kT} \qquad (2.30)$$

One thing that is of utmost importance that the reader can take from this knowledge is that the signal formed in a typical MRI experiment is related to the square of the field strength, B_0^2. Systems employed for routine clinical use generally are limited to 3.0 T and below; however, many research environments have human scanners (at the time of this writing) as high as 9.4 T (CMRR, University of Minnesota) and animal scanners employing field strengths up to 21 T and greater. The take home message here is

that with the advent of higher field strength scanners, the greater the potential signal can be. A heuristic explanation of why this is (in addition to the result of Equation 2.30) can be derived from the distribution of spins predicted from the Boltzmann distribution (Equation 2.12). From this formalism, it can be seen that at higher field, there will be a greater number of excess spins, which contribute to an increase in the available magnetization that can be manipulated for imaging purposes. As will be seen in future chapters, the more signal that can be obtained, the greater the opportunity for (1) increased speed of acquisition, (2) higher resolution, or (3) sensitivity to subtle tissue characteristics via spectroscopy and other methods.

One caveat to the reader is that while the signal that can be obtained in an MRI experiment goes with the square of the field strength, the noise in an experiment can vary linearly with the field strength. Therefore, the net signal-to-noise gain of increasing the field strength is actually linear, rather than quadratic. The remainder of signal formation will be discussed in Chapter 4.

2.8 Final Thoughts: From Signal to Image

The final section of this chapter deals with the basic relationship between the observed signal and the underlying density of protons in the tissue. Specific mathematical aspects of this relationship will be explained in Chapter 4. At this point, we have generated an observable MR signal through manipulation of the bulk proton magnetic moment and have written equations describing the time-varying signal changes. However, MRI is unique from conventional NMR in that an image can be formed where each image element (voxel in 3-space and pixel in 2-space) contains information about the underlying tissue. If a 90° B_1 RF field (so-called pulse) is applied at ω_o, the bulk magnetization will nutate to the transverse plane and will subsequently decay with T_2^*. A receiver set to record this signal will report a sinusoidal signal exponentially decaying with T_2^* and is called the free induction decay (FID). For a single proton, the FID is simply one frequency of oscillation and one decay constant T_2^*. For many protons distributed about the sample, the FID is a convolution of sines and cosines of multiple frequencies at different locations in 3-space (x,y,z); the result is complex. To gain information about the decay and ultimately an image of signal intensities, it is pragmatic to transform from time space used in conventional NMR to spatial frequency space. The spatial frequency dependence and intensity of a signal are related by the Fourier transform. The signal of an MRI experiment can be written as the sum over all proton densities, $\rho(x,y,z)$, distributed in the volume, Γ:

$$S = \iiint_\Gamma \rho(x,y,z)\,dx\,dy\,dz \qquad (2.31)$$

To manipulate signal intensity of the single water resonance spatially, linear magnetic field gradients are applied to the sample to "encode" their location in space, which can be back projected to form an image, which is related to the proton density, ρ, by the Fourier transform:

$$S(k_x, k_y, k_z) = \iiint_\Gamma \rho(x, y, z) e^{-i(k_x x + k_y y + k_z z)} \, dx \, dy \, dz \qquad (2.32)$$

where k_x, k_y, k_z are related to the integral of the gradient wave forms. From Equation 2.19, one can easily see that a set of intensities $S(k_x, k_y, k_z)$ can be recorded and located on a k-space grid that can then be inversely Fourier transformed to give a 3D image of the bulk proton signal $\rho(x,y,z)$ for each voxel.

2.9 Conclusion

The focus of this chapter was to describe the existence of nuclei in a sample, the nature of the influence of the external (B_0) and external (B_1) magnetic fields on the nuclei in a sample, and to provide a heuristic examination of the interaction (relaxation) of the nuclei after an applied perturbation. While this explanation is not complete in its scope, the goal is to provide the reader with the understanding of how nuclei evolve in a simple NMR or MRI experiment and how this evolution can be mapped into an observable quantity or signal. The following chapters will explore how these measurements are made, the hardware of an MRI system, and specific cases of spin evolution. For more information, the interested reader is asked to consult reference [7].

References

1. Bloch F, Hansen WW, Packard M. Nuclear induction. *Phys Rev* 1946;69:127.
2. Purcell EM, Torrey HC, Pound RV. Resonance absorption by nuclear magnetic moments in a solid. *Phys Rev* 1946;69:37.
3. Lauterbur P. Image formation by induced local interactions: Examples employing nuclear magnetic resonance. *Nature* 1973;242:190–191.
4. Mansfield P. Multi-planar image formation using NMR spin echoes. *J Phys C* 1977;10:L55.
5. Bushberg JT, Seibert JA, Leidholdt EM, Boone JM. *The Essential Physics of Medical Imaging*. Philadelphia: Lippincott, Williams & Wilkins; 2002.
6. Liboff RL. *Introductory Quantum Mechanics*. Reading, MA: Addison-Wesley; 1998.
7. Haacke EM, Brown RW, Thompson MR, Venkatesan R. *Magnetic Resonance Imaging: Physical Principles and Sequence Design*. New York: Wiley-Liss; 1999.
8. Bernstein MA, King KF, Zhou XJ. *Handbook of MRI Pulse Sequences*. Burlington: Elsevier; 2004.
9. Pathria RK. *Statistical Mechanics*. Oxford: Butterworth-Heinemann; 1999.
10. Bottomley PA, Foster TH, Argersinger RE, Pfeifer LM. A review of normal tissue hydrogen NMR relaxation times and relaxation mechanisms from 1–100 MHz: Dependence on tissue type, NMR frequency, temperature, species, excision, and age. *Med Phys* 1984;11:425–448.
11. Bottomley PA, Hardy CJ, Argersinger RE, Allen-Moore G. A review of 1H nuclear magnetic resonance relaxation in pathology: Are T1 and T2 diagnostic? *Med Phys* 1987;14:1–37.
12. Bloch F. The principle of nuclear induction. *Science* 1953;118(3068):425–430.
13. Damadian R, Zaner K, Hor D, Dimaio T. Human tumors by NMR. *Physiol Chem Phys* 1973;5:381–402.
14. Edden RAE, Smith SA, Barker PB. Longitudinal and multi-echo transverse relaxation times of normal breast tissue at 3 tesla. *J Magn Reson Imaging* 2010; in press.
15. Lu H, Clingman C, Golay X, van Zijl PC. Determining the longitudinal relaxation time (T1) of blood at 3.0 Tesla. *Magn Reson Med* 2004;52:679–682.
16. Stanisz GJ, Odrobina EE, Pun J, Escaravage M, Graham SJ, Bronskill MJ, Henkelman RM. T_1, T_2 relaxation and magnetization transfer in tissue at 3T. *Magn Reson Med* 2005;54:507–512.
17. Smith SA, Edden RA, Farrell JA, Barker PB, Van Zijl PC. Measurement of T1 and T2 in the cervical spinal cord at 3 tesla. *Magn Reson Med* 2008;60:213–219.
18. Parizel PM, Van Riet B, van Hasselt BA, Van Goethem JW, van den Hauwe L, Dijkstra HA, van Wiechen PJ, De Schepper AM. Influence of magnetic field strength on T2* decay and phase effects in gradient echo MRI of vertebral bone marrow. *J Comput Assist Tomogr* 1995;19:465–471.

3

Magnetic Resonance Imaging: Hardware and Data Acquisition

E. Brian Welch
Vanderbilt University

A system for performing magnetic resonance imaging requires several fundamental hardware components to generate, encode, and receive the signals that are transformed into an image. First, a strong magnet generates a uniform static magnetic field over a large volume of space in which the imaged object is placed. Second, a magnetic field gradient system (composed of gradient coils and gradient amplifiers) creates local variations in the magnetic field to encode spatial information. Finally, radiofrequency (RF) transmit hardware (composed of an RF amplifier and RF transmit coil) injects energy into the resonating nuclei within the imaged sample. After excitation, RF energy is emitted back from the imaged object and detected by RF receive hardware. The received analog RF signal is digitally sampled and stored by a computer-controlled spectrometer until enough data are acquired to reconstruct an image. In addition to the magnet, gradients, and RF components, a modern conventional MRI scanner includes other hardware components such as those shown schematically in Figure 3.1. Typically, three separate interconnected computers control the scanner (shown in gray boxes in Figure 3.1). The operator uses a high-powered computer workstation and viewing console for image display, scan planning, and other operator input of scanning control parameters that are sent/downloaded to the spectrometer. A modern spectrometer usually is itself a high-speed digital computer running a real-time operating system to allow precise pulse and waveform generation to output to the gradient and RF hardware, and to enable precise data acquisition timing. Acquired data samples are received by a high-speed image reconstruction computer with a large memory capacity to enable fast image generation even with large data matrices. Physiological monitoring hardware is frequently used to observe, record, or synchronize cardiac, respiratory, or other physiological signals coming from the imaged subject. Not only are the physiology signals displayed on the operator's console, but they may also be used by the spectrometer to trigger or synchronize the MRI scanning process [1]. Most scanners employ a mechanical patient table and a precise landmarking system, e.g., laser alignment cross hairs for patient positioning. Figure 3.1 also shows the components of an active magnetic shimming system, including shim amplifiers and shim coils, needed to improve the homogeneity of the static magnetic field on most high field scanners. For longer-term storage of images, the viewing console is typically connected to a mass storage device, archiving mechanism for burning of CD/DVD media, and a local area network (LAN) to reach a picture archiving and communication system (PACS). Other typical hardware components not depicted in Figure 3.1 include a cryogen cooling system that is required when using a superconducting magnet, magnetic shielding to decrease the effect of the fringe magnetic fields in the area surrounding the MRI magnet, and RF shielding of the MR scanner suite to isolate the system from external sources of RF interference.

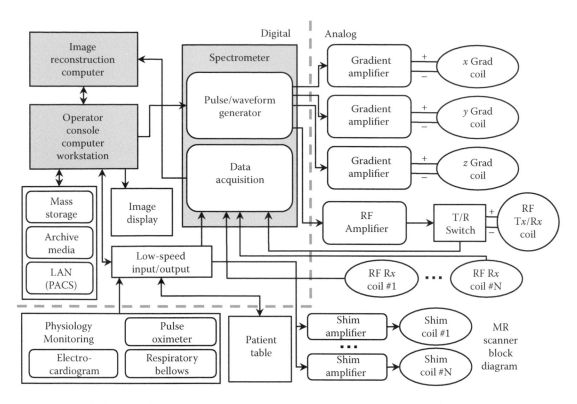

FIGURE 3.1 Hardware block diagram of a magnetic resonance imaging scanner. In a typical design, three communicating and precisely synchronized computer systems (gray boxes) are used to implement the control console, spectrometer, and image reconstruction computer. Of the three computer systems, the real-time operational specifications of the spectrometer computer are the highest because it must have extremely precise and high temporal resolution control (on the order of microseconds) of other hardware components. The spectrometer outputs analog pulses and waveforms to control gradient and RF components and receives analog RF signals. The control console is connected to a high-resolution image display for scan planning and evaluation. The console is often attached to one or more data archiving mechanisms such as mass storage hard drive(s), long-term archive media burner such as a CD or DVD writer, and a PACS accessible over a local area network connection. The console may also send and receive low speed control signals from other scanner hardware such as the patient table and physiological monitoring equipment. The dashed line in the diagram represents the boundary between digital and analog signals. While the computer systems are digital, the signals driving the RF, gradient, and shim amplifiers are analog as are the physiological signals. Three physical gradient coils for the *x*-, *y*-, and *z*-axes are used to spatially encode the MR signal. At least one RF amplifier is necessary to transmit. Sometimes a single RF coil can serve for both transmit and receive (T/R), but often dedicated receive coils are present to provide improved SNR and parallel imaging capabilities. Higher field systems may have dedicated shim coils (and amplifiers) to smooth the static magnetic field beyond the first order corrections achievable using just the imaging gradients. (Adapted from Atlas, S., *Magnetic Resonance Imaging of the Brain and Spine*, Raven Press, 1996.)

3.1 Magnets and Magnet Structures

In the field of MRI, static field strengths are most commonly stated in units of tesla (T), where 1 T is equal to 10,000 gauss (G), another common unit of magnetic field strength. The earth's naturally occurring magnetic field varies between 0.2 and 0.7 G. Commercial magnets for human imaging are available in field strengths ranging from 0.01 to 3.0 T for clinical applications and in ultra-high field strengths, 7.0 T and above, for research applications. Magnets are available as three distinct types: electromagnetic resistive magnets, permanent magnets, and superconducting magnets. A constant magnetic field can be generated by an electric current (electromagnet) used in resistive or superconducting magnets or by a permanent magnetic (ferromagnetic) material. The combination of large static magnetic field strength, high field homogeneity at the center of a magnet's volume (isocenter), and high field stability is challenging to achieve. Figure 3.2 presents schematic representations

of the basic magnet types. Designs of electromagnets and permanent magnets normally include an iron yoke connecting the magnetic poles in order to reduce the magnetic "resistance" of the return magnetic flux and can significantly improve the efficiency (tesla/ampere) of the magnet [2]. A magnet with one return yoke is known as a "C-arm magnet." A magnet with two return yokes is called an "H magnet." A C-arm magnet has the advantage that a human subject can approach from the side.

Resistive magnets use coils with many windings of wire conducting electrical current to create a magnetic field according to the Biot–Savart law expressed as follows:

$$\vec{B} = \int \frac{\mu_0 I}{4\pi |r|^3} \vec{r} \times \vec{ds}; \qquad (3.1)$$

where $\boldsymbol{B}$ is a vector representing the magnetic field as an integral of the electric current I passing through a current conducting

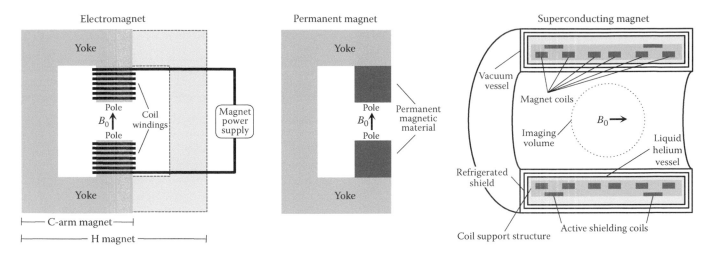

FIGURE 3.2 Schematic descriptions of three basic magnet types (resistive, permanent, and superconducting). An electromagnet (left) requires a power supply to conduct current through windings of wire positioned to create the two poles of the magnet. A strong magnetic field is generated between the poles. A yoke (often made of iron) mechanically separates the poles and provides a return path for the magnetic flux. Two form factors are common for yokes. One form factor creates what is known as a C-arm magnet (dark gray yoke) because of the overall shape of the magnet. The other form factor is known as an H-magnet (dark and light gray yoke) because of the shape of the open space within the yoke halves. A permanent magnet (middle) is similar to a resistive magnet in that it typically has two poles with a strong magnetic field created between the poles. Unlike a resistive magnet, a permanent magnet does not require a power supply because it employs a permanently magnetic material to generate the magnetic field. C-arm and H magnet form factors are common for permanent magnets. Most modern magnetic resonance imaging scanners use the third depicted magnet type, a cylindrical superconducting magnet (cross section shown on right). Loops of superconducting material are immersed in a cryogenically cooled vessel. Liquid helium is often used to maintain the low temperature necessary to create the superconducting state of the coil material. A power supply is necessary to energize the coil, but power is removed once the desired field strength has been achieved. Because of the zero resistance in the superconducting state, the current will continue flowing in the coil without the need for power. (Adapted from MacFall, J.R., *Categorical Course in Physics: The Basic Physics of MR Imaging—Hardware and Coils for MR Imaging,* in Riederer, S.J., Wood, M.L., editors, Chicago, IL, 1997, and Vlaardingerbroek, M., Boer, J., *Magnetic Resonance Imaging: Theory and Practice,* Springer, 1999.)

element of length and direction ds at a position r in space. The magnetic constant (permeability of free space) is denoted by μ_0 ($4\pi \times 10^{-7}$ T m²/A = 1.26×10^{-6} T m²/A). The magnitude of the magnetic field B generated at the center of a loop of current is given by Equation 3.2, where R is the radius of the loop.

$$B = \frac{\mu_0 I}{2R} \qquad (3.2)$$

To create a strong magnetic field, a resistive magnet requires many coil windings, strong electrical current, or a combination of both. The magnitude of the magnetic field B within a long solenoid of length L expressed in meters with N evenly spaced turns of wire carrying current I is given by

$$B = \mu_0 \frac{N}{L} I \qquad (3.3)$$

It is difficult to create a stable and homogeneous static magnetic field strength using a resistive magnet. Temperature must be controlled to preserve magnetic field stability. For example, the permeability of iron, causes the magnetic field to shift 2000 ppm for every 1 K change in temperature. Resistive magnet designs exist with both vertical and horizontal field alignments.

Permanent magnets require no electrical power to generate a magnetic field, but it is challenging to generate field strengths much larger than 0.3 T using a permanent magnet. Permanent magnets usually have a rectangular aperture with a vertical field. Similar to resistive magnets, temperature must be well controlled to maintain a stable magnetic field when using a permanent magnet. Also, permanent magnets are very heavy. The material comprising the permanent magnet is ferromagnetic and is characterized by a relationship between magnetic flux and magnetic field strength. Equation 3.4 describes the static magnetic field strength $B = |B_0|$ created by a permanent magnet, where $B_m H_m$ is the total magnetic energy per unit volume of the permanent magnetic material, V_m is the volume of the permanent magnetic material, and V_g is the volume of the gap between the magnet poles.

$$B = \sqrt{\mu_0 \frac{B_m H_m V_m}{V_g}} \qquad (3.4)$$

The total magnetic energy present per unit volume of a permanent magnetic material is a fixed property. To double the magnetic field, the amount of magnetic material must increase by a factor of four. The most common material used for permanent magnet MR systems is neodymium boron iron (NdBFe), which is expensive. Magnets made of NdBFe are limited by their weight and price to a maximum field strength of about 0.25 T, which should be compared to the price of the equivalent superconducting magnet. Also, the temperature must be kept constant because the magnetic field changes approximately 1000 ppm/K.

To achieve higher magnetic field strengths, superconducting magnets are commonly used. In a superconducting magnet, coil windings are immersed in a cryogenic fluid, such as liquid helium. At the low temperature of the liquid helium (4.3 K), the coil windings become superconducting; i.e., their electrical resistance becomes zero. Both helium and nitrogen can be used as cryogenic liquids to maintain superconductivity. Cryogens will slowly boil off and must be replenished regularly. The absence of electrical resistance means that once an external power supply achieves an inductive current in the windings, the power supply can be disconnected and the current (and the associated magnetic field) will be maintained. Superconducting magnets have excellent magnetic field stability and field homogeneity and typically come in the form of a cylinder with the static magnetic field oriented along the long axis of the cylindrical bore. The wires of superconducting magnets used in MRI systems are often made of niobium titanium or niobium tin alloy [3] filaments embedded in copper, which become superconductive at a temperature below 12 K. Figure 3.2 presents a cross section of a cylindrical superconducting magnet. Several sets of superconducting coils are used to create the strong static field at the center of the magnet bore. Superconducting active shielding coils are often used to reduce fringe fields and sit inside the same cryogen-cooled reservoir as the other superconducting coils.

In the low field strength range of 0.06 to 0.5 T, permanent and resistive magnets have a lower initial cost compared to superconducting magnet options. Permanent and resistive magnets also have low fringe field strengths because the iron yoke common in such designs confines the field and yields a relative safety and siting advantage. Air-core designs, both resistive and superconducting, have higher fringe field strengths.

3.1.1 B_0 Inhomogeneity Compensation

Ideally, the main magnetic field of the MR imaging hardware would be perfectly homogeneous, but in practice, there is always some level of inhomogeneity, ΔB_0. The change in magnetic field causes a change in the precessional frequency, Δf (Hz), according to Equation 3.5, where γ is the gyromagnetic ratio characteristic of the measured nuclei and $\gamma/2\pi = 4.257 \times 10^4$ Hz/mT for ^{1}H.

$$\Delta f = \Delta B_0 \left(\frac{\gamma}{2\pi} \right) \qquad (3.5)$$

If the field homogeneity varies spatially, the adverse effects will be similar to susceptibility artifacts and cause spatial distortion proportional to the amount of inhomogeneity. The magnitude of distortion is inversely proportional to the gradient strength applied during data readout in the MRI pulse sequence. Equation 3.6 provides the amount of spatial displacement Δx (mm) caused

by a static field inhomogeneity, Δf (Hz), in the presence of a given readout gradient of a strength equal to G_{read} (mT/m).

$$\Delta x = \frac{\Delta f}{(\gamma/2\pi)G_{read}} \times 10^3 \qquad (3.6)$$

If ΔB_0 is constant across the entire field of view, it will cause a new effective echo time for each frequency-encoded view. These shifts in the spatial frequency domain (Fourier domain or k-space, see Chapter 4) add phase ramps (according to the Fourier shift theorem) in the spatial domain image along the frequency encode direction. B_0 inhomogeneity can be reduced by appropriate maintenance and quality control on the MR imaging system by calibrating and applying "shim" gradients. Passive shimming refers to the process of placing pieces of iron within the bore of the magnet to optimize the homogeneity of the static magnetic field. Just the presence of an object inside the magnet bore affects the field homogeneity. Modern MRI systems include hardware and software methods that enable the adjustment of field homogeneity using the gradient coils or other dedicated active shimming coils. Commonly, a prescan procedure automatically optimizes applied shim gradients to make the static field as homogeneous as possible. However, the shimming procedure is typically run only once before scanning begins for a given imaging location. If the object moves during the scan or between similarly located scans, the homogeneity of the field will be disturbed and can cause image artifacts.

3.1.2 Siting and Magnetic Shielding

Magnetic shielding is necessary to protect the environment surrounding the MRI system from the effects of fringe magnetic fields, as other electronic and mechanical equipment (e.g., wristwatches, credit cards) can be damaged by a magnetic field. The effect of magnetic fields on implanted medical devices such as cardiac pacemakers, neurostimulators, and other similar devices is another important concern. To minimize safety hazards, the magnet must be isolated to an area into which individuals with magnetic contraindications will not be allowed to enter. Besides protecting the external environment from the magnet, it is also necessary to protect the MRI system from magnetic field disturbances that can be caused by passing cars, trains, or elevators. Magnets can be passively shielded using heavy iron plating to surround the magnet room. Many tons of iron are required to passively shield high field magnets; for example, approximately 20 tons for a 1.5 T magnet and 400 tons for a 7 T magnet may be used as a passive shield. Alternatively, a magnet can be actively shielded using a second outer superconducting coil set surrounding the inner superconducting coil set (see Figure 3.2) that produces the main magnetic field. The currents in the inner and outer coil sets flow in the opposite direction, which contains and reduces the magnetic fringe fields.

3.1.3 RF Shielding

The MRI signal is weak, and even small sources of external RF interference can degrade signal and image quality. Therefore, MR systems are shielded from external sources of RF energy by building RF shielding into the walls, floor, and ceiling of the MR scanner room. The RF shielding also prevents the signal generated by the MR system from disturbing RF-sensitive equipment outside of the MR suite.

3.2 Spectrometer

The spectrometer of the MR imaging system has three primary purposes: (1) reference frequency generation, (2) generation and transmission of desired gradient and RF waveform shapes, and (3) reception of RF signals. The reference frequency of the spectrometer must be as stable as the static magnetic field, e.g., on the order of 0.1 ppm/h. The reference frequency is typically generated using a temperature-controlled crystal oscillator. Spectrometers can be designed as analog or digital devices. Spectrometers may include a variety of RF components, including phase shifters, signal splitters, signal combiners, signal modulators, and signal mixers. Advanced users of MR imaging systems who need to develop custom gradient and RF waveform pulse sequences require a pulse programming system to control the spectrometer. Typically, a computer program that describes the exact timing and amplitudes of RF and gradient waveforms as well as the timing of data acquisition implements a pulse programming system. Pulse programming methods and languages vary across MR system manufacturers with some languages being more complicated or more flexible than others. Pulse programming systems need large waveform memories in order to manipulate the high-precision (16-bit) descriptors required to adequately represent RF and gradient waveforms.

3.3 Gradient System

MRI employs magnetic field gradients to spatially encode signals emitted from the imaged object according to the magnetic field and the associated resonant frequency existing at that spatial location. In general, the magnetic field gradients used by the MRI scanner are weak compared to the static magnetic field on which the gradient fields are superimposed. Conventional magnetic field gradients are designed to change linearly with position. Three sets of orthogonally positioned gradient coils create gradients in the x-, y-, and z-directions necessary to perform imaging of a 3D object. The strength of a gradient is expressed in units of tesla or gauss per centimeter (1 G/cm = 0.01 T/m = 10 mT/m). The maximum resolution of the MR imaging system is determined by the maximum gradient strength. The minimum spatial dimension Δx_{min} (mm) that can be encoded by the maximum gradient strength G_{max} (mT/m) in a sampling time $T_{readout}$ (s) is given by Equation 3.7:

$$\Delta x_{min} = \frac{(\pi/\gamma)}{T_{readout} G_{max}} \times 10^3 \tag{3.7}$$

where $\pi/\gamma = 1.175 \times 10^{-6}$ Hz/mT for ^{1}H.

The electrical currents passing through the gradient coils must be switched on and off many times during MR imaging, and this must be done in a reliable and rapid manner. The time that the current in a gradient coil requires to reach its maximum value is known as the "rise time," and the rate at which the current changes is called the "slew rate." State-of-the-art clinical MR systems in 2010 are available with maximum gradient strengths of 80 mT/m and slew rates of 200 T/m/s. For any given MR imaging sequence, the slew rate is usually set to the maximum allowed by the gradient hardware; however, it may be restricted if peripheral nerve stimulation (dependent on gradient orientation) is a concern [4]. Figure 3.3 presents schematics of prototypical current conduction pathways for x, y, and z

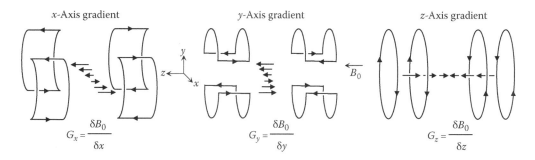

FIGURE 3.3 Prototypical current conduction pathways and generated magnetic fields for the fundamental x, y, and z physical axes of the magnetic resonance scanner's gradient subsystem. Solid black lines represent the current conduction pathways with arrowheads demarking the direction of current flow. Arrows of changing length at the center of the gradient coils represent the varying magnetic field strength created by the gradient coil. At the center of all three coils, the magnetic field contributed by the coil is zero. The designs for the x- and y-axis gradients employ complimentary saddle-shaped pairs of coils. Essentially, the x and y coils are rotated versions of the same coil. The z-axis gradient coil uses one or more Maxwell pairs of current loops flowing in opposite directions. The depicted current pathways are an oversimplification. Modern gradient coils are often designed using numerical optimization of the conducting patterns, which can create intricate patterns. The complicated patterns have caused such coils to be nicknamed "fingerprint" coils [5] to describe the conductor layout pattern. (Adapted from Rinck, P., *Magnetic Resonance in Medicine: The Basic Textbook of the European Magnetic Resonance Forum*, ABW-Wiss.-Verl., 2003, and Oppelt, A., *Imaging Systems for Medical Diagnostics: Fundamentals, Technical Solutions and Applications for Systems Applying Ionization Radiation, Nuclear Magnetic Resonance and Ultrasound*, Publicis, 2005.)

gradient coils. The arrows of varying magnitude at the center of the gradient coils represent the varying magnetic field strength created by the individual gradient coil. The magnetic field contributed by the coil at the geometric center (isocenter) of all three coils is zero. The solid black lines represent the current conduction pathways with arrowheads showing the direction of current flow. The designs for the x- and y-axis gradient coils use saddle-shaped pairs of coils with opposing directions of current in the opposing saddle. A common z-axis gradient coils uses one or more Maxwell pairs of circular current loops flowing in opposite directions. Modern gradient coils are often designed using numerical optimization of the conducting patterns, which can create intricate patterns such as so-called "fingerprint" coils [5].

3.3.1 Eddy Currents

The changing magnetic field gradients used by the MRI scanner generate eddy currents in other conductive parts of the scanner such as the metal shields of the magnet. The eddy currents can themselves produce unwanted gradient fields that cause problems such as image artifacts and additional helium consumption. The effects of eddy currents can be reduced by the use of actively shielded gradient coils in which an additional coil surrounds the gradient coil and counteracts the generated eddy current fields. When eddy currents are present, the onset of applied gradients can be delayed and the gradient amplitude may decrease. This can alter the timing and rate of the k-space trajectory (see Chapter 4) especially in sequences with a long

read-out time, such as echo planar imaging (EPI). In EPI, the k-space trajectory crosses k-space back and forth many times, and the misalignment in the frequency encode direction can cause inconsistencies in the orthogonal phase encode (PE) direction that leads to ghosting in the PE direction of the spatial domain image. The effects of eddy currents can be reduced with appropriate timing and amplitude tuning of the gradients from quality control measurements. Also, self-shielded gradient coils help reduce eddy current effects. Figure 3.4 presents plots of nominal and eddy current-compensated gradient coil driving currents and the associated output gradient waveforms adapted from references [6] and [7]. Without eddy current compensation, a perfectly rectangular bipolar driving current induces counteracting eddy currents in the conducting structures of the magnet that generate an undesirable gradient field (G_{eddy}) that superimposes on the desired gradient waveform to yield an imperfect total gradient (G_{total}). If the driving current is altered to include pre-emphasis to compensate for the generated eddy currents, the eddy currents will still exist, but their superposition with the nominal gradient waveform generated by the compensated driving current will yield the desired rectangular waveform shape for the total gradient.

3.4 RF System

The function of the RF system of the MR scanner is to accurately transmit the RF waveforms associated with the MR imaging pulse sequence and receive RF signals emitted from the imaged

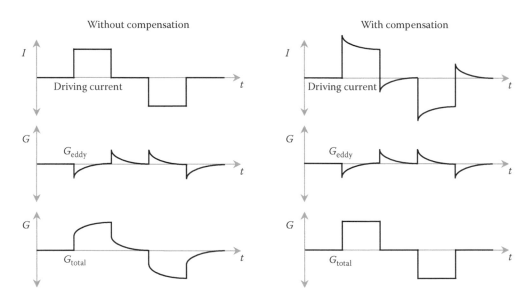

FIGURE 3.4 Plots of nominal and eddy current compensated driving currents and the associated output gradient waveforms. In the left column of plots, a rectangular bipolar driving current induces counteracting eddy currents in the conducting structures of the magnet that generate an undesirable gradient field (G_{eddy}) that superimposes on the desired gradient waveform to yield an imperfect total gradient (G_{total}). In the right column, the driving current is altered to include pre-emphasis to compensate for the generated eddy currents. The eddy currents still exist, but their superposition with the nominal gradient waveform generated by the compensated driving current yields the desired rectangular waveform shape for the total gradient. (Adapted from Sprawls, P., Bronskill, M., *American Association of Physicists in Medicine. The Physics of MRI: 1992 AAPM Summer School Proceedings*, American Institute of Physics, 1993, and Oppelt, A., *Imaging Systems for Medical Diagnostics: Fundamentals, Technical Solutions and Applications for Systems Applying Ionization Radiation, Nuclear Magnetic Resonance and Ultrasound*, Publicis, 2005.)

object. The primary components of the MR scanner's RF system include an RF waveform synthesizer, RF power amplifier, RF cabling and signal multiplexers, RF coil switching circuits, transmit and receive RF coils, and an RF power monitoring and protection system.

3.4.1 Requirements for RF Systems

The important characteristics of a power amplifier are its linearity, stability, duty factor, and noise level. The RF amplifier of an MR imaging system should be linear (within 1%) over the full dynamic range of both positive and negative amplitudes [8]. The RF output should not be contaminated by frequencies not included in the desired pulse. RF harmonic frequencies should be suppressed by at least 70 dB within the RF coil bandwidth. The amplifier should not produce noise when the amplifier input is equal to zero; otherwise, undesired spin excitation may occur. RF power requirements increase with the square of the static magnetic field strength. A typical 3 T clinical MR scanner needs an RF amplifier capable of 20 kW output power. RF pulse amplitude should not deviate by more than 0.1% from the desired amplitude over the course of an imaging acquisition. For MR systems used for exciting and detecting nuclei other than protons (see Chapters 15 and 22), a broadband RF amplifier that can be tuned to the frequencies of phosphorus, carbon, fluorine, sodium, deuterium, helium, xenon, etc., may be necessary.

3.4.2 RF Transmitters (B_1) Systems

The B_1 transmitter delivers precise amplitude and phase controlled RF waveform pulses to one or more transmit antennas. The RF transmit system is generally composed of a small amplitude signal synthesizer and an RF power amplifier. The maximum power, P, demanded in an MRI scan is proportional to the product of the square of the flip angle, α, and the square of the precessional frequency, ω_0, and is inversely proportional to the pulse duration, t_α, as expressed in Equation 3.8:

$$P \propto \frac{\alpha^2 \omega_0^2}{t_\alpha^2} \qquad (3.8)$$

RF waveforms generated by the small signal synthesizer must be amplified to kilowatt power levels by the RF power amplifier. Cables and transmit/receive (T/R) switches in the transmit signal pathway attenuate the power before it reaches the transmit RF coil. In the design of RF power amplifiers, several important requirements must be kept in mind, including maximum required pulse length, maximum allowable pulse droop, linearity, reproducibility, noise, and spurious signals. Droop describes the decrease in the RF amplifier power output over the time course of a long RF pulse. Poor linearity causes distortion of the RF waveform shape. Pulse reproducibility is often confounded by thermal drift, which causes amplifier gain variation. When not transmitting, the RF amplifier should not introduce noise. Finally, the RF amplifier should not introduce coherent

spurious signals that can manifest themselves as artifacts in the MR images.

3.4.3 Multitransmit Parallel RF Transmission Technology

A recent development in RF transmission technology is the simultaneous use of two or more RF transmit antennas to better control RF uniformity and power deposition in the imaged object [9]. The additional degrees of freedom provided by independent transmit RF amplitudes, phases, frequencies, and even waveform shapes enable "B_1 shimming" and minimization of locally absorbed RF power. MR imaging performed at high static field strengths (3 T and greater) is likely to benefit significantly from the introduction of multitransmit RF technology because of the RF uniformity challenges characteristic of higher static fields.

3.4.4 Sampling of Received Signals

When the RF power amplifier is not active, its output is disconnected to prevent its output noise from interfering from the relatively weak RF signals detected by the MR system. Diode switches are typically used to actively block (detune) RF receive circuitry and keep transmitted RF power from reaching the sensitive receive electronics. In a conventional receiver, the reference frequency is split and passed through a 90° phase shifter to provide 0° and 90° references. The original and shifted reference signals are separately mixed with the detected MR signal to produce the real (Q) and imaginary (I) channels. Signal receivers in an MR system typically have a digitally controlled gain stage to adjust the signal to an optimal amplitude range in the analog-to-digital conversion (ADC) process.

3.4.5 RF Coils

RF coils function as the transmitting antennas for the RF excitation field (B_1^+ field), and also serve as reception antennas for the emitted (B_1^- field) MR signal. When an object is placed inside or nearby an RF coil, impedance is added to the coil circuitry and is known as coil loading. At the high field strengths typical of MRI systems, the noise introduced by the imaged object dominates the inherent thermal noise generated by the RF coil circuitry. The signal-to-noise ratio (SNR) yielded by the coil is also a function of its quality (Q) factor (the sharpness of the coil's resonance), which, in general, should be as large as possible without sacrificing too much spatial uniformity in the coil's volume of excitation/detection [10]. Equation 3.9 defines the quality factor Q of an RF coil as the resonance center frequency ω_0 divided by $2\Delta\omega$, the full width at half maximum of the coil's spectral response.

$$Q = \frac{\omega_0}{2\Delta\omega} \qquad (3.9)$$

When a coil is loaded by the presence of the imaged object, Q decreases [6]. While it is important to generate a uniform RF excitation field for the best image quality, uniform receive sensitivity is generally not a requirement. Inhomogeneous receive sensitivity can in fact be advantageous when many receive coils are used together. In conventional MR imaging, signal transmission and reception are not performed simultaneously, and a single coil can perform both functions. Separate coils can be used for transmission and reception. In current MR imaging systems, a common strategy is to use a volume coil for RF transmission and an array of surface coils for signal reception. Figure 3.5 presented three common RF coil designs used in MRI. A common transmit and/or receive coil is the birdcage volume coil. A simple loop can be used to create a transmit and/or receive surface coil. Multiple surface coils can be combined to create a receive coil phased array. Uniform volume coils are most often constructed using a birdcage design. Solenoid volume coils can be used on systems with a vertical static field alignment or with horizontal field systems in cases where the solenoid coil axis can be aligned perpendicular to the horizontal axis. Surface coils can yield higher SNRs than volume coils. Typical SNR enhancements factors on the order of 3–5 can be achieved using surface coils. Phased array coils consist of multiple coils and can achieve high SNR over a larger volume than can be covered by a single coil without the geometric restrictions of a volume coil or smaller surface coil. The multiple signals received by the different coil elements of the array are combined to reconstruct one image. The SNR is not improved by simply adding the signals from the multiple coils. Separate preamplifiers, receiver channels, and more advanced reconstruction algorithms must be employed to capture the SNR gains available from a phased array of receive coils.

3.4.6 Linear versus Quadrature Coils

Quadrature excitation refers to the technique of exciting a transmit RF coil with two RF signals that are 90° out of phase. Unlike a linearly driven coil that creates a B_1 field oscillating in amplitude, quadrature excitation generates a circularly polarized field that is constant in amplitude. The use of a quadrature RF coil improves SNR. A quadrature transmit coil also reduces the

required RF power and the energy dissipated within the imaged object. In receive mode, the proper combination of the two outputs of a quadrature coil can yield a factor of $\sqrt{2}$ increase in SNR. For better signal detection, the imaged object should fill the majority of the detection coil. Typically, a filling factor of 70% or more is desired. Although the RF homogeneity of a quadrature volume coil is better than that of a wraparound surface coil, the increased filling factor possible with the surface coil can lead to a better SNR compared with a volume coil.

3.4.7 Receive-Only Coils

Figure 3.6 shows a prototypical circuit diagram of a simple receive-only MRI coil. A shielded coaxial cable connects the coil to the MRI scanner's spectrometer using a so-called balun (balanced–unbalanced) circuit in the cable, which acts as a type of electrical transformer to help match the impedance of the spectrometer and the coil. A low noise amplifier (LNA) is used to boost the weak MR signal detected by the coil's resonator circuit. The resonance frequency of the coil is determined by its fixed distributed capacitance in combination with several tunable capacitors. During RF transmission all receive-only coils are detuned to prevent transmitted RF power from entering the RF receive chain of the system. Detuning circuits may be active or passive. In the design shown in Figure 3.6, a passive detuning circuit is activated during RF transmit when voltages across the PIN diodes cause them to short-circuit and introduce an inductor into the coil's resonator circuit. The addition of the inductance into the coil's circuit shifts its center frequency far away from the transmitted RF frequency.

3.4.8 Parallel Imaging

Parallel imaging in MRI refers to methods to reduce scan time using data from multiple RF receive coils. Scan time is reduced by acquiring less data than is nominally required to reconstruct the entire field of view of interest. The reduced data acquisition leads to aliasing in the spatial domain image. However, under the proper conditions, the signals from the separate coils can

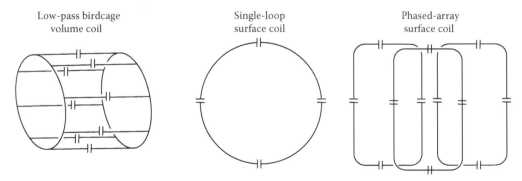

FIGURE 3.5 Three common RF coil designs used in magnetic resonance imaging. A common transmit and/or receive coil is the birdcage volume coil (left). A simple loop can be used to create a transmit and/or receive surface coil (middle). Multiple surface coils can be combined to create a receive coil phased array (right). (Adapted from Jin, J.-M., *Electromagnetic Analysis and Design in Magnetic Resonance Imaging*, CRC Press, Boca Raton, FL, 1999.)

Receive-only MRI coil

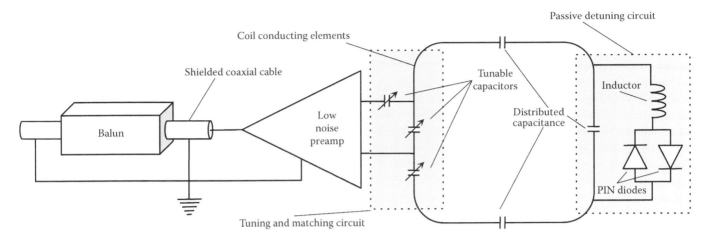

FIGURE 3.6 Circuit diagram of a simple receive-only MRI coil. The coil is connected to the MRI scanner's spectrometer using a shielded coaxial cable. A balun (balanced–unbalanced) circuit in the cable serves as a type of electrical transformer to help match the impedance of the spectrometer and the coil. A low noise figure preamplifier is used to amplify the weak MR signal detected by the coil's resonator circuit. The resonance frequency of the coil is controlled by several tunable capacitors. Other fixed distributed capacitors affect the resonant frequency of the coil. Receive-only coils must be detuned during the transmission of RF power by a different transmit coil. Detuning circuits may be active or passive. In the design shown here, a passive detuning circuit is activated during RF transmit when voltages across the PIN diodes cause them to short-circuit and introduce an inductor into the coil's resonator circuit to shift its center frequency far away from the transmitted frequency. (Adapted from Oppelt, A., *Imaging Systems for Medical Diagnostics: Fundamentals, Technical Solutions and Applications for Systems Applying Ionization Radiation, Nuclear Magnetic Resonance and Ultrasound*, Publicis, 2005.)

be used to regenerate the missing parts of the spatial frequency domain, also known as the Fourier domain, or *k*-space, in which MR imaging data are acquired. In a method known as SMASH (simultaneous acquisition of spatial harmonics) [11], a multi-element coil is designed with a specific geometry to allow for generation of unsampled spatial harmonics using the complex spatial domain sensitivity profiles of each coil element in the array. In a related method, known as GRAPPA (generalized autocalibrating partially parallel acquisition) [12], data are fully acquired near the center of *k*-space to enable regeneration of the missing data without the need for a separate coil sensitivity scan. Other methods operate in the spatial domain to "unfold" the aliased image such as the approach known as SENSE (sensitivity encoding) [13].

3.5 Safety and Security

Safety concerns around an MR imaging system include the strong static magnetic field, potential for RF burns, tissue heating from deposited RF energy, peripheral nerve stimulation, acoustic noise, and the dangers associated with a magnet quench, a sudden loss of superconductivity, which causes very rapid boil off of liquid cryogen (see Section 3.5.6).

3.5.1 Static Magnetic Field

The primary hazard associated with the static magnetic field is ferromagnetic attraction in which a ferrous object will experience a strong force if brought close to the magnet [14]. The force

can be strong enough to turn the ferrous object into a projectile. Modern computer equipment can tolerate up to 20 G. Shielded computer monitors, magnetic data storage devices, credit cards, and other shielded medical equipment (CT scanners, nuclear cameras, etc.) can tolerate up to 10 G. Neurostimulation and other biostimulation devices can tolerate up to 5 G.

3.5.2 RF Burns from Current Loops

Conductive leads or cables that are exposed to the time-varying RF field of the MR scanning environment may have high currents induced within them that could transfer heat to the imaged subject [15]. A typically reported RF-related injury involves conductive leads or cables placed against the subject's bare skin. Often the leads are part of the physiological monitoring hardware, electrocardiogram or pulse oximeter, or may be part of the RF coil. To reduce the risk of RF burns, operators must ensure that conductive cables do not form loops and do not come into contact with bare skin. Practices to avoid the problem include cable pathway guides and thermal and electrical insulating pads and mattresses. RF burns can also occur when the detuning/decoupling circuitry fails in a receive-only RF coil. As a precaution, such receive coils should be insulated from the subject's skin. Current conducting loops can occur within the subject's body without any external conductors. For example, a subject touching her hands together can create a conductive loop. All skin-to-skin contact points should be avoided with the use of linens or other insulating materials (pads, cushions, or coil mattresses). Finally, some conductive materials applied to the skin, such as tattoos, tattooed eyeliner, and some cosmetics can result in burns [16].

3.5.3 Specific Absorption Rate

Biologic tissue exposed to time-varying electromagnetic RF fields in the megahertz range will experience heating. The RF power absorbed per kilogram of tissue (watts/kilogram) is known as the specific absorption rate (SAR). Equation 3.10 shows that SAR is a function of the frequency-dependent tissue conductivity $\sigma(\omega)$, tissue density ρ, and electric field E.

$$SAR = \frac{\sigma(\omega)E^2}{\rho} \qquad (3.10)$$

In the context of magnetic resonance imaging, SAR is proportional to the product of the square of the precessional frequency ω_0 and the square of the transmit RF amplitude B_1 as well as to the product of the square of the flip angle α and the square of the amplitude of the static magnetic field B_0 as stated in Equation 3.11.

$$SAR \propto \omega_0^2 B_1^2 \propto \alpha^2 B_0^2 \qquad (3.11)$$

The quadratic relationship in Equation 3.11 reveals why SAR becomes a limiting factor in high field MR scanning. International standards [17] set global limits on SAR for the whole-body and head as well as local SAR limits for the head/trunk region and the extremities. Different operating levels are defined, including normal, first level controlled, and second level controlled. The normal operating mode SAR limits for the whole global body, global head, local head/trunk, and local extremities are 2, 3.2, 10, and 20 W/kg, respectively. Limits for first and second level controlled operating mode remain the same except that the limit for global whole-body SAR increases to 4 W/kg. In subjects with an unimpaired thermoregulatory capacity, whole-body SAR of 4 W/kg will not increase the core body temperature by more than 1°C, a temperature rise for which no adverse health effects have been reported. The actual heating experienced by tissue is a result of the absorbed RF energy, the tissue heat conductance, and the tissue's thermoregulatory abilities (vascular dilation, cardiac rate, metabolic rate, perspiration, etc.). Certain tissues with reduced blood circulation (e.g., the lens of the eye, testicles) are more sensitive to RF energy absorption and require special attention, especially when employing surface transmit coils that have spatially inhomogeneous B_1 fields.

3.5.4 Peripheral Nerve Stimulation

Rapidly switching magnetic field gradients create dB/dt that can lead to peripheral nerve stimulation (PNS) in imaged subjects. The nerve stimulation is a result of the electric field induced in tissue. The effect ranges from tingling sensations to muscle twitching to pain. Modeling of the effect is challenging because of the heterogeneity of the body's electrical conductivity, but it is widely accepted that the likelihood and magnitude of stimulation increases with a larger peak field generated inside the body [18]. To evaluate the peak field, all gradient coil contributions (x, y, and z) must be considered. International standards [17] set limits on the time-rate of change for magnetic fields at three operating levels (normal, first level controlled, and second level controlled). For a 200-μs rise time, the limits are 45, 56, and 56 T/s, respectively. For a 400-μs rise time the limits are 30, 38, and 38 T/s, respectively.

3.5.5 Acoustic Noise

The electrical currents passing through the gradient coils are on the order of several hundred amperes. Because the current flows in the presence of a large magnetic field, large forces are exerted on the gradient coils and their mechanical housings. The gradient structures can oscillate during gradient switching and produce large amounts of acoustic noise. The acoustic vibration can be alleviated to some extent with the use of dampening material between the coil and patient bore, but this approach is limited by the spatial constraints of the magnet design. Ear protection in the form of earplugs or headphones is typically provided to imaged subjects. Also, gradient coils can be designed to balance Lorentz forces within the coil structure to address directly the cause of the acoustic noise [5].

3.5.6 Quenches

When there is a sudden loss of superconductivity, a quench occurs. If the wires within a superconducting magnet become resistive, heat is generated that further increases the resistance of the wires. This unstable process ultimately leads to the dissipation of all of the stored energy within the magnet and to the boiling off of the liquid helium in a process known as quenching. A quench can be a violent process, and constructional precautions are necessary, such as overpressure safety valves and venting channels for gaseous helium. If not properly controlled, a quench can be destructive to the static magnetic field coil, and passive circuits are often used to limit damage. These circuits are activated by the start of a quench and help spread the quench throughout the coil to distribute the destructive energy [7]. To quickly eliminate the static magnetic field in the case of a safety emergency, an operator may deliberately activate an emergency quench. The deliberate mechanism uses a controlled heating of the coil to avoid damage to the system and can operate in a matter of seconds. All quenches result in a large loss of cryogens (e.g., helium), which must be safely vented to avoid suffocation by anyone in the area. After a quench, the magnet must be thoroughly checked, refilled with cryogens, and ramped back up to the normal field strength—a time-consuming and costly consequence.

3.6 Summary

In this chapter, we have discussed the fundamental hardware components to generate, encode, and receive the signals that are transformed into an image. The following chapter discusses how the signals acquired can be used to generate images.

References

1. Bernstein MA, King KF, Zhou ZJ. *Handbook of MRI Pulse Sequences.* Burlington, Massachusetts: Elsevier Academic Press; 2004. pp. 443–490.

2. Vlaardingerbroek M, Boer J. *Magnetic Resonance Imaging: Theory and Practice.* New York: Springer-Verlag; 1999.

3. Partain C. *Magnetic Resonance Imaging*: Philadelphia: W.B. Saunders Co.; 1988.

4. King KF, Schaefer DJ. Spiral scan peripheral nerve stimulation. *J Magn Reson Imaging* 2000;12:164–170.

5. Mansfield P, Chapman BL, Bowtell R, Glover P, Coxon R, Harvey PR. Active acoustic screening: reduction of noise in gradient coils by Lorentz force balancing. *Magn Reson Med* 1995;33:276–281.

6. Sprawls P, Bronskill M, *American Association of Physicists in Medicine. The Physics of MRI: 1992 AAPM Summer School Proceedings.* 1993. Published for the American Association of Physicists in Medicine by the American Institute of Physics.

7. Oppelt A. *Imaging Systems for Medical Diagnostics: Fundamentals, Technical Solutions and Applications for Systems Applying Ionization Radiation, Nuclear Magnetic Resonance and Ultrasound.* Erlangen, Germany: Publicis; 2005.

8. MacFall JR. *Categorical Course in Physics: The Basic Physics of MR Imaging—Hardware and Coils for MR Imaging.* In: Riederer SJ, Wood ML, editors; 1997; Oak Brook, Illinois: Radiological Society of North America. pp. 41–57.

9. Katscher U, Bornert P. Parallel RF transmission in MRI. *NMR Biomed* 2006;19:393–400.

10. Cho ZH, Jones J, Singh M. *Foundations of Medical Imaging.* New York: Wiley; 1993.

11. Sodickson DK, Manning WJ. Simultaneous acquisition of spatial harmonics (SMASH): fast imaging with radiofrequency coil arrays. *Magn Reson Med* 1997;38:591–603.

12. Griswold M, Jakob P, Heidemann R, Nittka M, Jellus V, Wang J, Kiefer B, Haase A. Generalized autocalibrating partially parallel acquisitions (GRAPPA). *Magn Reson Med* 2002;47:1202–1210.

13. Pruessmann K, Weiger M, Scheidegger M, Boesiger P. SENSE: Sensitivity encoding for fast MRI. *Magn Reson Med* 1999;42:952–962.

14. McRobbie D. *MRI from Picture to Proton.* New York: Cambridge University Press; 2003.

15. Westbrook C. *Handbook of MRI Technique.* Malden, Massachusetts: Blackwell Science; 1999.

16. Carr JJ. Danger in performing MR imaging on women who have tattooed eyeliner or similar types of permanent cosmetic injections. *AJR Am J Roentgenol* 1995;165: 1546–1547.

17. IEC. *Medical Electrical Equipment. Part 2-33: Particular Requirements for the Safety of Magnetic Resonance Equipment for Medical Diagnosis.* Volume IEC 601-2-33 Ed. 2.0. Geneva, Switzerland: International Electrotechnical Commission; 2002.

18. Schmitt F, Stehling MK, Turner R, Bandettini PA. *Echo-Planar Imaging: Theory, Technique, and Application.* New York: Springer-Verlag; 1998.

19. Atlas S. *Magnetic Resonance Imaging of the Brain and Spine.* Philadelphia: Lippincott-Raven Press; 1996.

20. Rinck P. *Magnetic Resonance in Medicine: The Basic Textbook of the European Magnetic Resonance Forum* (including MR Image Expert, version 2.5; Dynalize, version 1.0 demo; with 45 tables). Boston: Blackwell Wissenschafts-Verlag; 2003.

21. Jin J-M. *Electromagnetic Analysis and Design in Magnetic Resonance Imaging.* Boca Raton, FL: CRC Press; 1999.

Image Formation

David R. Pickens
Vanderbilt University

4.1 Introduction

Creation of useful image information from the nuclear magnetic resonance (NMR) phenomenon used for chemical analysis began with the work of Paul Lauterbur. In 1973, he reported on the successful creation of a 2D image of a pair of test tubes using a projection methodology and the careful modification of a static magnetic field by means of an applied linear gradient field [1]. Lauterbur's insight was in determining that a Fourier decomposition of the signals received could relate frequency to position in the presence of the linear gradient. By this manipulation, a collection of 1D projections was produced. A reconstruction process similar to computed tomography using filtered back projection yielded a transverse image of the test tubes.

Shortly after Lauterbur's seminal publication, others presented methods to produce image information from live animals placed in a magnetic field. One such approach, called FONAR, for field focusing nuclear magnetic resonance, created a resonance region in the body of an animal by manipulating the shape of the static magnetic field and a radiofrequency (RF) field [2]. This field, reported to be spherical and about a millimeter in diameter, could be scanned through the object to be studied. The development of the scanning field approach led to development of sensitive point and other scanning methods, but the technique pioneered by Lauterbur became the basis for the imaging methods used today.

For the phenomenon of resonance to occur in MR imaging, the frequency of the RF pulse must match the precessional frequency of the nuclear species that are present in the tissues of interest according to the Larmor equation:

$$\omega = \gamma B_0 \qquad (4.1)$$

where ω is the precessional frequency in radians per second, γ is the gyromagnetic ratio in radians per second per tesla, and B_0 is the field strength of the static magnetic field. By recalling that $\omega = 2\pi f$, the Larmor equation can be expressed as

$$f = \frac{\gamma}{2\pi} B_0 \qquad (4.2)$$

where f is the precessional frequency in hertz (Hz) and $\gamma/2\pi$ is expressed in hertz per tesla (Hz/T).

This equation is fundamental to magnetic resonance imaging because it is the manipulation of B_0 and the observation that the change in frequency and phase will provide the necessary information to recover signals related to the location and numbers of different groups of precessing nuclei, a concept from Lauterbur's original work. These locations can be expressed as contrast differences in 2D and 3D image sets that can be produced by the methods to be discussed.

A variation of a few millitesla will cause a detectable variation in the precessional frequency of nuclei placed in a static magnetic field. For the discussions that follow, protons will be the nuclei of primary concern. The gyromagnetic ratio for protons is 42.58 MHz/T. If one assumes a magnet has a B_0 field strength of 1.5 T, then the precessional frequency would be 63.87 MHz. A change of 1 mT (0.001 T) at 1.5 T would cause a change in precessional frequency of 63,870 Hz. If one can cause small variations in the experienced magnetic field and know precisely how this is done and what the local field strength is within a part of a sample, then changes in the local static field produced by the addition of gradient fields will cause protons at different locations to precess at different frequencies. If a signal is composed of many frequencies, then a way to decode the signal is to use the Fourier transform to find the individual frequencies and relate them to a known position, leading to the process of image formation.

4.2 Description of Image Formation

The goal of MRI is to produce an image of the distribution of protons within the object of interest, which is often the human body. To do this, the system must somehow detect the resonance signal from these protons and determine the location from which this signal originates. This process is known as spatial localization of the signal information, which will require the system to select the slice or slab of interest, excite protons within the slice, receive the signals from the excited protons, and then perform a reconstruction of the signals from these protons to produce a useable image.

Images that are formed depend on three characteristics of the tissues. The first characteristic is the density of the nuclear species, ρ. The other two characteristics, discussed in Chapter 2, are the relaxation times T_1 and T_2. All contrast between tissues that will be found in the resulting images is based on differences between these characteristics and the timing of the application of RF and gradient fields to interrogate these characteristics.

In Chapter 2, the signal induced in a loop of wire positioned in the transverse plane was described by Equation 2.30. The signal that results is known as a free induction decay (FID), which can be shown to decay with a time constant T_2, where

$$M_{xy} = M_0 e^{-t/T_2} \tag{4.3}$$

and M_{xy} is the transverse magnetization, M_0 is the equilibrium magnetization, and t is the time. However, due to the imperfections of the main magnetic field B_0, the decay of the FID follows a more rapid T_2^* decay. As the spins release energy and return to equilibrium, the longitudinal magnetization, M_z, recovers, following the form

$$M_z = M_0(1 - e^{-t/T_1}) \tag{4.4}$$

where the regrowth of magnetization in the z direction follows the T_1 relaxation time constant.

In real tissues, this signal is a convolution of contributions from protons experiencing different local magnetic environments, which affect their Larmor frequency slightly [3]. Although the signal that is produced is a complicated mix of frequencies, it does not contain information that can be used to decode the position and density of protons in three dimensions. It is this decoding and reconstruction aspect of the signal that leads to useful maps of the location of these protons that are displayed as gray-scale images. The process for producing these images can take a number of forms that are discussed below.

Pulse sequences specify the application of RF pulses, including the time and intensity, the switching on and off of linear gradient fields, the switching on and off of the receiver and digitizer systems to acquire information, as well as how often these events occur. Pulse sequences can be described in timing diagrams,

called pulse sequence diagrams (PSDs), that represent the events graphically and show how to manipulate a tissue's magnetization in a specific way. A pulse sequence is really a specification for the timing of a large number of events that ultimately lead to image information being collected for reconstruction processing. Among the timing components specified in a PSD are the TE, the time to the maximum amplitude of an echo, which is described by Equation 4.3, and the TR, the repetition time for the pulse sequence, as well as the timing for various other events, such as switching on and off of gradients and the receiver/analog-to-digital converter (ADC) subsystems.

4.3 Echoes

An echo is different from a FID in that it is a rephasing of signal and originates from one of two mechanisms. The shape of the FID is a decaying sinusoidal wave induced in a receiver coil as the transverse magnetization goes to 0. An echo, on the other hand, has a sinusoidal form that comes from the refocusing of the transverse magnetization so that the decaying FID is reestablished (refocused), causing the sinusoidal wave to increase to a peak and then begin to decay as a result of continuing dephasing of the transverse magnetization. Echoes can be produced using RF pulses or can be produced using the reversal of magnetic field gradients.

4.3.1 RF Echoes

RF echoes were first described by Hahn in 1950 [4]. These involve the application of multiple RF pulses to a spin system in order to produce a FID and then force an echo to evolve using a refocusing RF pulse that follows the decay of the FID signal. It is this echo that can be acquired and converted into image information during the process of image formation. A typical application of RF pulses consists of the application of a 90° excitation pulse followed after a time τ by a 180° rephasing pulse as follows:

$$90° - \tau - 180° \tag{4.5}$$

The signal at time 2τ that results from this application of two RF pulses is known as a spin echo (SE). The time 2τ is the TE time, the time to the maximum amplitude of the echo.

In an SE pulse sequence, where the 90° RF pulse moves the net magnetization M_0 into the xy plane, the echo is induced by application of a 180° RF refocusing pulse. After application of the 90° pulse, the net magnetic vector begins to evolve according to the solution to the Bloch equations:

$$M_z(t) = M_0(1 - e^{-t/T_1}), \quad \text{for } t = TR, \text{ the repetition time} \tag{4.6}$$

which describes the evolution of the vector component of M_0 in the z direction. The equation

$$M_{xy}(t) = M_{xy}(0)e^{-t/T_2}, \quad \text{for } t = TE, \text{ the echo time} \tag{4.7}$$

describes the dispersal of the magnetization in the xy plane due to the influences of varying magnetic fields from interactions with other spinning protons (spins). From a practical standpoint, groups of spins (isochromats) that are precessing in synchrony at the end of the 90° pulse immediately begin to spread from each other or dephase due to the effects of the particular tissue environment in which they exist. This dispersion means that some spins lag behind because their precessional frequency is slightly different from other spins. This lagging decreases the M_{xy} vector component of magnetization that will eventually become zero unless additional perturbation of the system occurs.

Application of a 180° RF pulse at some time τ, however, forces these dispersing spins to come back into "focus" or synchronization. Because the dispersing spins are in the xy plane, application of a 180° pulse forces the spins to rotate 180° across the xy plane, but on the other side of the circular path. This rotation effectively reverses the precessional order so that fast-moving spins are now able to overtake slow-moving spins, causing the signal to come back into focus. As these spins refocus at TE, the M_{xy} magnetization returns to nearly the level seen right after the 90° pulse was turned off. However, the amplitude is somewhat less due to true T_2 relaxation effects in the tissues of interest. Remember that the M_z component of M_0 is evolving as well, albeit more slowly than the changes in the M_{xy} component.

Upon the appearance of the refocused signal that peaks at time TE, the induced current that is present in the receive coil is converted to digital form for later processing. This is accomplished by the process of analog-to-digital conversion where the analog, time-varying signal is converted to digital form. The ADC is turned on for a sampling time, T_s, which is determined by the number of steps necessary to characterize the refocused signal and the frequencies to be recorded from the signal.

There are many other possible ways to interact with the spin system using RF echoes, including fast SE pulse sequences and various inversion recovery pulse sequences. The inversion recovery SE pulse sequence permits the measurement of T_1 characteristics. This sequence applies an initial inverting 180° pulse followed at time TI, the inversion time, by a 90° pulse, and a 180° pulse with evolution of an echo at time TE. After the initial 180° pulse, the M_z magnetization now has a negative value:

$$M_z(0) = -M_0 \qquad (4.8)$$

Since the time TI is the time delay between the initial 180° pulse and the following 90° pulse, the evolution of the M_z is

$$M_z(t) = M_0(1 - 2e^{-t/T_1}), \text{ for } 0 \le t < \text{TI} \qquad (4.9)$$

Following a second 180° pulse, the longitudinal magnetization is tipped into the transverse plane so that it can produce a signal for readout. The M_{xy} signal then follows the form

$$M_{xy}(t) = \left| M_0(1 - 2e^{-T_I/T_1}) \right| e^{-(t-TI)T_2^*}, \quad t > T_I \qquad (4.10)$$

for a single application of 180°–90°–180° sequence. The ADC is turned on to capture the maximum refocusing of the transverse magnetization, which in this case permits the readout of the longitudinal magnetization signal that has been rotated into the xy plane during the inversion time TI.

4.3.2 Gradient Echoes

4.3.2.1 Gradients

Gradient are implemented in all three dimensions on imaging systems following the form

$$\vec{G} = \frac{\partial B_z}{\partial x}\hat{x} + \frac{\partial B_z}{\partial y}\hat{y} + \frac{\partial B_z}{\partial z}\hat{z} \equiv G_x\hat{x} + G_y\hat{y} + G_z\hat{z} \qquad (4.11)$$

The gradient vector $\vec{G}$ is expressed here as the unit vectors of the Cartesian coordinate system with G_x, G_y, and G_z as the orthogonal components of $\vec{G}$ [5]. The gradient in any direction produces a field in the z direction according to

$$\vec{B} = \left(B_0 + G_x x + G_y y + G_z z \right)\hat{z} \qquad (4.12)$$

For SE imaging, gradients are turned on and off sequentially, but gradient switching can also be used for producing echoes.

Gradient echoes are echoes produced by application of a field gradient rather than a 180° RF pulse to induce rapid dispersion of spins in the xy plane followed by a reversal of the gradient polarity that results in dispersing spins coming back into focus to produce an echo. In the generation of gradient echoes, a linear gradient is applied during the evolution of the FID. At some time, t, after the linear gradient has caused dephasing of the spins, the gradient is reversed and remains on for a length of time sufficient to equal the period when the initial gradient was on. While the linear rephasing gradient is on, the spins come back into phase in the transverse plane at time TE, the echo time. The gradient remains on until the signal decays completely, with the timing usually selected so that the echo occurs in the middle of the rephasing gradient turn-on period. Because the echoes are dephased and rephased in the same direction as the magnetic field and there is no 180° refocusing pulse, there is no cancellation of the effect of inhomogeneities and tissue susceptibility effects, unlike SEs that do result in cancellation of these effects. The transverse magnetization decays with a T_2^* characteristic, which is much more rapid a loss of phase than is T_2 decay.

In SE imaging, the flip angle α, the angle achieved from application of the RF pulse between the z direction and the xy plane, is usually either 90° for an SE pulse sequence or 180° followed by a 90° pulse for an inversion recovery SE pulse sequence. In gradient echo imaging, the flip angle is often less than 90° because the flip angle can have a significant effect on the recorded signal amplitude and, ultimately, upon the contrast in reconstructed images. Furthermore, a small flip angle allows the pulse sequence to use a very short TR, which results in quicker image acquisitions. With a very short TR, a flip angle of a few degrees may be desirable to achieve the contrast desired. As the TR becomes longer (>200 ms) and the flip angle becomes greater than 45°, the contrast approaches that of an SE sequence, except that the T_2^* decay characteristic is present rather than the T_2 characteristic.

4.4 Image Formation

The goals of the imaging process are to determine the density of the spins present in a given spatially located volume element (voxel) and to determine the relaxation characteristics of those spins within the voxel. The characteristics of these spins will be used in a reconstruction process to produce 2D- and 3D maps in the form of image information. Among the requirements for this to be done accurately are the requirements to select a slice of the body to image and to interrogate each volume element in the slice to find out the characteristic differences based on different tissues distributed through the volume of interest.

While there are many ways to employ gradients to make images, the primary approach to imaging uses slice selection, followed by frequency and phase encoding. The frequency and phase of the spin isochromats contain sufficient information to enable placing the signal from these isochromats in the proper position relative to others within an *x,y,z* coordinate system in time or signal space, also known as *k*-space. This information is subsequently converted from the signal (*k*-space) domain to the frequency or image domain by means of a Fourier transformation in two dimensions or three dimensions. Through this mechanism, frequency information from signals induced in the receiver coil is transformed to spatial information to be reconstructed as a gray-scale image.

4.4.1 Slice Selection

To find the distribution of spins within a slice of tissue of a given thickness, it is necessary to select an appropriate slice position prior to applying imaging gradients that will encode within-slice information. This process of slice selection uses the capabilities of a spatially selective gradient that is turned on in the presence of the excitation RF field. In reality, depending on the type of imaging that will be performed, the selected slice can be more like a thick slab of tissue. The basic idea for slice selection involves the transmission of an RF pulse having a specified frequency range or bandwidth around the Larmor frequency dictated by the static field. This band-limited RF pulse is applied to the imaging volume during the time that a slice-select gradient is turned on at a specified amplitude. The shape of the RF pulse is in the form of a sync function, which specifies the range of frequencies in the pulse. Coupled with the gradient strength of the slice-select gradient, the combination of band-limited sync pulse and gradient strength produce a range of Larmor frequencies, which cause spins to resonate at the particular frequency dictated by the gradient amplitude at that point. This range defines the thickness of the slice that has been selected as

$$\Delta z = \frac{2\pi\Delta f}{\gamma G_z} \qquad (4.13)$$

where Δz is the thickness of the slice, in this case in the *z* direction, Δf is the bandwidth of the RF excitation pulse, and G_z is the gradient [5]. For this formulation, the assumption is that the axial slice is located at the isocenter of the gradient, G_z, which is centered around the Larmor frequency of the system with no gradients turned on. However, to shift the slice location, it is a simple matter to change the carrier frequency offset from the Larmor frequency or change the gradient strength so that the excited slice is offset from the central location of the slice to a new slice position. With this approach, it is easy for the system to turn on the slice select gradient and apply an RF pulse of the specified bandwidth necessary to select a slice of a desired thickness and position along the gradient direction.

One of the issues that must be addressed when the slice-select gradient is turned on is the issue of transverse magnetization phase dispersal. This dispersal reduces the signal available in the transverse plane across the slice, so it is usually necessary to have a rephasing gradient lobe as part of the slice selection process. This lobe is of opposite polarity to the slice selection part of the gradient and is designed to restore the lost phase due to the gradient being on during the RF pulse. The rephasing gradient lobe is equal to one half of the area of the slice-select gradient.

4.4.2 Frequency Encoding

The process of frequency encoding involves the application of a gradient in any direction that produces a spatial encoding of the position of isochromats. The encoding process follows the previously described variation of the local magnetic field in a linear fashion so that the resulting time-domain signals exhibit a range of frequencies where the frequencies are related to the position information in one direction. This approach has the advantage that the resulting signal, after decoding of position with Fourier transform processing, produces a 1D projection that can be used in filtered back projection reconstruction as well as in more elaborate reconstruction approaches using Fourier transforms.

For the frequency-encoding process to work properly, there must be two components in typical pulse sequences that provide both a prephasing function and a rephasing function. These separate components or lobes can be found in various locations within the frequency-encoding timing diagram, based in part on the type of pulse sequence that is in use and how it is implemented. Pulse sequences based on the SE technique typically place the prephasing part of the gradient in front of the 180° pulse, although it can follow the 180° pulse but with opposite polarity. The readout portion of the gradient follows the 180° rephasing pulse and is usually centered around the echo.

For gradient echo pulse sequences, the prephasing lobe can be negative or positive and often is adjacent to the readout lobe, which has the opposite polarity, thereby providing the gradient reversal that results in the readout signal. The gradient reversal rephases the out-of-phase isochromats, causing them to come into phase to produce an echo signal that can be digitized and used for image reconstruction.

The development of frequency encoding follows from the Larmor equation. In the development by Bernstein et al., for the rotating reference frame with a frequency of $\omega = \gamma B_0$,

$$\omega_x = \gamma G_{xp}(t)x \qquad (4.14)$$

where $G_{xp}(t)$ is the prephasing gradient for an isochromat in the x direction [5]. The prephasing gradient causes the accumulation of phase, φ_{xp}, while it is activated, leading to the formulation

$$\varphi_{xp} = \int_0^T G_{xp}(t)dt \qquad (4.15)$$

where T is the "on" time for the prephasing gradient. From the phase relationship in Equation 4.15, the signal in the transverse plane can be found by the summation of all of the accumulated vectors of the isochromats as a weighted function of their spin densities, $\rho_{x1}, \rho_{x2} \cdots \rho_{xn}$. The signal, S_{xp}, is approximated by the continuous distribution of spin densities and phase dispersion as

$$S_{xp} \cong \int_{-\infty}^{+\infty} \rho(x)e^{-i\varphi_p(x)}\,dx \qquad (4.16)$$

For an SE pulse sequence using a 180° RF refocusing pulse, the phase dispersion is eliminated by changing the sign, yielding

$$S'_{xp} \cong \int_{-\infty}^{+\infty} \rho(x)e^{i\varphi_p(x)}\,dx \qquad (4.17)$$

Following this, a readout gradient lobe, $G_x(t)$, is applied, resulting in additional phase dispersion and leading to the signal

$$S(t) = \int_{-\infty}^{+\infty} \rho(x)e^{-i(\varphi(x,t)-\varphi_p(x))}\,dx \qquad (4.18)$$

At a time, t_{echo}, where the phase dispersal described in Equation 4.15 becomes 0, it can be said that all of the isochromats are in phase and an echo is formed. This time point defines the center of k-space, which is ordinarily sampled symmetrically while the readout is taking place, so that the period for acquisition is defined over a period from $-T/2$ to $+T/2$, centered around the echo. The readout gradient area to the left of the t_{echo} time is the same as the prephasing gradient lobe. Negative k-space is sampled during this time, while positive k-space is sampled during the right-of-center period covered by the time from t_{echo} until the end of the period $+T/2$ [5].

Gradient echo pulse sequences use the gradient polarity reversal to cause the production of an echo, so the prephasing gradient is negative with respect to the readout gradient. The same symmetrical approach to sampling as described for SEs can be used for gradient echo systems as well, although asymmetrical sampling approaches are described in the literature for specialized applications requiring very short echo times [5,6].

4.4.3 Phase Encoding

Phase encoding implements a way of obtaining enough information to produce the second dimension of data that can be reconstructed into an image. The information obtained from frequency encoding produces 1D signals, which can be processed into projections that can be used for image reconstruction, mentioned earlier. However, projection reconstruction can take more time than directly reconstructing 2D data. Phase encoding the signal during collection permits implementation of 2D reconstruction methods.

The basic concept of phase encoding is that a gradient can be turned on for a short period after the spins have been rotated into the transverse plane by the initial RF pulse. Different areas under the phase-encoding gradient cause different phase shifts, which can be decoded to produce a linear range of phase shifts related to the direction of the applied gradient. Following each of these phase shifts, frequency-encoding occurs, as described previously. By sequentially incrementing the phase-encoding gradient over multiple TR periods, it is possible to fill 2D k-space so that subsequent application of a 2D Fourier transform produces an image representing the distribution of spins in the tissue of interest.

Phase encoding is implemented as follows from Bernstein et al. [5]: a gradient, G_y, applied in the y direction as a phase-encoding gradient, will produce an angular frequency, ω, in the rotating frame

$$\omega = \gamma G_y y \qquad (4.19)$$

When the phase-encoding gradient is turned on and then returns to zero, the phase, φ_y, in the transverse plane is

$$\varphi_y = y\gamma \int_0^T G_y(t')dt' \qquad (4.20)$$

where T is the time duration of the phase-encoding gradient and S is the induced signal from the magnetization in the transverse plane for one dimension, which becomes

$$S(k_y) = \int M_\perp(y)e^{-i\varphi(y)}\,dy \qquad (4.21)$$

where $M_\perp = M_x + iM_y$, the magnetization in the transverse plane, and k_y represents spatial frequency elements with units of cycles/distance. Since the phase-encoding gradients are applied discretely in time in a repetitive fashion, the signal, $S(k_y)$, is approximated by the summation

$$S(k_y) = \sum_{n=0}^{N-1} M_\perp(n\Delta y)e^{-2\pi i(n\Delta y)k_y} \qquad (4.22)$$

where $y = n\Delta y$, Δy is the pixel size, and N is the number of pixels to be encoded. The repeated application of G_y produces N

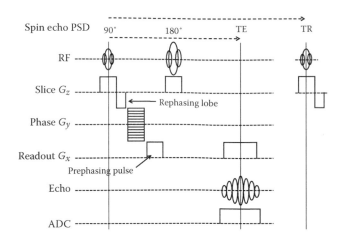

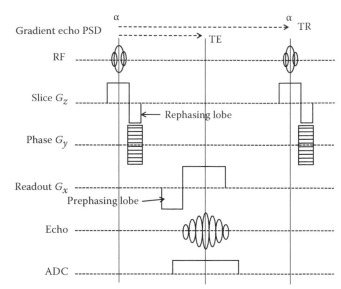

FIGURE 4.1 SE PSD. The events and relative timing necessary to implement a simple SE pulse sequence are shown for a single TR period. However, multiple TR periods would be necessary to collect enough phase encoded steps to reconstruct a 2D image. These steps are shown as the "ladder" in the diagram. Of particular note is the presence of the rephasing lobe in the slice selection gradient G_z and the prephasing gradient lobe in the readout gradient G_x. Also note that the echo evolves at time TE after the rephasing 180° pulse at TE/2.

FIGURE 4.2 Gradient echo PSD. In this PSD, there is no rephasing pulse to cause formation of an echo. Here the biphasic readout gradient G_x causes evolution of a gradient echo that is digitized by the receiver/ADC system. Note that as in the SE PSD, there is the repetition of multiple TR periods to record enough phase-encoded echoes to reconstruct a 2D image. Also note that instead of a 90° RF pulse, a pulse α of less than 90° would ordinarily be used. The gradient echo pulse sequence would execute more quickly than the SE sequence.

different phase steps that can be used to reconstruct the second dimension in a 2D data set. The magnitude of the steps between different applications of the phase-encoding gradient, G_y, control the resolution in k-space. Additionally, the maximum gradient strength determines the FOV, $N\Delta y$, in the y direction. Considerations for sampling, FOV, and pixel size all dictate the size of the phase-encoding gradient [5,6].

Figure 4.1 shows a PSD for a simple SE pulse sequence. Note that the diagram shows a single TR period, the number of which would be determined in part by the number of phase-encoding steps necessary to define the phase encoded direction in the collected data set. The gradient switching is specified for slice selection as well. In Figure 4.2 the PSD shows a gradient echo pulse sequence. The TR period would ordinarily be much shorter than in an SE pulse sequence so the data could be collected much more quickly. Of particular note is the lack of the 180° rephasing pulse seen in Figure 4.1. The echo in the gradient echo pulse sequence is produced by the biphasic readout gradient.

4.4.4 Extensions to Three Dimensions

There are basically two different ways to acquire 3D information. The first involves collecting multiple slices positioned so that each 2D slice is formed as described earlier, with slices positioned in such a way as to produce a stack covering the three dimensions, often known as an anisotropic acquisition because the slice thickness dimension is larger than the in-plane dimensions. This approach to 3D imaging is used clinically because the coverage can be adequate to answer a clinical question even with gaps between slices. Additionally, the acquisition can be performed rapidly relative to full 3D acquisitions.

In true 3D imaging, the acquisition encodes all three dimensions and the reconstruction is done in three dimensions as well. In this case the slice that is selected is actually a "slab" of thickness approximating the field of view in the other two dimensions. The approach is to use a broadband RF pulse to excite the entire volume during each TR period and apply an additional phase-encoding gradient to encode the third dimension. Thus, the excitation volume is the same during each TR period, but the phase-encoding step encoding the third dimension is used in addition to the phase-encoding step for in-plane encoding. Therefore, two sets of phase-encoding steps are required to be completed to encode fully 3D data in k-space. The great advantage to this approach is that the slices in the z-direction can be made very thin, since the slice thickness is controlled by the phase-encoding step in the z dimension, allowing for the possibility of isotropic image sets, those that have the same spatial resolution in all three dimensions. Additionally, there can be improvements in signal-to-noise ratios over the more conventional multislice imaging approach. However, the trade-off for using a true 3D acquisition is that the imaging times can increase substantially compared to a multislice acquisition.

4.5 *k*-Space and Pulse Sequences

4.5.1 *k*-Space Description

The data collected by manipulating gradients, RF energy, and timing of various events produce signals that are recorded in a specific way. The "raw" or "data" space is called "*k*-space," and mathematically; it is the conjugate of the image space, having units of inverse distance, and is related to the image by the Fourier transform. *k*-Space for a 2D image is a 2D space of signal information, the central area of which consists of low spatial frequency information that describes the contrast and shape of the object that is being imaged. In the center is the spatial frequency of 0. As one moves out from the center of *k*-space, the more distant locations represent the higher spatial frequencies that help define the edges within the images that result from the reconstruction process. A diagram of *k*-space is shown in Figure 4.3.

The received signal is the Fourier transform of the density of the spins in the object that is being studied. This is shown for three dimensions in the following development from Haake et al.

$$s(k_x, k_y, k_z) = \iiint \rho(x, y, z)e^{-i2\pi(k_x x + k_y y + k_z z)}\, dx\, dy\, dz \quad (4.23)$$

where ρ is the spin density and k is the spatial frequency in the direction x, y, or z such that $k = k(t)$, having dimensions of cm^{-1} [6]. It can be readily seen that this relationship relates the signal in *k*-space to the spin density by the Fourier transform as shown by

$$s(k_x, k_y, k_z) = \mathcal{F}[\rho(x, y, z)] \quad (4.24)$$

This is commonly called the imaging equation in three dimensions. The time dependence of $k(t)$ is related to the integral of

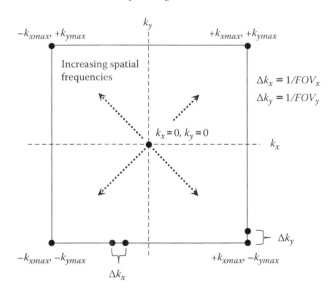

k-Space diagram

FIGURE 4.3 *k*-Space diagram. *k*-Space is an array of discrete points that represent spatial frequency information with 0 spatial frequency, sometimes called "DC," at the center. Filling *k*-space involves stepping through the range of spatial frequencies from positive to negative in both k_x and k_y directions in some sort of systematic way. The particular pulse sequence design dictates how this space is filled. Once *k*-space is filled, the 2D Fourier transform reconstructs grayscale images of the distribution of spins in the imaged slice.

the appropriate gradient, so that for each of the directions x, y, and z

$$k_x(t) = \gamma \int_0^t G_x(t')dt',\ k_y(t) = \gamma \int_0^t G_y(t')dt',\ k_z(t) = \gamma \int_0^t G_z(t')dt' \quad (4.25)$$

Note that the range of integration is over the beginning of the applied gradient to time t at the end of the gradient-on period. For a 2D acquisition, recall that the slice selection is done by manipulation of the slice gradient and tailoring the frequency spectrum of the RF energy to excite the selected spins. The imaging equation for two dimensions is reduced to

$$s(k_x, k_y) = \iiint \rho(x, y, z)e^{-i2\pi(k_x x + k_y y)}\, dx\, dy\, dz \quad (4.26)$$

where filling of *k*-space is accomplished in some systematic fashion.

The recording of *k*-space information is a discrete process because of the need to digitize the signal for reconstruction processing and because of the application of phase encoding in discrete steps. This means that *k*-space is a collection of discrete values that are an approximation of a continuous space. Because the analog-to-digital conversion process is used to convert the signal into discrete form, the number of points sampled must obey the Nyquist sampling criterion to avoid aliasing problems in the signal.

In a 2D acquisition, often *k*-space is sampled so that the FOV, Δk_x, and Δk_y are the same. In the Δk_x direction

$$\Delta k_x = \frac{\gamma}{2\pi}G_x \Delta t = \frac{1}{FOV} \quad (4.27)$$

where Δt is the sample time for each complex point in *k*-space [5]. This period, known as the "dwell time" is equal to the inverse of the readout bandwidth.

4.5.2 Filling *k*-Space

Filling *k*-space is a function of the pulse sequence implemented, the parameters used to digitize the received signals into samples, and the goals of the pulse sequence. *k*-Space can be filled in a linear fashion, a radial fashion, variations of a diagonal scanning fashion, partial fashion, spiral fashion, and so on. Many techniques have been developed to increase the speed of acquisition or to achieve a particular result using nonrectangular strategies for filling *k*-space. This section will discuss two ways to fill *k*-space using rectangular sampling strategies.

4.5.2.1 SE Pulse Sequence

In an SE pulse sequence of conventional design, each line in the k_x direction is filled by the digitized sample of the signal produced by the receiving system with the center of the signal centered at the middle of k-space. Δk_x, the distance between sample points along the k_x axis, can be shown to be inversely proportional to the FOV in the x direction. The distance from the negative maximum $-k_{xmax}$ through the center of k-space to $+k_{xmax}$ produces a set of spatial frequencies, which after reconstruction will contribute to the image. It can be shown that the pixel size in the x direction is inversely proportional to two times k_{xmax}. Likewise, Δk_y is inversely proportional to the field of view in the y direction. The distance from $-k_{ymax}$ to $+k_{ymax}$ moves in steps of Δk_y, where the pixel size in the y direction is inversely proportional to two times the k_{ymax} [7]. The goal of the imaging process is to use the chosen pulse sequence to fill this space as efficiently as possible.

The SE pulse sequence uses two RF pulses as described previously. The 180° rephasing pulse causes the evolution of an echo with the maximum amplitude occurring at the time TE. However, in order to sample the signal that evolves, the readout gradient, G_x, is turned on prior to the maximum amplitude of the echo signal. In a standard SE pulse sequence, the area under the gradient prior to the central point of k-space must be equal to the area of the prephasing pulse that occurs before the 180° rephasing pulse. Thus, sampling begins at $-k_{xmax}$ and proceeds through the maximum of the echo to $+k_{xmax}$. The area under the readout gradient is symmetric with the area preceding the center of k-space so that the entire k_x range is sampled. Figure 4.4 illustrates the trajectory in k-space that is possible when the prephasing readout gradient is located before the 180° refocusing pulse and the phase-encoding gradient begins the readout at $+k_{ymax}$.

For a simple SE pulse sequence, the phase-encoding gradient, G_y, typically is turned on after the first RF pulse at maximum amplitude for a fixed period. For each TR period, the phase-encoding gradient will be turned on at a stepped amplitude decreasing from positive to negative values through successive TR periods so that the coverage of k-space proceeds from $+k_{ymax}$ to $-k_{ymax}$. This approach that fills k-space from the outer edge at maximum G_y to the opposite edge at minimum G_y is known as a "top-to-bottom sequential" filling scheme [5]. There are, however, numerous other ways to use the phase-encoding gradients to acquire the necessary k_y data to fill k-space in a different order simply by controlling the amplitude of the phase-encoding gradient at a fixed turn-on time [5].

4.5.2.2 Gradient Echo Pulse Sequence

A gradient echo pulse sequence uses a single RF pulse followed by application of gradients necessary to fill k-space. Because the echo is formed by a gradient reversal, the initial polarity of the gradient can be either negative or positive, followed by the reversal and readout of the signal. Phase-encoding steps manipulate the phase of the system, as in the case of SE pulse sequences, so that k-space will be covered in an appropriate manner. A simple 2D gradient echo sequence using phase-encoding and gradient echo readout traverses k-space in a fashion very similar to that seen in the simple SE example. The application of the gradient, G_y, after an RF pulse with slice selection establishes the beginning of the movement through k-space. If the pulse sequence

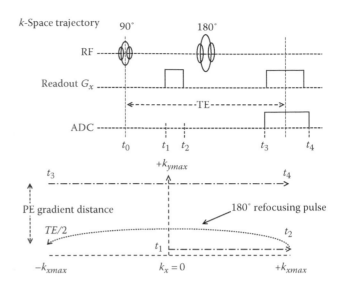

FIGURE 4.4 k-Space trajectory. For a conventional SE pulse sequence, where the prephasing gradient occurs before the 180° refocusing pulse, the trajectory of the signal location to be filled in k-space is controlled by the turn-on and turn-off times of the readout and phase-encoding gradients. Starting at t_1 when the first readout gradient prephasing pulse begins, the trajectory moves from t_1 to t_2 as shown in the bottom of the figure. At t_2, the 180° refocusing pulse moves the magnetization to the position across k-space to $-k_{xmax}$ at time TE/2. Depending on the phase-encoding gradient strength, the readout of the signal begins at t_3 and proceeds through k-space from $-k_{xmax}$ to $+k_{xmax}$ at t_4, when the readout of the signal is complete. The next TR period uses a different phase-encoding step, which will position the k_x line to be readout in a different part of k-space along the k_y direction, repeating the readout process just described.

starts with the maximum amplitude of the G_y gradient, then the k_{ymax} position will be the start followed by the readout from $-k_{xmax}$ to $+k_{xmax}$ to cover the first line of k-space at the highest spatial frequency. The next TR period would apply the G_y gradient at the next lower amplitude, effectively moving the position of the line to be read out toward the center of k-space. The application of the biphasic G_x gradient reads out the line of k-space, this time from $+k_{xmax}$ to $-k_{xmax}$. This phase-stepping process with alternating readout directions continues through the number of phase-encoding steps specified by the pulse sequence until k-space is fully covered.

4.5.2.3 Other Ways to Fill k-Space

The trajectories in k-space can be used to provide advantages in immunity to certain types of artifacts or improvements in the speed at which the pulse sequence is able to cover the applicable k-space. In true 3D imaging, using either the SE or the gradient echo approaches, it is fully possible to dispense with the single slice selection gradient applied during the RF pulses and to incorporate an approach using a broadband RF pulse that excites a volume plus the addition of a second phase-encoding gradient, such that phase steps cover two of the three directions needed to describe a volume of *k*-space. The readout approach remains the same in this arrangement.

Several other methods of coverage can be used as well. Many of these methods are applicable to clinically important pulse sequences. *k*-Space can be covered in such a way that radial lines are formed through the center of *k*-space. This approach is sometimes referred to as the projection imaging method. Somewhat similar in approach is a so-called spiral sequence where spiral trajectories originate from the center of *k*-space outward to cover *k*-space with either a single RF pulse or with multiple RF pulses. The spiral sequence has the advantage that it does not so sparsely sample the lower spatial frequency areas of *k*-space as happens with radial pulse sequences where there can be a loss of low frequency information. Other approaches can involve incomplete filling of *k*-space in which zero-padding fills unsampled parts of *k*-space with zeros to decrease the acquisition time for the sequence. Other methods incorporate acquisitions that use the property of symmetry for the spatial frequency data to omit collection of a complete set of *k*-space data. In this design, slightly more than half of *k*-space plus data at zero spatial frequency are collected and the missing data synthesized from the collected data. The trade-off is that the signal-to-noise level is inferior to a full *k*-space acquisition. These and other approaches have been developed to move beyond the simple pulse sequence methods [5].

4.6 Reconstruction

Reconstruction is the process of taking the data sampled into *k*-space and converting it into maps of the distribution of isochromats. These maps are the grayscale images that one is familiar with in MR imaging, formed from 2D or 3D *k*-space data that are translated to image space. The most common method used now is to apply the 2D or 3D Fourier transform to the 2D or 3D *k*-space data that have been acquired in a rectilinear fashion to reconstruct the image information. However, other methods of reconstruction can be used to address particular ways in which *k*-space has been filled. One example is the use of projection reconstruction techniques similar to those used in computed tomography when the *k*-space data appear as a series of radial lines in *k*-space. All of these methods are of interest, but most reconstruction takes place using the Fourier transform. In some cases an interpolation process is used when the acquisition is not rectilinear to create a rectilinear mapping of *k*-space from the original nonrectilinear data. However the reconstruction process is performed, the key information is that the reconstruction algorithm should be appropriate for the way the data were originally placed into *k*-space.

4.6.1 Fourier Reconstruction

The *k*-space data are related to the image information by a Fourier transform in two or three dimensions (see Equations 4.23 and 4.24). In the case of rectilinearly filled *k*-space, a series of 1D Fourier transforms can be applied in the form of the fast Fourier transform (FFT), a computer implementation of the discrete Fourier transform [8]. Other ways of filling *k*-space, such as the radial path described earlier can use a filtered back projection method for reconstruction, if some interpolation is performed to place the data in rectangular form. The regridding process to interpolate data onto a rectilinear grid is necessary using conventional versions of the FFT.

The procedure for implementing the reconstruction requires that there be some preparations to the data before the actual transformation step. Some of these preparations are related to the use of the FFT. In particular, it is often necessary to pad the *k*-space data with zeros to bring the array to be processed by the FFT up to a power of 2, which often is required by the version of the FFT. Usually, this is performed in a symmetric fashion [5]. In some situations, padding with zeros extends to powers of 2 beyond what is necessary for proper operation of the FFT. This extended padding can enhance the resolution of the resulting reconstructed image by providing a smoothing of the data in the image due to a reduction in the partial volume effect [5,9].

Another prereconstruction step involves phase shifting to address the issue that the FFT typically will expect that the zero frequency time point is at the beginning of the sequence of values. Since data are typically collected with a range of $-k_x$ to $+k_x$ or reverse depending on the gradient polarity with $k_x = 0$ located in the center, application of a phase-correction step correctly shifts the data after the FFT has been applied. Using the Fourier shift theorem, this phase shift can be implemented by alternating the sign of the FFT output [5]. A similar approach properly places the DC point in the center of the array by applying a similar phase shift for each *k*-space sample prior to application of the FFT.

Another processing step that is also performed is using a windowing function to control the presence of ringing and data overshoot artifacts due to truncation of the *k*-space data. This truncation artifact is often called the Gibbs artifact. It is equivalent to multiplying the *k*-space data by a rectangular function, which leads to a characteristic artifact because of the sharp transition from data at the edge of *k*-space to zero. These effects can be reduced using a nonrectangular window function, a filter that reduces the higher spatial frequencies in some sort of smooth fashion at the edges of the defined *k*-space data. The process is called apodization and can use a variety of window functions to reduce the Gibbs artifact at the expense of some loss of resolution [5].

Other processing steps accommodate different issues with the *k*-space data. In many cases, rectangular fields of view are used to shorten the acquisition time by reducing phase-encoding steps. A zero-filling process makes up for fewer phase-encoding steps followed by an interpolation in image space to achieve a square matrix. In some systems it is necessary to accommodate the fact that the gradients that are supposed to be linear over their range are not, leading to some geometric and intensity changes that have to be corrected. Additionally, there often are some scaling processes so that the images that are stored fall within the range of the display system. Other processing can include more advanced methods to correct for baseline shifts related to the DC level, although the phase cycling process typically eliminates the problem.

4.6.2 Parallel Imaging and Reconstruction

Parallel imaging methods use multiple coils to acquire signal information from the imaging field. Each of these multiple coils is coupled to a separate receiver so that the excitation and reception processes are separated in what is termed a "phased array." Using an array of coils, some of the temporal requirements for spatial encoding are reduced by reducing the need to switch multiple phase-encoding gradients, resulting in improvements in acquisition speed without affecting image spatial resolution and contrast [10]. This technique has been known for some time when it was introduced as a way to improve the coverage of large areas of anatomy while improving the signal-to-noise ratios over that of a large single element coil [11]. The resulting images can be assembled using a square-root-of-the-sum-of-the-squares formulation to create magnitude images from the multiple coils. While this type of multiple coil acquisition expanded the ability of MR imaging to efficiently cover large areas for imaging regions of the body such as the spine, more recent developments of parallel imaging provided the imaging community with the ability to increase the speed of acquisition using multiple coils and much more sophisticated processing of the signals from each coil.

There are two basic approaches that have been developed for using coil arrays to improve acquisition speed. The idea of using multiple coils to increase the speed of acquisition is to take fewer lines of k-space in the phase-encoding direction by increasing the distance between the phase-encoding lines. This distance is sometimes given the name R to indicate a "reduction factor" or "acceleration factor" [5]. Increasing the distance between phase-encoding lines reduces the FOV, typically resulting in a wrap-around error (aliasing) if the object extends outside of the FOV. Since it is possible to determine the RF field (B1) characteristics of each coil element and the resulting sensitivities at all points in the B1 field of each coil, such information can be used by specialized reconstruction methods to either remove or prevent the occurrence of the aliasing effect. The two approaches introduced to do this with phased arrays are known as SENSE (sensitivity encoding) and SMASH (simultaneous acquisition of spatial harmonics) introduced by Sodickson and Manning in 1997 [12] and Pruessmann et al. in 1999 [13], respectively.

4.6.2.1 SENSE

SENSE is a parallel imaging technique that unfolds aliased image information from multiple receiver coils in the image domain to produce an acceptable image. In SENSE imaging for rectilinear sampled k-space, the approach is to acquire undersampled data in k-space by reducing the number of phase-encoding steps by a factor R, the reduction factor. After the FFT, the result is an aliased image from each coil due to the reduced FOV, but with a speed-up in the acquisition by R times.

This is illustrated in Figure 4.5 for a six-channel SENSE coil. The reconstruction process takes these aliased images and prescan data from the coil sections to reconstruct the correct image. The prescan data come from a 3D low resolution acquisition, which can be run very quickly, that provides the spatial sensitivity maps for all of the coils with the patient in place. All of the processing to produce the final image is performed after application of the FFTs to reconstruct the k-space data into the image domain [14].

In SENSE, each pixel in a reduced FOV image from each of the coils in the array contains information from R numbers of pixels in the final full FOV image. From Blaimer et al., the signal at a pixel in location x, y in the kth coil image I_k is described as

$$I_k(x,y) = C_k(x,y_1)\rho(x,y_1) + \cdots + C_k(x,y_R)\rho(x,y_R) \qquad (4.28)$$

where C is the coil sensitivity at the corresponding location in the image with full FOV and y_R is the maximum number of pixel locations at the R-specified phase-encoding steps [10].

If the index k goes from 1 to N_c, where N_c refers to the number of coil elements in the array, one can rewrite Equation 4.28 in the following form

$$I_k = \sum_{l=1}^{R} C_{kl}\rho_l \qquad (4.29)$$

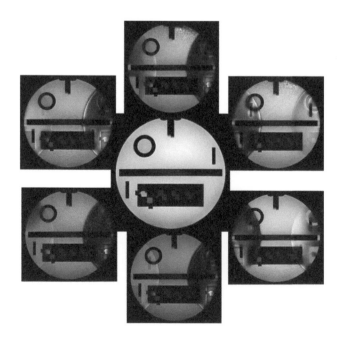

FIGURE 4.5 SENSE with six coils. A six-channel receiver system was used to acquire the individual aliased coil images from a six-channel SENSE head coil for an acquisition speed-up (SENSE factor) of about 1.5. The images were collectively reconstructed to form the SENSE image in the center of the figure.

for $l = 1$ to R, which gives the locations of the pixels. Blaimer et al. provide the following formulation, which leads to a set of linear equations with R unknowns (in matrix notation):

$$\vec{I} = \hat{C} \cdot \vec{\rho} \tag{4.30}$$

such that the $\vec{I}$ represents the value in the complex coil image for a chosen pixel over the number of coils in use [10]. $\hat{C}$ is the coil sensitivity matrix for each coil at the R superimposed positions with dimensions of number of coils times R. $\vec{\rho}$ is the vector of R pixels in the full FOV image for which the solution is found using an inverse of the sensitivity matrix [10,13,14]:

$$\vec{\rho} = \left(\hat{C}^H \hat{C} \right)^{-1} \hat{C}^H \cdot \vec{I} \tag{4.31}$$

In this development, no noise correlation is taken into account [10]. More details of the full derivation can be found in Pruessmann et al. [13]. The ability to perform this inversion is based in part on there being fewer R pixels than the number of coils in the array. Additionally, the equation above must be repeated for each pixel in the reduced FOV image to reconstruct the full FOV image. An added consideration is the ease with which the inversion can be performed, which is based on the geometry of the coil array in use during the acquisition. This factor, called the geometry factor, or g-factor, is a measure of the difficulty of inverting the matrix in Equation 4.31 above. The penalty in SNR for using SENSE is given by

$$SNR_{SENSE} = \frac{SNR_{Full}}{g\sqrt{R}} \tag{4.32}$$

A detailed example of the development of a parallel imaging SENSE coil for head imaging can be found in an article by de Zwart et al. [15].

4.6.2.2 SMASH

SMASH refers to a k-space technique that uses estimated coil sensitivities to produce k-space phase-encoding data that would normally be collected if all of the phase-encoding steps were actually performed [10,12,14,16]. The general idea is to approximate the complex exponential functions that would have been present in the missing phase-encoding lines of k-space using linear combinations of coil sensitivity information. As in SENSE, SMASH provides a significant speed-up in acquisition by making use of the coil array information to replace the missing lines of data normally present in complete (100%) phase encoding.

SMASH uses coil sensitivity values, $C_k(x,y)$, and linear weights, $n_k^{(m)}$, in order to produce composite sensitivity profiles having sinusoidal spatial variations of order m where

$$C_m^{cp}(x,y) = \sum_{k=1}^{N_c} n_k^{(m)} C_k(x,y) \cong e^{im\Delta k_y} \tag{4.33}$$

and C_m^{cp} is the composite c sensitivity profile. In this formulation from Blaimer et al., $\Delta k_y = 2\pi/FOV$, k is an index from 1 to N_c, where N_c is the number of coils in an array, and m represents the integer order of the spatial harmonic produced [10]. Since this is a linear formulation with the linear weights, $n_k^{(m)}$, as the unknowns, a linear least-squares fitting approach can be used to fit the coil sensitivity profiles, C_k, to the mth order spatial harmonic represented by $e^{im\Delta k_y}$. The 1D signal, $S_k(k_y)$, from the phase-encoding direction in a coil k can be shown to be the Fourier transform of the spin density with coil sensitivity profile weighting:

$$S_k(k_y) = \int \rho(y)C_k(y)e^{ik_y y}\, dy \tag{4.34}$$

By combining Equations 4.33 and 4.34, the basic SMASH equation is formed for k-space lines that are shifted as $S(k_y + m\Delta k_y)$, based on the weighted combinations of the coil signals measured during the acquisition process [10]:

$$\sum_{k=1}^{N_c} n_k^{(m)} \cdot S_k(k_y) = \int \rho(y) \sum_{k=1}^{N_c} n_k^{(m)} C_k(y)e^{ik_y y}\, dy$$

$$\cong \int \rho(y)e^{im\Delta k_y y}e^{iky} = S^{cp}(k_y + m\Delta k_y) \tag{4.35}$$

The original version of SMASH suffers from the limitation that it is only valid for certain coil geometries in which it is possible to approximate spatial harmonics relatively accurately. This is not always easy to do with newer coil designs, so other strategies have been developed that supplant the original SMASH approach and add some efficiencies in the reconstruction process. The most successful of these is known as GRAPPA, for generalized auto calibrating partially parallel acquisition, that was proposed by Griswold et al. [17]. A primary difference in GRAPPA and SMASH is the use of reference data in the form of calibration signals rather than the entire coil sensitivity data set as in SMASH. GRAPPA attempts to create missing lines in k-space from combinations of neighboring lines [16,18]. Extra calibration lines acquired during data acquisition from each coil are used as a reference to determine what combination of adjacent lines in data from each coil most closely approximates the calibration line in each single coil. In this approach, a series of coefficients are developed that can be used to generate the missing k-space lines in data from each coil. Once this is accomplished, each full k-space data set for each coil of the array can be Fourier-transformed into image space where all of the coil images are combined.

Parallel imaging using image space measurements (SENSE) and k-space methods (SMASH and GRAPPA) are widely available on commercial imaging systems. In some commercial systems, both SENSE and SMASH are available for use when one or the other method will yield the better performance. It should be noted that SENSE uses accurate sensitivity maps for full reconstruction of the data, while SMASH and GRAPPA, because they both use a least-squares approximation to spatial harmonics in *k*-space, are less accurate in the final reconstruction of the images. However, both *k*-space and image space implementations of parallel acquisition technology have been highly successful, with continuing development of systems having larger numbers of parallel channels demonstrating concomitant increases in imaging performance.

References

1. Lauterbur PC. Image formation by induced local interactions: Examples employing nuclear magnetic resonance. *Nature* 1973;242:190–191.

2. Damadian R, Minkoff L, Goldsmith M, Stanford M, Koutcher J. Field focusing magnetic resonance (FONAR): Visualization of a tumor in a live animal. *Science* 1976; 194(4272):1430–1432.

3. Liang Z-P, Lauterbur PC. *Principles of Magnetic Resonance Imaging.* Piscataway, NJ: IEEE Press; 2000.

4. Hahn EL. Spin echoes. *Phys. Rev.* 1950;80:580–594.

5. Bernstein MA, King KF, Zhou XJ. *Handbook of MRI Pulse Sequences.* Boston: Elsevier; 2004.

6. Haake EM, Brown RW, Thompson MR, Venkatesan R. *Magnetic Resonance Imaging: Physical Principles and Sequence Design.* New York: Wiley-Liss; 1999.

7. Paschal CB, Morris HD. K-space in the clinic. *JMRI* 2004;19:145–159.

8. Gonzalez RC, Woods RE. *Digital Image Processing.* Upper Saddle River, NJ: Prentice Hall; 2001.

9. Du YP, Parker DL, Davis WL, Cao G. Reduction of partial-volume artifacts with zero-filed interpolation in three-dimensional MR angiography. *JMRI* 1994;4:733–741.

10. Blaimer M, Breuer F, Mueller M, Heidemann RM, Giswold MR, Jakob PM. SMASH, SENSE, PILS, GRAPPA: How to choose the optimal method. *Top Magn Reson Imaging* 2004;15(4):223–236.

11. Roemer PB, Edelstein WA, Hayes CE, Souza SP, Mueller OM. The NMR phased array. *Magn Reson Med* 1990;16:192–225.

12. Sodickson DK, Manning WJ. Simultaneous acquisition of spatial harmonics (SMASH): Fast imaging with radiofrequency coil arrays. *Magn Reson Med* 1997;38:591–603.

13. Pruessmann KP, Weiger M, Scheidegger MB, Boesiger P. SENSE: Sensitivity encoding for fast MRI. *Magn Reson Med* 1999;46:952–962.

14. Pruessmann KP. Encoding and reconstruction in parallel MRI. *NMR Biomed* 2006;19:288–299.

15. de Zwart JA, Ledden PJ, Kellman P, van Gelderen P, Duyn JH. Design of a SENSE-optimized high-sensitivity MRI receive coil for brain imaging. *Magn Reson Med* 2002;47:1218–1227.

16. Larkman DJ, Nunes RG. Parallel magnetic resonance imaging. *Phys Med Biol* 2007;52:R15–R55.

17. Griswold MA, Jakob PM, Heidemann RM, Nittka M, Jellus V, Wang J, Kiefer B, Haase A. Generalized autocalibrating partially parallel acquisitions (GRAPPA). *Magn Reson Med* 2002;47:1202–1210.

18. Hashemi RH, Bradley WG Jr, Lisanti CJ. *MRI: The Basics,* Third Edition. Philadelphia: Wolters Kluwer Lippincott, Williams & Wilkins; 2010.

Characterizing Tissue Properties with Endogenous Contrast Mechanisms

Quantitative Measurement of T_1, T_2, T_2^*, and Proton Density

Richard D. Dortch
Vanderbilt University

Mark D. Does
Vanderbilt University

5.1 Introduction

While the majority of clinical MRI today is qualitative, ^{1}H NMR in tissue began with quantitative evaluations of relaxation times. Even before the development of MRI, Belton et al. [1–3], Hazelwood et al. [4,5], Damadian et al. [6–9], and a number of other investigators [10–15] characterized ^{1}H relaxation times in a variety of normal and diseased tissues. In fact, the observation of T_1 and T_2 differences between normal and cancerous tissue by Damadian et al. [6] was, arguably, the primary motivating factor for the development of practical signal localization with gradients, leading to the development of contemporary MRI. With this development, however, methods for robust measurement of tissue water proton relaxation became more complicated and, consequently, somewhat overlooked by the majority of the MRI community. This is particularly true of those involved in the application of MRI to clinical diagnostics. In fact, the vast majority of clinical cancer MRI performed today, including scans for tumor diagnostics, aims to simply maximize tissue contrast in order to visualize tumors and perhaps measure their size. Nonetheless, an ever-growing body of literature is developing to warrant the application of MRI protocols that quantitatively measure tissue water proton relaxation characteristics. In the context of cancer imaging, the most relevant clinical example is the use of dynamic measurements of T_1 following contrast

agent injection to characterize tumor vasculature via dynamic contrast enhanced (DCE) MRI (see Chapter 10).

In this chapter, this and other quantitative T_1, T_2, and T_2^* measurement protocols of interest are discussed. Emphasis is placed on optimization and practical implementation. The advantages and disadvantages of each technique are discussed within this framework. Prior to these more quantitative discussions, the existing literature is reviewed and the proposed biophysical basis for relaxation time changes in cancerous tissue is discussed.

5.2 NMR Relaxation in Cancerous Tissue

In this section, we review *in vivo* and *ex vivo* studies aimed at characterizing NMR relaxation times in tumor. For a more thorough review of this subject, especially with respect to earlier works, the reader is referred to reference [16].

As previously stated, elevated proton T_1 and T_2 relaxation times in excised cancerous rat tissue were first observed by Damadian et al. [6] in 1971. Similar findings were subsequently reported in a number of animal models [4,10,11,14,15] and excised human tumors [7–9,12,13]. Together, this body of literature prompted the suggestion that relaxation data may contain quantitative information for differentiating normal from cancerous and even benign from malignant tumors in humans _in vivo_.

This promise, along with the advent of MRI, prompted investigators to develop methods for imaging tumor relaxation times _in vivo_. The earliest of these studies [17–19] simply sought to generate contrast between normal and cancerous tissue via T_1- and/or T_1-weighted imaging—this practice still dominates clinical cancer MRI today. Subsequently, more sophisticated quantitative MRI protocols for measuring T_1 and T_2 were developed and applied in human tumors [20–24] and in animal models [25–28] _in vivo_. Varying degrees of success (with respect to diagnostic specificity) have been reported, which can be attributed, at least in part, to the experimental difficulties associated with imaging relaxation times _in vivo_ (see Sections 5.4–5.6).

5.2.1 Biophysical Basis of Relaxation Time Changes

In this section, we discuss the biophysical basis for relaxation time changes that occur in cancerous tissue. We limit the discussion to T_1 and T_2 measurements. Given that T_2 is a component of T_2^*, many of the points regarding T_2 also apply for T_2^*. It should be noted, however, that T_2^* is additionally affected by magnetic field inhomogeneities, which can be of physiological interest in some cases (e.g., when caused by changes in blood oxygenation).

A number of physiological and microanatomical parameters are altered in tumors that may affect relaxation times. This list includes variations in pO_2, pH, water content, and macromolecular content as well as the presence (or absence) of irregular vasculature, necrotic tissue, and/or densely packed proliferating cells [15,29–32]. The effect of each of these with respect to changes in T_1 and/or T_2 in cancerous tissue is discussed below in more detail.

Increased tissue relaxation times in tumors may be attributable to increased water content, decreased macromolecular content, hypoxia (low pO_2), and/or changes in pH. For example, it has been known for some time that the relaxation times of viable tumor are elevated relative to normal tissue and that this increase correlates with increased water content [15]. Additionally, decreases in macromolecular content, which affects the observed mobile water pool via magnetization transfer (MT; see Chapter 8), also likely play a significant role [33]. Changes in pO_2 and pH may also play a role with respect to T_2. Previous studies in skeletal muscle [34] have indicated that hypoxia-induced intracellular acidosis results in increased tissue T_2 times via pH-modulated chemical exchange between water and titratable protons of protein side groups [35]. Given that viable tumor cells are commonly hypoxic and acidic, these may also be underlying factors with respect to T_2 changes in tumors.

Although relaxation times are typically elevated in tumors, several factors may result in decreased tissue relaxation times, including the presence of densely packed proliferating cells, irregular tumor vasculature, and/or necrotic tissue. Previous work [26] has shown that relaxation times and cellularity are inversely related, most likely attributed to the fact that macromolecular content increases and water content decreases with increased cellularity. Additionally, the presence of irregular tumor vasculature (e.g., dead ends, shunts that are commonly found in tumor associated vasculature [29]) may result in the accumulation of paramagnetic deoxyhemoglobin and a corresponding decrease in relaxation times for adjacent tissue. Furthermore, necrotic tissue has been shown to exhibit decreased water content relative to normal tissue in some instances [36], which may explain reported [37] decreases in relaxation times concurrent with the onset of necrosis.

Thus, it can be said that tissue relaxation times are sensitive to a wide array of parameters that can be altered in tumors. Because many of these factors occur concurrently and exhibit competing effects, questions regarding the specificity of relaxation time measurements in tumors remain. One proposed method to improve this specificity is multiexponential characterization of tumor relaxation times, which is the subject of the following section.

5.2.2 Multiexponential Relaxation

Given the heterogeneous nature of tumors, one might expect to observe multiexponential relaxation in tumors. In tumors, multiexponential T_1 [33,38,39] and T_2 have been observed; however, multiexponential T_1 is most likely attributable to MT between the observed mobile water protons and protons that are closely associated with relatively immobile macromolecules [40]. As a result, we limit our discussion to multiexponential T_2 in this section.

Multiexponential T_2 relaxation has been observed in animal models (both _in vivo_ [28,38] and _ex vivo_ [37]), in surgical samples _ex vivo_ [36,39], and in humans _in vivo_ [23]. Of these previous studies, only two [23,28] attempted to characterize multiexponential T_2 relaxation in tumors using an imaging approach others used a nonimaging [41,42] approach. The reported findings from these studies have been inconsistent, which is presumably a product of differences in tumor types, models, and staging, as well as differences in the data acquisition and analysis techniques used (see Sections 5.4.4–5.4.5). Nonetheless, these studies do indicate that (i) T_2 is multiexponential in a number of tumors, and (ii) these measurements may contain more specific information about the tumor microenvironment and progression compared to conventional single component T_2 measurements. For example, a recent study [28] observed two T_2 components in a rat glioblastoma model claimed to be derived from necrotic and viable tumor regions in order of increasing relaxation time (single component characterization of T_2 in the same model was unsuccessful in characterizing this heterogeneity). Identifying the physiological origins of the observed T_2 components is an ongoing area of research; however, this approach may allow for more specific quantitative characterization of the tumor microenvironment _in vivo_ on a subvoxel scale.

5.2.3 Applications

Applications involving NMR relaxation measurements in cancer, both in humans and in animal models, fall into one of two categories: (i) relative measurements and (ii) absolute measurements. The most relevant example of a relative NMR relaxation measurement in cancer imaging is DCE-MRI (see Chapter 10). In this approach, a series of postcontrast T_1-weighted signal images are acquired, yielding a measure of ΔT_1. Fitting the resulting ΔT_1 timecourse to an appropriate pharmacokinetic model yields physiological parameters that relate to tissue perfusion, microvascular vessel wall permeability, and extracellular volume (n.b., more rigorous approaches involve a precontrast T_1 measurement to convert ΔT_1 to an absolute T_1). Dynamic susceptibility (DSC; see Chapter 11) MRI, a similar technique based upon T_2 changes, provides additional estimates of blood flow, blood volume, and mean transit time. Studies involving absolute NMR relaxation measurements seek to determine if these measurements can directly serve as a reliable biomarker for tumor detection, staging, and treatment response. Because previous reports have called into question the specificity of relaxation measurements, there are many who remain skeptical about this prospect. Nonetheless, there is sufficient promise and an ever-growing body of literature to warrant further investigation. For example, recent work [43] has shown that a decrease in tumor T_1 reflects hypocellularity in a number of experimental tumor models and may serve as a generic marker of chemotherapeutic treatment response.

The long-term success of such studies depends on the ability of the chosen approach to accurately and robustly quantify the NMR parameter (T_1, T_2, T_2^*, or M_0) of interest, which in turn depends on a number of experimental factors (e.g., chosen pulse sequence, sensitivity to $\mathbf{B}_1$, and $\mathbf{B}_0$ variations). In the following sections, we will discuss methods for measuring these NMR parameters in this context. Prior to this, the general pulse sequences used are introduced and the corresponding signal equations are derived.

5.3 Pulse Sequences and Signal Equations

In general, any number of pulse sequences can be used to measure NMR parameters (T_1, T_2, T_2^*, or M_0): one simply needs a signal equation that relates the observed signal to independent pulse sequence parameters (e.g., TE, TR). The observed signal can then be regressed against the independent sequence parameter(s) to fit the model parameter(s). The effectiveness of this fitting depends on the chosen pulse sequence, its sensitivity to $\mathbf{B}_1$ and $\mathbf{B}_0$ field variations, the chosen independent sequence parameter values, and, of course, the signal-to-noise ratio (SNR). For any sequence, the signal equation can be derived using the Bloch equations (Equation 2.22), usually ignoring relaxation during radiofrequency (RF) pulses and neglecting off-resonance terms.

Consider, for example, a spin echo pulse sequence with a 90° excitation pulse (Figure 5.1). In this case, it is intuitive to see that the Bloch equations for transverse magnetization solve to

$$M_T(t) = M_z(t = 0^-)\exp(-t/T_2) \qquad (5.1)$$

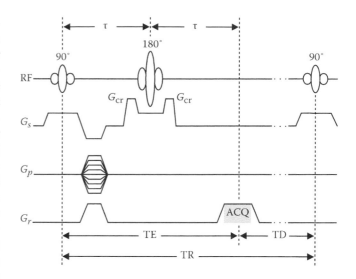

FIGURE 5.1 Spin echo imaging pulse sequence. A 180° refocusing pulse is applied after a delay, τ, forming the acquired (ACQ) spin echo after an additional delay, τ. The resulting signal is described by Equation 5.5. Crusher gradients (G_{cr}) are applied about the 180° refocusing pulse to ensure that the acquired signal arises from spins excited by the 90° excitation pulse. $G_{s,p,r}$ = gradient in slice, phase, and read directions.

where $M_z(t = 0^-)$ is the longitudinal magnetization immediately prior to the excitation pulse. Solving the Bloch equations for longitudinal relaxation can determine this term with respect to pulse sequence timings, TR and TE, and equilibrium magnetization, M_0, yielding

$$M_T(TR,TE) = M_0[1 - 2\exp(-(TR - TE/2)/T_1) + \exp(-TR/T_1)]$$
$$\exp(-TE/T_2). \qquad (5.2)$$

In many cases, $TE \ll TR$ and T_1, and Equation 5.2 is written as

$$M_T(TR,TE) \approx M_0[1 - \exp(-TR/T_1)]\exp(-TE/T_2) \qquad (5.3)$$

which is convenient because terms containing TR and T_1 are now independent of terms with TE and T_2. Alternatively, one can make note that $M_z(TE) \approx 0$ and solve the Bloch equations from this point to give

$$M_T(TD,TE) \approx M_0[1 - \exp(-TD/T_1)]\exp(-TE/T_2) \qquad (5.4)$$

which does not require assumptions about the relative size of TE, TR, and T_1. By convention, we use the symbol M when describing magnetization, but use S to describe the signal we observe in an NMR experiment, which is proportional to M_T but also includes instrument factors and additive noise. The signal equation for a spin echo sequence as defined in Figure 5.1 is then

$$S(TD,TE) = S_0[1 - \exp(-TD/T_1)]\exp(-TE/T_2) + v \qquad (5.5)$$

where v is drawn from a Gaussian distribution. Note that when the signal S is derived from a magnitude image, it is then drawn

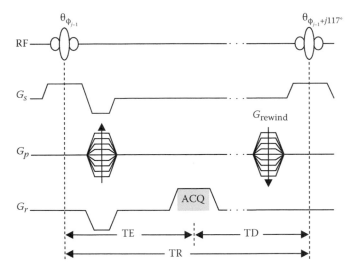

FIGURE 5.2 Spoiled gradient echo imaging pulse sequence. A gradient echo is acquired (ACQ) following an excitation flip angle of θ. The resulting steady-state signal is described by Equation 5.6. The phase encode rewind gradients (G_{rewind}) ensures that the phase accrual is the same for each TR period. This allows one to RF spoil—ensuring the acquired signal arises from spins excited during the current TR period—by simply incrementing the phase ϕ_j of each consecutive RF pulse by linearly incrementing the phase from one period to the next. $G_{s,p,r}$ = gradient in slice, phase, and read directions.

from a Rician distribution, which complicates analysis (see Section 5.8). Using a similar approach and solving in the steady state, the signal equation for a spoiled gradient echo sequence as defined if Figure 5.2 is

$$S(\theta, TR, TE) = S_0 \sin\theta \frac{\left(1 - \exp\left(-TR/T_1\right)\right)}{\left(1 - \cos\theta \exp\left(-TR/T_1\right)\right)} \exp\left(-TE/T_2^*\right) + v \quad (5.6)$$

Again, note that the terms containing TE and T_2^* are independent of the terms containing TR, θ, and T_1.

5.4 Spin–Spin Relaxation (T_2)

5.4.1 Spin Echo

Spin echo sequences, in one form or another, are most commonly used to measure T_2. From Equation 5.5, above, it is apparent that repeated signal acquisition using a conventional spin echo sequence with varied TE and constant TD (or constant TR if $TE \ll TR$) allows for a simple fitting of the signal to an exponentially decaying function. In this case, the T_1-weighting term in Equation 5.4 is constant and thus can be incorporated into the apparent proton density, S_a, leaving

$$S_i \equiv S(TE_i) = S_a \exp(-TE_i/T_2) + v_i \quad (5.7)$$

for the ith signal. Given a vector **s** of observations of S_i, $i = 1$ to N, T_2 (and S_a) can be estimated using standard nonlinear regression

methods (e.g., Levenberg–Marquardt) to find a least sum square solution

$$\underset{\hat{T}_2, \hat{S}_a}{\arg\min} \left\| \mathbf{s} - \hat{S}_a \exp\left(-\mathbf{t}/\hat{T}_2\right) \right\|_2 \quad (5.8)$$

where **t** is a vector of echo times corresponding to the vector of observations **s** and $\|\cdot\|_2$ is the Euclidian norm.

The drawbacks of nonlinear least squares fitting include (i) the need to supply initial parameter guesses and (ii) their increased computational expense relative to linear least squares fitting. An alternative fitting approach is to linearize the signal equation through a log-transformation

$$\log\left[S_i\right] \approx \log\left[S_a\right] - TE_i/T_2 \quad (5.9)$$

where the noise term has been ignored for simplicity because it is no longer additive. This approach avoids the drawbacks associated with nonlinear regression, but the log-transformation introduces systematic errors because Equation 5.9 is only a good approximation when the additive noise in Equation 5.7 is very small. The resulting fitted parameters $\left(\hat{S}_a, \hat{T}_2\right)$ will be biased estimates of the true model parameters. For this reason, when accuracy is more important than processing time, nonlinear regression is recommended. Note, however, that even with nonlinear regression, biased estimates of model parameters can result from the Rician nature of the noise (see Section 5.8).

5.4.2 Optimization of Acquisition Parameters

The choice of appropriate acquisition parameters (TR and TE) is crucial given the limited SNR typically available for clinical MRI. Generally, one would like to choose the TR that optimizes the SNR of the images per unit acquisition time. The total acquisition time is proportional to the product of TR and the number of acquisitions (N_{acq})

$$T_{total} \propto N_{acq} TR \quad (5.10)$$

The signal magnitude is directly proportional to N_{acq}

$$signal \propto N_{acq} \left(1 - e^{-TR/T_1}\right) \quad (5.11)$$

and the standard deviation of the noise is proportional to the square-root of N_{acq}

$$\sigma_v \propto \sqrt{N_{acq}} \quad (5.12)$$

Hence, the SNR per unit acquisition time, or SNR efficiency, is

$$SNR_{eff} \propto \frac{1}{\sqrt{TR}} \left(1 - e^{-TR/T_1}\right) \quad (5.13)$$

Note that SNR_{eff} is maximized by $TR_{opt} \approx 1.25T_1$. In practice, for a fixed total scan time, this TR may not be possible because N_{acq} is restricted to integer values. Close to optimal SNR_{eff} is established by rounding the optimal N_{acq} down to the next integer value and increasing TR to meet the total scan time.

Choice of the optimal TE is less straightforward. Previous work using Cramér–Rao lower bound theory [44]—a method that provides the lower variance bound for fitted experimental parameters, assuming they are determined from an unbiased estimator—has shown the optimal sampling pattern involves acquiring 22% of the acquisitions at the minimum possible TE and 78% of the acquisitions at $1.28T_2$. Two issues arise with these optimization protocols: (1) they require a priori knowledge of T_2 and (2) they only yield optimal results for a narrow range of T_2s. To adequately sample the T_2 decay curve from all tissues of interest, one typically needs to acquire a relatively large number of spin echoes, resulting in prohibitively long acquisition times for the single spin echo sequence considered to this point. Fortunately, multiple spin echo sequences have been developed that allow one to efficiently acquire a large number of spin echoes following a single acquisition. This is the topic of Section 5.4.4. Prior to this discussion, the effect variation in $\mathbf{B}_0$ and $\mathbf{B}_1$ on the observed signal in single and multiple spin echo sequences is discussed.

5.4.3 $\mathbf{B}_0$ and $\mathbf{B}_1$ Imperfection

The spin echo sequence refocuses phase accumulation due to variation in $\mathbf{B}_0$ and thus produces images free of $\mathbf{B}_0$-induced artifacts as long as the receiver bandwidth per voxel is large compared to the resonant frequency variation across each voxel. However, variations in $\mathbf{B}_0$ will impact quantitative estimates of T_2 through their effect on the RF refocusing pulse. For single echo sequences, as discussed heretofore, if the effective flip angle of the refocusing pulse $\theta_2 \neq 180°$ (and care is taken to only measure signal from the spin echo, as opposed to signal excited from equilibrium by θ_2), then the spin echo magnitude is reduced by a factor $\sin^2(\theta_2/2)$. This factor can be incorporated into S_a, and the estimate of T_2 remains accurate. However, when multiple spin echoes are acquired using multiple RF refocusing pulses, variation of this flip angle significantly from 180° will have significant effect on the measurement of T_2, as discussed below.

5.4.4 Multiple Spin Echoes

The Carr–Purcell–Meiboom Gill (CPMG) sequence [41,42] shown in Figure 5.3 is commonly used to acquire high fidelity nonlocalized T_2 decay data. By applying a series of refocusing (180°) pulses at odd integers of a delay (τ = echo spacing/2) and collecting signal at even integers of τ, signal loss due to $\mathbf{B}_0$ inhomogeneities is refocused, forming a so-called spin echo that decays according to T_2. Generally, sequences that apply a large number of RF pulses exhibit artifacts associated with $\mathbf{B}_0$ and $\mathbf{B}_1$ inhomogeneities. In the CPMG sequence, the relative phase of the excitation and refocusing pulses is designed to minimize this effect.

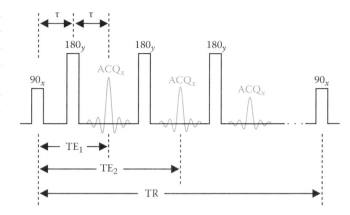

FIGURE 5.3 CPMG pulse sequence for nonlocalized measurement of T_2. The relative phase of the excitation (90_x) and refocusing (180_y) pulses is designed to reduce sensitivity to $\mathbf{B}_0$ and $\mathbf{B}_1$ inhomogeneities. Only the first three spin echoes (ACQ$_x$) are shown; however, one typically acquires thousands of echoes following each excitation.

The multiple spin echo sequence (Figure 5.4) developed by Poon and Henkelman [45] is commonly used to acquire high fidelity multiple spin echo imaging data. Again, by applying a series of refocusing pulses at odd integers of a delay (τ = esp/2) and collecting signal at even integers of τ, multiple echoes that decay according to T_2 can be acquired following a single excitation RF pulse. Unfortunately, the imaging gradients (specifically the phase encode gradients G_p in Figure 5.4) employed in this sequence render the CPMG phase cycling scheme ineffective in reducing artifacts associated with $\mathbf{B}_0$ and $\mathbf{B}_1$ inhomogeneities. To minimize these artifact contributions, broadband composite refocusing pulses are commonly applied between pairs of crusher gradients of alternating and descending magnitude. The broadband composite pulses serve to correct flip angle errors due

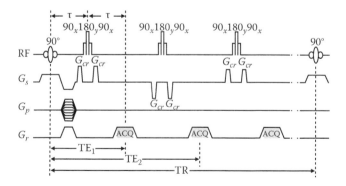

FIGURE 5.4 Multiple spin echo imaging pulse sequence for measuring T_2. Only the first three echoes are shown; however, typically 32 echoes or more are acquired per excitation. The combination of broadband composite refocusing pulses ($90_x180_y90_x$) and crusher gradients (G_{cr}) placed about each refocusing pulse (in an alternating and descending fashion) serve to eliminate signal from unwanted coherence pathways (e.g., stimulated echoes). $G_{s,p,r}$ = gradient in slice, phase, and read directions.

to $\mathbf{B}_1$ and/or $\mathbf{B}_0$ inhomogeneity; $90_x 180_y 90_x$ [46] pulses, which are designed to correct for $\mathbf{B}_1$ errors, are most commonly employed. The arrangement of crusher gradients is optimized to dephase any remaining signal from unwanted coherence pathways.

The use of broadband refocusing pulses makes multislice imaging with this sequence difficult. For example, Maier et al. [47] implemented a multislice version of a multiple spin echo sequence by using slice-selective refocusing pulses. Their results indicated that the accuracy of T_2 measurements derived from such a sequence were significantly affected by the presence of stimulated echoes (i.e., introduction of T_1 contrast) and off-resonance effects from adjacent slice-selective pulses (i.e., introduction of MT contrast). One way to extend the volumetric coverage of this sequence is to convert it to a 3D sequence—this can be achieved by replacing the slice-selective excitation pulse with a nonselective broadband pulse or, in some cases, a slab-selective pulse. This, however, can result in acquisition times that are unsuitable for clinical imaging. Several investigators have developed novel approaches to get around this issue. Does and Gore [48] proposed acquiring multiple gradient echoes about each refocusing pulse, with each of these gradient echoes encoding a different line of k-space. Oh et al. [49] proposed using a multiple echo T_2 preparation period followed by a multislice spiral readout as a means to rapidly acquire volumetric T_2 data. One particular advantage of this approach is that it allows one to optimally sample the T_2 decay curve using a logarithmic echo spacing [50,51], potentially reducing the total number of echoes one need acquire.

5.4.5 Multiexponential T_2

The observed NMR signal has been shown to exhibit multiexponential T_2 (MET$_2$) decay in a number of tissues, often arising from a combination of microanatomical water compartition (e.g., intracellular and extracellular water) and slow intercompartmental exchange. In these tissues, the signal equation for spin echoes must be generalized to include the sum of signals with different T_2 time constants. Considering K such signal components, the signal equation becomes

$$S_i = \sum_{j=1}^{K} S_{a,j} \exp\left(-TE_i / T_{2,j}\right) + v_i \qquad (5.14)$$

Given a vector $\mathbf{s}$ of observations $\mathbf{s} = [S_1, \cdots, S_K]^T$, one then wishes to estimate $T_{2,j}$ and $S_{a,j}, j = 1$ to K; however, this is a difficult numerical problem. For values of K as low as 3 and realistic values of SNR, estimates of $T_{2,j}$ and $S_{a,j}, j = 1$ to K, may be highly dependent on the initial guesses needed for nonlinear regression.

An alternative approach, which does not require initial parameter guesses, is to rewrite Equation 5.14 into a linear system of equations, spanning all possible T_2 values,

$$\mathbf{s} = \mathbf{A}\mathbf{d} + \mathbf{v} \qquad (5.15)$$

where $\mathbf{v} = [v_1, \cdots, v_K]^T$ is a vector of additive noise, $\mathbf{A}$ is a matrix of known decaying exponentials defined by

$$\mathbf{A} = \begin{bmatrix} e^{-TE_1/T_{2,1}} & \cdots & e^{-TE_1/T_{2,K}} \\ \vdots & \ddots & \vdots \\ e^{-TE_N/T_{2,1}} & \cdots & e^{-TE_N/T_{2,K}} \end{bmatrix} \qquad (5.16)$$

and $\mathbf{d}$ is a vector of K unknown apparent proton densities, which can be thought of as a spectrum of T_2 component amplitudes (also known as a T_2 spectrum). With a relatively large value of K (≈ 100), this becomes a discrete Laplace transform, the inversion of which is well known to be an ill-posed problem because matrix $\mathbf{A}$ is rank-deficient. There are numerous approaches to regularizing this problem, but the most common approach is to solve Equation 5.15 by Lawson and Hanson's nonnegative least square (NNLS) algorithm [52], which solves for $\mathbf{d}$ subject to the constraint $d_i \geq 0$, $i = 1$ to K. This constraint makes physical sense for a MET$_2$ measurement, but the resulting T_2 spectrum tends to be composed of a few isolated nonzero components.

In many cases, a smooth spectrum is easier to interpret and thought to be more physically meaningful. One can incorporate regularization into Equation 5.15 by augmenting $\mathbf{A}$ and $\mathbf{s}$. For example, augmenting $\mathbf{A}$ with a diagonal $K \times K$ matrix,

$$\mathbf{A}_r = \begin{bmatrix} \mathbf{A} \\ \hline \begin{matrix} \mu & & & & \\ & \mu & & & \\ & & \ddots & & \\ & & & \mu & \\ & & & & \mu \end{matrix} \end{bmatrix}, \quad \mathbf{s}_r = \begin{bmatrix} \mathbf{s} \\ \hline 0 \\ 0 \\ \vdots \\ 0 \\ 0 \end{bmatrix} \qquad (5.17)$$

results in the incorporation of a minimum energy constraint because the lower K rows contribute an energy term

$$\mu \sum_{i=1}^{K} d_i^2 \qquad (5.18)$$

weighted by a regularizer term, μ, into to the sum square error in the least-square solution of Equation 5.15 [53]. As the regularizer term is adjusted closer to zero, the least-square fit dominates and a spectrum with isolated delta functions is obtained. Conversely, as the regularizer term is increased, smoother spectra that are less sensitive to noise are obtained at the cost of data misfit. Multiple strategies exist to set μ, including using a value that results in a fixed increase in χ^2 misfit (typically 1%–2%) relative to the unregularized, least-squares solution [54] or using the generalized cross-validation approach (GCV) [55]. Alternatively, one may simply adjust this term "by eye" until a suitably conservative spectrum results, then use this same μ value for all MET$_2$ measurements to be compared.

This process is demonstrated in Figure 5.5 for a T_2 decay curve with two components: (1) a short-lived component with a $T_2 = 20$ ms that represents 40% of the signal and (2) a long-lived

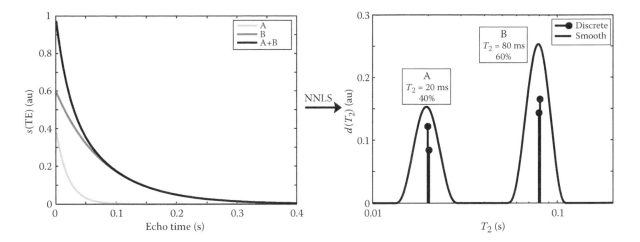

FIGURE 5.5 Example MET$_2$ decay (left) and corresponding T_2 spectrum (right). The biexponential decay (A + B) is decomposed into signal components A and B using the NNLS algorithm. Adding constraints to regularize the solution results in a smooth spectrum that is less sensitive to noise at the cost of data misfit. Note, the discrete T_2 distribution is scaled in for display purposes.

component with a $T_2 = 80$ ms that represents the remaining signal fraction. Note the two discrete components observed in the unregularized solution are replaced by smooth peaks in the regularized solution. These spectra were fitted using a minimum curvature constraint with regularization set by GCV using a freely available MERA Toolbox (http://vuiis.vanderbilt.edu/~doesmd/MERA/MERA_Toolbox.html) for MATLAB® (Natick, MA).

5.5 Effective Spin–Spin Relaxation (T_2^*)

5.5.1 Spoiled Gradient Echo

Spoiled gradient echo sequences, both single (Figure 5.2) and multiple echo (Figure 5.6), are most commonly used to quantify T_2^*, although asymmetric spin echo sequences [56] can be

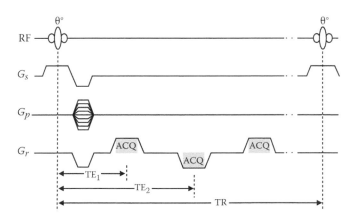

FIGURE 5.6 Multiple gradient echo imaging pulse sequence for measuring T_2^*. Multiple gradient echoes are acquired (ACQ) following a single excitation by alternating the polarity of the read gradient. This results in a signal that decays according to T_2^*. $G_{s,p,r}$ = gradient in slice, phase, and read directions.

employed as well. In either case, the signal equation from these sequences can be reduced to

$$S_i = S_a \exp\left(-TE_i/T_2^*\right) + v_i \qquad (5.19)$$

where TE_i is gradient echo time or the time relative to the center of the spin echo for asymmetric spin echo sequences. As before, the T_1-weighting terms from either the spin echo or spoiled gradient echo signal equations are constant with respect to TE and can be combined with S_0 to create the apparent proton density term, S_a.

5.5.2 Optimization of Acquisition Parameters

The discussion with regard to fitting and optimization of T_2 data (Section 5.4.2) holds true for T_2^* data as well. Where T_2^* quantification differs from T_2 is in the practical issues that arise, specifically the sensitivity of the gradient and asymmetric spin echo signal to macroscopic background magnetic field gradients that arise due to susceptibility differences between tissues.

5.5.3 B_0 and B_1 Imperfection

Unlike measurements of T_2, measurements of T_2^* using multiple gradient echoes are insensitive to variations in B_1, but highly sensitive to spatial variation in B_0—also known as background field gradients. In the presence of a background field gradient, the signal equation from a gradient or asymmetric spin echo sequence can be altered to include an additional term, $f_{\Delta\phi}$, that models the fractional signal loss due to background field gradients

$$S_i = S_a \exp\left(-TE_i/T_2^*\right) f_{\Delta\phi}\left(TE_i\right) + v_i \qquad (5.20)$$

The phase dispersion across each dimension can be estimated from the first derivative of the magnetic field

$$\Delta\phi_j\left(TE_i\right) = \gamma\Delta r_j \int_0^{TE_i} dt\,\frac{\partial B_0\left(\mathbf{r}\right)}{\partial r_j} \qquad (5.21)$$

where $\mathbf{r}$ is the voxel location, the partial term is the gradient of the magnetic field in the jth direction, and Δr_j is the width of the voxel in the same direction. The term $f_{\Delta\phi}$ can then be estimated from

$$f_{\Delta\phi}\left(TE_i\right) = \prod_{j=1}^{3}\left|\mathrm{sinc}\left(\frac{\Delta\phi_j\left(TE_i\right)}{2}\right)\right| \qquad (5.22)$$

assuming the spin density within each voxel is uniform [57]. To accurately quantify T_2^* in the presence of a significant background field gradient, one needs to either correct for this term prospectively or account for this term retrospectively. Given the voxel dimension dependence of this effect, in-plane contributions are often neglected in these methods as in-plane voxel dimensions are typically much smaller than the slice thickness for 2D sequences.

5.5.3.1 Prospective Correction of Background Field Gradients

The most straightforward way to remove background field gradients is through shimming. A number of automated techniques are available is order to optimize shim values (see reference [57] and references therein). Recent advances in dynamic shimming [57] offer further improvement by allowing the user to update shims on a slice-by-slice basis. Despite these advances, complete removal of susceptibility-induced background field gradients is often not possible, as they tend to be highly nonlinear. This is especially true at high static magnetic fields, where these susceptibility gradients are more pronounced. As a result, additional techniques are needed to accurately quantify T_2^* in the presence of susceptibility gradients.

Background gradient associated signal loss can be minimized by designing tailored RF pulses whose through-slice phase profile is the inverse of the phase profile generated by the background gradients themselves [58]. Unfortunately, these RF pulses are only optimized for a single echo time, making it difficult to implement such a technique for quantifying T_2^*. The phase profile across the slice can also be modulated via the slice refocus gradient — this method is often referred to as "z-shimming" (Figure 5.7). To correct for a given background gradient, G_b, at time TE_i, one can modulate the slice refocus gradient by a factor of ΔG_{sr}

$$\Delta G_{sr} = \frac{-G_b TE_i}{t_{sr}} \qquad (5.23)$$

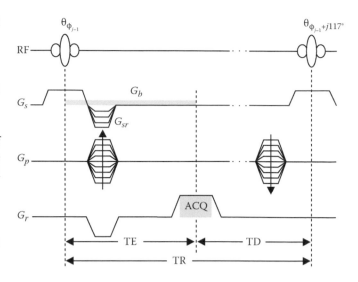

FIGURE 5.7 *z*-Shimming imaging pulse sequence. To correct for a given background gradient G_b at time TE, the slice refocus gradient G_{sr} is modulated according to Equation 5.23.

where t_{sr} is the slice refocus gradient duration. However, because this correction factor will vary both spatially and temporally, one needs to acquire a number of images with different slice refocus amplitudes to apply this correction at all positions and TEs. Three-dimensional sequences, which typically have much smaller through-plane dimensions than 2D sequences, can also be used to minimize the signal loss due to background gradients. An obvious tradeoff with this approach is the decreased SNR associated with the decreased voxel dimensions. However, the SNR can be recovered by averaging the magnitude of adjacent partitions in the image volume (after 3D Fourier transforming the acquired *k*-space data).

5.5.3.2 Retrospective Correction of Background Field Gradients

To retrospectively correct for background gradients in T_2^* measurements, one first needs to estimate the main magnetic field offset $\Delta\mathbf{B}_0$ over a 3D volume. This can be determined from two gradient echo (or asymmetric spin echo) phase images acquired at two different TEs

$$\Delta\mathbf{B}_0\left(\mathbf{r}\right) = \frac{\Delta\phi\left(\mathbf{r}\right)}{\gamma\Delta TE} \qquad (5.24)$$

where $\Delta\phi(\mathbf{r})$ is the phase accrual between images over the echo time difference ΔTE. The gradient of the magnetic field offset can then be determined numerically (e.g., center differencing), which can then be used to estimate the phase accrual at time TE_i via Equation 5.25 and the corresponding signal loss, $f_{\Delta\phi}$.

Fernández-Seara and Wehrli [59] suggested implicitly calculating the through slice background gradient G_b by including it as a free parameter in the fit. Combining Equations 5.20–5.22 and ignoring in-plane contributions yields the following signal equation

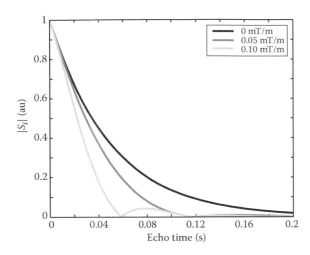

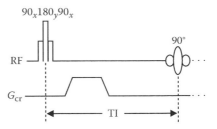

FIGURE 5.9 Inversion recovery preparation. An inversion pulse, in this case a broadband composite pulse, is followed by an inversion recovery delay *TI*. The resulting signal is described by Equation 5.28. The crusher gradient (G_{cr}) spoils any signal that is excited by the inversion pulse.

this case, prior to excitation, magnetization is inverted using a 180° RF pulse and a variable delay (*TI*) is allowed during which longitudinal magnetization relaxes toward thermal equilibrium (M_0). As before, if we use the good approximation $M_z(TE) \approx 0$ and incorporate T_2 weighting into the apparent proton density, the resultant signal equation for this sequence is

$$S_i \equiv S(TI_i, TD) = S_a[1 - 2\exp(-TI_i/T_1) + \exp(-TD/T_1)] + v_i \quad (5.28)$$

from which T_1 and S_a can be estimated similarly to Equation 5.27.

5.6.2 Variable Flip Angle Gradient Echo

An alternate and popular method for measuring T_1 is through the acquisition of multiple spoiled gradient echo images acquired with variable flip angle as described by Equation 5.6. Once again, the T_2^* weighting term can be incorporated into the apparent proton density (or simply neglected if $TE \ll T_2^*$)

$$S_i = S_a \sin\theta_i \frac{\left(1 - \exp\left(-TR/T_1\right)\right)}{\left(1 - \cos\theta_i \exp\left(-TR/T_1\right)\right)} + v_i \quad (5.29)$$

and nonlinear regression of S_i versus θ_i yields estimates $\hat{S}_a$ and $\hat{T}_1$. This approach is attractive because the flip angles can be chosen to generate signal variation in S_i over a wide range of *TR* values. That is, *TR* can be reduced to arbitrarily short values (unlike saturation and inversion recovery), limited only by the SNR and instrument performance. Therefore, when short acquisition times and/or large volumetric coverage are needed, the variable flip angle approach is often employed. However, as noted below in Section 5.6.4, in most practical circumstances this method requires both a spatial mapping of $\mathbf{B}_1$ and, for 2D acquisitions, corrections for through-plane variation in $\mathbf{B}_1$.

5.6.3 Optimization of Acquisition Parameters

Quantification of T_1 via inversion and saturation recovery methods typically requires sampling data at long *TI* and/or *TD* values, resulting in sequences with prohibitively long acquisition

FIGURE 5.8 T_2^* decay data in the presence of background magnetic field gradients. Plots were generated using Equation 5.25 and the following parameters: $T_2^* = 50$ ms; $G_b = 0$, 0.05, and 0.1 mT/m; and $\Delta z = 2$ mm.

$$S_i = S_a \exp\left(-TE_i/T_2^*\right)\text{sinc}\left(\frac{\gamma TE_i G_b \Delta z}{2}\right) + v_i \quad (5.25)$$

where Δz is the slice thickness. Example multiple gradient echo data (in the presence of a background gradient) are shown in Figure 5.8.

5.6 Spin-Lattice Relaxation (T_1)

5.6.1 Saturation and Inversion Recovery

T_1 can be readily measured using either spin-echo or gradient echo pulse sequences, as is apparent from Equations 5.5 and 5.6, both of which can be separated into independent T_1- and T_2-weighting terms. A standard spin echo sequence can be repeated with constant *TE* and varied *TD* (or varied *TR* if $TE \ll TR$) in a method known as saturation recovery because recovery of longitudinal magnetization is measured from $M_z = 0$. In this case, the T_2-weighting term in Equation 5.5 can be incorporated into the apparent proton density, leaving

$$S_i \equiv S(TD_i) = S_a[1 - \exp(-TD_i/T_1)] + v_i \quad (5.26)$$

for the *i*th signal. Note that the apparent proton density, S_a, is not the same as that in Equation 5.7. As for the previous case of fitting T_2, given a vector **s** of observations of S_i, $i = 1$ to N, T_1 (and S_a) can be estimated using nonlinear regression to find a least sum square solution

$$\underset{\hat{T}_2, \hat{S}_a}{\arg\min} \left\| \mathbf{s} - \hat{S}_a\left[1 - \exp\left(-\mathbf{t}/\hat{T}_1\right)\right] \right\|_2 \quad (5.27)$$

A common variation on the saturation recovery method, which provides twice the dynamic range of signal amplitude, is the inversion recovery spin echo pulse sequence (Figure 5.9). In times. As such, there is a significant body of literature dedicated to optimizing these methods (see reference [60] and references therein). Generally speaking, a logarithmic progression of *TI* (or *TR*) has been shown to be optimal for inversion (or saturation) recovery with the maximum *TI* (or *TR*) at thermal equilibrium ($\approx 5T_1$). There are, however, more elaborate schemes that offer increased performance under certain scenarios (e.g., for a fixed acquisition time, for a range of T_1 values). For example, Li et al. [61] noted that varying both *TI* and *TD* in the inversion recovery experiment yields significant efficiency gains over varying *TI* alone, allowing one to quantify T_1 from as few as three images.

Even with optimization, saturation and inversion recovery sequences are often too long for many clinical applications. One way to decrease acquisition times is to incorporate fast imaging readouts (e.g., echo planar) to acquire multiple lines of *k*-space per shot. Alternatively, or in conjunction with these fast readouts, one can use the Look–Locker method [62,63]. This approach involves sampling M_z repeatedly during a single inversion recovery period, using small flip angle (θ) uniformly spaced excitations followed by a gradient echo signal acquisition (Figure 5.10)—somewhat analogous to acquiring multiple spin echoes per *TR* period for T_2 mapping. This sampling reduces M_z concomitantly with its recovery due to longitudinal relaxation. Therefore, Equation 5.28 must be altered to account for this effect, resulting in the following expression

$$S_i = S_a \left[1 - \beta \exp\left(-TI_i / T_1^* \right) \right] + v_i \qquad (5.30)$$

where β can be fitted as a free parameter and

$$T_1^* = \frac{T_1}{1 + \left(T_1 / \tau \right) \ln \left[\cos \alpha \right]} \qquad (5.31)$$

with τ equal to the spacing between θ pulses.

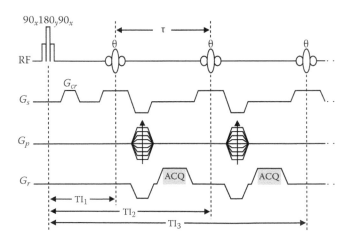

FIGURE 5.10 Look–Locker imaging pulse sequence for rapid measurement of T_1. The inversion preparation in Figure 5.9 is followed by a series of low flip angle (θ) gradient echo acquisitions, each separated by a delay of τ. The resulting recovery of longitudinal magnetization is described by Equations 5.30 through 5.31.

5.6.4 B_0 and B_1 Imperfection

The measurement of T_1 with either saturation or inversion recovery depends upon knowing the state of longitudinal magnetization at the beginning of the variable delay period (*TD* or *TI*, for saturation or inversion recovery, respectively), and this depends upon knowing the effective flip angle, θ, of the excitation or inversion RF pulse. Likewise, and more obviously, T_1 measurements using variable flip-angle spoiled gradient echo depend on accurate knowledge of the effective flip angles. As noted above, variations in both $\mathbf{B}_0$ and $\mathbf{B}_1$ will alter θ. In many cases, especially when imaging, one cannot assume to know θ with a high degree of accuracy.

Equation 5.4 and, subsequently, Equation 5.26 were derived assuming that the excitation RF pulse rotates magnetization by exactly 90°. If this assumption is not correct when measuring T_2, only the estimate $\hat{S}_a$ is biased, but if this assumption is not correct when measuring T_1, the estimate of T_1 itself is biased. There are RF pulse designs, such as the hyperbolic secant pulse [64] and the aforementioned broadband composite pulses (e.g., $90_x 180_y 90_x$), which can mitigate the effect of $\mathbf{B}_0$ and $\mathbf{B}_1$ variations on the effective flip angle. If such pulses cannot be employed or uncertainty in θ remains, the signal equation can be modified to account for this additional unknown parameter.

Equation 5.26 modified to include θ as a free parameter is

$$S_i \equiv S(TD_i) = S_a [1 - (1 - \cos \theta) \exp(-TD_i / T_1)] + v_i. \qquad (5.32)$$

Thus, the least square fit of this equation to the observed data involves three unknown terms

$$\underset{\hat{T}_2, \hat{S}_a, \hat{\theta}}{\arg \min} \left\| \mathbf{s} - \hat{S}_a \left[1 - \left(1 - \cos \hat{\theta} \right) \exp \left(-\mathbf{t} / \hat{T}_1 \right) \right] \right\|_2 \qquad (5.33)$$

Similarly for inversion recovery, we cannot assume that the prescribed 180° RF pulse exactly inverts the magnetization, and Equation 5.28 can be extended to include an arbitrary flip angle prior to the *TI* delay,

$$S_i \equiv S(TI_i, TD) = S_a [1 - (1 - \cos \theta) \exp(-TI_i / T_1) + \exp(-TD / T_1)] + v_i \qquad (5.34)$$

Similar to Equation 5.33, T_1, θ, and S_a are estimated by

$$\underset{\hat{T}_2, \hat{S}_a, \hat{\theta}}{\arg \min} \left\| \mathbf{s} - \hat{S}_a \left[1 - \left(1 - \cos \hat{\theta} \right) \exp \left(-\mathbf{t} / \hat{T}_1 \right) - \exp \left(-TD / \hat{T}_1 \right) \right] \right\|_2. \qquad (5.35)$$

When measuring T_1 using variable flip angle spoiled gradient echo, it is not possible to simply make θ an additional unknown model parameter because the signal itself is regressed versus flip angle. Thus, in addition to the variable flip angle measurements, it is also necessary to acquire an independent measure of the applied flip angle, which involves a mapping of $\mathbf{B}_1$ across the image space. Also, for 2D imaging using slice-selective excitation pulses, a through-plane variation in flip angle is unavoidable and is not linear with flip angle beyond the regime that is well-described by the low-flip angle approximation ($\theta < 30°$).

5.7 Proton Spin Density (M_0)

All aforementioned measurements involve fitting an apparent proton density, S_a, which is equal to S_0 and some additional relaxation-weighting depending on the sequence being used. Thus, to measure S_0, one must minimize or correct for the additional relaxation weighting in S_a. For a spin echo sequence, the T_1-weighting term in Equation 5.5 can be removed if $TD \gg T_1$, but this dictates a relatively long measurement time. Alternatively, the T_2^* weighting can be removed from a spoiled gradient echo sequence using $TE \ll T_2^*$, which is practical for many tissues and imposes no acquisition time constraint on the measurement. Ultimately, S_a is still a relative measure of proton density, but if it can be compared to a simultaneously acquired measure of S_a from a sample with known proton concentration, it may be converted to absolute units of moles ^{1}H. In doing so, one may also need to account for spatial variations in the effective flip angle (from spatial variation in $\mathbf{B}_0$ and/or $\mathbf{B}_1$) and receive fields (from spatial variation in receptivity).

5.8 Rician Noise

A significant problem in quantitative analysis of MRI signals results from the fact that the noise in a magnitude MR image is not governed by Gaussian noise, but rather by Rician noise [65]. For most measurements, this difference is inconsequential because a Rician noise distribution is approximately Gaussian when the SNR $\gtrsim$ 10. However, when measuring transverse relaxation, for example, this can bias data samples near the noise floor (i.e., at long *TE*s). More specifically, these points will appear artifactually larger than they would in the presence of Gaussian noise, thereby biasing the estimate of T_2 to a larger value (Figure 5.11). Several strategies exist to correct for this effect [65,66]. A simple correction is to subtract off the baseline offset (or incorporate it into the signal equation). This, however, will also bias the resulting fit. A more appropriate approach to correct for Rician-induced bias is based upon the maximum likelihood estimate of the measured signal, $E[M]$, in the presence of Rician noise [66]. This expectation value for the Rician distribution is given by

$$E[M] = \sigma\sqrt{\frac{\pi}{2}}\exp(-\alpha)[(1+2\alpha)I_0(\alpha) + 2\alpha I_1(\alpha)] \quad (5.36)$$

where I_i is the ith-order modified function of the first kind, and $\alpha = A^2/4\sigma^2$. Here, A the expectation value of the underlying Gaussian distribution, which is the parameter of interest. It can be estimated numerically by finding $\hat{A}$ that minimizes the squared

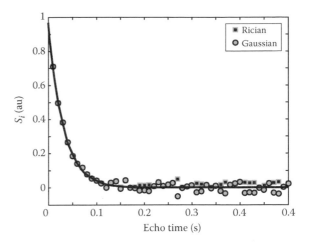

FIGURE 5.11 T_2 decay curve in the presence of Rician and Gaussian noise (SNR = 50, T_2 = 30 ms). Note the bias in the Rician data samples near the noise floor (i.e., at long *TE*s). These points are artifactually larger than the same samples in the presence of Gaussian noise which can bias the resulting estimate to T_2.

difference between the observed signal, M, and its expectation in Equation 5.36.

References

1. Belton PS, Jackson RR, Packer KJ. Pulsed NMR studies of water in striated muscle. I. Transverse nuclear spin relaxation times and freezing effects. *Biochim Biophys Acta* 1972;286(1):16–25.
2. Belton PS, Packer KJ. Pulsed NMR studies of water in striated muscle. 3. The effects of water content. *Biochim Biophys Acta* 1974;354(2):305–314.
3. Belton PS, Packer KJ, Sellwood TC. Pulsed NMR studies of water in striated muscle. II. Spin-lattice relaxation times and the dynamics of non-freezing fraction of water. *Biochim Biophys Acta* 1973;304(1):56–64.
4. Hazlewood CF, Chang DC, Medina D, Cleveland G, Nichols BL. Distinction between the preneoplastic and neoplastic state of murine mammary glands. *Proc Natl Acad Sci U S A* 1972;69(6):1478–1480.
5. Hazlewood CF, Chang DC, Nichols BL, Woessner DE. Nuclear magnetic resonance transverse relaxation times of water protons in skeletal muscle. *Biophys J* 1974;14(8):583–606.
6. Damadian R. Tumor detection by nuclear magnetic resonance. *Science* 1971;171(976):1151–1153.
7. Damadian R, Zaner K, Hor D, Dimaio T. Human tumors by NMR. *Physiol Chem Phys* 1973;5(5):381–402.
8. Damadian R, Zaner K, Hor D, DiMaio T, Minkoff L, Goldsmith M. Nuclear magnetic resonance as a new tool in cancer research: Human tumors by NMR. *Ann N Y Acad Sci* 1973;222:1048–1076.

9. Damadian R, Zaner K, Hor D, DiMaio T. Human tumors detected by nuclear magnetic resonance. *Proc Natl Acad Sci U S A* 1974;71(4):1471–1473.

10. Frey HE, Knispel RR, Kruuv J, Sharp AR, Thompson RT, Pintar MM. Proton spin-lattice relaxation studies of non-malignant tissues of tumorous mice. *J Natl Cancer Inst* 1972;49(3):903–906.

11. Weisman ID, Bennett LH, Maxwell LR, Sr, Woods MW, Burk D. Recognition of cancer *in vivo* by nuclear magnetic resonance. *Science* 1972;178(67):1288–1290.

12. Cottam GL, Vasek A, Lusted D. Water proton relaxation rates in various tissues. *Res Commun Chem Pathol Pharmacol* 1972;4(3):495–502.

13. Hollis DP, Economou JS, Parks LC, Eggleston JC, Saryan LA, Czeister JL. Nuclear magnetic resonance studies of several experimental and human malignant tumors. *Cancer Res* 1973;33(9):2156–2160.

14. Kiricuta IC, Jr., Demco D, Simplaceanu V. State of water in normal and tumor tissues. *Arch Geschwulstforsch* 1973; 42(3):226–228.

15. Kiricuta IC, Jr., Simplaceanu V. Tissue water content and nuclear magnetic resonance in normal and tumor tissues. *Cancer Res* 1975;35(5):1164–1167.

16. Bottomley PA, Hardy CJ, Argersinger RE, Allen-Moore G. A review of ^{1}H nuclear magnetic resonance relaxation in pathology: Are T_1 and T_2 diagnostic? *Med Phys* 1987;14(1):1–37.

17. Damadian R, Minkoff L, Goldsmith M, Stanford M, Koutcher J. Field focusing nuclear magnetic resonance (FONAR): Visualization of a tumor in a live animal. *Science* 1976;194(4272):1430–1432.

18. Damadian R, Minkoff L, Goldsmith M, Stanford M, Koutcher J. Tumor imaging in a live animal by focusing NMR (FONAR). *Physiol Chem Phys* 1976;8(1):61–65.

19. Lauterbur PC, Lai CM, Frank JA, Dulcey CS. *In vivo* zeugmatographic imaging of tumors. *Physics Can* 1976;32: abstract 33.11.

20. Smith FW, Mallard JR, Reid A, Hutchison JM. Nuclear magnetic resonance tomographic imaging in liver disease. *Lancet* 1981;1(8227):963–966.

21. Naruse S, Horikawa Y, Tanaka C, Hirakawa K, Nishikawa H, Yoshizaki K. Significance of proton relaxation time measurement in brain edema, cerebral infarction and brain tumors. *Magn Reson Imaging* 1986;4(4):293–304.

22. Kjaer L, Thomsen C, Iversen P, Henriksen O. *In vivo* estimation of relaxation processes in benign hyperplasia and carcinoma of the prostate gland by magnetic resonance imaging. *Magn Reson Imaging* 1987;5(1):23–30.

23. Schad LR, Brix G, Zuna I, Harle W, Lorenz WJ, Semmler W. Multiexponential proton spin–spin relaxation in MR imaging of human brain tumors. *J Comput Assist Tomogr* 1989;13(4):577–587.

24. Kjaer L, Thomsen C, Gjerris F, Mosdal B, Henriksen O. Tissue characterization of intracranial tumors by MR imaging. *In vivo* evaluation of T_1- and T_2-relaxation behavior at 1.5 T. *Acta Radiol* 1991;32(6):498–504.

25. Liu YH, Hawk RM, Ramaprasad S. *In vivo* relaxation time measurements on a murine tumor model—Prolongation of T_1 after photodynamic therapy. *Magn Reson Imaging* 1995;13(2):251–258.

26. Jakobsen I, Lyng H, Kaalhus O, Rofstad EK. MRI of human tumor xenografts *in vivo*: Proton relaxation times and extracellular tumor volume. *Magn Reson Imaging* 1995;13(5):693–700.

27. Gambarota G, Veltien A, van Laarhoven H, Philippens M, Jonker A, Mook OR, Frederiks WM, Heerschap A. Measurements of T_1 and T_2 relaxation times of colon cancer metastases in rat liver at 7 T. *MAGMA* 2004;17(3–6):281–287.

28. Dortch RD, Yankeelov TE, Yue Z, Quarles CC, Gore JC, Does MD. Evidence of multiexponential T_2 in rat glioblastoma. *NMR Biomed* 2009;22(6):609–618.

29. Bergers G, Benjamin LE. Angiogenesis: tumorigenesis and the angiogenic switch. *Nat Rev Cancer* 2003;3(6):401–410.

30. Gatenby RA, Gillies RJ. Why do cancers have high aerobic glycolysis? *Nat Rev Cancer* 2004;4(11):891–899.

31. Krohn KA, Mankoff DA, Eary JF. Imaging cellular proliferation as a measure of response to therapy. *J Clin Pharmacol* 2001;96S–103S.

32. Padhani AR, Krohn KA, Lewis JS, Alber M. Imaging oxygenation of human tumours. *Eur Radiol* 2007;17(4): 861–872.

33. Bakker CJ, Vriend J. Multi-exponential water proton spin–lattice relaxation in biological tissues and its implications for quantitative NMR imaging. *Phys Med Biol* 1984;29(5):509–518.

34. Damon BM, Gregory CD, Hall KL, Stark HJ, Gulani V, Dawson MJ. Intracellular acidification and volume increases explain R(2) decreases in exercising muscle. *Magn Reson Med* 2002;47(1):14–23.

35. Zhong JH, Gore JC, Armitage IM. Relative contributions of chemical exchange and other relaxation mechanisms in protein solutions and tissues. *Magn Reson Med* 1989;11(3):295–308.

36. Shioya S, Haida M, Ono Y, Fukuzaki M, Yamabayashi H. Lung cancer: Differentiation of tumor, necrosis, and atelectasis by means of T_1 and T_2 values measured *in vitro*. *Radiology* 1988;167(1):105–109.

37. Kovalikova Z, Hoehn-Berlage MH, Gersonde K, Porschen R, Mittermayer C, Franke RP. Age-dependent variation of T_1 and T_2 relaxation times of adenocarcinoma in mice. *Radiology* 1987;164(2):543–548.

38. Kroeker RM, Stewart CA, Bronskill MJ, Henkelman RM. Continuous distributions of NMR relaxation times applied to tumors before and after therapy with X-rays and cyclophosphamide. *Magn Reson Med* 1988;6(1):24–36.

39. Fantazzini P, Sarra A. A comparison of the proton relaxation in human epithelial tumors and associated uninvolved tissue. *MAGMA* 1994;2(3):405–407.

40. Edzes HT, Samulski ET. Cross relaxation and spin diffusion in the proton NMR or hydrated collagen. *Nature* 1977;265(5594):521–523.

41. Carr HY, Purcell EM. Effects of diffusion on free precession in nuclear magnetic resonance experiments. *Phys Rev* 1954;94(3):630.

42. Meiboom S, Gill D. Modified spin-echo method for measuring nuclear relaxation times. *Rev Sci Instrum* 1958;29(8):688–691.

43. McSheehy PM, Weidensteiner C, Cannet C, Ferretti S, Laurent D, Ruetz S, Stumm M, Allegrini PR. Quantified tumor T_1 is a generic early-response imaging biomarker for chemotherapy reflecting cell viability. *Clin Cancer Res* 2010;16(1):212–225.

44. Jones JA, Hodgkinson P, Barker AL, Hore PJ. Optimal sampling strategies for the measurement of spin–spin relaxation times. *J Magn Res B* 1996;113(1):25–34.

45. Poon CS, Henkelman RM. Practical T_2 quantitation for clinical applications. *J Magn Reson Imaging* 1992;2(5):541–553.

46. Levitt MH, Freeman R. Compensation for pulse imperfections in NMR spin-echo experiments. *J Magn Reson* 1981;43(1):65–80.

47. Maier CF, Tan SG, Hariharan H, Potter HG. T_2 quantitation of articular cartilage at 1.5 T. *J Magn Reson Imaging* 2003;17(3):358–364.

48. Does MD, Gore JC. Rapid acquisition transverse relaxometric imaging. *J Magn Reson* 2000;147(1):116–120.

49. Oh J, Han ET, Lee MC, Nelson SJ, Pelletier D. Multislice brain myelin water fractions at 3T in multiple sclerosis. *J Neuroimag* 2007;17(2):156–163.

50. Shrager RI, Weiss GH, Spencer RG. Optimal time spacings for T_2 measurements: Monoexponential and biexponential systems. *NMR Biomed* 1998;11(6):297–305.

51. Does MD, Gore JC. Complications of nonlinear echo time spacing for measurement of T_2. *NMR Biomed* 2000;13(1):1–7.

52. Lawson CL, Hanson RJ. *Solving Least Squares Problems.* Englewood Cliffs, NJ: Prentice-Hall; 1974.

53. Whittall KP, Mackay AL. Quantitative interpretation of NMR relaxation data. *J Magn Reson* 1989;84(1):134–152.

54. Graham SJ, Stanchev PL, Bronskill MJ. Criteria for analysis of multicomponent tissue T_2 relaxation data. *Magn Reson Med* 1996;35(3):370–378.

55. Golub GH, Heath M, Wahba G. Generalized cross-validation as a method for choosing a good ridge parameter. *Technometrics* 1979;21(2):215–223.

56. Stables LA, Kennan RP, Gore JC. Asymmetric spin-echo imaging of magnetically inhomogeneous systems: theory, experiment, and numerical studies. *Magn Reson Med* 1998;40(3):432–442.

57. Zhao YS, Anderson AW, Gore JC. Computer simulation studies of the effects of dynamic shimming on susceptibility artifacts in EPI at high field. *J Magn Reson* 2005;173(1):10–22.

58. Chen NK, Wyrwicz AM. Removal of intravoxel dephasing artifact in gradient-echo images using a field-map based RF refocusing technique. *Magn Reson Med* 1999;42(4):807–812.

59. Fernandez-Seara MA, Wehrli FW. Postprocessing technique to correct for background gradients in image-based R_2^* measurements. *Magn Reson Med* 2000;44(3):358–366.

60. Ogg RJ, Kingsley PB. Optimized precision of inversion-recovery T_1 measurements for constrained scan time. *Magn Reson Med* 2004;51(3):625–630.

61. Li K, Zu Z, Xu J, Janve VA, Gore JC, Does MD, Gochberg DF. Optimized inversion recovery sequences for quantitative T_1 and magnetization transfer imaging. *Magn Reson Med* 2010;64(2):491–500.

62. Look DC, Locker DR. Time saving in measurement of NMR and EPR relaxation times. *Rev Sci Instrum* 1970;41(2):250–251.

63. Crawley AP, Henkelman RM. A comparison of one-shot and recovery methods in T_1 imaging. *Magn Reson Med* 1988;7(1):23–34.

64. Silver MS, Joseph RI, Chen CN, Sank VJ, Hoult DI. Selective population inversion in NMR. *Nature* 1984;310(5979):681–683.

65. Gudbjartsson H, Patz S. The Rician distribution of noisy MRI data. *Magn Reson Med* 1995;34(6):910–914.

66. Bonny JM, Zanca M, Boire JY, Veyre A. T_2 maximum likelihood estimation from multiple spin-echo magnitude images. *Magn Reson Med* 1996;36(2):287–293.

6

Arterial Spin Labeling Techniques

Bruce M. Damon
Vanderbilt University

Christopher P. Elder
Vanderbilt University

6.1 Introduction

Tumors, like other tissues, require oxygen and nutrient delivery to sustain high rates of growth and development. The demand for oxygen and nutrients is met by increased perfusion, which we define as the tissue mass-normalized rate of blood delivery to an organ or tumor via its capillary network. The high perfusion requirement makes angiogenesis a critical event in the growth of a tumor; angiogenesis, in turn, is the logical target of many forms of anticancer therapy. These considerations, along with the spatially heterogeneous patterns of blood flow that are characteristic of tumors, indicate that using tomographic imaging to quantify tumor blood flow could help significantly in tumor characterization or in determining a tumor's responsiveness to therapy.

There are two broad classes of MRI methods available for measuring perfusion. Both are based on the application of well-established diffusible tracer methods [1]. Dynamic contrast-enhanced MRI and dynamic susceptibility contrast MRI employ exogenous molecules (MRI contrast agents) as the tracers; these topics are considered in Chapters 10 and 11, respectively. In this chapter, we consider arterial spin labeling (ASL), in which the water in blood serves as an endogenous tracer. This is made possible by magnetically labeling the blood using radiofrequency (RF) pulses.

In Section 6.2, we provide a brief overview of ASL. In Section 6.3, we describe ASL methods in detail. Section 6.4 deals with ASL quantification of perfusion. In the final section, we describe some of the studies in which ASL has been applied to studies of tumor biology.

6.2 Overview of ASL

ASL treats the tissue of interest as a system having two compartments: tissue parenchyma and blood. The parenchyma is considered to consist largely of static (unmoving) spins, but there is exchange between the compartments. Consistent with the idea that water is freely diffusible through capillary membranes, the exchange is considered to be fast enough that the compartments are well mixed on the timescale of the ASL experiment.

To conduct an ASL experiment, two types of images are acquired. The first is a "control" image, in which the longitudinal magnetization of the inflowing arterial blood is unperturbed. Therefore, its longitudinal magnetization equals its equilibrium magnetization. There is exchange between blood and tissue water, but because the equilibrium magnetization values of the tissue and blood are approximately equal when expressed per unit mass, the longitudinal magnetization of the tissue is not affected by the exchange process. The second type of image is a "tagged" image, in which the magnetization of the spins flowing into the imaging slice plane has been inverted (inversion pulses are described in Chapter 5). Consequently, the inflowing blood has lower longitudinal magnetization than it does in the control condition. As the labeled blood perfuses the tissue of interest and exchanges with the tissue water, the longitudinal magnetization of the tissue parenchyma is reduced. The longitudinal magnetization difference between the tagged and control conditions gives rise to a signal difference between the two conditions. The signal difference is very small (a few percent) and is proportional to the rate of blood perfusion to the tissue, as well as other quantities. As we will see below, mathematical expressions exist that allow the computation of quantitative maps of perfusion in both healthy and cancerous tissue based on ASL data.

Often, ASL data are acquired as part of a dynamic time series. For example, the data might be acquired from the tissue of interest during some physiological event (such as from a muscle during exercise or from the brain during a cognitive task). Or, the data could be acquired to test the responsiveness of the local vasculature to the infusion of a vasodilator. Because ASL is a difference technique (perfusion is encoded as the signal difference between two images), it is very sensitive to motion. Therefore, motion insensitive, rapid imaging methods such as echo-planar imaging (EPI) are frequently used to form the image. While the image formation process itself in EPI may be motion insensitive, interimage subject motion may still cause registration errors between images. Therefore, to reduce the effects of motion artifacts in dynamic time series acquisitions further, the control and tagged images are acquired in an interleaved fashion; each sequential image pair can then be used to calculate perfusion images. Because each image in the pair is acquired only several seconds apart, the amount of interimage motion is typically fairly small and can be corrected using image registration techniques, if necessary.

The many different ASL techniques can be broadly classified as continuous ASL (CASL) or pulsed ASL (PASL). CASL was the first ASL technique described, with the first works published in the early 1990s [2,3]. There are several basic variants of PASL [4–7], each having been originally described in the mid-1990s. CASL and PASL are alike in that inflowing spins are labeled, perfuse the imaging slice plane, and affect the image signal intensity in the manner described above. The techniques differ in that in CASL, the magnetization in a bolus of blood flowing into a distal tissue of interest is inverted at a single position for a prolonged period; in contrast, PASL methods invert the magnetization in a large region of blood vessels upstream of the tissue of interest for a very brief period.

The description of ASL presented above is conceptually fairly simple: a control image is acquired; a perfusion-sensitized image is made by magnetically labeling the inflowing blood in either a continuous or pulsed fashion; the signal difference between the images tells us something about the perfusion to the tissue of interest. However, the quantitative accuracy of ASL rests critically on the extents to which four key assumptions are satisfied. Assumptions 1 and 2 relate to the definition and formation of the control image

1. The static spins in the slice plane contribute equal amounts of signal under the control and tagged imaging conditions.
2. There is no labeling of the inflowing blood in the control condition.

Assumptions 3 and 4 pertain to the generation of the tagged image. They are

3. The inversion of the inflowing spins' magnetization is uniform and complete.
4. The time required for the labeled blood to perfuse and mix with spins in the slice plane of interest is spatially and temporally uniform, the latter condition applying mainly to dynamic types of acquisitions.

The degrees to which Assumptions 1 and 2 are satisfied vary depending on the specific ASL method. We will therefore discuss the validity of these assumptions in the specific contexts of each ASL method. As for Assumption 3, we will see that this assumption is well met for some methods and not well met for others. As we describe the ASL pulse sequences, we will discuss methods (such as adiabatic inversion pulses) that address this issue. We will also show that incomplete inversion can be accounted for quantitatively. The issue described in Assumption 4—regarding the time required for blood to pass from its point of labeling to the imaging slice plane—provides for a great deal of uncertainty in ASL. While the actual transmit time for the blood is unknown, the assumption concerning its spatial and temporal homogeneity is typically assumed to be well met for most healthy tissues. However, this assumption may not be valid in certain pathologies. We will revisit this issue when we discuss ASL quantification. The detailed presentation of the various ASL methods is the subject of the next section of this chapter.

6.3 ASL Techniques

We will start with a qualitative discussion of CASL. We will then describe the two basic PASL approaches qualitatively, along with some of their variants.

6.3.1 Continuous ASL

As noted above, CASL was the original ASL technique introduced. Detre et al. first implemented CASL by saturating the inflowing magnetization proximal to the imaging plane using a train of 90° RF pulses [2]; Williams et al. [3] subsequently used inversion labeling of the inflowing spins. In the inversion recovery approach, the signal difference between the control and tagged images, ΔS, is proportional to $(M_{0,A} - [-M_{0,A}]) = 2M_{0,A}$, where $M_{0,A}$ is the equilibrium magnetization of the arterial blood; in the saturation approach, ΔS is proportional to $M_{0,A}$. The dependence of ΔS on $2M_0$ reflects a doubled labeling efficiency. Because of this, the inversion approach is preferred and we discuss this method only. In particular, CASL inverts the inflowing spins' magnetization using a technique called flow-induced adiabatic inversion [8]. We begin by presenting this method and then describe its implementation in the CASL sequence.

6.3.1.1 Flow-Induced Adiabatic Inversion

An adiabatic inversion pulse is used to invert the magnetization homogeneously throughout the entire region of tissue experiencing the pulse. The adiabatic passage principle states that the magnetization will always respond to the effective magnetic field (B_{Eff}), given as the vector sum of the B_0 field and an applied B_1 field, so long as the direction of B_{Eff} does not change appreciably during a single precession of the magnetization about B_{Eff}. For a pulse applied initially along the x-axis, this condition (the "adiabatic condition") is met as long as the following inequality is satisfied:

$$\left|\frac{d\varphi}{dt}\right| \ll \frac{\gamma}{2\pi}\left|B_{Eff}\right| \qquad (6.1)$$

where $\varphi = \arctan\left(\dfrac{B_x(t)}{B_z(t)}\right)$, B_x, and B_z refer to the X and Z components of B_{Eff}, respectively, t is time, and γ is the gyromagnetic ratio. The left-hand side of this inequality represents the rate of change in the direction of B_{Eff} and the right-hand side gives the precessional frequency about B_{Eff}. Experimentally, the adiabatic condition is generated by simultaneously modulating the amplitude and frequency of the RF pulse. This is done by sweeping the RF pulse's frequency from a value greater than the resonance frequency to a value lower than the resonance frequency; simultaneously, the RF amplitude is modulated from zero to its maximum and back to zero. These modulation patterns can be generated using pairs of functions that vary in appropriate manners, such as sine and cosine functions or hyperbolic tangent and hyperbolic secant functions. As long as the pulse is long enough and the B_1 amplitude is large enough everywhere in the sample, the magnetization will nutate such that it lies along B_{Eff}, regardless of the B_1 amplitude. This homogeneous inversion eliminates the confounding influence of the variation in B_1 amplitude that normally occurs even in volume excitation coils, such as birdcage resonators. As the spatial variation in B_1 can be ±10% or more, adiabatic inversion pulses can improve inversion uniformity considerably and increase the degree to which Assumption 3 is satisfied.

Flow-induced adiabatic inversion is a variant of this method, having originally been introduced for angiography applications [8]. In this technique, the labeling RF pulse is modulated in neither frequency nor amplitude. So that the inflowing blood will experience an effectively frequency-modulated pulse, a magnetic field gradient G_{Tag} is applied along the direction of flow in the vessel of interest (the term G_{Tag} is used because it corresponds to the RF pulse used to label or "tag" the inflowing spins). A labeling plane is defined in the vessel of interest at a distance Δx from the imaging plane, calculated as

$$\Delta x = \omega_{Tag}/(\gamma G_{Tag}) \qquad (6.2)$$

where ω_{Tag} is the frequency of the tagging RF pulse. The spins flowing into the labeling plane experience a frequency modulated pulse: as blood flows along G_{Tag}, the resonant frequencies of the spins are initially lower than ω_{Tag}, then equal to ω_{Tag}, and finally greater than ω_{Tag}. Because of the changing temporal relationship between the spins' resonant frequencies and ω_{Tag} and the similarity of this relationship to the condition described above, an adiabatic inversion is achieved. This will be true for any blood velocity v that satisfies the following inequalities:

$$R_{2,A} \ll \frac{Gv}{\left|B_{Eff}\right|} \ll \frac{\gamma}{2\pi}\left|B_{Eff}\right| \qquad (6.3)$$

where $R_{2,A}$ is the transverse relaxation rate of the arterial blood. The left-hand inequality expresses the idea that the blood must

be flowing quickly enough to reach the imaging plane before its signal has decayed due to R_2 effects (for spin-echo images; we would use R_2^* for gradient-echo images). The right-hand inequality is analogous to Equation 6.1; i.e., it defines the adiabatic condition. Because the tagging pulse is off-resonance to the static spins in the imaging plane, these spins are not directly affected by the inversion pulse. We now place this adiabatic inversion scheme in the context of CASL.

6.3.1.2 CASL Scheme and Pulse Sequence

Figure 6.1 graphically illustrates the CASL sequence. Figure 6.1a depicts the tagged image acquisition and Figure 6.1b describes the control image acquisition. In each panel, the diagrams across the top show the pulse sequence and the diagrams across the bottom of the panel schematically illustrate the effect of the pulse sequence on the blood vessel network of a hypothetical tissue. Both the pulse sequence and tissue diagrams evolve in time from left to right. In the pulse sequence diagrams, the top row illustrates the RF pulses and the bottom row illustrates the applications of the slice selection gradient (G_{Slice}) only. In the tissue diagrams, the proximal-to-distal direction (arterial-to-venous) is from bottom to top.

The sequence can be divided into three sections. The first is a preparation (labeling) period, which occurs as follows. G_{Tag} is applied in the slice selection direction. After the gradient has ramped up to its intended value, an RF pulse is applied with a duration Δt and a frequency ω_{Tag}, generating the flow-induced adiabatic inversion condition described above. In the tissue diagram, the gray bar represents the labeling plane and the dashed bar represents the imaging plane; as noted, the labeling plane is proximal to the imaging plane. The leftmost tissue diagram corresponds to the start of the tagging pulse. To label a sufficient bolus of inflowing blood, a tagging duration (Δt) of several seconds is required. However, a continuous pulse of such duration may exceed the hardware capabilities of the RF amplifier and/or may deposit excessive power into the subject, causing local tissue heating. For these reasons, a continuous inversion pulse is approximated using a series of short, hard pulses; a duty cycle of 70%–90% is typically used. This is represented by the boxed, vertically striped character of the tagging RF pulse. G_{Tag} and ω_{Tag} must be set considering both the desired location of the labeling plane (typically, several centimeters proximal to the imaging plane) and the expected range of blood velocities to satisfy the inequality described by Equation 6.3. The middle tissue diagram depicts the state of the blood magnetization at the end of the tagging pulse. As suggested by comparing the leftmost and middle tissue diagrams in Figure 6.1a, by the end of the tagging pulse, a large bolus of blood will have been labeled.

The final two phases of the pulse sequence include a delay following inversion (characterized by the inversion time, TI) and image acquisition. The first event during TI is that a spoiling gradient is applied. This gradient dephases any coherent transverse magnetization that may have resulted from an imperfect tagging pulse and is represented in the pulse sequence diagram as the large trapezoidal gradient waveform. During the remainder of TI (represented by the dashed portion of the RF and G_{Slice}

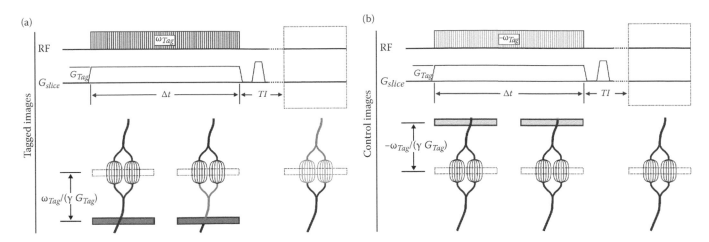

FIGURE 6.1 CASL pulse sequence and schematic illustration of labeling scheme in tissue. (a) Pulse sequence and a diagram illustrating its effect on the tissue for the tagged image. (b) Pulse sequence and tissue diagram for the control image. In the tagged image, the tagging gradient G_{Tag} and an RF pulse (at frequency ω_{Tag}) are applied, corresponding to a proximal (to the slice plane) position Δx. Flow in the direction of the gradient creates an adiabatic inversion of the inflowing blood in the labeling plane, indicated by the gray bar in the tissue diagram. This occurs for a prolonged period (Δt), labeling a large bolus of blood. During the inversion delay TI, the labeled spins perfuse imaging slice of interest, as indicated by the light gray shading of the vessels in the tissue diagram; finally, an image is acquired. In the control image, the tagging RF pulse is applied at the frequency $-\omega_{Tag}$. Because the sign of the tagging gradient is maintained, the labeling plane now occurs distally to the imaging slice, and the image is not flow-sensitized. However, the reversal of the sign of ω_{Tag} allows for proper control over MT effects. For further details, the reader is referred to the main text.

timelines in the pulse sequence diagram), labeled spins flow into and perfuse the imaging plane; this is indicated by the light gray shaded vessels in the tissue diagram. Note that CASL requires that TI be greater than the time required for the leading edge of the labeled arterial blood bolus to arrive at the distal edge of the imaging slice (an important quantity called the transit time, t_a) and less than the time required for the trailing edge of the labeled arterial blood bolus to reach the proximal edge of the imaging slice (=$t_a + \Delta t$). This condition ensures that there is labeled blood perfusing all portions of the imaging plane at the time of image acquisition. The final portion of the CASL sequence is to acquire an image in the slice of interest. This is represented by the dashed gray box in the pulse sequence diagram. In the tissue diagram, the imaging slice is represented by the dashed boxes through the capillary bed. As noted previously, a rapid imaging sequence such as EPI is typically used to read out the image, although other techniques may be used as well.

We turn our attention now to the control image acquisition. As noted above, the static spins in the imaging plane are not directly affected by the inversion pulse. However, they are indirectly affected by this pulse. This occurs by way of incidental magnetization transfer (MT). MT is described in more detail in Chapter 8; for our purposes, a brief description will suffice. MT results from the fact that the solid phase of the tissue (containing the macromolecules) has an extremely short T_2 and therefore a very broad resonance peak [9]. Because of the breadth of this peak, an RF pulse applied off-resonance for water can saturate the magnetization of the protons in the solid tissue phase. This saturation can be transferred to the nearby free water by way of chemical exchange and via through-space magnetic interactions. This

transfer of magnetization reduces the signal from the static spins in the imaged slice; the signal losses can be quite large (up to tens of percent in some tissues and for some offset frequencies of the saturating RF field). If these effects are not properly accounted for, Assumption 1 above would be violated, as the tissue magnetization would be modulated by a factor other than perfusion.

Figure 6.1b shows how the CASL sequence controls for this effect. Here, the tagging pulse is shown to occur at a frequency $-\omega_{Tag}$; because G_{Tag} is unchanged, the labeling plane occurs on the distal (and presumably, the venous or downstream) side of the imaging slice. Assuming that all arterial flow occurs in the proximal-to-distal direction, the imaging slice is not flow-sensitized. MT effects are often assumed to be generally symmetric about the water resonance. In this case, the static spins in the imaging slice experience approximately the same MT effect in the control condition as in the tagged condition. Note that another way to place the control labeling plane in this location would be to reverse the sign of G_{Tag} rather than ω_{Tag}.

In the preceding paragraph, we noted that a general symmetry of the MT effect is often assumed. In fact, there is some MT asymmetry about the water resonance frequency (on the order of a few percent [10]). While this may be small compared to the overall MT effect, it is similar to the signal difference between the control and tagged conditions; because of this, asymmetric MT effects can impair ASL quantification. In the paragraph above, we also noted the two options for defining a downstream labeling plane in the control image. The option in which the sign of G_{Tag} is changed controls for MT asymmetry better than changing the sign of ω_{Tag}, because in each case the saturating RF pulse is applied at the same frequency. For imaging readouts such as EPI, changing the sign of ω_{Tag} controls for eddy current–induced distortions in the image better than changing the sign of G_{Tag}, because the induced gradient is the same in both types of images.

It may be the case that the labeling plane in the control image occurs entirely outside of the body. But when the tissue geometry is such that labeling occurs in a tissue distal to the imaging slice, there will be some labeling of venous blood that occurs; such inflow may affect the signal in the control image, impacting the validity of Assumptions 1 and 2. Also, if there is irregular vascular geometry resulting in arterial inflow to the slice from both directions (such as may exist in a tumor), then this flow would not be properly accounted for in this unidirectional labeling scheme. Other sequences, discussed below, address these issues.

This concludes the qualitative description of CASL. We move on now to a qualitative discussion of PASL methods.

6.3.2 Pulsed ASL

As noted above, PASL is another inversion labeling method; but in contrast to CASL, which inverts the magnetization in a small upstream region of tissue for a long period, PASL inverts the magnetization in a large, upstream region of tissue for a short period. Two basic PASL approaches exist, which differ in terms of the spatial extent of the inversion pulse and the manner in which the control image is generated. In the EPISTAR method (EPI and signal targeting with alternating RF [4]) and its variants, the magnetization is inverted proximally to the imaging slice only. in the other technique (flow-sensitive alternating inversion recovery [FAIR] [5,7]), the magnetization is inverted globally about the slice.

6.3.2.1 Proximal Inversion Methods: EPISTAR and Its Variants

The EPISTAR pulse sequence is depicted in Figure 6.2, with Figure 6.2a showing the tagged image acquisition and Figure 6.2b showing the control image acquisition. Similar notations are used in the pulse and tissue diagrams to those of Figure 6.1. Beginning with the tagged image, the pulse sequence first shows a slice-selective 90° pulse. We will delay the discussion of this pulse and first describe the tagging approach itself. Like CASL, there are three main parts to the sequence: a labeling phase, an inversion delay, and image read-out. To initiate the labeling portion of the sequence, an adiabatic inversion pulse (typically a hyperbolic secant pulse) is applied with spectral width $\Delta\omega$ and center frequency ω_{Tag} in the presence of a gradient G_{Tag}. G_{Tag} is weak compared to the slice selection gradient, and so a large band of tissue (often referred to as a slab because of its thickness—on the order of 10 cm) experiences the RF pulse. $\Delta\omega$, ω_{Tag}, and G_{Tag} are set such that the inversion slab is placed proximal to the imaging plane and so that a gap of approximately 1 cm exists between the slab and the imaging plane. During the time *TI* (about 1 s), the inverted blood perfuses the imaging plane, exchanging with the tissue spins. Finally, an image is acquired using EPI or another rapid imaging technique. To form the control image, the inversion slab is shifted to the distal side of the imaging plane. As illustrated in Figure 6.2b, this can be accomplished by reversing the sign of ω_{Tag}. The inversion slab can also be shifted by reversing the sign of G_{Tag}. These practices have similar advantages and disadvantages with regard to asymmetric incidental MT effects and eddy current–induced distortions as discussed above for CASL.

We conclude our discussion of EPISTAR with two technical notes. The first relates to the so-called slice profile (the

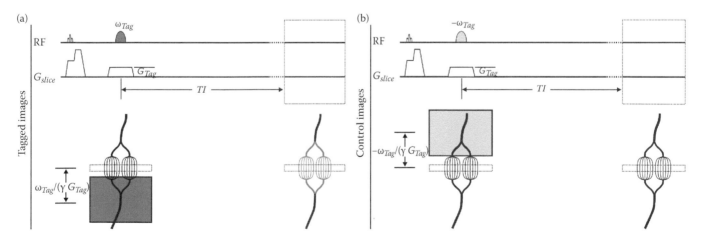

FIGURE 6.2 EPISTAR pulse sequence and schematic illustration of labeling scheme in tissue. (a) Pulse sequence and a diagram illustrating its effect on the tissue for the tagged image. (b) Pulse sequence and tissue diagram for the control image. In the tagged image, the tagging gradient G_{Tag} is applied and a hyberbolic secant inversion pulse is applied with a spectral width $\Delta\omega$ and at frequency ω_{Tag}. This inverts the magnetization in a large, proximal region of tissue, indicated by the gray box in the tissue diagram. During the inversion delay *TI*, the labeled spins perfuse imaging slice of interest, as indicated by the light gray shading of the vessels in the tissue diagram; finally, an image is acquired. In the control image, the tagging RF pulse is applied at the frequency $-\omega_{Tag}$. Because the sign of the tagging gradient is maintained, the labeling band now occurs distally to the imaging slice, and the image is not flow-sensitized. However, the reversal of the sign of ω_{Tag} allows for proper control over MT effects. For further details, the reader is referred to the main text.

completeness and uniformity of in-slice and out-of-slice excitation). The RF pulses used for slab and slice selection are intended to excite none of the spins outside of the selected plane and fully and equally excite the spins inside the selected plane. However, this is never so. Some out-of-slice excitation always occurs; in-slice excitation is more homogeneous with adiabatic pulses than sinc pulses, but still is not perfect. Therefore, the ~1 cm gap between the inversion slabs and the slice plane is necessary so that the magnetization in the imaging plane is not disturbed by the inversion pulses, and vice versa. The presence of this gap creates the transit delay for labeled blood to reach the slice plane (the PASL equivalent of the quantity t_a referenced earlier), the value of which is unknown. As we will see below, the existence of an unknown t_a complicates quantification. Also, the 90° pulse at the beginning of the pulse sequence helps to control for such effects by saturating the magnetization in the imaging plane. Because the magnetization in the imaging plane is already saturated, any additional RF excitation (such as that caused by out-of-slab effects in the inversion pulse) is ineffective. Note that we have also indicated a spoiler gradient at the end of the saturation pulse's slice-selection gradient, which will dephase any coherent transverse magnetization in the imaging plane caused by the saturation pulse.

There are several variants of EPISTAR. Each of them addresses the issue of distal labeling of venous and (potentially) arterial blood described earlier for CASL. Proximal inversion with correction for off-resonance effects (PICORE) [6] uses the same tagging sequence as EPISTAR, but defines the control image differently (Figure 6.3). Instead of applying the inversion pulse at frequency $-\omega_{Tag}$ in the presence of a gradient, the pulse is simply applied at frequency ω_{Tag}, but without any gradient at all. This practice controls for asymmetric incidental MT effects. However, eddy current–induced geometric distortions in the EPI data are not as well controlled for in PICORE as in EPISTAR, because the

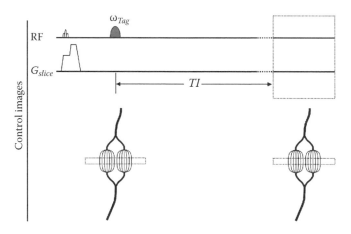

FIGURE 6.3 Formation of the control image in the PICORE pulse sequence and schematic illustration of labeling scheme in tissue. The tagged image is formed in an identical manner to EPISTAR. In the control image, the tagging RF pulse is applied at the frequency ω_{Tag}, but without the tagging gradient. This allows for proper control over MT effects. For further details, the reader is referred to the main text.

gradient waveforms differ. Another variant of EPISTAR is TILT (Transfer Insensitive Labeling Technique; [11]). Here, the single 180° inversion pulse is approximated with two 90° pulses in rapid succession. In the tagged image acquisition, the pulses are applied with the same phase, and so the magnetization nutates by 180°. In the control image, the 90° pulses oppose each other in phase. For example, a 90° pulse along X is followed by a 90° pulse along $-X$. Thus, in the control image, the magnetization experiences no net nutation; however, MT effects are well compensated for (the 90° pulses all occur at the same frequency).

6.3.2.2 Global Inversion Methods: FAIR

The FAIR pulse sequence is depicted in Figure 6.4. To begin, we note that some authors reverse the definitions of the tagged and control images in their description of the FAIR technique from those that we have presented here. Here, we will maintain our previous definition of the tagged image as one in which inflowing blood has been inverted and that of the control image in which no inversion of the inflowing blood has occurred. Also, for reasons that will soon become apparent, many investigators describe the two image types acquired in FAIR using the terms "selective" and "nonselective."

Figure 6.4a shows the tagged (or nonselective inversion) image pulse sequence and tissue diagram. In the labeling phase of the sequence, a hyperbolic secant inversion pulse is applied, as indicated in the pulse sequence diagram by the shaded, semioval RF pulse shape. This RF pulse occurs on resonance for water. Because there is not a gradient being applied at this time, the magnetization of all of the spins within the sensitive volume of the RF coil is inverted. This is indicated by the large shaded region in the tissue diagram below. During the inversion delay, a spoiling gradient may be applied; this is indicated by the dashed, trapezoidal gradient waveform in the pulse sequence diagram. During this delay, labeled spins perfuse the slice of interest. Finally, the image is read out.

Figure 6.4b shows the pulse sequence and tissue diagrams for the control (or selective inversion) image acquisition. The RF portion of the sequence is identical to that for the tagged image. However, the inversion pulse now occurs in the presence of a gradient G_{Tag}, which is applied in the same direction as the slice-selection gradient. Because G_{Tag} is present during the inversion pulse, only a select region of tissue experiences the pulse. This region has the same center position as the imaging slice, but is about twice as thick as the imaging slice. Because of its thickness, the region is sometimes referred to as the inversion "slab." By extending the inverted slab to a thicker volume than the imaging slice, there is uniform excitation within the imaging plane itself; this practice corrects for the imperfect slice-profile effects discussed above. During TI, the labeled spins flow out of the slice plane. As a final comment about the pulse sequence itself, we note that G_{Tag} may be entirely absent during the tagged image acquisition, as discussed above, or played out at a different time during the pulse sequence. The latter option would account for eddy current–induced distortions in the images better than the former option. The advantages of FAIR with respect to EPISTAR

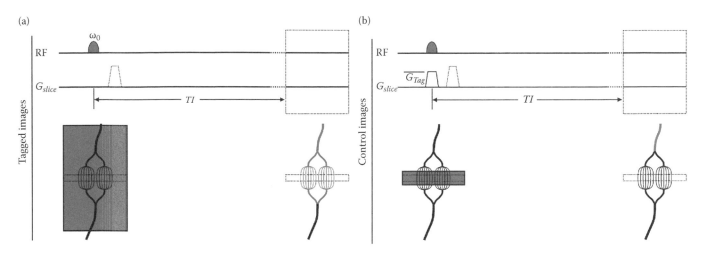

FIGURE 6.4 FAIR pulse sequence and schematic illustration of labeling scheme in tissue. (a) Pulse sequence and a diagram illustrating its effect on the tissue for the tagged image. (b) Pulse sequence and tissue diagram for the control image. In the tagged image, a hyberbolic secant inversion pulse is applied on resonance (i.e., at the frequency ω_0). Because no gradient is applied during the labeling pulse, the magnetization within the entire sensitive volume of the RF coil is inverted. During the time *TI*, the inverted spins perfuse the imaging slice, as indicated by the light gray shading of the vessels in the tissue diagram; finally, an image is acquired. During the control image acquisition, a tagging gradient is applied during the inversion pulse, and a volume slightly thicker than the slice only is affected. During *TI*, the labeled spins flow out of the slice plane. For further details, the reader is referred to the main text.

and its variants are, first, that the assumption of unidirectional arterial flow is not required and, second, that there is no incidental MT, because all pulses are applied on resonance. The former property is quite relevant to tumors, and the latter property is advantageous because it facilitates multislice imaging.

6.4 ASL Quantification

We start with a quantitative description of the effects of flow on the apparent T_1 of a tissue. We follow this with a discussion of a quantitative, general kinetic model for ASL. Finally, we discuss some important issues related to ASL quantification and present the methodological developments that have addressed these issues.

6.4.1 Description of Quantitative ASL Models

6.4.1.1 Effects of Flow on the Apparent T_1 of the Tissue

As defined in Chapter 5, T_1 is an exponential time constant describing the growth of longitudinal magnetization. This growth results largely from through-space magnetic interactions between magnetic dipoles rotating or translating at the Larmor frequency or twice the Larmor frequency [12]. The Bloch equation for Z magnetization describes the rate of change in the longitudinal magnetization, M_Z, as

$$\frac{dM_Z}{dt} = \frac{M_0 - M_Z}{T_1} \tag{6.4}$$

To describe only the inherent relaxation properties of the tissue parenchyma, we could rewrite Equation 6.4 using the symbols

$M_{Z,T}$, $M_{0,T}$, and $T_{1,T}$ to replace M_Z, M_0, and T_1, respectively; the subscript T refers to the tissue. However, perfusion will cause the tissue's Z magnetization to change at a different rate, because Z magnetization is delivered to and removed from the tissue by way of flow. Assuming rapid exchange of water across the capillary walls on the timescale of the ASL experiment and equal blood and tissue T_1s, the rate at which Z magnetization is delivered to the tissue by way of flow (f) will equal

$$fM_{Z,A} \tag{6.5}$$

The rate at which Z magnetization is removed from the tissue by way of flow equals

$$-\frac{f}{\lambda} M_{Z,T} \tag{6.6}$$

where λ is the tissue–blood partition coefficient for water, defined as

$$\lambda = \frac{g \, Water/g \, Tissue}{g \, Water/g \, Blood} \tag{6.7}$$

As noted by Detre et al. [2], Equations 6.4 through 6.7 can be combined to describe the rate of change in $M_{Z,T}$ in a perfused tissue as

$$\frac{dM_{Z,T}}{dt} = \frac{M_{0,T} - M_{Z,T}}{T_{1,T}} + fM_{Z,A} - \frac{f}{\lambda} M_{Z,T} \tag{6.8}$$

We can go on to describe further an apparent T_1 (T_1') that is influenced by both the natural relaxation and flow-dependent processes and is defined as

$$\frac{1}{T_1'} = \frac{1}{T_{1,T}} + \frac{f}{\lambda} \qquad (6.9)$$

Kwong et al. [7] have also written this equation to include a "mixed" T_1 in place of $T_{1,T}$, i.e., the mixture of $T_{1,A}$ and $T_{1,T}$—their average weighted by the blood volume fraction.

Equation 6.9 suggests that by varying the *TI* systematically under both control and tagged image conditions, respectively, one could estimate T_1 and T_1'; by further assuming a value for λ (~0.90 being typical for the brain [13]), one could calculate the perfusion, f. This is indeed possible, although it is a lengthy protocol to implement and requires that *TI* always be intermediate to t_a and ($t_a + \Delta t$). The main difficulty in this last statement lies in how well known t_a is. Such considerations have created the need for a more general formulation of the quantitative model of the ASL experiment.

6.4.1.2 A General Model for ASL

There have been several quantitative models of the ASL experiment presented. The above discussion is similar to models presented by Detre et al. [2] for CASL and by Kwong et al. [7] and Kim et al. [5] for PASL. Building on these and other [6,14,15] works, Buxton et al. [16] presented a general, quantitative kinetic model of the ASL experiment. We briefly summarize the Buxton model here and refer the reader to the original work for the full description and derivation of the model and a complete presentation of those authors' sensitivity analyses.

As noted, there is a longitudinal magnetization difference ($\Delta M_Z(t)$) between the tagged and control images in ASL, defined formally as

$$\Delta M_Z(t) = M_{Z,L}(t) - M_{Z,C}(t) \qquad (6.10)$$

where subscripts L and C refer to the labeled (i.e., tagged) and control images, respectively. In developing their model, Buxton et al. make Assumption 1, above. In this context, this means that they assume $M_{Z,C}$ to be constant during both types of image acquisition. They also assume that a proper control image has been made (i.e., Assumption 2, above).

Buxton et al. develop their model by noting that the quantity $\Delta M_Z(t)$ is determined quantitatively by the magnitude of inverted magnetization (=$2M_{0,A}$), f, and the rates of the delivery, exchange, and recovery of magnetization. The delivery function $C(t)$ describes the time dependence of the arrival of labeled magnetization and considers both transit time and relaxation effects. Upon its arrival to the perfused tissue, arterial water may exchange with water of the tissue parenchyma, causing some of the labeled water molecules to remain in the tissue at the time of image acquisition. The persistence of some of the labeled magnetization in the slice plane is described by a residue function, $R(t)$.

Finally, some portion of the labeled magnetization, now in the tissue, will undergo longitudinal relaxation; this is described by a magnetization relaxation function, $M(t)$.

The collective action of these functions can be described mathematically by defining $\Delta M_Z(t)$ as the convolution of $C(t)$ with $[R(t) \cdot M(t)]$. Considering also the magnetization inversion and flow terms, we have

$$\Delta M_Z(t) = 2M_{0,A}f\{C(t) * [R(t) \cdot M(t)]\} \qquad (6.11)$$

where * indicates the convolution operation. We now consider the specific forms of $C(t)$, $R(t)$, and $M(t)$.

In forming the delivery function $C(t)$, Buxton et al. assume magnetization to arrive in the voxel in a uniform, plug–flow fashion (i.e., an instantaneous transition exists between nonlabeled and labeled water molecules at both the leading and trailing edges of the labeled arterial bolus). Under this condition, the labeled arterial magnetization will perfuse the imaging plane uniformly and between the times t_a and ($t_a + \Delta t$) only. Also, we note that the arterial magnetization will begin to undergo longitudinal relaxation immediately following the inversion pulse. Considering both of these factors along with the "inversion efficiency," α (a term ranging from zero to one describing the effectiveness of the inversion pulse and accounting for the fact that the images may not be acquired under fully relaxed conditions [14,15]), $C(t)$ can be written as

$$C(t) = \begin{cases} 0 & 0 < t < t_a \\ \alpha e^{-t/T_{1,A}} & t_a \leq t < (t_a + \Delta t) \\ 0 & (t_a + \Delta t) \leq t \end{cases} \qquad (6.12a)$$

for PASL and

$$C(t) = \begin{cases} 0 & 0 < t < t_a \\ \alpha e^{-t_a/T_{1,A}} & t_a \leq t < (t_a + \Delta t) \\ 0 & (t_a + \Delta t) \leq t \end{cases} \qquad (6.12b)$$

for CASL. Note that t is taken to be the *TI* and that α accounts for any failure to achieve complete inversion (Assumption 3). As noted in the description of CASL, Δt is the duration of the inversion pulse, but for PASL, Δt depends also on the spatial extent of the tagging band and is not known a priori.

The residue function $R(t)$ is formed using the following considerations. In a given amount of time t', a certain number of labeled water molecules will arrive in an imaging voxel. Buxton et al.'s residue function quantifies the fraction of those molecules that persist in that voxel at a time t. Assuming that arterial and tissue water exchange rapidly, this fraction will be constant over time. As long as the physiological state of the tissue is not changing during the measurement period, the residue function can then be written as a function of t only:

$$R(t) = e^{-ft/\lambda} \qquad (6.13)$$

for both PASL and CASL.

Buxton et al. use the magnetization relaxation function $M(t)$ to describe the longitudinal magnetization in the labeled water that persists in the voxel at time t as a fraction of the original longitudinal magnetization tag. Assuming that a water molecule entering into the tissue from the vasculature will immediately adopt the T_1 of the tissue, and again assuming that the physiological state of the tissue does not change during the measurement period, then

$$M(t) = e^{-t/T_{1,T}} \qquad (6.14)$$

again for both PASL and CASL.

Putting all of this together, $\Delta M_Z(t)$ can be expressed in the following manner for the PASL experiment:

$$\Delta M_Z(t) = \begin{cases} 0 & 0 < t < \Delta t \\ 2\alpha M_{0,A} f(t - t_a) e^{-t/T_{1,A}} Q_P(t) & \Delta t < t < (t_a + \Delta t) \\ 2\alpha M_{0,A} f t_a e^{-t/T_{1,A}} Q_P(t) & (t_a + \Delta t) \le t \end{cases}$$

$$(6.15a)$$

We note that $M_{0,A} = \dfrac{M_0}{\lambda}$. Also, $Q_P(t)$ is a dimensionless quantity defined as

$$Q_P(t) = \begin{cases} \dfrac{e^{kt}(e^{-kt_a} - e^{-kt})}{k(t - t_a)} & \Delta t < t < (t_a + \Delta t) \\ \dfrac{e^{kt}(e^{-kt_a} - e^{-k(t_a + \Delta t)})}{kt_a} & (t_a + \Delta t) \le t \end{cases}$$

$$(6.15b)$$

with

$$k = \frac{1}{T_{1,A}} - \frac{1}{T_1'} \qquad (6.15c)$$

and T_1' having the meaning ascribed in Equation 6.9. The function $Q_P(t)$ approaches a value of one when $T_{1,A}$ and T_1' are very similar and under most conditions can be ignored. Equation 6.15a can then be simplified by eliminating the $Q_P(t)$ term. For CASL, we have

$$\Delta M_Z(t) = \begin{cases} 0 & 0 < t < \Delta t \\ 2\alpha M_{0,A} f T_1' e^{-t_a/T_{1,A}} Q_C(t) & \Delta t < t < (t_a + \Delta t) \\ 2\alpha M_{0,A} f T_1' e^{-t_a/T_{1,A}} e^{-(t - t_a - \Delta t)/T_1'} Q_C(t) & (t_a + \Delta t) < t \end{cases}$$

$$(6.16a)$$

with

$$Q_C(t) = \begin{cases} 1 - e^{-(t - t_a)/T_1'} & \Delta t < t < (t_a + \Delta t) \\ 1 - e^{-\Delta t/T_1'} & (t_a + \Delta t) < t \end{cases}$$

$$(6.16b)$$

Like $Q_P(t)$, $Q_C(t)$ is a dimensionless quantity; it approaches a value of one when t and Δt are much larger than T_1'. For this reason, a long delay time is used in CASL [15]; under this condition, Equation 6.16a can be simplified by eliminating $Q_C(t)$.

Equations 6.15a and 6.16a can be used to calculate quantitative perfusion images, on a pixel by pixel basis. Because raw MR images reflect arbitrary signal intensity and not the magnetization per se, ΔS is divided by the equilibrium signal intensity S_0 and used to represent $\Delta M_Z/M_0$. $T_{1,A}$ can be measured or obtained from the literature [17,18]. The inversion efficiency α can be calculated theoretically or determined experimentally and typically has values >0.95 for PASL; for CASL, α is lower and highly blood velocity dependent [19]. However, t_a is not known a priori; therefore (together with f), there are two unknowns in Equations 6.15a and 6.16a. Therefore, $\Delta S/S_0$ must be measured using at least two values of TI. When two values of TI are used, f and t_a can be calculated by solving the system of two equations; if more than two values of TI are used, a curve-fitting approach can be used.

While this general model accounts for a large number of effects, there are some persistent issues with regard to quantification. These can be considered to have technical and physiological bases. The discussion of these issues and the technical advancements that have been made to address them is the focus of the remainder of this section of the chapter.

6.4.2 Issues Affecting Quantification: Technically Based

6.4.2.1 Multislice Imaging

The sequences presented thus far have described only single-slice imaging. However, tumors are three-dimensional structures and it is desirable to characterize the perfusion to the entire tumor. There are two issues that hinder the implementation of multislice perfusion methods, however. The first is that the transit times for the blood to the different slices will vary. The technical advancements to address this issue are discussed below. The other issue, which pertains to CASL and to a lesser extent proximal inversion PASL methods, relates to incidental MT. The specific issue with regard to multislice imaging is that the center frequency of each slice differs, but ω_{Tag} is constant. Therefore, how far off-resonance the labeling RF pulse is from the water in the imaging plane depends on the spatial position of the imaging slice. This affects the validity of Assumption 1 for multislice image acquisitions, because changing the sign of ω_{Tag} or G_{Tag} results in a suitable control for incidental MR effects in the center slice only. For CASL, this has been addressed by applying the labeling pulse using a small, additional coil [20]. Because the B_1 field of a small coil has a limited spatial extent, there will be a very small amount of incidental MT, and so multislice imaging is more feasible. However, not all imagers have the hardware configuration required for the second coil.

6.4.2.2 Radiation Damping

One difficulty that can arise in FAIR is radiation damping, which describes the induction of current in the RF coil by the precessing magnetization. Current running through the coil generates small B_1 fields that perturb the magnetization in the coil, creating an addition relaxation pathway. The magnitude of this effect is determined in part by the proportion of the coil's volume occupied by the object being imaged. As noted by Zhou et al., for FAIR this includes the total volume of tissue within the coil experiencing the inversion pulse [21]. This will affect the accuracy of the perfusion estimates if not properly accounted for. Several variants of FAIR have been developed to address this problem. Zhou et al. introduced FAIRER (FAIR Excluding Radiation damping), which uses small gradient pulses to dephase the residual coherent transverse magnetization caused by radiation damping [21]. Helpern et al. [22] and Tanabe et al. [23] have described uninverted FAIR (UNFAIR), in which each single inversion pulse is replaced by two consecutive inversion pulses. In the tagged image, one pulse is slice-selective and the other is not. Thus, the spins in the imaging plane are rotated by 360° (no net effect) and the inflowing spins are inverted. In the control image, both pulses are nonselective, meaning that there is no net effect on any of the spins. The 360° rotation in the control image makes this image insensitive to inversion time and T_1 differences between the blood and tissue, helping to reduce radiation damping effects.

6.4.3 Issues Affecting Quantification: Physiologically Based

6.4.3.1 Assumption of Unidirectional Flow

We have discussed this issue previously, but for completeness, we refer to it here also. Unidirectional flow in the proximal-to-distal direction is a reasonable assumption for normal tissues, but this assumption may not be well met in tissues in which the vascular geometry is highly irregular, such as tumors. Therefore, the quantitative accuracy of CASL and proximal inversion methods relies on the degree to which the assumptions that venous return is minimal and that the arterial flow is unidirectional are met.

6.4.3.2 Labeling of Blood That Will Not Ultimately Perfuse the Imaging Plane

When the inversion tag is applied, some of the labeled blood will exist in larger, upstream arteries. Some of this blood will perfuse tissues distal to the slice plane, but when imaged will appear as perfusion to the slice of interest. These signals can be reduced using a small amount of diffusion weighting gradients but, in general, can be better dealt with using the methods described below that sharply demarcate the temporal extent of the labeled blood bolus.

6.4.3.3 Unknown Transit Time Delays

This is perhaps the most important single issue affecting ASL quantification and has generated a large amount of technical innovation. As just discussed, the inability to assume t_a a priori, and the likelihood that t_a varies spatially in heterogeneous tissues such as tumors, can be circumvented experimentally by measuring ΔS at multiple values of *TI*. However, this makes perfusion quantification very time consuming; also, reproducibility,

temporal stability, and temporal sampling frequency concerns come into play for dynamic acquisitions. Therefore, several pulse sequences have been devised to reduce or eliminate the sensitivity of ASL to t_a. Several of these techniques have evolved to provide better delineation of the labeled arterial blood bolus. These include quantitative imaging of perfusion using a single subtraction in its various forms (QUIPPS, QUIPPS-II, and Q2-TIPS), inflow-turbo sampling EPI-FAIR (ITS-FAIR), and quantitative STAR labeling of arterial regions (QUASAR).

QUIPPS, QUIPP-II, and Q2-TIPS [6,24] have in common the application of additional saturation pulses during the inversion time. In QUIPPS, these are applied at the imaging slice and in QUIPPS-II, and Q2-TIPS, these are applied in the tagging region. These saturation pulses create a well defined edge along the labeled blood bolus. Because they occur at a known point in time, the unknown value t_a in Equation 6.15a can be replaced with an experimentally specified pulse sequence parameter. This method works well in tissues with normal and uniform values for t_a but may fail when there is a broad distribution in t_a. As with the application of the general kinetic model of ASL, this problem can be circumvented by acquiring data at multiple *TIs*.

In the ITS-FAIR sequence, Günther et al. [25] presented a solution to the problem of the long time required to acquire ASL data at multiple *TIs*. They used an inversion recovery sampling scheme known as a Look–Locker acquisition. In this approach, the recovery following a single inversion pulse is sampled repeatedly using a large number of small nutation angle RF pulses, each of which occurs at a different *TI*. By appropriately adapting the quantitative model of ASL to account for the perturbation of the inversion recovery process caused by the repeated RF pulses, Günther et al. were able to provide quantitative perfusion maps. A similar approach was implemented using TILT by Hendrikse et al. [26]. Petersen et al. [27] subsequently proposed that if the shape of the arterial input function (AIF = $2M_{0,A} C(t)$) was known, then a model could be developed that does not require assumptions about the duration of the labeled blood bolus, λ, or the number of tissue compartments contributing to the ASL signal. The technique they introduced, QUASAR, labels spins according to an EPISTAR-like scheme. Additional saturation pulses are applied in the manner of QUIPPS and its variants. As with ITS-FAIR, a Look–Locker acquisition is used to sample the recovery from inversion. A subset of the images is obtained using a different nutation angle, which allows for B_1 inhomogeneity to be compensated for quantitatively. Finally, another subset of images is obtained with spoiling gradients; these eliminate the signal from within the vessels, allowing the AIF to be characterized. In sum, QUASAR represents a model-free approach to ASL that facilitates multislice imaging and improves quantification.

Another approach to eliminating transit time sensitivity is velocity-selective ASL (VS-ASL). As its name implies, VS-ASL labels blood based on its velocity, rather than its spatial position. The VS tag consists of a series of RF and gradient pulses that label only blood flowing above a specified velocity. Moreover, the imaging readout is configured such that it acquires data only from spins traveling below the cutoff velocity. Because arterial blood generally decelerates and venous blood accelerates, VS-ASL is thus able to select specifically for arterial blood. In principal, this permits labeling to occur in regions very close to the imaging slice, meaning also that flow through large vessels that will not ultimately perfuse the imaging slice is not labeled.

6.5 ASL in Cancer Biology Studies

The previous section outlined many different approaches to ASL that are available. Effective application of these techniques in oncology practice is based on the ability to detect, characterize, and follow the treatment response of tumors based on changes in perfusion. These techniques may be pertinent because angiogenesis is a key process in the growth and development of tumors, needed to overcome the limitations to tumor size that result from diffusion-limited nutrient delivery. In this final section of the chapter, we will provide examples illustrating the current state of application of ASL techniques in oncology practice, focusing first on diagnosis and characterization, then on following response to treatment and correlations with clinical outcomes.

6.5.1 Tumor Diagnosis and Characterization

Cancer can occur in virtually any part of the body. Throughout a tumor's growth and development, its characteristics can differ internally and also from those of other tumors. ASL is specifically applicable to cancerous tissue diagnosis and characterization where tumors can be distinguished from healthy tissue and each other based on an increase, decrease, or spatial change in their blood flow patterns. We will discuss some preliminary diagnostic ASL applications in thyroid adenomas, animal models of renal cell carcinoma in the kidney, and specific grading of brain gliomas in patients.

Many presentations of pathology can be detected with standard MRI contrast techniques, but some thyroid pathologies such as granulomatous thyroiditis, Hashimoto's thyroiditis, and Riedel's struma have been related to changes in thyroid perfusion. Ultrasound and gamma scintigraphy are the standard imaging technologies used for studying thyroid pathology, but these do not provide absolute measures of blood flow. ASL is being studied as a possible quantitative method for detecting the presence and severity of thyroid pathology, including cancer. One preliminary study [28] was conducted in normal volunteers and a single patient with a thyroid adenoma confirmed by a standard scintigraph. The investigators used a FAIR sequence with a TrueFISP image readout at 1.5 T to measure a mean perfusion between 400 and 500 ml 100 g^{-1} min^{-1} over multiple slices in the thyroid. The image acquisition for the entire thyroid required 17–31 min and the adenoma had reduced perfusion relative to normal thyroid tissue in the patient. In addition, cysts demonstrating no perfusion were detected in the thyroid gland in five of the healthy volunteers, and a significant difference in flow was found between the right and left lobes of the thyroid. Though

these results are preliminary, this study demonstrates the potential diagnostic application of ASL in the thyroid gland and particularly for detection of thyroid adenoma.

Once a tumor is discovered, treatment decisions depend on accurate characterization of the extent of the tumor's development, its potential for continued aggressive growth, and its potential resistance to therapy. In the kidney, the model of renal cell carcinoma has been used to demonstrate the potential for ASL to characterize tumors. To do so, Schor-Bardach et al. engineered tumors with different patterns of resistance to sorafenib [29], a multikinase inhibitor that blocks signaling through the vascular endothelial growth factor (VEGF) receptor. Sorafenib has shown good initial responses in patients, but most develop resistance within 5–10 months of beginning treatment. By implementing a FAIR technique with background suppression and a single slice through the largest diameter of the tumors, the investigators determined that in the sorafenib-resistant portions of the tumor, the flow at pretreatment baseline was 10.2 ml 100 g^{-1} min^{-1}, while the sorafenib-sensitive tumor flow was 80.1 ml 100 g^{-1} min^{-1}. A third tumor, with an early sensitivity and late resistance, had a baseline flow of 75.1 ml 100 g^{-1} min^{-1}. Also, the spatial distribution of flow in these tumors corresponded to their histological appearance, suggesting the possibility for internal tumor characterization. These results are promising and suggest the possibility of using ASL to predict resistance to treatment regimes, but remain to be repeated and extended in patient trials.

Gliomas are tumors arising from the glial cells of the brain and are some of the most frequently diagnosed primary brain tumors. Characterizing the development and potential for continued aggressive growth in glioma is normally based on biopsy-based observation of tissue characteristics, such as cellularity, mitosis, morphology, and evidence of necrosis. Data from histological examination are used to classify gliomas using World Health Organization (WHO) grades I–IV. Development of new vascular structures is another marker of glioma development, and this leads to increased tumor blood flow that may be assessed by perfusion MRI methods. Gaa et al. [30], the first to report application of ASL in brain tumors, were able to distinguish among meningiomas, high (III–IV)- and low (I–II)-grade astrocytomas, and lymphomas based on different flow values expressed relative to healthy white matter.

The ability to detect grading and types of brain tumors with ASL techniques has progressed along with improvements in methodology. Weber et al. [31] compared the diagnostic qualities of ASL (including Q2-TIPS and ITS-FAIR techniques), exogenous susceptibility contrast agent–based determination of blood volume, and MR spectroscopy in a sample of 79 brain tumor patients. The ASL techniques were able to distinguish among grades II–IV of glioma with a range of 92%–97% accuracy and 33%–50% specificity. Differentiation between grade IV glioma and lymphoma and between grade IV glioma and metastasis was achieved with 97% and 100% accuracy, respectively, and with respective specificities of 80% and 71%. The authors noted that the perfusion MRI techniques were preferable over spectroscopy for clinical applications because of shorter acquisition time and better predictive value for differential diagnosis. This study demonstrates the potential utility of ASL techniques in characterizing brain tumors prior to any invasive surgery. It should be noted that because of baseline differences in absolute flow at rest, these characterizations do not rely on absolute perfusion or flow; rather, their greatest value occurs when measured relative to healthy brain tissue or adjusted for patient age.

These examples demonstrate the development of successful applications of ASL techniques in different organs for the detection and characterization of cancerous tissue. They highlight the quantitative and noninvasive nature of ASL compared to other techniques. These types of characterizations may ultimately be able to lead physicians toward specific treatment strategies. The next section will discuss applications to monitoring treatment response and highlight two more aspects of ASL techniques: their ability to follow treatment responses over time and their predictive value for patient outcomes.

6.5.2 Tumor Treatment

Once a tumor has been diagnosed and characterized, physicians will determine a treatment strategy. Patients demonstrate variable responses to treatment, and it is often advantageous to discontinue one treatment in favor of another. These decisions are best made based on longitudinal data collected from tumor sites; earlier detection of responses means necessary changes can be made efficiently. The response to treatment is most commonly assessed by measuring changes in tumor size using MRI or computed tomography. More recently, specific antiangiogenic therapies have been developed, which affect the vasculature before a change in size can be detected. ASL techniques are promising for monitoring the flow consequences of these vascular changes. Tumor may show clear responses of reduced blood flow from one time to the next or may demonstrate a slowing progression and a lack of increase in blood flow. In this section, we will discuss the monitoring of treatment responses in renal cell carcinoma and the use of posttreatment measures in brain gliomas to predict treatment outcomes.

The monitoring of treatment responses in renal cell carcinoma using ASL was first reported by de Bazelaire et al. [32]. Their more recent study [33] is an excellent example of the application of ASL to monitoring treatment responses. Patients with metastatic renal cell carcinoma were given antiangiogenic therapy; blood flow was measured in the largest cross section of tumors using a single slice FAIR sequence before and after 1 month of treatment and tumor size with structural MRI after 4 months. Patients who had reduced tumor blood flow after 1 month presented with stable disease after 4 months, while those with unchanged flow after 1 month showed disease progression after 4 months. The authors pointed out that their 40-s long ASL image acquisition could be repeated even within the same exam, whereas sequences depending on an exogenous contrast agent cannot be performed twice in the same day (i.e., time must be allowed for renal elimination of the contrast agent).

Many tumors that begin in a specific organ or tissue can metastasize through blood or lymph in the later stages of their development. The brain is a common site for the deposition of metastases, affecting 20%–40% of all cancer patients. Stereotactic radiosurgery is a specifically targeted radiation therapy for such metastases that has shown 82%–96% local control rates. Weber et al. [34] investigated the use of ASL and DSC measurements of relative regional blood flow to predict treatment outcome earlier than standard morphological MRI techniques. For the ASL acquisition, a five-slice Q2-TIPS sequence was employed to measure flow prior to and 6, 12, and 24 weeks following stereotactic radiosurgery. Tumor size was then assessed 6 months following treatment. The change in flow from baseline to 6 weeks after treatment predicted patient outcome of remission, stable disease, or disease progression at 6 months. Blood flow in metastases relative to gray matter demonstrated 100% specificity, sensitivity, and positive and negative predictive values for treatment outcome. As with tumor characterization in the brain, these promising results are based on measures of flow relative to healthy gray matter. These results are impressive but are limited to a small number of patients. Further evidence in larger trials will be necessary to fully support implementation of these ASL techniques in practice.

6.6 Summary

The preceding examples demonstrate the current state of application of ASL techniques in oncology. We have presented progress in diagnosis, characterization, and the following of responses to treatment in thyroid, kidney, and brain. The common themes thus far are that ASL techniques are not yet independent of standard, more invasive, or exogenous contrast-based techniques because they cannot provide the breadth of detailed data that physicians receive from histological analysis, for example. However, it has been shown in both animal models and limited patient trials that ASL techniques do show promise for providing unique information about blood flow that may allow completely noninvasive characterization of high-flow tumors, as well as accurate identification of treatment responses earlier than standard MRI morphological techniques. All of these applications are preliminary in the sense that none have progressed to the point of large randomized clinical trials. As ASL techniques continue to progress along with the understanding of tumor biology and treatments, we will perhaps see greater application of ASL over a broader range of clinical practice and a broader range of organ and tissue types.

References

1. Kety SS, Schmidt CF. The nitrous oxide method for the quantitative determination of cerebral blood flow in man; theory, procedure and normal values. *J Clin Invest* 1948; **27**: 476–483.
2. Detre JA, Leigh JS, Williams DS, Koretsky AP. Perfusion imaging. *Magn Reson Med* 1992; **23**: 37–45.

3. Williams DS, Detre JA, Leigh JS, Koretsky AP. Magnetic resonance imaging of perfusion using spin inversion of arterial water. *Proc Natl Acad Sci USA* 1992; **89**: 212–216.

4. Edelman RR, Siewert B, Darby DG, Thangaraj V, Nobre AC, Mesulam MM, Warach S. Qualitative mapping of cerebral blood flow and functional localization with echo-planar MR imaging and signal targeting with alternating radio frequency. *Radiology* 1994; **192**: 513–520.

5. Kim SG. Quantification of relative cerebral blood flow change by flow-sensitive alternating inversion recovery (FAIR) technique: Application to functional mapping. *Magn Reson Med* 1995; **34**: 293–301.

6. Wong EC, Buxton RB, Frank LR. Implementation of quantitative perfusion imaging techniques for functional brain mapping using pulsed arterial spin labeling. *NMR Biomed* 1997; **10**: 237–249.

7. Kwong KK, Chesler DA, Weisskoff RM, Donahue KM, Davis TL, Ostergaard L, Campbell TA, Rosen BR. MR perfusion studies with T1-weighted echo planar imaging. *Magn Reson Med* 1995; **34**: 878–887.

8. Dixon WT, Du LN, Faul DD, Gado M, Rossnick S. Projection angiograms of blood labeled by adiabatic fast passage. *Magn Reson Med* 1986; **3**: 454–462.

9. Wolff SD, Balaban RS. Magnetization transfer contrast (MTC) and tissue water proton relaxation in vivo. *Magn Reson Med* 1989; **10**: 135–144.

10. Jun H, Craig KJ, Jaishri B, Seth AS, Peter CMvZ, Jinyuan Z. Quantitative description of the asymmetry in magnetization transfer effects around the water resonance in the human brain. *Magn Reson Med* 2007; **58**: 786–793.

11. Golay X, Stuber M, Pruessmann KP, Meier D, Boesiger P. Transfer insensitive labeling technique (TILT): Application to multislice functional perfusion imaging. *J Magn Reson Imaging* 1999; **9**: 454–461.

12. Bloembergen N, Purcell E, Pound R. Relaxation effects in nuclear magnetic resonance absorption. *Physical Reviews* 1948; **73**: 679–712.

13. Herscovitch P, Raichle ME. What is the correct value for the brain–blood partition coefficient for water? *J Cereb Blood Flow Metab* 1985; **5**: 65–69.

14. Zhang W, Williams DS, Koretsky AP. Measurement of rat brain perfusion by NMR using spin labeling of arterial water: In vivo determination of the degree of spin labeling. *Magn Reson Med* 1993; **29**: 416–421.

15. Alsop DC, Detre JA. Reduced transit-time sensitivity in noninvasive magnetic resonance imaging of human cerebral blood flow. *J Cereb Blood Flow Metab* 1996; **16**: 1236–1249.

16. Buxton RB, Frank LR, Wong EC, Siewert B, Warach S, Edelman RR. A general kinetic model for quantitative perfusion imaging with arterial spin labeling. *Magn Reson Med* 1998; **40**: 383–396.

17. Spees WM, Yablonskiy DA, Oswood MC, Ackerman JJ. Water proton MR properties of human blood at 1.5 Tesla: Magnetic susceptibility, T_1, T_2, T_2^*, and non-Lorentzian signal behavior. *Magn Reson Med* 2001; **45**: 533–542.

18. Lu H, Clingman V, Golay X, van Zijl P. Determining the longitudinal relaxation time (T_1) of blood at 3.0 Tesla. *Magn Reson Med* 2004; **52**: 679–682.

19. Wong EC, Buxton RB, Frank LR. A theoretical and experimental comparison of continuous and pulsed arterial spin labeling techniques for quantitative perfusion imaging. *Magn Reson Med* 1998; **40**: 348–355.

20. Zhang W, Silva AC, Williams DS, Koretsky AP. NMR measurement of perfusion using arterial spin labeling without saturation of macromolecular spins. *Magn Reson Med* 1995; **33**: 370–376.

21. Zhou J, Mori S, van Zijl PC. FAIR excluding radiation damping (FAIRER). *Magn Reson Med* 1998; **40**: 712–719.

22. Helpern JA, Branch CA, Yongbi MN, Huang NC. Perfusion imaging by un-inverted flow-sensitive alternating inversion recovery (UNFAIR). *Magn Reson Imaging* 1997; **15**: 135–139.

23. Tanabe JL, Yongbi M, Branch C, Hrabe J, Johnson G, Helpern JA. MR perfusion imaging in human brain using the UNFAIR technique. Un-inverted flow-sensitive alternating inversion recovery. *J Magn Reson Imaging* 1999; **9**: 761–767.

24. Luh WM, Wong EC, Bandettini PA, Hyde JS. QUIPSS II with thin-slice TI1 periodic saturation: A method for improving accuracy of quantitative perfusion imaging using pulsed arterial spin labeling. *Magn Reson Med* 1999; **41**: 1246–1254.

25. Günther M, Bock M, Schad LR. Arterial spin labeling in combination with a look-locker sampling strategy: Inflow turbo-sampling EPI-FAIR (ITS-FAIR). *Magn Reson Med* 2001; **46**: 974–984.

26. Hendrikse J, Lu H, van der Grond J, Van Zijl PC, Golay X. Measurements of cerebral perfusion and arterial hemodynamics during visual stimulation using TURBO-TILT. *Magn Reson Med* 2003; **50**: 429–433.

27. Petersen ET, Lim T, Golay X. Model-free arterial spin labeling quantification approach for perfusion MRI. *Magn Reson Med* 2006; **55**: 219–232.

28. Schraml C, Boss A, Martirosian P, Schwenzer NF, Claussen CD, Schick F. FAIR true-FISP perfusion imaging of the thyroid gland. *J Magn Reson Imaging* 2007; **26**: 66–71.

29. Schor-Bardach R, Alsop DC, Pedrosa I, Solazzo SA, Wang X, Marquis RP, Atkins MB, Regan M, Signoretti S, Lenkinski RE, Goldberg SN. Does arterial spin-labeling MR imaging-measured tumor perfusion correlate with renal cell cancer response to antiangiogenic therapy in a mouse model? *Radiology* 2009; **251**: 731–742.

30. Gaa J, Warach S, Wen P, Thangaraj V, Wielopolski P, Edelman RR. Noninvasive perfusion imaging of human brain tumors with EPISTAR. *Eur Radiol* 1996; **6**: 518–522.

31. Weber MA, Zoubaa S, Schlieter M, Juttler E, Huttner HB, Geletneky K, Ittrich C, Lichy MP, Kroll A, Debus J, Giesel FL, Hartmann M, Essig M. Diagnostic performance of spectroscopic and perfusion MRI for distinction of brain tumors. *Neurology* 2006; **66**: 1899–1906.

32. de Bazelaire C, Rofsky NM, Duhamel G, Michaelson MD, George D, Alsop DC. Arterial spin labeling blood flow magnetic resonance imaging for the characterization of metastatic renal cell carcinoma. *Acad Radiol* 2005; **12**: 347–357.

33. de Bazelaire C, Alsop DC, George D, Pedrosa I, Wang Y, Michaelson MD, Rofsky NM. Magnetic resonance imaging-measured blood flow change after antiangiogenic therapy with PTK787/ZK 222584 correlates with clinical outcome in metastatic renal cell carcinoma. *Clin Cancer Res* 2008; **14**: 5548–5554.

34. Weber MA, Thilmann C, Lichy MP, Günther M, Delorme S, Zuna I, Bongers A, Schad LR, Debus J, Kauczor HU, Essig M, Schlemmer HP. Assessment of irradiated brain metastases by means of arterial spin-labeling and dynamic susceptibility-weighted contrast-enhanced perfusion MRI: Initial results. *Invest Radiol* 2004; **39**: 277–287.

Diffusion-Weighted MRI

Lori R. Arlinghaus
Vanderbilt University

Thomas E. Yankeelov
Vanderbilt University

7.1 Introduction

One of the hallmarks of cancer is the unlimited potential for cell replication [1], which leads to changes in tissue microstructure, such as increased cellularity and decreased extracellular volume fraction (ECVF), compared with healthy tissue. Successful treatment with radiation therapy and chemotherapy results in increased ECVF, decreased cellularity, and a loss of cell membrane integrity [2–4]. These changes can be detected indirectly with diffusion-weighted magnetic resonance imaging (DW-MRI), which provides information about tissue microstructure by means of a noninvasive magnetic resonance (MR) imaging sequence that is sensitive to the motion of water molecules. In this chapter, the biological basis for changes observed with DW-MRI and the imaging mechanism behind the technique will be described, along with practical considerations. Evidence for its potential as a valuable imaging biomarker (Biomarkers Definitions Working Group 2001) for the detection, staging, and monitoring of therapy in cancer will be presented, and, finally, future directions for the technique will be discussed.

7.2 Biological Basis of DW-MRI

DW-MRI [5–7] is based upon the observations of Hahn [8] and Carr and Purcell [9] that spin-echo (SE) MR sequences are sensitive to the diffusion of molecules within the sample. This random motion, also known as Brownian motion, is due to thermal energy and has been well described by Einstein [10]. For freely diffusing molecules, the distribution of displacements that the particles will experience over a given amount of time (t) is described by a Gaussian function. The standard deviation of the distribution describes the mean-squared displacement of the molecules in one dimension over time t:

$$\langle x^2 \rangle = 2Dt \tag{7.1}$$

where D is the diffusion coefficient of the molecules (see Figure 7.1). The diffusion coefficient describes the rate of diffusion and is dependent upon temperature, the size of the particles, and the viscosity of the medium.

Free diffusion of water molecules at normal body temperature (37°C), with a diffusion coefficient of 3×10^{-3} mm²/s, would result in an average displacement of 17–25 μm in one dimension over diffusion times of 50–100 ms. However, water molecules within tissue are not freely diffusing on this time scale. They encounter macromolecules, cell membranes, and other tissue microstructures that hinder diffusion, causing a reduction in the average diffusion displacement, compared with that of free water. This effective, or apparent diffusion coefficient (ADC), can be measured with DW-MRI.

The complex interactions between water and tissue microstructures that lead to the observed ADC are not entirely understood. However, it is currently recognized that cellularity, extracellular volume fraction (ECVF), membrane permeability, and tortuosity play an important role. A study modifying cell density and extracellular volume fraction in human glial and red blood cells *ex vivo* revealed an inverse relationship between cell density and ADC value and that ADC increased when cells shrank (increased ECVF) and decreased with cellular swelling (decreased ECVF) [11]. Several groups have demonstrated that ADC values are inversely correlated with tumor cellularity in

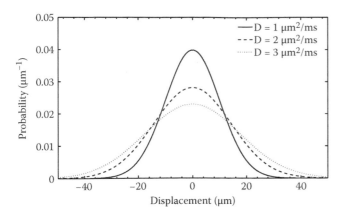

FIGURE 7.1 The Gaussian diffusion displacement distribution for different diffusion coefficients with a diffusion time of 50 ms. As the diffusion coefficient increases, the distribution broadens, indicating increased motility for the given diffusion time.

vivo [12–14]. Latour et al. [15] demonstrated that ADC measurements were sensitive to changes in cellular permeability in erythrocytes, and Lang et al. [16] observed increased ADC measurements associated with necrosis and the associated breakdown of the cell membranes. As diffusing water molecules interact with macromolecules and cell structures, their paths of motion are forced to deviate from a straight line, resulting in a more tortuous environment. In a study of an ischemia model in neonatal rat brain, it was shown that an increase in tortuosity resulted in a reduction in ADC [17]. The effects of changes in cellularity, ECVF, membrane permeability, and tortuosity on the measured ADC can be seen in Figure 7.2.

7.3 Diffusion-Weighted MRI

Many of the diffusion imaging sequences used today are based upon the pulsed gradient spin-echo (PGSE) scheme [18] shown in Figure 7.3. In this sequence, two gradient pulses are applied, one on either side of the 180° radiofrequency (RF) pulse and each with amplitude G and duration δ. The net effect of these gradient pulses is the labeling of spins, such as protons in water, which move by self-diffusion of molecules during the time Δ between the applications of each gradient.

To demonstrate the effect of these gradients on the magnetization of the spins, take the case of the pulsed gradients applied along the z-axis. When the first pulse is applied, the transverse magnetization of the spins acquires a phase shift ϕ_1 that varies linearly according to the spins' location along the z-axis so that

$$\phi_1 = \gamma \int_0^\delta Gz_1 \, dt = \gamma Gz_1 \delta \qquad (7.2)$$

where γ is the gyromagnetic ratio and z_1 is the location of the spin, which is assumed to remain constant over the pulse duration δ. The 180° RF pulse reverses the phase of the transverse

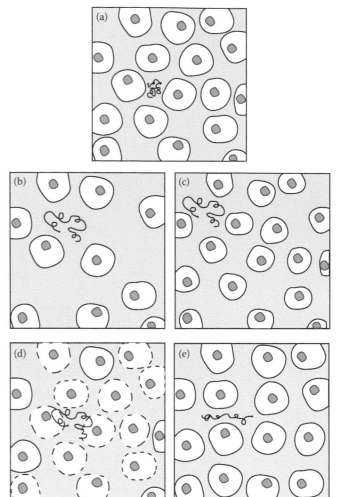

FIGURE 7.2 Examples of cellular changes leading to an increase in ADC. (a) Example of typical cell organization and the path of an extracellular water molecule for a given amount of time. (b) Reduction in cell density leads to increased ECVF, giving water more room to move. (c) Reduction in cell size, with cell density remaining constant, also gives the water more room to move. (d) Increased membrane permeability of cells reduces the restrictions they face. (e) Reduced tortuosity reduces the number of turns a water molecule must make to get from one location to another.

magnetization so that it is now $-\phi_1$. When the second gradient pulse is applied, the transverse magnetization of the spin, now located at position z_2, acquires a second phase shift ϕ_2:

$$\phi_2 = \gamma \int_0^\delta Gz_2 \, dt = \gamma Gz_2 \delta \qquad (7.3)$$

Thus, the net effect of the gradients on the phase of a spin's magnetization is the sum of the two phase shifts:

$$\phi_2 - \phi_1 = \gamma G\delta(z_2 - z_1) \qquad (7.4)$$

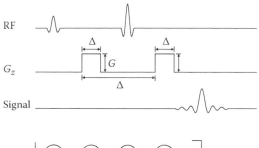

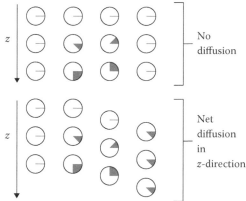

FIGURE 7.3 The pulse sequence diagram for a PGSE experiment (top) and schematics demonstrating the effects of the sequence on spins that do not diffuse (middle) and spins that do diffuse (bottom) during the diffusion encoding time Δ. The spins initially have no phase, then the first diffusion sensitizing gradient (G) applies a location-dependent phase to each spin. The spins are allowed to diffuse, and a 180° RF pulse is applied, reversing the phase of each spin. The spins diffuse for a total time period Δ, when a second diffusion sensitizing gradient (G) is applied. If the spins have not moved from their original locations (middle), they have no net phase at the time of the signal readout. If the spins have moved (bottom), they will have a net phase change that is related to the net distance they have moved along the direction of the diffusion sensitizing gradients.

If the spin does not change position along the z-axis between the application of the two gradients, then $z_1 = z_2$, and there is no net effect of the gradient pulses on the phase. In practice, however, molecules move about randomly (Brownian motion), so that for any given spin, z_1 may not be equal to z_2. This results in a phase shift of the spin's magnetic moment m:

$$m = m_0 \exp(i\gamma G\delta(z_2 - z_1)) \quad (7.5)$$

where m_0 is the magnitude of the magnetic moment in the case of $G = 0$ and the exponential term is the phase factor. The observed signal S in an MR experiment is due to the bulk magnetization, which is the sum of the magnetization vectors of individual spins, whose locations are not known. The random motion of the spins leads to a random distribution of displacements, and the bulk signal attenuation is given by

$$\frac{S}{S_0} = \int\int_{-\infty}^{\infty} P(z_2|z_1,\Delta)\exp(i\gamma G\delta(z_2 - z_1))dz_2\,dz_1 \quad (7.6)$$

where S_0 is the intrinsic signal, and $P(z_2|z_1,\Delta)$ is the probability that a spin initially located at z_1 is located at z_2 after time Δ. For unrestricted, isotropic diffusion in one dimension, the distribution $P(z_2|z_1,\Delta)$ can be assumed to be a normalized Gaussian:

$$P(z_2|z_1,\Delta) = \frac{1}{\sqrt{4\pi D\Delta}}\exp\left(\frac{-(z_2 - z_1)^2}{4D\Delta}\right) \quad (7.7)$$

where the diffusion coefficient D is related to the mean square displacement of a molecule by Equation 7.1 with $t = \Delta$. Substituting Equation 7.7 into Equation 7.6 and evaluating the resulting integral gives an attenuation factor

$$\frac{S}{S_0} = \exp(-\gamma^2 G^2 \delta^2 D\Delta) \quad (7.8)$$

Hardware limitations prevent δ from being negligible as compared with Δ. In this case, the equation describing the signal attenuation can be derived from the Bloch–Torrey equations [19]. For unrestricted, isotropic diffusion, the signal attenuation at the time of the echo (*TE*) depends upon the diffusion coefficient D and an attenuation factor b [7]:

$$\frac{S(TE)}{S_0} = \exp(-bD) \quad (7.9)$$

where

$$b = \int_0^{TE} \mathbf{k}(t)\cdot\mathbf{k}(t)dt \quad (7.10)$$

$$\mathbf{k}(t) = \gamma\int_0^t \mathbf{G}(t')dt' \quad (7.11)$$

and

$$\mathbf{G}(t) = (G_x(t),G_y(t),G_z(t))^T \quad (7.12)$$

(The interested reader is referred to reference [20] for a detailed derivation.) For the PGSE sequence, Equation 7.10 takes the form

$$b = \gamma^2\delta^2 G^2\left(\Delta - \frac{\delta}{3}\right) \quad (7.13)$$

The b factor (or b value) defines the sensitivity of the imaging sequence to diffusion.

The value of each variable, except D, in Equations 7.9 and 7.13 is known. By obtaining two images with different b values (e.g.,

$b_1 = 0$ s/mm^2 and $b_2 \geq 500$ s/mm^2), Equations 7.9 and 7.13 can be used to calculate the value of D:

$$D = \frac{\ln\left(S_1/S_2\right)}{b_2 - b_1} \qquad (7.14)$$

where S_1 is the signal acquired with b value b_1 and S_2 is the signal acquired with b value b_2. In tissue, D represents the ADC, measured in one spatial direction. Typically, the spatially invariant mean ADC (*mADC*) value is calculated by applying diffusion gradients along three orthogonal directions (x,y,z), calculating the ADC value for each direction (ADC_x, ADC_y, ADC_z) according to Equation 7.14, and then calculating the mean value:

$$mADC = \frac{ADC_x + ADC_y + ADC_z}{3} \qquad (7.15)$$

7.4 Practical Considerations

7.4.1 Diffusion Models

Equation 7.9 assumes the relationship between the signal and the diffusion coefficient is monoexponential, which may be an oversimplification of the complex environment within tissue and its effects on diffusion. For example, intracellular and extracellular water molecules may experience very different diffusion environments. Several groups have investigated multiexponential model of diffusion in the brain [21–23]. For a biexponential model, Equation 7.9 takes the following form:

$$\frac{S(TE)}{S_0} = f \exp\left(-bD_1\right) + \left(1-f\right)\exp\left(-bD_2\right) \qquad (7.16)$$

where f is the fraction of water molecules with diffusion coefficient D_1, and D_2 is the diffusion coefficient of the remaining water molecules. Bennett et al. [24] proposed a stretched exponential model to investigate the distribution of diffusion coefficients measured in the brain. In this case, it is assumed that a continuum of diffusion coefficients are being experienced by the sample, and Equation 7.9 takes the form of

$$\frac{S(TE)}{S_0} = \exp\left(-\left(b \cdot DDC\right)^{\alpha}\right) \qquad (7.17)$$

where DDC is the distributed diffusion coefficient and α is the intravoxel heterogeneity index. For a monoexponential model, $\alpha = 1$, and for a multiexponential model, $\alpha < 1$.

While multiexponential models may be able to provide a more detailed picture of the diffusion properties of the underlying

tissue, they are not without their limitations. For example, multiple images with b values extending to very high values are required for accurate characterization of these models. This results in long scan times, and as will be discussed in more detail in the following section, background noise begins to dominate the signal at high b values.

In general, it is reasonable to assume monoexponential behavior in most DW-MRI applications in cancer. However, if multiexponential models are explored, the results should be interpreted with caution. There are many sources of potential error, including perfusion effects, that can make signal that is truly monoexponential appear to be multiexponential. (The effects of perfusion on the measured signal will be discussed in more detail in the following section.) There is also debate over the interpretation of data exhibiting biexponential behavior. The two diffusion coefficients are often referred to as fast (or large) and slow (or small) and are often suggested to be related to diffusion of extracellular and intracellular water, respectively, e.g., references [21,25–27]. However, the relative contribution of each diffusion coefficient does not correlate with the measured volume fractions of intracellular and extracellular water [22], and biexponential signal has been observed in studies of intracellular water alone [28].

7.4.2 Choice of b Values

When designing a DW-MRI study, several factors affect the choice of optimum b values. The appropriate optimization method depends on the degree of anisotropy in the tissue of interest and on the parameter of interest—signal intensity (i.e., for qualitative assessment) or ADC (i.e., for quantitative assessment). Additional factors that affect the calculation of optimum b values include the ADC of the tissue of interest, the amount of perfusion in the tissue, and the amount of background noise.

While the focus of this chapter is on quantitative applications of DW-MRI in cancer, the inherent contrast between different tissue types in DW-MRIs can be useful for qualitative purposes, including tumor localization. Maximizing the contrast between the tissue of interest and other tissue types requires optimizing the b values. Assuming isotropic diffusion and that S_{0A} and S_{0B} are the signal intensities in regions A and B when no diffusion weighting is applied ($b = 0$ s/mm^2), the contrast between regions A and B with $D_A > D_B$ is maximized when [29]

$$b_{\Delta S_{max}} = \frac{\ln\left(\dfrac{D_A}{D_B}\right) - \ln\left(\dfrac{S_{0B}}{S_{0A}}\right)}{D_A - D_B}. \qquad (7.18)$$

Optimization of DW-MRIs for quantitative use requires the minimization of the effects of noise on the measurements. Starting with Equation 7.9 and using a propagation of errors calculation, it can be shown that σ_{ADC}, the standard deviation of the

ADC measured with N independent b values, is described by the following equation:

$$\sigma_{ADC} = \frac{D_0}{SNR_0} \sqrt{\frac{\sum\limits_n \left(b_n N - \sum\limits_n b_n \right)^2 e^{2b_n D_0}}{N \sum\limits_n b_n^2 - \left(\sum\limits_n b_n \right)^2}} \qquad (7.19)$$

where D_0 is the true ADC of the tissue of interest and SNR_0 is the signal-to-noise ratio (SNR) of the image when $b = 0$ s/mm² [30]. Minimizing σ_{ADC} with respect to b_n provides the optimum set of b values.

It is clear from Equations 7.18 and 7.19 that the true ADC values of the tissue being imaged play an important role in the optimization of b values, regardless of the application. Prior knowledge of the true ADC values of tumors and healthy tissue are not always available and reported values can vary greatly. Padhani et al. [31] published a list of suggested b values for various tissues. However, it is often best to acquire images with multiple b values, if possible.

As mentioned in the previous section, factors other than random diffusion, such as the bulk flow of water and noise, affect the measured signal. Blood flow in the capillaries causes an artificial decrease in signal, and the amount of signal attenuation is dependent on the b value. This effect is related to the perfusion fraction, which is the fraction of total water located within the capillaries in the tissue [32], and it affects the optimum value of the low b value (b_1 in Equation 7.14) used for DW-MRI. For example, in healthy brain tissue, where the perfusion fraction is relatively low (around 5%–10% [33–35]), the effect of perfusion on the measured ADC is minimal, and a low b value of 0 s/mm² is commonly used. Tumors, on the other hand, exhibit sustained angiogenesis [1] and a broad range of perfusion factors (20–80%) [36,37], increasing the effects of perfusion on the signal measured at low b values. Therefore, values of 50–200 s/mm² are typically used for b_1 (Equation 7.14) in cancer applications. At higher b values (e.g., ≥3000 s/mm²), there is greater signal attenuation due to diffusion and the measured signal is artificially increased by noise as the noise floor is reached [38]. The rate at which the noise floor is approached is related to the ADC value: the higher the ADC value, the faster the signal is attenuated at a lower b value. These effects are illustrated in Figure 7.4. When selecting b values, one should also consider the fact that the difference between b_1 and b_2 must be large enough to produce adequate contrast between the two images.

7.4.3 Anisotropic Diffusion

The highly organized structure of some tissues, such as white matter in the brain, renal tubules in the kidneys, and fibromuscular tissue in the prostate, results in highly anisotropic diffusion, which may be altered by tumors. Although most tumors exhibit relatively isotropic diffusion, measures of diffusion anisotropy may be useful for tumor delineation and detection of tumor infiltration in tissues with highly structured organization.

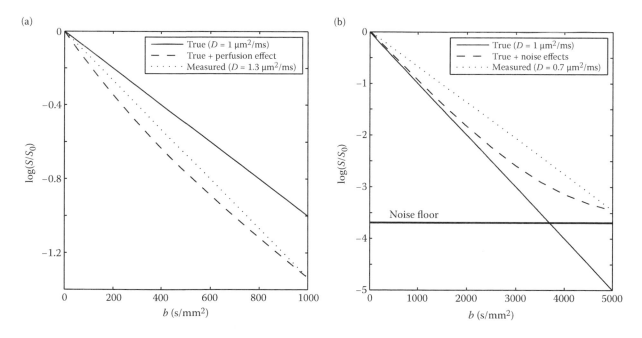

FIGURE 7.4 Simulated signal attenuation plotted as a function of the b value to illustrate the effects of (a) perfusion and (b) the noise floor. (a) Perfusion mimics diffusion at a diffusion rate much faster than the true diffusion rate. This leads to a nonmonoexponential decay of the signal (dashed line). However, if the ADC value is calculated with signal attenuation values at $b = 0$ and 1000 s/mm² (dotted line), the resulting ADC will be higher than the true value. (b) At higher b values, background noise will artificially increase the measured amount of signal. This also leads to nonmonoexponential decay. However, in this case, if the ADC value is calculated using $b = 0$ and 5000 s/mm², the resulting ADC will be lower than the true value.

In these tissues, the dependence of the measured ADC on the direction along which the diffusion gradient is applied can be taken advantage of to reveal information related to the tissue structure in three dimensions. This requires a variant of DW-MRI referred to as diffusion tensor imaging (DTI). In DTI, diffusion gradients are applied along a minimum of six noncollinear directions not all in the same plane, the ADC value is now represented as a tensor matrix **D** [39], and the corresponding *b* values are also represented as a matrix **b**. Equation 7.9 now takes the form

$$\frac{S(TE)}{S_0} = \exp\left(-\sum_{i=x,y,z}\sum_{j=x,y,z} b_{ij}D_{ij}\right) \tag{7.20}$$

where b_{ij} and D_{ij} are the *ij*th components of **b** and **D**, respectively. **D** is a symmetric matrix:

$$\mathbf{D} = \begin{bmatrix} D_{xx} & D_{xy} & D_{xz} \\ D_{yx} & D_{yy} & D_{yz} \\ D_{zx} & D_{zy} & D_{zz} \end{bmatrix} \tag{7.21}$$

where

$$D_{xy} = D_{yx}$$
$$D_{xz} = D_{zx} \tag{7.22}$$
$$D_{yz} = D_{zy}$$

The *b*-matrix (**b**) is also symmetric, with

$$b_{ij} = \gamma^2\delta^2 G_i G_j\left(\Delta - \frac{\delta}{3}\right) \tag{7.23}$$

where *i,j* = (*x,y,z*). The diffusion tensor has six independent elements, which can be solved for using multivariate linear regression [39]. Each tensor can then be diagonalized to obtain the eigenvectors $(\vec{e}_1,\vec{e}_2,\vec{e}_3)$ and eigenvalues $(\lambda_1,\lambda_2,\lambda_3)$, which characterize the direction and magnitude, respectively, of the 3D diffusion displacement distribution within a given voxel.

Several rotationally invariant scalar values can be derived from the tensor. The most commonly used values are the trace of the diffusion tensor [39], mean diffusivity, and fractional anisotropy (FA) [40]. The trace of **D**, Tr(**D**), is the sum of the eigenvalues, giving a measure of the total diffusivity:

$$\text{Tr}(\mathbf{D}) = \lambda_1 + \lambda_2 + \lambda_3 \tag{7.24}$$

The mean diffusivity $\bar{\lambda}$, comparable to the mean ADC value calculated in DW-MRI, is the average of the effective diffusivities:

$$\bar{\lambda} = \frac{\text{Tr}(\mathbf{D})}{3} \tag{7.25}$$

FA is an index of the variation of diffusion over measurement directions. It is defined as

$$FA = \sqrt{\frac{3}{2}}\frac{\sqrt{(\lambda_1-\bar{\lambda})^2 + (\lambda_2-\bar{\lambda})^2 + (\lambda_3-\bar{\lambda})^2}}{\sqrt{\lambda_1^2 + \lambda_2^2 + \lambda_3^2}} \tag{7.26}$$

and it ranges from 0 for isotropic diffusion to 1 for completely anisotropic diffusion restricted to one direction. Although only a minimum of six diffusion-encoding directions is required to calculate the diffusion tensor, it has been shown for tissue exhibiting anisotropic diffusion that a minimum of 20 unique directions is necessary for accurate estimation of diffusion anisotropy and a minimum of 30 unique directions is necessary for accurate estimation of tensor orientation and mean diffusivity [41].

These scalar diffusion parameters provide indirect measures of changes in tissue microstructure. For example, $\bar{\lambda}$ decreases as cellularity increases or ECVF decreases. FA is high in highly structured tissue, such as muscle fibers and white matter in the brain, and low in tissue where structures are randomly organized, such as gray matter in the brain. It should be noted, however, that regions of relatively low FA could exist even within regions of highly organized tissue. For example, if a voxel contains signal from two intersecting white matter tracts, partial volume averaging will lead to a decreased FA value compared with the value that would be measured if the voxel contained a single white matter tract.

7.4.4 Image Acquisition Methods

Diffusion-weighting gradients can be inserted into several different MR image acquisition schemes, and they can be applied in the readout, phase, and slice-selection gradients simultaneously at varying magnitudes to alter the net direction of the diffusion-encoding. DW-MRI is most commonly applied to spin-echo sequences because the contribution of T_2-weighting is relatively simple. The most common of these schemes will be described here and are illustrated in Figure 7.5.

7.4.4.1 Spin Echo

The PGSE sequence described by Stejskal and Tanner [18] is the most straightforward implementation of DW-MRI since it is an extension of the familiar spin-echo experiment. An excitation pulse, typically 90°, is applied, followed by the first diffusion-weighting gradient. Then, the 180° RF refocusing pulse is applied, and the second diffusion-weighing gradient is applied. Finally, the signal is acquired so that the sampling of *k*-space is done so that the sampling of the center and the peak of the spin echo coincide. The PGSE method is sensitive to bulk subject motion and, since it is a spin-echo technique, requires a comparatively long time to acquire the data, since each line in *k*-space is acquired with a single RF excitation. This technique is useful in preclinical applications where the animals are anesthetized and immobilized. Phase errors induced by residual motion can

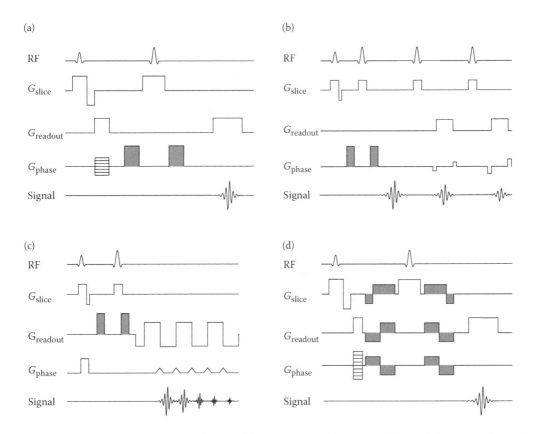

FIGURE 7.5 Common pulse sequences used for DWI: (a) PGSE, (b) fast spin-echo, (c) EPI, and (d) trace diffusion weighting. The diffusion gradients (filled) can be applied in combination on all three axes for diffusion weighting in various directions.

be corrected with the addition of navigators or navigator echoes [42], which are partial sets of *k*-space data used to track motion between diffusion weightings.

7.4.4.2 Fast Spin Echo

The fast spin-echo (FSE) method, also referred to as rapid acquisition with relaxation enhancement (RARE) or turbo spin echo (TSE), is a modified version of the spin-echo sequence that allows for significantly faster scan times with less sensitivity to motion than traditional SE sequences [43]. In this method, a single RF excitation pulse is followed by a series of RF refocusing pulses, producing a spin echo for each refocusing pulse. Each echo is spatially encoded so that multiple lines of *k*-space can be sampled after a single excitation pulse, reducing the total scan time by the number of echoes acquired after the excitation pulse.

The major advantage of FSE acquisitions is that they are relatively insensitive to B_0 field inhomogeneities and susceptibility artifacts. Although scan times are greatly reduced compared with traditional spin-echo sequences, FSE acquisitions are slower than EPI because data cannot be sampled during the application of the RF pulses. Additionally, the amount of RF power deposited in the subject is increased in FSE compared with EPI because of the application of multiple refocusing RF pulses over a relatively short amount of time. Care must be taken not to exceed normal safety guidelines, particularly at fields 3 T or higher, as RF power

increases with the square of field strength [44]. Similar to SE, FSE sequences are sensitive to motion. Additional disadvantages to FSE methods are blurring, edge enhancement, and ghosting. DW-FSE sequences violate the Carr–Purcell–Meiboom–Gill (CPMG) conditions required for FSE sequences [45], leading to phase errors between the spin echoes and the stimulated echoes, and care must be taken to avoid the effects of this violation.

7.4.4.3 Echo Planar Imaging

The most common method used to acquire DW-MRIs is single-shot echo planar imaging (EPI) [46]. In a single-shot EPI acquisition, the entire image is acquired after one RF excitation, hence the name "single shot." The sequence is designed so that the spin-echo forms just as the center of *k*-space is reached. This removes gross motion artifacts caused by dephasing. Often, only part of *k*-space is acquired, which will save even more time. The major advantages of the single-shot EPI method are that it is fast and that it is relatively immune to bulk motion artifacts. This is particularly advantageous for neuroimaging applications, where full head coverage in a very short amount of time is desired.

Unfortunately, EPI acquisitions are prone to severe image artifacts, including eddy current distortion and susceptibility artifacts. These distortions occur along the phase encoding direction in an EPI acquisition because the long readout time in the phase encoding direction results in a low effective bandwidth.

The rapid switching of the gradients induces eddy currents in the electrically conductive structures of the MRI scanner. This leads to additional magnetic fields that cause linear distortions characterized by translation, scaling, and shearing in the phase-encoding direction. Large differences in magnetic susceptibility produce local magnetic field gradients that cause nonlinear distortions in the images. The effects are most severe at interfaces between tissue and air, such as in the brain tissue surrounding the sinuses, and even between different types of tissue, such as adipose and glandular tissue in the breast. These distortions worsen as field strength increases, making it challenging to apply at field strengths greater than 1.5 T.

7.4.4.4 Trace Diffusion Weighting

DTIs and measurements of anisotropy can be obtained by acquiring DWIs in multiple directions, with diffusion weighting in a single direction per acquisition. However, if only the ADC value is needed, an alternative method, trace diffusion-weighted imaging, may be used [47,48]. In trace imaging, instead of acquiring three images with diffusion-weighting gradients played out in separate orthogonal directions for each, equal diffusion weighting is applied with bipolar gradients along three orthogonal directions during the diffusion time, Δ. In this case, Equation 7.13 takes the form

$$b = n\gamma^2 G^2 \delta^2 \left(\Delta - \frac{\delta}{3} \right) \qquad (7.27)$$

where n is the number of individual bipolar gradients applied during diffusion weighting.

The primary application for trace diffusion weighting has been in preclinical applications. For example, Moffat et al. [49] describe a scheme for implementing trace diffusion weighting in a spin-echo sequence for high throughput of animals on a 7 T scanner. EPI at high field strengths is prone to severe image distortions, and spin-echo sequences have long acquisition times. Trace diffusion weighting with a spin-echo sequence allows for a significantly shorter acquisition time (on the order of 20 min in reference [49]) without the distortions associated with EPI.

7.4.5 Fat Suppression

Regardless of the image acquisition scheme used, it is necessary to suppress the signal from protons in fat surrounding the tumor or tissue of interest. The purpose of this is twofold. First, the ADC of fat is very low compared with muscle, organs, and tumors. Contributions from fat may average with other tissues, resulting in an artificially low ADC measurement. For example, van Rijswijk et al. [50] reported an ADC of 0.2×10^{-3} mm^2/s for fat and 1.5×10^{-3} mm^2/s for soft tissue lesions. Second, the protons within fat precess at a lower frequency than water protons. This results in a shift of the signal from fat in the frequency encoding direction for most image acquisition schemes and in

the phase encoding direction for EPI acquisitions, potentially contaminating voxels within the tissue of interest. Several different techniques exist for fat suppression, and the most common techniques used in DW-MRI are short T_1 inversion recovery (STIR), spectral inversion recovery (SPIR), and spectral attenuated inversion-recovery (SPAIR).

STIR fat suppression takes advantage of the fact that the T_1 of fat is shorter than that of water. A 180° inversion RF pulse is applied prior to the image excitation pulse. This results in an inversion of the longitudinal magnetization of both water and fat. The excitation pulse is then applied at the time when the longitudinal magnetization of fat is at zero. Because the relaxation rate of water is slower, the resulting signal will be from water only. The main advantage of this technique is that it is relatively insensitive to B_0 field inhomogeneities. This method results in a decrease in the SNR because the longitudinal magnetization of water is not fully recovered at the time of the excitation. It also results in images with inverted T_1 contrast: tissues with a long T_1 will appear brighter than tissues with a short T_1.

The SPIR fat suppression technique also implements an inversion pulse prior to excitation; however, the pulse is spectrally selective to invert fat only. As in STIR, the excitation pulse is applied when the longitudinal magnetization of fat is at zero so that only water is excited. This technique is also sensitive to B_0 inhomogeneities; however it does not affect image contrast as STIR does. SPAIR is very similar to SPIR except that an adiabatic RF pulse is used to selectively invert fat. The advantage of the adiabatic pulse is that it is relatively insensitive to B_1 (RF) inhomogeneities.

7.4.6 Image Analysis Methods

The ultimate goal of acquiring DW-MRIs is often to perform a quantitative comparison of the measured diffusion parameters between two or more groups. Common examples include comparisons of patients versus healthy controls, comparisons between treatment populations, and comparisons within patient over time. The most common methods for quantitative comparison of diffusion parameters are described below. For simplicity, each method will be described in terms of comparing ADC values; however, any quantitative measure, such as FA or $\mathrm{Tr}(D)$, can be substituted.

7.4.6.1 Region of Interest

The most basic form of analysis is to calculate an average ADC value within a manually selected region of interest (ROI) and then compare that value across subject populations. This method is illustrated in Figure 7.6. The advantages of this method are that it does not require coregistration of pretreatment and posttreatment DW-MRIs for comparison and it potentially increases the SNR in the ADC calculation. This method has been used in both preclinical [51–53] and clinical studies [54–57]. Some groups have taken care to define ROIs in an attempt to avoid contamination from areas of necrosis or cysts in both preclinical [58,59] and clinical studies [60–62]. However, this method does

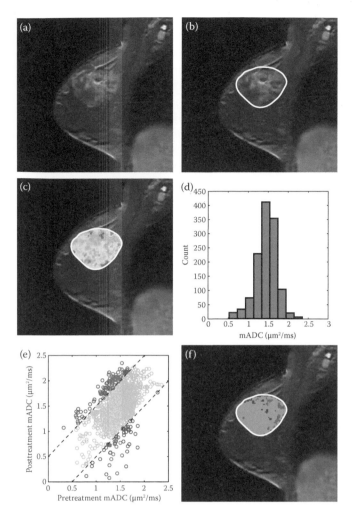

FIGURE 7.6 (See color insert.) Examples of common analysis methods in ADC of breast cancer. (a–b) Before performing any quantitative comparisons, the ROI must be defined, either manually (b) or with an automated algorithm. Typically, an anatomical image, such as a contrast enhanced T_1-weighted image (a), is used to define the ROI. (c) For a ROI analysis, the ROI is applied to the ADC map, and the statistical properties (e.g., mean value) of the ADC values within the ROI are then compared either to another ROI of healthy tissue within the same patient, across time points, or between subject populations. (d) In a histogram analysis, the ADC values within the ROI are grouped into bins, and the number of voxels is plotted as a function of the ADC range for the bin. The shape and statistical characteristics of the histograms are then compared. (e–f) fDM provides a means of comparing ADC values across time points with spatial specificity. ADC maps acquired at two different time points (e.g., pretreatment, posttreatment) first have to be aligned. Then, ADC values within the ROI at the second time point are plotted against the corresponding values at the first time point. 95% confidence intervals are then used to determine voxels within the ROI demonstrating significant increases (red) and decreases (blue) in ADC value between the two time points. This information can then be translated back to the parametric map to display where the changes occurred within the ROI.

not take tumor heterogeneity into account, potentially providing misleading results.

7.4.6.2 Histogram Analysis

Even with the careful definition of ROIs to avoid regions of necrosis and cysts, it is difficult to consistently perform this across subjects and across time points. Also, it may be of interest to know if necrotic regions or cysts have changed in volume over time. Comparison of histograms of the ADC values from ROIs between time points and/or groups is an alternative analysis method that allows for this. This method is illustrated in Figure 7.6. The frequency of a given ADC value within the ROI is plotted in order of increasing ADC value. Changes in the shape of the histogram, ADC value of the maximum frequency, and the mean ADC represent changes in the distribution of ADC values within the ROI. This method has been used in both preclinical [63–65] and clinical applications [64,62,66]. While this method provides information about the heterogeneity of the tissue of interest, one of its limitations is that there is no spatial specificity.

7.4.6.3 Functional Diffusion Mapping

A more sophisticated method was introduced by Moffat et al. [67] to analyze heterogeneous brain tumors. DW-MRIs acquired prior to and after treatment are coregistered so that similar anatomical features are located within the same voxels in each data set. ADC values are then compared on a voxelwise basis within a manually defined ROI. The posttreatment ADC value for each voxel is plotted against its pretreatment value. An additional ROI is defined in the healthy-appearing, contralateral tissue. Pretreatment and posttreatment ADC values of each voxel in this second ROI are correlated with a linear least squares analysis, providing 95% confidence intervals and standardized residuals. These values are then used to determine which voxels in the tumor ROI demonstrated a significant change in ADC value after treatment, and the voxels are then color-coded by their change: red for an increase in ADC, blue for a decrease in ADC, and green for no change. The number of voxels in each category is then normalized by the total number of voxels in the tumor ROI to give the parameters V_R, V_B, and V_G corresponding to the percentage of voxels with increased, decreased, and unchanged ADC, respectively. Differences in these parameters can then be used to discriminate patients who ultimately respond differently to therapy. This method has been applied to both preclinical [68,69] and clinical studies [70]. The major challenge in applying this technique is the need for accurate image registration across time points. This is relatively straightforward in the brain, which is surrounded by a rigid structure and where several anatomical landmarks exist. In applications outside the brain, nonlinear deformation of the tissue and a relative lack of standard anatomical landmarks make this a much more difficult task. The major advantage of this method is that it is not limited by heterogeneity within the tumor; rather, it incorporates changes in that heterogeneity to assess response.

7.4.7 Timing

Dynamic changes in the tumor environment may require accurate timing of DW-MRI acquisition to maximize its sensitivity in detection of treatment response. If cells initially swell prior to necrosis, increasing intracellular water and decreasing the ECVF extracellular water, then a reduction in ADC would be expected. Immediately after necrosis, it would be expected that ADC would increase because of the breakdown of cell walls and increase in extracellular water. Then as healthy cells repopulate the area or as fibrosis occurs, the extracellular volume fraction may decrease again leading to a decrease in ADC.

A biphasic response in tumor ADC to photodynamic therapy (PDT) on human prostate adenocarcinoma xenografts implanted in nude mice was reported by Plaks et al. [71]. Seven hours after treatment, tumors showed a significant ($p < .005$) decrease in mean ADC. ADC measurements at 24 and 48 h posttreatment showed a significant ($p < .005$) increase in mean ADC, compared with the pretreatment value. Histology revealed that tumors treated with PDT contained viable neoplastic cells with cytoplasmic vacuolation and exhibited edema 7 h posttreatment. Twenty-four hours posttreatment, tumors contained both apoptotic and necrotic cells, which led to massive necrosis in tumors 48h posttreatment.

Clinical data demonstrating changes in ADC values within 48 h of therapy are limited and conflicting. In a small study of three patients with malignant gliomas being treated with convection-enhanced taxol, Mardor et al. [54] reported an initial increase in mean tumor ADC values compared with pretherapy values just 1 to 2 days after treatment. After 2 days, mean ADC values increased and remained increased compared with baseline in the two subjects who ultimately responded to therapy. In a more robust study, Kamel et al. [72] measured tumor ADC in 24 men with hepatocellular carcinoma at six time points: prior to treatment and 24 h and 1, 2, 3, and 4 weeks after treatment. Compared to the pretreatment value, tumor ADC was increased at each subsequent time point. However, the increase was only significant ($p = .004$) at weeks 1 and 2 after treatment, and the ADC values measured at weeks 2, 3, and 4 were lower than the values measured during the previous week. The authors suggested that this transient change was the result of cellular changes occurring in response to necrosis [73]. In practice, most clinical studies acquire diffusion-weighted images 1 to 2 weeks after the start of treatment.

7.5 Applications

7.5.1 Preclinical Applications

7.5.1.1 Early Detection of Treatment Response

The usefulness of DW-MRI in the assessment of treatment response was first demonstrated by Zhao et al. [51]. They showed that ADC values in radiation-induced fibrosarcomas significantly increased prior to decrease in tumor size in response to treatment with cyclophosphamide. Chenevert et al. [58] treated intracranial 9L gliomas in rats with BCNA (Carmustine) and observed a significant increase in ADC within the tumor over the course of treatment, while the volume of the tumor continued to increase at a reduced rate, compared with untreated tumors. A 90% cell kill was achieved with the dosage administered. T_1 and T_2 values were measured pretreatment and 6–8 days posttreatment, as well. ADC increased by 50%, T1 increased by 16%, and T2 increased by 27%. Histopathologic comparison of untreated and treated tumors at 6 days posttreatment revealed increased extracellular space and a large number of apoptotic cells. Similar results have been reported in a variety of tumor models with both cytotoxic and cytostatic chemotherapeutic agents. Galons et al. [65] treated human breast cancer xenografts that were sensitive (MCF7/S) and resistant (MCF7/D40) to doxorubicin in SCID (severe combined immunodeficient) mice with Paclitaxel. Two days posttreatment, the rate of tumor growth was reduced only in the taxol-sensitive tumors, and the ADC values were increased compared with pretreatment. There was no significant difference in tumor volume between the treated and untreated tumors (in either case, sensitive or resistant). The authors concluded that the ADC value can be used to identify treatment response to taxols and discriminate between drug-sensitive and drug-resistant tumors at an earlier stage than tumor volume. Jennings et al. [65] treated LnCaP tumors derived from human prostatic adenocarcinomas with docetaxel and found increased ADC values on days 2 and 4 after the initiation of treatment. The authors were able to measure dose-dependent changes in ADC, which preceded changes in tumor volume.

7.5.1.2 Dose Dependence

The treatment response observed by Zhao et al. [51] was dependent upon dose size. Mice were divided into two treatment groups: 150 mg/kg cyclophosphamide and 300 mg/kg cylophosphamide. While tumors in both groups exhibited an increase in ADC, the increase was larger and more prolonged for the 300 mg/kg treatment group. The 300 mg/kg treatment group exhibited a maximum increase in tumor ADC of almost 100% compared with pretreatment ADC values and maintained an increase of 50% or more for days 2–9 after treatment. Whereas the 150 mg/kg group had a maximum increase in tumor ADC of over 50%, which was only maintained for days 3 and 4 after treatment.

Ross et al. have further investigated the ability of DW-MRI to detect dose-dependent responses to treatment in rat 9L glioma models with two different analysis methods [64,68]. In both studies, intracerebral 9L glioma tumors were induced in male Fischer 344 rats. The animals were then divided into three treatment groups and a control group. The three treatment groups received a single dose of BCNU on day 0 in the amounts of 6.65 mg/kg ($0.5 \times LD_{10}$), 13.3 mg/kg ($1.0 \times LD_{10}$), and 26.6 mg/kg body weight ($2.0 \times LD_{10}$), where LD_{10} is the dose at which 10% of the animals died. In the first study, mean tumor ADC values were measured prior to treatment and then every 2 days for up to 34 days depending upon survival [64]. Evaluation of the percent change in mean tumor ADC value for each group compared with its baseline value revealed that both magnitude and duration of treatment response were dose-dependent. Mean tumor ADC values did not significantly change for the control group in the days following treatment in the other groups, whereas all three treatment groups exhibited an increase in mean tumor ADC, with significant changes in ADC appearing 4 days posttreatment and no corresponding change in tumor volume. Mean tumor ADC values for the 6.65 mg/kg treatment group peaked 4 days posttreatment with an increase of over 10% compared with baseline values and then gradually decreased over days 6–12 to a value slightly lower than the baseline ADC for the group. Mean tumor ADC values for the 13.3 and 26.6 mg/kg treatment groups peaked 6 days posttreatment with an increase of over 50% compared with baseline. Mean tumor ADC then gradually decreased for the 13.3 mg/kg group on days 8–22 to values at or slightly below baseline ADC values for the group. The 26.6 mg/kg group maintained their peak ADC value for 4 days; then, values gradually decreased from days 10 to 32 to the baseline ADC value for the group. Changes in cell density, measured from histological sections acquired from a separate group of animals treated with a 26.6 mg/kg dose of BCNU, were inversely correlated with mean tumor ADC ($p = .041$). In the second study, the use of an alternative analysis method, functional diffusion mapping (fDM) [67], to measure preclinical dose escalation effects was evaluated [70]. Animals were imaged prior to treatment and every day for up to 14 days after treatment. Each posttreatment ADC map was coregistered to its corresponding pretreatment ADC map. Each voxel within the tumor was then classified into one of three groups: significantly increased ADC, significantly decreased ADC, or no change in ADC; and normalized tumor volume for each group was calculated by dividing the number of voxels in each group by the total number of voxels in the tumor. The normalized tumor volume that exhibited increased ADC was found to correlate with drug dose ($p = 5.8 \times 10^{-6}$) and with overall survival ($p = .002$).

7.5.2 Clinical Applications

7.5.2.1 Early Detection of Treatment Response

The first reported application of DW-MRI for treatment assessment in humans was in two patients with primary CNS tumors treated with both radiation and adjuvant chemotherapy [64]. Similar to results from preclinical studies, changes in tumor ADC values occurred before changes in tumor volume. Tomura et al. [57] reported that tumor ADC values were significantly increased ($p < .05$) 2–4 weeks after stereotactic radiation therapy and that the changes in ADC preceded changes in tumor size by 4 weeks. Huang et al. [73] studied 21 patients with brain metastases treated with site-specific radiation and found a significant increase ($p = .009$) in mean tumor ADC value 1 week after treatment, which was well before changes in tumor size. The authors also reported that tumors with radiation-induced central necrosis exhibited significantly higher ($p < .001$) *mADC* values than tumors with noncentral necrosis.

These results are not unique to brain tumors treated with radiation therapy. As predicted by preclinical animal studies, several types of tumors have shown increased ADC prior to changes in

volume after treatment with chemotherapy, as well. For example, a recent longitudinal study of patients with locally advanced breast cancer reported that the mean ADC of the tumors significantly ($p < .002$) increased after just one round of neoadjuvant chemotherapy and that the increase was greater in patients who ultimately responded to treatment than in those who did not [74]. Tumor volume decreased by similar amounts in both responders and nonresponders after the first cycle of treatment, leading the authors to suggest that ADC might be a valuable tool for differentiating the two groups early in therapy. Additional studies have also demonstrated increases in tumor ADC that preceded changes in tumor size after neoadjuvant treatment of breast cancer after the first cycle [62] and after completion of the full course of therapy [62].

Theilmann et al. [56] observed decreased ADC values in patients who ultimately responded to therapy 4 to 11 days after administration of chemotherapy for treatment of breast cancer metastases in the liver. These changes preceded changes in tumor volume. Lee et al. [75] followed a patient with metastatic prostate cancer in the bone initially treated with a combination of androgen blockade and bicalutamide and goserelin acetate and then an additional palliative radiation treatment 2 weeks later. Increased ADC was observed in each lesion 2 weeks after the initiation of therapy, with an increased effect 8 weeks after initiation, and without a significant change in tumor volume.

7.5.2.2 Tissue Characterization

Tumor classification and grading currently require an invasive biopsy to collect a tissue sample that is then investigated for pathological changes. This can result in tissue damage that may complicate future imaging studies. Therefore, a noninvasive procedure to report on tissue status, such as DW-MRI is highly desirable. However, the extent of the usefulness of DW-MRI may be limited because factors that cannot be measured with DW-MRI, such as the rate of cell division and abnormalities in cell structure, are also involved in tumor classification and grading. Also, structural heterogeneity makes it difficult to classify and grade tumors by ADC values alone. In spite of these shortcomings, there are examples in the literature of using ADC to successfully differentiate benign and malignant tumors.

ADC values have been shown to be useful in distinguishing malignant from benign lesions in the breast [76,61]. Rubesova et al. [61] performed a clinical prospective study of 110 breast lesions in 78 patients. Twenty-three lesions were excluded from analysis because they were too small (less than 0.7 cm), they were cystic lesions, pathology was not obtained for them, or image artifacts prevented accurate analysis. Of the remaining 87 lesions, 25% were found to be benign and 75% were found to be malignant according to pathology. Mean ADC values of viable tumor (excluding necrotic regions) were found to be significantly ($p < .001$) lower in malignant tumors (mean $\pm$ SEM = $0.95 \pm 0.021 \times 10^{-3}$ mm^2/s) than benign tumors (mean $\pm$ SEM = $1.51 \pm 0.068 \times 10^{-3}$ mm^2/s). The authors reported that a threshold value 1.13×10^{-3} mm^2/s allowed differentiation of malignant and benign breast tumors with a sensitivity of 84% and a specificity of 86%, similar to values reported by Guo et al. [76]. Although there was a significant difference between the mean ADC values of the two tumor groups, there was considerable overlap in the range of values for benign (0.95–2.36 × 10^{-3} mm^2/s) and malignant (0.65–1.49 × 10^{-3} mm^2/s) lesions.

An additional potential use for DW-MRI is the differentiation of normal tissue, necrotic tissue, and recurrent tumor. A study of 36 men with rising PSA levels, indicating recurrence, after definitive external beam radiation therapy for treatment of prostate cancer reported a significant ($p < .01$) difference in ADC values of biopsy-proven tumors (0.98 ± 0.23 × 10^{-3} mm^2/s) and benign tissue (1.60 ± 0.21 × 10^{-3} mm^2/s) [77]. Although, again, an overlap between the range of values observed for each group was observed. The study also showed that DW-MRI in combination with T_2-weighted MRI increased the sensitivity and specificity of predicting locally recurrent cancer after radiation therapy.

7.5.2.3 Outcome Prediction

As survival is the ultimate goal of therapy, a method that can accurately predict how response to therapy at an early stage correlates with clinical outcome is required. Such a method would potentially allow for changes to be made early in treatment, if necessary. The most convincing examples of the use of DW-MRI to predict clinical outcome have been fDM studies of brain tumors [67,78]. Using fDM, Moffat et al. [67] were able to correctly classify the response of 20 patients with a variety of brain tumors as partial response (PR), stable disease (SD), or progressive disease (PD) 3 weeks after the initiation of treatment with 100% specificity and sensitivity. The standard radiographic response used to determine patient response was the "crossed diameter product" of the tumor measured 4 weeks after completion of the entire course of therapy. This means that the fDM method allowed prediction of treatment response two months earlier than the standard radiographic response measurement. This study led to a subsequent investigation of treatment response in a group of 29 patients with malignant gliomas [70]. Treatment response was evaluated 3 weeks after the initiation of treatment and compared with radiographic response 10 weeks after the start of treatment. Patient responses predicted with fDM parameters 3 weeks posttreatment initiation demonstrated 75% sensitivity and 93% specificity and were shown to be a better predictor of progressive disease response than mean ADC or cross-diameter product at the same time point. The authors also found that changes in fDM parameters correlated with overall survival and time to progression. Another study of 60 patients with high-grade gliomas found that combining fDM analysis early in the course of treatment with traditional radiological response measurements provided a more accurate prediction of patient survival than either measurement alone [78].

7.6 Future Directions

As we have seen in this chapter, DW-MRI has several favorable characteristics for cancer imaging applications. It is a noninvasive imaging technique that does not require an exogenous contrast agent. It provides quantitative measures indirectly related to changes in the cellular environment related to tumor growth and treatment. Also, changes in diffusion parameters associated with effective treatment can be detected prior to gross morphological changes.

Major areas for growth in both the preclinical and clinical settings include developing a better understanding of the relationship between observed changes in diffusion parameters and changes occurring at the molecular level and improvement of and standardization of DW-MRI acquisition, image processing, and image analysis techniques.

7.6.1 Preclinical

In the preclinical setting, it is possible to apply advanced DW-MRI techniques requiring scan times that are not clinically feasible. Advancements in these techniques are allowing further investigation of the complex diffusion environment in tissue and provide a better understanding of the relationship between changes observed with DW-MRI and cellular mechanisms involved in tumor growth and treatment. Another advantage of DW-MRI in the preclinical setting is that it is compatible with high-throughput, longitudinal studies necessary in the drug development process. As drug development becomes increasingly targeted, it is important to have an imaging technique that is sensitive to changes induced by effective therapies.

7.6.1.1 Intracellular Diffusion

Recall from Equation 7.1 that the mean squared distance covered by a molecule depends upon the amount of time it is left to diffuse. In typical preclinical and clinical studies, the smallest possible diffusion time (Δ) is typically restricted by hardware limitations, so that only distances larger than the typical size of a cell are probed. However, changes in diffusion on a much smaller scale may provide additional information to our understanding of changes in intracellular organization that are occurring during tumor development and treatment. Higher b values are necessary to probe the smaller diffusion distances; however, hardware limitations prevent achievement of the necessary b values. One way to overcome that challenge is to use an oscillating gradient spin-echo (OGSE) sequence [79,80]. In the OGSE sequence, sinusoidally varying gradients of frequency f replace the pulsed gradients of the PGSE sequence, as seen in Figure 7.7. The effective diffusion time is reduced to less than the period of one oscillation [80], allowing much shorter diffusion times.

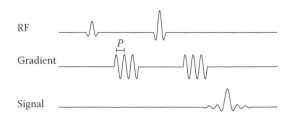

FIGURE 7.7 Instead of block-shaped pulses, oscillating gradients are applied. This allows for effective increase in the b value. The b value is related to the period of the oscillating gradients.

Recall from Equation 7.1 that a reduction of the diffusion time *t* results in a reduced diffusion distance.

The hardware necessary to produce oscillating gradients with high *b* values is readily available for animal scanners, and the potential for the use of OGSE methods to probe tumors has been demonstrated *in vivo* in a glioblastoma model in rats [81]. In this study, ADC maps calculated from a traditional PGSE acquisition with a diffusion time of 15 ms were compared with ADC maps calculated from OGSE acquisitions with effective diffusion times of 8.3, 1.7, and 1 ms. As the diffusion times decreased, the heterogeneity of tumor ADC values increased. The increase in ADC contrast within the tumors provides information that cannot be detected with conventional methods that only probe extracellular diffusion.

7.6.1.2 Drug Development

DW-MRI is an ideal candidate for use in the high-throughput, longitudinal studies that are critical for expediting the drug development process. As we have seen in this chapter, DW-MRI is sensitive to changes in tumor ADC prior to changes in tumor volume in successful treatment with traditional cytotoxic therapies, such as chemotherapy [58,63,65] and radiation [51]. However, increasing focus is being placed upon the development of molecularly targeted drugs with cytostatic [82] or antiangiogenic [83] actions as alternatives to traditional cytotoxic therapies. Successful treatment with these drugs will not necessarily produce a decrease in tumor volume. Therefore, imaging methods, such as DW-MRI, that are sensitive to changes occurring at the molecular level will be even more important in the development of these drugs.

7.6.2 Clinical

DW-MRI holds great promise for use in clinical applications, such as tumor detection, classification, and treatment assessment. However, before it becomes incorporated into standard practice, there is still much work that needs to be done, including standardization of image acquisition and analysis and the performance of large-scale, multicenter trials [31]. As these goals are actively being pursued, the research field is continuing to develop new applications and take advantage of hardware advances.

7.6.2.1 Diffusion-Weighted Whole-Body Imaging with Background Body Signal Suppression

An alternative use of DW-MRI for evaluating the entire body for tumors is diffusion-weighted whole-body imaging with background body signal suppression (DWIBS) [88]. DW-MRIs are acquired either over the entire body or a substantial ROI with relatively high diffusion weighting (e.g., 500–1000 s/mm²) to suppress tissue other than structures with low ADC. Maximum intensity projections are then computed at multiple projection angles and inverted to make low ADC structures appear hypointense to resemble the scintigraphic images acquired with positron emission tomography (PET). Fat suppression is extremely important, as lipids with low ADC may result in false positive findings. This method is exciting in that it appears to provide images comparable to those acquired with PET but without ionizing radiation.

7.6.2.2 Extensions to Fields >3 T

As the field strength of human MR scanners continues to increase beyond 3 T, the use of DW-MRI at these field strengths is beginning to be investigated. The primary advantage of increased field strength is increased SNR. However, some of the challenges of performing DW-MRI at lower field strengths are exacerbated at higher strengths.

SNR increases linearly with the square of the strength of the main magnetic field (B_0) [85]. With increased SNR, it is possible to obtain improved spatial and/or temporal resolution. Increased spatial resolution allows for better detection of small tumors and characterization of tissues with inhomogeneous diffusion properties. Improved temporal resolution allows for shorter scan times and reduces the potential for patient motion during the scan. It should be noted that because of tradeoffs between spatial and temporal resolution, it is not possible to take full advantage of increased SNR in both at the same time. It should also be noted that shortened T_2 relaxation times might counteract some of the SNR gain at higher field strengths.

The primary challenges of imaging at higher magnetic fields are related to field inhomogeneities. Recall from Section 3.4.3 that susceptibility-induced distortions in DW-EPIs worsen as field strength increases. Parallel imaging [86] may help reduce some of the susceptibility-induced distortion by effectively increasing the bandwidth in the phase encoding direction. Development of alternative image acquisition schemes that are not as sensitive to these distortions is an active area of research [87]. Inhomogeneities in the RF signal strength (B_1) are also a problem at higher field strengths. Methods for retrospective correction of image intensity variation exist, e.g., reference [88], and hardware advances now allow for transmission of multiple RF pulses tailored to produce a more homogeneous field [89].

Acknowledgments

We thank the National Institutes of Health for funding through NCI 5R01CA129961 and NCI 1U01CA142565.

References

1. Hanahan D, Weinberg RA. The hallmarks of cancer. *Cell* 2000;100(1):57–70.
2. Peterson HI, Appelgren L, Kjartansson I, Selander D. Vascular and extravascular spaces in a transplantable rat tumour after local X-ray irradiation. *Z Krebsforsch Klin Onkol Cancer Res Clin Oncol* 1976;87(1):17–25.
3. Braunschweiger PG. Effect of cyclophosphamide on the pathophysiology of RIF-1 solid tumors. *Cancer Res* 1988;48(15):4206–4210.

4. Morse DL, Galons JP, Payne CM, Jennings DL, Day S, Xia G, Gillies RJ. MRI-measured water mobility increases in response to chemotherapy via multiple cell-death mechanisms. *NMR Biomed* 2007;20(6):602–614.

5. Merboldt KD, Hanicke W, Frahm J. Self-diffusion NMR imaging using stimulated echoes. *J Magn Reson* 1985;64(3): 479–486.

6. Taylor DG, Bushell MC. The spatial mapping of translational diffusion coefficients by the NMR imaging technique. *Phys Med Biol* 1985;30(4):345–349.

7. Le Bihan D, Breton E, Lallemand D, Grenier P, Cabanis E, Laval-Jeantet M. MR imaging of intravoxel incoherent motions: Application to diffusion and perfusion in neurologic disorders. *Radiology* 1986;161(2):401–407.

8. Hahn EL. Spin echoes. *Phys Rev* 1950;80(4):580–594.

9. Carr HY, Purcell EM. Effects of diffusion on free precession in nuclear magnetic resonance experiments. *Phys Rev* 1954;94(3):630–638.

10. Einstein A. Über die von der molekularkinetischen Theorie der Wärme geforderte Bewegung von in ruhenden Flüssigkeiten suspendierten Teilchen. *Ann Phys* 1905;322:549–560.

11. Anderson AW, Xie J, Pizzonia J, Bronen RA, Spencer DD, Gore JC. Effects of cell volume fraction changes on apparent diffusion in human cells. *Magn Reson Imaging* 2000;18(6):689–695.

12. Sugahara T, Korogi Y, Kochi M, Ikushima I, Shigematu Y, Hirai T, Okuda T, Liang L, Ge Y, Komohara Y, Ushio Y, Takahashi M. Usefulness of diffusion-weighted MRI with echo-planar technique in the evaluation of cellularity in gliomas. *J Magn Reson Imaging* 1999;9(1):53–60.

13. Gauvain KM, McKinstry RC, Mukherjee P, Perry A, Neil JJ, Kaufman BA, Hayashi RJ. Evaluating pediatric brain tumor cellularity with diffusion-tensor imaging. *AJR Am J Roentgenol* 2001;177(2):449–454.

14. Kono K, Inoue Y, Nakayama K, Shakudo M, Morino M, Ohata K, Wakasa K, Yamada R. The role of diffusion-weighted imaging in patients with brain tumors. *AJNR Am J Neuroradiol* 2001;22(6):1081–1088.

15. Latour LL, Svoboda K, Mitra PP, Sotak CH. Time-dependent diffusion of water in a biological model system. *Proc Natl Acad Sci U S A* 1994;91(4):1229–1233.

16. Lang P, Wendland MF, Saeed M, Gindele A, Rosenau W, Mathur A, Gooding CA, Genant HK. Osteogenic sarcoma: Noninvasive in vivo assessment of tumor necrosis with diffusion-weighted MR imaging. *Radiology* 1998;206(1): 227–235.

17. van der Toorn A, Sykova E, Dijkhuizen RM, Vorisek I, Vargova L, Skobisova E, van Lookeren Campagne M, Reese T, Nicolay K. Dynamic changes in water ADC, energy metabolism, extracellular space volume, and tortuosity in neonatal rat brain during global ischemia. *Magn Reson Med* 1996;36(1): 52–60.

18. Stejskal EO, Tanner JE. Spin diffusion measurements— Spin echoes in presence of a time-dependent field gradient. *J Chem Phys* 1965;42(1):288–292.

19. Torrey HC. Bloch equations with diffusion terms. *Phys Rev* 1956;104:563–565.

20. Le Bihan D. *Diffusion and Perfusion Magnetic Resonance Imaging: Applications to Functional MRI*. New York: Raven Press; 1995.

21. Niendorf T, Dijkhuizen RM, Norris DG, van Lookeren Campagne M, Nicolay K. Biexponential diffusion attenuation in various states of brain tissue: Implications for diffusion-weighted imaging. *Magn Reson Med* 1996;36(6):847–857.

22. Mulkern RV, Gudbjartsson H, Westin CF, Zengingonul HP, Gartner W, Guttmann CR, Robertson RL, Kyriakos W, Schwartz R, Holtzman D, Jolesz FA, Maier SE. Multicomponent apparent diffusion coefficients in human brain. *NMR Biomed* 1999;12(1):51–62.

23. Clark CA, Le Bihan D. Water diffusion compartmentation and anisotropy at high b values in the human brain. *Magn Reson Med* 2000;44(6):852–859.

24. Bennett KM, Schmainda KM, Bennett RT, Rowe DB, Lu H, Hyde JS. Characterization of continuously distributed cortical water diffusion rates with a stretched-exponential model. *Magn Reson Med* 2003;50(4):727–734.

25. Stanisz GJ, Szafer A, Wright GA, Henkelman RM. An analytical model of restricted diffusion in bovine optic nerve. *Magn Reson Med* 1997;37(1):103–111.

26. Buckley DL, Bui JD, Phillips MI, Zelles T, Inglis BA, Plant HD, Blackband SJ. The effect of ouabain on water diffusion in the rat hippocampal slice measured by high resolution NMR imaging. *Magn Reson Med* 1999;41(1):137–142.

27. Bui JD, Buckley DL, Phillips MI, Blackband SJ. Nuclear magnetic resonance imaging measurements of water diffusion in the perfused hippocampal slice during N-methyl-D-aspartate–induced excitotoxicity. *Neuroscience* 1999;93(2): 487–490.

28. Sehy JV, Ackerman JJ, Neil JJ. Evidence that both fast and slow water ADC components arise from intracellular space. *Magn Reson Med* 2002;48(5):765–770.

29. Kingsley PB. Introduction to diffusion tensor imaging mathematics: Part III. Tensor calculation, noise, simulations, and optimization. *Conc Magn Reson A* 2006;28A(2):155–179.

30. Bito Y, Hirata S, Yamamoto E. *Optimal Gradient Factors for ADC Measurements*. Berkeley, CA: ISMRM; 1995. p. 913.

31. Padhani AR, Liu G, Mu-Koh D, Chenevert TL, Thoeny HC, Takahara T, Dzik-Jurasz A, Ross BD, Van Cauteren M, Collins D, Hammoud DA, Rustin GJ, Taouli B, Choyke PL. Diffusion-weighted magnetic resonance imaging as a cancer biomarker: Consensus and recommendations. *Neoplasia* 2009;11(2):102–125.

32. Le Bihan D, Breton E, Lallemand D, Aubin ML, Vignaud J, Lavaljeantet M. Separation of diffusion and perfusion in intravoxel incoherent motion MR imaging. *Radiology* 1988;168(2):497–505.

33. Weiss HR, Buchweitz E, Murtha TJ, Auletta M. Quantitative regional determination of morphometric indexes of the total and perfused capillary network in the rat brain. *Circ Res* 1982;51(4):494–503.

34. Turner R, Le Bihan D, Maier J, Vavrek R, Hedges LK, Pekar J. Echo-planar imaging of intravoxel incoherent motion. *Radiology* 1990;177(2):407–414.

35. Chenevert TL, Pipe JG. Effect of bulk tissue motion on quantitative perfusion and diffusion magnetic-resonance-imaging. *Magn Reson Med* 1991;19(2):261–265.

36. Bernsen HJJA, Rijken PFJW, Oostendorp T, Vanderkogel AJ. Vascularity and perfusion of human gliomas xeno-grafted in the athymic nude-mouse. *Br J Cancer* 1995;71(4): 721–726.

37. Riches SF, Hawtin K, Charles-Edwards EM, de Souza NM. Diffusion-weighted imaging of the prostate and rectal wall: Comparison of biexponential and monoexponential modelled diffusion and associated perfusion coefficients. *NMR Biomed* 2009;22(3):318–325.

38. Dietrich O, Heiland S, Sartor K. Noise correction for the exact determination of apparent diffusion coefficients at low SNR. *Magn Reson Med* 2001;45(3):448–453.

39. Basser PJ, Mattiello J, LeBihan D. MR diffusion tensor spectroscopy and imaging. *Biophys J* 1994;66(1):259–267.

40. Basser PJ, Pierpaoli C. Microstructural and physiological features of tissues elucidated by quantitative-diffusion-tensor MRI. *J Magn Reson B* 1996;111(3):209–219.

41. Jones DK. The effect of gradient sampling schemes on measures derived from diffusion tensor MRI: A Monte Carlo study. *Magn Reson Med* 2004;51(4):807–815.

42. Ehman RL, Felmlee JP. Adaptive technique for high-definition MR imaging of moving structures. *Radiology* 1989;173(1):255–263.

43. Hennig J, Nauerth A, Friedburg H. Rare imaging—A fast imaging method for clinical MR. *Magn Reson Med* 1986;3(6):823–833.

44. Collins CM, Smith MB. Signal-to-noise ratio and absorbed power as functions of main magnetic field strength, and definition of "90 degrees" RF pulse for the head in the birdcage coil. *Magn Reson Med* 2001;45(4):684–691.

45. Meiboom S, Gill D. Modified spin-echo method for measuring nuclear relaxation times. *Rev Sci Instrum* 1958;29(8):688–691.

46. Mansfield P. Multi-planar image-formation using NMR spin echoes. *J Phys C: Solid State Phys* 1977;10(3): L55–L58.

47. Mori S, Vanzijl PCM. Diffusion weighting by the trace of the diffusion tensor within a single scan. *Magn Reson Med* 1995;33(1):41–52.

48. Wong EC, Cox RW, Song AW. Optimized isotropic diffusion weighting. *Magn Reson Med* 1995;34(2):139–143.

49. Moffat BA, Hall DE, Stojanovska J, McConville PJ, Moody JB, Chenevert TL, Rehemtulla A, Ross BD. Diffusion imaging for evaluation of tumor therapies in preclinical animal models. *MAGMA* 2004;17(3–6):249–259.

50. van Rijswijk CS, Kunz P, Hogendoorn PC, Taminiau AH, Doornbos J, Bloem JL. Diffusion-weighted MRI in the characterization of soft-tissue tumors. *J Magn Reson Imaging* 2002;15(3):302–307.

51. Zhao M, Pipe JG, Bonnett J, Evelhoch JL. Early detection of treatment response by diffusion-weighted 1H-NMR spectroscopy in a murine tumour in vivo. *Br J Cancer* 1996;73(1):61–64.

52. Lemaire L, Howe FA, Rodrigues LM, Griffiths JR. Assessment of induced rat mammary tumour response to chemotherapy using the apparent diffusion coefficient of tissue water as determined by diffusion-weighted 1H-NMR spectroscopy in vivo. *MAGMA* 1999;8(1):20–26.

53. Chinnaiyan AM, Prasad U, Shankar S, Hamstra DA, Shanaiah M, Chenevert TL, Ross BD, Rehemtulla A. Combined effect of tumor necrosis factor–related apoptosis-inducing ligand and ionizing radiation in breast cancer therapy. *Proc Natl Acad Sci U S A* 2000;97(4):1754–1759.

54. Mardor Y, Roth Y, Lidar Z, Jonas T, Pfeffer R, Maier SE, Faibel M, Nass D, Hadani M, Orenstein A, Cohen JS, Ram Z. Monitoring response to convection-enhanced taxol delivery in brain tumor patients using diffusion-weighted magnetic resonance imaging. *Cancer Res* 2001;61(13):4971–4973.

55. Mardor Y, Pfeffer R, Spiegelmann R, Roth Y, Maier SE, Nissim O, Berger R, Glicksman A, Baram J, Orenstein A, Cohen JS, Tichler T. Early detection of response to radiation therapy in patients with brain malignancies using conventional and high b-value diffusion-weighted magnetic resonance imaging. *J Clin Oncol* 2003;21(6):1094–1100.

56. Theilmann RJ, Borders R, Trouard TP, Xia G, Outwater E, Ranger-Moore J, Gillies RJ, Stopeck A. Changes in water mobility measured by diffusion MRI predict response of metastatic breast cancer to chemotherapy. *Neoplasia* 2004;6(6):831–837.

57. Tomura N, Narita K, Izumi J, Suzuki A, Anbai A, Otani T, Sakuma I, Takahashi S, Mizoi K, Watarai J. Diffusion changes in a tumor and peritumoral tissue after stereotactic irradiation for brain tumors: Possible prediction of treatment response. *J Comput Assist Tomogr* 2006;30(3):496–500.

58. Chenevert TL, McKeever PE, Ross BD. Monitoring early response of experimental brain tumors to therapy using diffusion magnetic resonance imaging. *Clin Cancer Res* 1997;3(9):1457–1466.

59. McConville P, Hambardzumyan D, Moody JB, Leopold WR, Kreger AR, Woolliscroft MJ, Rehemtulla A, Ross BD, Holland EC. Magnetic resonance imaging determination of tumor grade and early response to temozolomide in a genetically engineered mouse model of glioma. *Clin Cancer Res* 2007;13(10):2897–2904.

60. Pickles MD, Gibbs P, Lowry M, Turnbull LW. Diffusion changes precede size reduction in neoadjuvant treatment of breast cancer. *Magn Reson Imaging* 2006;24(7):843–847.

61. Rubesova E, Grell AS, De Maertelaer V, Metens T, Chao SL, Lemort M. Quantitative diffusion imaging in breast cancer: A clinical prospective study. *J Magn Reson Imaging* 2006;24(2):319–324.

62. Yankeelov TE, Lepage M, Chakravarthy A, Broome EE, Niermann KJ, Kelley MC, Meszoely I, Mayer IA, Herman CR, McManus K, Price RR, Gore JC. Integration of

quantitative DCE-MRI and ADC mapping to monitor treatment response in human breast cancer: Initial results. *Magn Reson Imaging* 2007;25(1):1–13.

63. Galons JP, Altbach MI, Paine-Murrieta GD, Taylor CW, Gillies RJ. Early increases in breast tumor xenograft water mobility in response to paclitaxel therapy detected by non-invasive diffusion magnetic resonance imaging. *Neoplasia* 1999;1(2):113–117.

64. Chenevert TL, Stegman LD, Taylor JM, Robertson PL, Greenberg HS, Rehemtulla A, Ross BD. Diffusion magnetic resonance imaging: An early surrogate marker of therapeutic efficacy in brain tumors. *J Natl Cancer Inst* 2000;92(24):2029–2036.

65. Jennings D, Hatton BN, Guo J, Galons JP, Trouard TP, Raghunand N, Marshall J, Gillies RJ. Early response of prostate carcinoma xenografts to docetaxel chemotherapy monitored with diffusion MRI. *Neoplasia* 2002;4(3):255–262.

66. Pope WB, Kim HJ, Huo J, Alger J, Brown MS, Gjertson D, Sai V, Young JR, Tekchandani L, Cloughesy T, Mischel PS, Lai A, Nghiemphu P, Rahmanuddin S, Goldin J. Recurrent glioblastoma multiforme: ADC histogram analysis predicts response to bevacizumab treatment. *Radiology* 2009;252(1):182–189.

67. Moffat BA, Chenevert TL, Lawrence TS, Meyer CR, Johnson TD, Dong Q, Tsien C, Mukherji S, Quint DJ, Gebarski SS, Robertson PL, Junck LR, Rehemtulla A, Ross BD. Functional diffusion map: A noninvasive MRI biomarker for early stratification of clinical brain tumor response. *Proc Natl Acad Sci U S A* 2005;102(15):5524–5529.

68. Moffat BA, Chenevert TL, Meyer CR, McKeever PE, Hall DE, Hoff BA, Johnson TD, Rehemtulla A, Ross BD. The functional diffusion map: An imaging biomarker for the early prediction of cancer treatment outcome. *Neoplasia* 2006;8(4):259–267.

69. Lee KC, Sud S, Meyer CR, Moffat BA, Chenevert TL, Rehemtulla A, Pienta KJ, Ross BD. An imaging biomarker of early treatment response in prostate cancer that has metastasized to the bone. *Cancer Res* 2007;67(8):3524–3528.

70. Hamstra DA, Chenevert TL, Moffat BA, Johnson TD, Meyer CR, Mukherji SK, Quint DJ, Gebarski SS, Fan X, Tsien CI, Lawrence TS, Junck L, Rehemtulla A, Ross BD. Evaluation of the functional diffusion map as an early biomarker of time-to-progression and overall survival in high-grade glioma. *Proc Natl Acad Sci U S A* 2005;102(46):16759–16764.

71. Plaks V, Koudinova N, Nevo U, Pinthus JH, Kanety H, Eshhar Z, Ramon J, Scherz A, Neeman M, Salomon Y. Photodynamic therapy of established prostatic adenocarcinoma with TOOKAD: A biphasic apparent diffusion coefficient change as potential early MRI response marker. *Neoplasia* 2004;6(3):224–233.

72. Kamel IR, Liapi E, Reyes DK, Zahurak M, Bluemke DA, Geschwind JF. Unresectable hepatocellular carcinoma: Serial early vascular and cellular changes after transarterial chemoembolization as detected with MR imaging. *Radiology* 2009;250(2):466–473.

73. Huang CF, Chou HH, Tu HT, Yang MS, Lee JK, Lin LY. Diffusion magnetic resonance imaging as an evaluation of the response of brain metastases treated by stereotactic radiosurgery. *Surg Neurol* 2008;69(1):62–68.

74. Sharma U, Danishad KK, Seenu V, Jagannathan NR. Longitudinal study of the assessment by MRI and diffusion-weighted imaging of tumor response in patients with locally advanced breast cancer undergoing neoadjuvant chemotherapy. *NMR Biomed* 2009;22(1):104–113.

75. Lee KC, Bradley DA, Hussain M, Meyer CR, Chenevert TL, Jacobson JA, Johnson TD, Galban CJ, Rehemtulla A, Pienta KJ, Ross BD. A feasibility study evaluating the functional diffusion map as a predictive imaging biomarker for detection of treatment response in a patient with metastatic prostate cancer to the bone. *Neoplasia* 2007;9(12):1003–1011.

76. Guo Y, Cai YQ, Cai ZL, Gao YG, An NY, Ma L, Mahankali S, Gao JH. Differentiation of clinically benign and malignant breast lesions using diffusion-weighted imaging. *J Magn Reson Imaging* 2002;16(2):172–178.

77. Kim CK, Park BK, Lee HM. Prediction of locally recurrent prostate cancer after radiation therapy: Incremental value of 3T diffusion-weighted MRI. *J Magn Reson Imaging* 2009;29(2):391–397.

78. Hamstra DA, Galban CJ, Meyer CR, Johnson TD, Sundgren PC, Tsien C, Lawrence TS, Junck L, Ross DJ, Rehemtulla A, Ross BD, Chenevert TL. Functional diffusion map as an early imaging biomarker for high-grade glioma: Correlation with conventional radiologic response and overall survival. *J Clin Oncol* 2008;26(20):3387–3394.

79. Parsons EC, Does MD, Gore JC. Modified oscillating gradient pulses for direct sampling of the diffusion spectrum suitable for imaging sequences. *Magn Reson Imaging* 2003;21(3–4):279–285.

80. Parsons EC, Does MD, Gore JC. Temporal diffusion spectroscopy: Theory and implementation in restricted systems using oscillating gradients. *Magn Reson Med* 2006;55(1):75–84.

81. Colvin DC, Yankeelov TE, Does MD, Yue Z, Quarles C, Gore JC. New insights into tumor microstructure using temporal diffusion spectroscopy. *Cancer Res* 2008;68(14):5941–5947.

82. Pietras RJ, Pegram MD, Finn RS, Maneval DA, Slamon DJ. Remission of human breast cancer xenografts on therapy with humanized monoclonal antibody to HER-2 receptor and DNA-reactive drugs. *Oncogene* 1998;17(17):2235–2249.

83. Folkman J. Angiogenesis: An organizing principle for drug discovery? *Nat Rev Drug Discov* 2007;6(4):273–286.

84. Takahara T, Imai Y, Yamashita T, Yasuda S, Nasu S, Van Cauteren M. Diffusion weighted whole body imaging with background body signal suppression (DWIBS): Technical improvement using free breathing, STIR and high resolution 3D display. *Radiat Med* 2004;22(4):275–282.

85. Hoult DI, Richards RE. Signal-to-noise ratio of nuclear magnetic-resonance experiment. *J Magn Reson* 1976;24(1):71–85.

86. Sodickson DK, Manning WJ. Simultaneous acquisition of spatial harmonics (SMASH): Fast imaging with radiofrequency coil arrays. *Magn Reson Med* 1997;38(4):591–603.

87. Heidemann RM, Porter DA, Anwander A, Feiweier T, Heberlein K, Knosche TR, Turner R. Diffusion imaging in humans at 7T using readout-segmented EPI and GRAPPA. *Magn Reson Med* 2010;64(1):9–14.

88. Meyer CR, Bland PH, Pipe J. Retrospective correction of intensity inhomogeneities in MRI. *IEEE Trans Med Imaging* 1995;14(1):36–41.

89. Vernickel P, Roschmann P, Findeklee C, Ludeke KM, Leussler C, Overweg J, Katscher U, Grasslin I, Schunemann K. Eight-channel transmit/receive body MRI coil at 3T. *Magn Reson Med* 2007;58(2):381–389.

8

Magnetization Transfer and Chemical Exchange Saturation Transfer Imaging in Cancer Imaging

Daniel F. Gochberg
Vanderbilt University

Martin Lepage
Université de Sherbrooke

The signal underlying magnetic resonance imaging (MRI) comes primarily from the proton spin magnetic moments in freely rotating water molecules. However, in addition to water, tissue is a complicated mixture of metabolites, proteins, and large macromolecules. Since these tissue components typically have low concentration or fast relaxation, direct detection is difficult or impossible. Fortunately, molecular environment affects the water spins' relaxation rates, and the resulting image contrast is what makes conventional MRI useful. There are, however, more direct and specific techniques for probing these interactions between water and its complimentary tissue components. Magnetization transfer (MT) and chemical exchange saturation transfer (CEST) are two such methods, and in this chapter, we will examine their foundations and application to cancer imaging.

8.1 Magnetization Transfer

MT imaging has found wide application since its beginnings in 1989 [1], most notably, in detecting and characterizing multiple sclerosis [2] and as a contrast mechanism between normal tissues and blood [3] or postcontrast-enhanced tissues. It is in this last application that MT has had the greatest, although still limited, impact on cancer imaging. In this section, we begin with a qualitative introduction to MT before proceeding to a quantitative description of the technique. We then examine several illustrative applications to cancer imaging and discuss its relation to similar MRI methods.

8.1.1 Qualitative Description of MT

Typical MRI sequences do not effectively encode spatial information for several milliseconds after the excitation radiofrequency (RF) pulse. Even ultra short echo time acquisition sequences typically take hundreds of microseconds before acquiring data on clinical imaging systems. Solid macromolecular proton samples, however, have T_2 values of roughly 10 μs, meaning that the signals from these samples will have disappeared before any acquisition begins. These proton spins are not, however, completely invisible. Via MT, these solid proton spins affect the imaged water proton spins. The pool of solid proton spins is not directly imaged. Instead, the solid proton spins are indirectly detected via their effect on the liquid proton spins.

Figure 8.1 diagrams the simple two-pool model that underlies MT. Although it is only ~10% the size of the free water pool, the solid macromolecular pool can have a pronounced effect if the pulse sequence is designed to maximize sensitivity. Figure 8.2 gives an example of how selective off-resonance saturation of the solid pool indirectly affects the liquid pool. The solid pool has a wide bandwidth due to its short T_2. Off-resonance saturation has minimal direct effect on the liquid pool but can significantly reduce the solid pool magnetization. Due to the coupling between the pools, as the solid pool is irradiated, the liquid pool magnetization decays with a time constant T_{1sat}. Such a sequence is used to generate MT contrast or to calculate the magnetization transfer ratio (MTR), a semiquantitative measure of the MT of the sample:

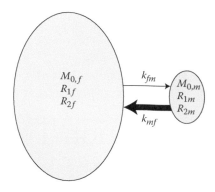

FIGURE 8.1 The MT model. The f and m subscripts represent free water and macromolecular protons, respectively. M_0, R_1, and R_2 are the sizes, longitudinal relaxation rates, and transverse relaxation rates. k_{fm} is the coupling rate from the free water to the macromolecular protons. k_{mf} is the rate in the reverse direction. Typically, $M_{0,f} \gg M_{0,m}$ and $k_{mf} \gg k_{fm}$, which are represented graphically by the size of the pools and connecting arrows, respectively.

$$MTR = \frac{M_0 - M_{sat}}{M_0} \quad (8.1)$$

where M_{sat} and M_0 are the signals with and without off-resonance saturation. MTR provides a number between 0 and 1 that is related to the size of the solid pool. This metric has found widespread use in assessing myelin content in white matter, but only limited use in cancer studies, as discussed below. MTR is only semiquantitative since its value depends on several sample relaxation and exchange rate characteristics. The relative signal weighting of each sample parameter depends on the frequency and power of the saturation pulse, rather than solely on the sample. Often, this saturation is achieved via short pulses due to amplifier and acquisition time limitations, rather than a long continuous wave saturation, and then the shape and repetition time of the pulses further complicate the MTR dependence on the underlying biology. These issues can be partially overcome by careful adherence to a standard imaging protocol across imaging sites [4] and correcting for variations in the applied field strength [5].

However, even if the same pulse sequence is uniformly applied, MTR is a function of all the relaxation and exchange parameters listed in Figure 8.1 and, in that sense, is not a specific measure of only the MT aspects of the tissue. The dependency

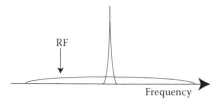

FIGURE 8.2 An MT experiment. The wide macromolecular resonance is saturated with RF irradiation. The narrow free water protons are indirectly saturated via their coupling to the macromolecules.

can be simplified in the ideal case of complete saturation of the macromolecular pool and no direct saturation of the free liquid pool, which is extremely difficult to achieve experimentally. In this limiting case,

$$MTR \rightarrow \frac{k_{fm}}{k_{fm} + R_{1f}} \quad (8.2)$$

where k_{fm} is the magnetization exchange rate from the free water pool to the macromolecular solid pool, k_{mf} is the rate in the reverse direction, and R_{1f} is the longitudinal relaxation rate of the free water pool if there were no MT. When k_{fm} is large, MTR approaches 1, whereas when R_{1f} is large, MTR approaches zero. As has been noted previously [6] and as Equation 8.2 makes clear, it is the relative size of these two rates that dictates the MT effect of off-resonance saturation on the measured signal in the ideal case.

8.1.2 Quantitative Description of MT

To separate the contrast due to MT vs. relaxation, a more quantitative (and potentially more pathologically specific) approach to tissue modeling is necessary. The MT tissue model illustrated in Figure 8.1 is equivalent to a set of coupled differential equations:

$$\frac{d}{dt}\frac{M_{x,f}}{M_{0,f}} = -R_{2f}\frac{M_{x,f}}{M_{0,f}} - \Delta M_{y,f} \quad (8.3)$$

$$\frac{d}{dt}\frac{M_{y,f}}{M_{0,f}} = -R_{2f}\frac{M_{y,f}}{M_{0,f}} + \Delta M_{x,f} + \omega_1\frac{M_{z,f}}{M_{0,f}} \quad (8.4)$$

$$\frac{d}{dt}\frac{M_{z,f}}{M_{0,f}} = -R_{1f}\left(\frac{M_{z,f}}{M_{0,f}} - 1\right) - k_{fm}\left(\frac{M_{z,f}}{M_{0,f}} - \frac{M_{z,m}}{M_{0,m}}\right) - \omega_1\frac{M_{y,f}}{M_{0,f}} \quad (8.5)$$

$$\frac{d}{dt}\frac{M_{z,m}}{M_{0,m}} = -R_{1m}\left(\frac{M_{z,m}}{M_{0,m}} - 1\right) - k_{mf}\left(\frac{M_{z,m}}{M_{0,m}} - \frac{M_{z,f}}{M_{0,f}}\right)$$
$$- W\left(\Delta, \omega_1, R_{2m}\right)\frac{M_{z,m}}{M_{0,m}} \quad (8.6)$$

The notation follows that used in Figure 8.1, where the subscripts f and m refer to the free water and solid macromolecular proton pools, respectively. R_1 and R_2 are the longitudinal and transverse relaxation rates, respectively, and M_0 is the equilibrium magnetization. ω_1 and Δ are the irradiation amplitude and the offset frequency, respectively. The free water pool dynamics is described by three equations for the x, y, and z magnetization components. The macromolecular pool dynamics are described by a single equation for the z magnetization, since its rapid transverse signal decay makes computation of its x and y magnetization components unnecessary. The macromolecular saturation effects, which normally would be implicitly modeled by the transverse magnetization equations, are instead incorporated into the z component equation by adding a saturation term W. In addition to reducing the number of equations, this inclusion has the added benefit of allowing flexible, more realistic macromolecular pool lineshapes, such as Gaussian or super Lorentzian, which better model the frequency dependence of the irradiation effect on the solid pool.

The sample parameters can be individually determined by numerically fitting the parameter values to a model of long continuous wave [7] or short, repetitive pulsed [8] irradiation at a range of powers and frequency offsets or the transient response to a selective inversion pulse [9]. Such methods are referred to as quantitative MT (qMT). For example, the ratio of the sizes of pools $= M_{0,m}/M_{0,f} = k_{fm}/k_{mf}$ has been determined *in vivo* using such methods and utilized as a more specific measure of demyelination than conventional relaxation measures (e.g., reference [10]). We discuss below the limited application of such methods to cancer imaging, where the underlying basis for changes in the sample parameters is not fully understood.

8.1.3 Magnetization Transfer versus Relaxation

Magnetization transfer is the coupling between the proton spins in the liquid and solid components of tissue. Such coupling may also play a significant role in relaxation. For example, the liquid spins in white matter are likely strongly coupled to the solid spins via the cholesterol, cerebrosides, and phosphatidylcholine [11,12], and this coupling causes both greater MT and faster relaxation times, in comparison to gray matter. In particular, M_{sat}/M_0 MT-weighted contrast is similar to T_2-weighted contrast since MT and transverse relaxation have a similar dependence on molecular motions.

There is, however, a fundamental difference between MT and relaxation. MT refers to the cross-relaxation effect: the dynamics of the water proton spins depends on the current state of the solid proton spins. This dependence of the liquid spin evolution on the current state of the solid spins differentiates MT from simple relaxation. It is also what makes MT useful. T_1 and T_2 are not sequence dependent. MT effects, on the other hand, can be manipulated. For example, they can be effectively turned on and off by saturating far off-resonance or not, and it is the ratio of the images with and without the MT effect that forms the basis of MTR.

8.1.4 MT in Cancer Imaging

MT has found its greatest application in studies of demyelinating diseases of the central nervous system, such as multiple sclerosis. It is not surprising, then, that cancer imaging studies employing MT have primarily been targeted at brain cancer, with additional notable work in breast cancer.

8.1.4.1 Breast Cancer

Imaging breast cancer with MRI is complicated by the inherent heterogeneity of breast tissue. Imaging based on single exponential T_1 and T_2 relaxation times have difficulty distinguishing normal tissue from lesions. Instead, it is the shape and time evolution of signal enhancement due to a gadolinium-based contrast injection that has generated the most interest, and such methods are discussed in Chapters 9–12. The primary use of MT imaging of breast cancer has been to aid in delineating areas of postcontrast enhancement [13,14]. The enhancement due to the gadolinium-based paramagnetic contrast agents is due to a shortening of T_1, and this T_1 shortening mechanism is independent of MT. In addition, since the size of the MT effect due to off-resonance saturation is inversely related to R_1 (see Equation 8.2), the increased R_1 value in enhanced regions will mitigate any MT effects. Fibroglandular tissue, on the other hand, can be suppressed by up to 50% by off-resonance saturation MT pulses [15]. These effects combine to increase the contrast between the enhanced region and surrounding healthy appearing tissue when applying off-resonance saturation. The increase in relative contrast agent enhancement can be utilized for simple contrast studies, dynamic studies, or, as in Leong et al. [16], studies of morphology. They utilized a 3D sequence that combined water selectivity (without fat excitation) with MT solid pool saturation postcontrast agent injection to determine lesion morphology. Their improved measures were 94% specific and 54% sensitive for malignancy.

Even with these benefits, MT methods are not currently widely used in clinical breast imaging [17]. A contributing factor is the rise of sequences like Rotating Delivery of Excitation Off Resonance (RODEO) [18], a fat-suppressed T_1-weighted method developed by largely the same group that earlier initiated MT breast imaging. They found that RODEO, in comparison to MT weighted sequences, had greater T_1-weighting, shorter acquisition times, and avoided unspecified artifacts ascribed to the MT pulse. Note that inadvertent MT effects still occur in simple gradient echo sequences (such as RODEO) [19], although likely of smaller magnitude than in sequences with off-resonance saturation pulses.

In addition to increasing postinjection contrast, MT may be able to characterize lesions. This characterization may complement dynamic measures of contrast agent signal enhancement, since such measures have difficulty distinguishing malignant lesions from, for example, benign fibroadenomas [20]. Callicott et al. [21] demonstrated *in vitro* that T_{1sat} discriminates neoplastic from normal breast tissue better than T_1, although it still cannot discriminate between benign and malignant disease. Bonini et al. [22], on the other hand, found a statistically significant difference in the MTR of benign and malignant lesions *in vivo*. They found that malignant lesions have a smaller MTR, which they ascribed to biochemical changes, possibly in proteolytic and enzyme inhibitor activity. This result hints that MT may yet play a larger role in lesion characterization.

8.1.4.2 Brain Cancer

Similarly to MT imaging of breast cancer, MT imaging of brain cancer is employed as a visualization aid in injected contrast studies. It is more rarely used as a means to characterize tissue. MT pulses suppress the signal from all brain tissue, but it suppresses signal from tumors to a smaller degree [23], hence increasing normal-tumor contrast [24,25], potentially to a greater degree than is the case for purely relaxation-based imaging [26]. Kurki et al. [27] initiated MT contrast studies in intracranial tumors at 0.1 T, finding that the application of an MT pulse increased the contrast-to-noise ratio (CNR) for both precontrast and postcontrast images. The greatest improvement in CNR occurred in lesions with the greatest postcontrast enhancement, indicating that it may be the increase in R_1 more than the MT rate k_{fm} that is responsible for the CNR improvement. Likewise, the same group [28] demonstrated that MTR could distinguish the grade of malignancy in astrocytomas and that the MTR value correlated strongly with collagen content. However, R_1 also distinguishes grades of malignancy and, while Lundbom [28] suggests that the magnetization transfer contrast (MTC) has a higher potential than R_1 in differentiating malignancy grades, it is unclear if MTC changes, in fact, are independent of these R_1 changes. Since relaxation times have limited ability to distinguish tumors [29–31], distinguishing MT from relaxation effects is essential.

qMT imaging (which can distinguish MT and relaxation effects) has had very limited application to cancer studies. Quesson et al. [32] applied qMT to rats with implanted C6 glioma cells. They found that the resulting tumors had a smaller solid proton pool size than neighboring brain regions. This result was unexpected, given the increased cell density in the tumor. One possible explanation is that only a small fraction of the cancer cell macromolecules are efficiently coupled to the free water. Quesson et al. also found that the coupling rate and solid pool transverse relaxation rates are more rapid than neighboring tissues, possibly indicating a more immobilized macromolecular structure. Although any conclusions are tentative, similar coupling rate and solid proton pool size results were found in a human glioma study [33].

8.2 Chemical Exchange Saturation Transfer

CEST and MT employ similar tissue models and experimental protocols. Both CEST and MT are based on two-pool models of tissue: free water and solid macromolecular protons for MT and free water and labile metabolite protons for CEST. Experimentally, both CEST and MT primarily use off-resonance irradiation in order to selectively saturate the second pool while measuring the resulting decrease in the coupled free water signal. In this way, the signal reflects the magnitude of the coupling between the pools. There are, however, significant differences between CEST and MT, and in this section, we give qualitative and quantitative explanations of CEST and examine its application to cancer imaging.

8.2.1 Qualitative Description of CEST

Since CEST imaging contrast reflects exchanging metabolite protons, it is a form of molecular imaging and is, in this respect, similar to localized magnetic resonance (MR) spectroscopy [34].

MR spectroscopy is a well-developed field with widespread applications, most significantly in metabolic disorders and oncology (see Chapter 9). However, conventional MR spectroscopic imaging suffers from low sensitivity, necessitating large imaging voxels and long acquisition times. CEST is a more recent development (almost entirely in the last decade) that provides a means for dramatically increased sensitivity to metabolites via their exchange of magnetization with water. This indirect detection can increase sensitivity by up to five orders of magnitude over direct MR spectroscopy [35] and has already been employed in studies of gliomas [36–40] and ischemia [41–44]. In addition, this detection via chemical exchange allows for the determination of exchange rates in addition to metabolite concentration, facilitating unique applications such as pH imaging [41–43,45–47]. In light of these initial applications, CEST imaging has the potential for great significance in research and clinical cancer studies.

CEST consists of measuring the remaining water MR signal after off-resonance saturation at several frequency offsets. CEST contrast is the normalized difference in the water signal when irradiating at the metabolite resonance and on the opposite side of the water peak. For example, if the metabolite resonates at a frequency 200 Hz less than the water peak, the CEST "contrast" is the normalized water signal difference when irradiating at 200 and –200 Hz relative to the water peak. This approach depends on the metabolite having a narrow peak (unlike the MT solid pool), so that the metabolite is unaffected by irradiation on the opposite side. In practice, a range of offsets are acquired to ensure capturing the metabolite and water resonant frequencies, given that static B_0 field inhomogeneities make these frequencies spatially varying.

CEST experiments apply irradiation between 1 and ~50 ppm from the water peak (roughly 60 Hz to 3 kHz at 1.5 T) depending on the resonant frequency of the targeted metabolite exchanging sites, such as hydroxyls, amines, and amides. The greatest frequency shifts are for paramagnetic agents that shift the exchanging site resonance by tens of parts per million. Diamagnetic agents, including endogenous metabolites, have shifts of a few parts per million. (This compares to an offset of a couple kilohertz in typical MT experiments, independent of field strength.)

The simplest and most spectrally selective acquisition method uses long (several seconds) irradiation pulses. However, such pulses are not possible on clinical imaging systems due to specific absorption rate (SAR) and RF amplifier limitations, and this limitation is typically dealt with by using a series of short (tens of milliseconds) RF pulses.

As noted above, tissues are most often modeled as two coupled pools, a single metabolite and free water [48], and direct saturation of the water pool is also often ignored or simplified (see, for example, the review of methods by Zhou and van Zijl [49]). However, this simple model does not account for additional overlapping metabolites, uncertainties in the free water resonance [50], or background macromolecular solid-like protons, such as those measured in MT experiments. Asymmetry in this background macromolecular pool with respect to the free water resonance is particularly problematic, appearing in some studies [51], but not in others [44].

8.2.2 Quantitative Description of CEST

A goal of quantitative CEST studies is to isolate the underlying biophysical changes that occur in pathology. There have been several CEST studies that quantify metabolite concentration and/or exchange by taking signal ratios of multiple CEST agents [45,52] or assuming values for the bulk of the tissue parameters [36,53]. A general fit is difficult due to the large number of free parameters. (For example, a tissue with only a single exchanging metabolite and nonsymmetric coupled solid-state protons is characterized by 14 parameters: 6 relaxation rates, 4 exchange rates, 3 frequencies, and 1 magnetization magnitude. Additionally, the applied field strength B_1 needs to be experimentally determined, which, along with the static field B_0, may be highly inhomogeneous, especially at high fields.) Oftentimes in quantitative studies, the background macromolecular pool is ignored [54], approximated, or fit in isolation from its effects on metabolite parameter determination [51]. Finally, there has been limited validation of these existing partially quantitative approaches.

As noted above, modeling of the CEST signal is similar to that of MT, and the MT tissue model illustrated in Figure 8.1 can also serve for CEST. However, while the exchanging macromolecular pool in MT is roughly 10% the size of the free water pool, the exchanging protons in CEST are typically several orders of magnitude less concentrated. The metabolites also have relatively long T_2 values (on the order of tens of milliseconds), meaning that their transverse magnetization cannot be ignored. Hence quantitative models are based on six coupled differential equations, three for the water (similar to Equations 8.3 through 8.5) and three for the coupled metabolite. An additional three coupled differential equations are required for each additional metabolite, and a single additional equation (like Equation 8.6) is required to model the background solid pool.

8.2.3 CEST in Cancer Imaging

8.2.3.1 Amide Proton Transfer

Amide proton transfer (APT) imaging is a type of CEST imaging that focuses on water exchange with endogenous amides in mobile proteins and peptides with a 3.5-ppm frequency offset from water. Amides are a defining feature of peptides and proteins. Noninvasive *in vivo* detection of amides is, therefore, related to the total protein concentration and may provide a coarse means for proteomic-like and cell-density information. This sensitivity has made applications to cancer one of the first areas of APT application [36,37,39,40], with a focus on determining tumor extent. Zhou et al. [36] imaged rat 9L gliosarcomas, but the amide exchange effect in tumors was only apparent when taking the difference in the APT signal of the tumor and normal regions. Jones et al. [37] extended APT to human brain tumor imaging at 3 T, with a focus on separating tumor and edematous regions. Zhou et al. [39] also examined human brain tumors, with a focus on improving the signal to noise ratio (SNR) efficiency and imaging robustness in the presence of B_0

inhomogeneities. Wen et al. [40] focused on the heterogeneity of human tumors with corresponding histopathology, separating tumor cores, peritumoral edema, necrotic regions, cystic cavities, and normal appearing white matter.

To simplify the data interpretation, variations in APT contrast have typically been ascribed to a single dominant source. In rat brain 9L gliosarcomas [36], these changes are ascribed to variations in amide concentration. (In rat brain ischemia [41,55], by contrast, these changes are ascribed to variations in amide exchange.) More likely, a spectrum of changes occur, reflecting differing biomechanical processes, such as edema, vascularization and tissue composition changes, and biochemical changes, such as variations in lactic acid and bicarbonate levels, with corresponding effects on pH [56]. The relative contribution of each biophysical process to the CEST contrast is not settled.

8.2.3.2 Paramagnetic CEST

Generation of CEST contrast places restrictions on the agents that can be used to obtain contrast through this mechanism. The requirement to saturate the spins associated with the agent implies that the exchange site must be chemically shifted from that of water so that direct saturation of water is not a significant contribution to signal loss. Additionally, the exchanging proton must reside on the agent for a time scale amenable to RF saturation. This essentially requires the chemical exchange rate to be slower than the chemical shift difference (in Hz) between the proton while it is on the agent and the proton while it is on water. It is, therefore, advantageous to maximize this chemical shift difference. To do this, some groups [57–59] have modified typical CEST agents with lanthanide ions (other than the efficient T_1 agent, Gd^{3+}) to impart a paramagnetic shift on the proton while it resides on the agent. The additional hyperfine shift increases the range of chemical exchange rates than may be probed using these (paramagnetic CEST, or PARACEST) agents.

In conjunction with the recent trend toward higher field MR systems, CEST and PARACEST techniques offer uniquely complementary means to obtain physiologically based MR contrast. The majority of the CEST and PARACEST agents currently described in the literature are always "on" and will generate contrast when the requisite RF conditions are met. However, Yoo et al. [59] have recently demonstrated the design of a "smart" PARACEST agent that responds to enzyme activity showing that chemical environments or enzyme action may be used to modulate accessibility of the exchange site or the proximity of the lanthanide ion to the exchange site in a predictable manner. Likewise, PARACEST agents have been developed that respond to environmental factors such as pH, temperature, zinc, glucose (reviewed in references [60,61]), and nitric oxide [62]. For example, noncovalent binding of a PARACEST agent to L-lactate shifts the exchangeable proton frequency [63]. A recent interesting development is the inclusion of small gadolinium complexes into liposomes. Whereas water exchange for the coordination sphere of the gadolinium complex would be too fast in solution, the encapsulation inside a liposome restricts the exchange to water molecules contained within the liposome, such that the liposome becomes a PARACEST agent, called a LIPOCEST agent [64].

While these methods have not yet been directly applied to cancer studies, their physiologic sensitivity has exciting potential.

8.3 Summary

In summary, MT and CEST are two ways to image the effects of magnetization exchange on the bulk tissue water signal. MT is sensitive to magnetization exchange with the macromolecular components of tissue, while CEST is sensitive to exchange with rapidly tumbling endogenous and exogenous agents. While the MT applications to cancer imaging are promising, the field is much less developed than, say, MT application to central nervous system pathologies, specifically multiple sclerosis. Protocols have not been optimized or fully tested. More importantly, the underlying biological basis of changes in MT parameters in cancer is unclear. CEST, on the other hand, is a more recent, but more rapidly developing area than MT. Its connection to molecular, and in particular protein, imaging makes CEST cancer applications especially promising.

References

1. Wolff SD, Balaban RS. Magnetization transfer contrast (MTC) and tissue water proton relaxation in vivo. *Magn Reson Med* 1989;10:135–144.
2. Ropele S, Fazekas F. Magnetization transfer MR imaging in multiple sclerosis. *Neuroimaging Clin North Am* 2009;19:27–36.
3. Mehta RC, Pike GB, Enzmann DR. Magnetization transfer magnetic resonance imaging: A clinical review. *Topics Magn Reson Imaging* 1996;8:214–230.
4. Silver NC, Barker GJ, Miller DH. Standardization of magnetization transfer imaging for multicenter studies. *Neurology* 1999;53:S33–S39.
5. Ropele S, Filippi M, Valsasina P, Korteweg T, Barkhof F, Tofts PS, Samson R, Miller DH, Fazekas F. Assessment and correction of B-1–induced errors in magnetization transfer ratio measurements. *Magn Reson Med* 2005;53:134–140.
6. Henkelman RM, Stanisz GJ, Graham SJ. Magnetization transfer in MRI: A review. *NMR Biomed* 2001;14:57–64.
7. Henkelman RM, Huang X, Xiang Q, Stanisz GJ, Swanson SD, Bronskill MJ. Quantitative interpretation of magnetization transfer. *Magn Reson Med* 1993;29:759–766.
8. Sled JG, Pike GB. Quantitative imaging of magnetization transfer exchange and relaxation properties in vivo using MRI. *Magn Reson Med* 2001;46:923–931.
9. Gochberg DF, Gore JC. Quantitative imaging of magnetization transfer using an inversion recovery sequence. *Magn Reson Med* 2003;49:501–505.
10. Narayanan S, Francis SJ, Sled JG, Santos AC, Antel S, Levesque I, Brass S, Lapierre Y, Sappey-Marinier D, Pike GB, Arnold DL. Axonal injury in the cerebral normal-appearing white matter of patients with multiple sclerosis is related to concurrent demyelination in lesions but not to concurrent demyelination in normal-appearing white matter. *Neuroimage* 2006;29:637–642.
11. Kucharczyk W, Macdonald PM, Stanisz GJ, Henkelman RM. Relaxivity and magnetization transfer of white matter lipids at MR imaging: Importance of cerebrosides and pH. *Radiology* 1994;192:521–529.
12. Koenig SH. Cholesterol of myelin is the determinant of gray-white contrast in MRI of brain. *Magn Reson Med* 1991;20:285–291.
13. Pierce WB, Harms SE, Flamig DP, Griffey RH, Evans WP, Hagans JE. 3-Dimensional gadolinium-enhanced MR imaging of the breast—Pulse sequence with fat suppression and magnetization transfer contrast—Work in progress. *Radiology* 1991;181:757–763.
14. Flamig DP, Pierce WB, Harms SE, Griffey RH. Magnetization transfer contrast in fat-suppressed steady-state 3-dimensional MR images. *Magn Reson Med* 1992;26:122–131.
15. Santyr GE, Kelcz F, Schneider E. Pulsed magnetization transfer contrast for MR imaging with application to breast. *J Magn Reson Imaging* 1996;6:203–212.
16. Leong CS, Daniel BL, Herfkens RJ, Birdwell RL, Jeffrey SS, Ikeda DM, Sawyer-Glover AM, Glover GH. Characterization of breast lesion morphology with delayed 3DSSMT: An adjunct to dynamic breast MRI. *J Magn Reson Imaging* 2000;11:87–96.
17. Kuhl C. The current status of breast MR imaging—Part I. Choice of technique, image interpretation, diagnostic accuracy, and transfer to clinical practice. *Radiology* 2007;244:356–378.
18. Harms SE, Flamig DP, Hesley KL, Meiches MD, Jensen RA, Evans WP, Savino DA, Wells RV. MR-imaging of the breast with rotating delivery of excitation off resonance—Clinical-experience with pathological correlation. *Radiology* 1993; 187:493–501.
19. Ou XW, Gochberg DF. MT effects and T-1 quantification in single-slice spoiled gradient echo imaging. *Magn Reson Med* 2008;59:835–845.
20. Stelling CB, Powell DE, Mattingly SS. Fibroadenomas—Histopathologic and MR imaging features. *Radiology* 1987;162:399–407.
21. Callicott C, Thomas JM, Goode AW. The magnetization transfer characteristics of human breast tissues: An in vitro NMR study. *Physics in Medicine and Biology* 1999;44:1147–1154.
22. Bonini RHM, Zeotti D, Saraiva LAL, Trad CS, Filho JMS, Carrara HHA, de Andrade JM, Santos AC, Muglia VF. Magnetization transfer ratio as a predictor of malignancy in breast lesions: Preliminary results. *Magn Reson Med* 2008;59:1030–1034.
23. Mehta RC, Pike GB, Haros SP, Enzmann DR. Central nervous system tumor, infection, and infarction—Detection with gadolinium-enhanced magnetization transfer MR imaging. *Radiology* 1995;195:41–46.
24. Finelli DA, Hurst GC, Gullapalli RP. T1-weighted three-dimensional magnetization transfer MR of the brain: Improved lesion contrast enhancement. *Am J Neuroradiol* 1998;19:59–64.

25. Finelli DA, Hurst GC, Gullapali RP, Bellon EM. Improved contrast of enhancing brain lesions on postgadolinium, T1-weighted spin-echo images with use of magnetization-transfer. *Radiology* 1994;190:553–559.

26. Lemaire L, Franconi F, Saint-Andre JP, Roullin VG, Jallet P, Le Jeune JJ. High-field quantitative transverse relaxation time, magnetization transfer and apparent water diffusion in experimental rat brain tumour. *NMR Biomed* 2000;13:116–123.

27. Kurki TJI, Niemi PT, Lundbom N. Gadolinium-enhanced magnetization transfer contrast imaging of intracranial tumors. *J Magn Reson Imaging* 1992;2:401–406.

28. Lundbom N. Determination of magnetization transfer contrast in tissue—An MR imaging study of brain tumors. *Am J Roentgenol* 1992;159:1279–1285.

29. Just M, Thelen M. Tissue Characterization with T1, T2, and proton density values—Results in 160 patients with brain tumors. *Radiology* 1988;169:779–785.

30. Komiyama M, Yagura H, Baba M, Yasui T, Hakuba A, Nishimura S, Inoue Y. MR imaging—Possibility of tissue characterization of brain tumors using T1 and T2 values. *Am J Neuroradiol* 1987;8:65–70.

31. Kjaer L, Thomsen C, Gjerris F, Mosdal B, Henriksen O. Tissue characterization of intracranial tumors by MR imaging—In vivo evaluation of T1-relaxation and T2-relaxation behavior at 1.5 T. *Acta Radiol* 1991;32: 498–504.

32. Quesson B, Bouzier AK, Thiaudiere E, Delalande C, Merle M, Canioni P. Magnetization transfer fast imaging of implanted glioma in the rat brain at 4.7 T: Interpretation using a binary spin-bath model. *J Magn Reson Imag* 1997;7:1076–1083.

33. Yarnykh VL. Pulsed Z-spectroscopic imaging of cross-relaxation parameters in tissues for human MRI: Theory and clinical applications. *Magn Reson Med* 2002;47: 929–939.

34. van der Graaf M. In vivo magnetic resonance spectroscopy: Basic methodology and clinical applications. *Eur Biophys J* 2010;39:527–540.

35. Goffeney N, Bulte JW, Duyn J, Bryant LH, Jr., van Zijl PC. Sensitive NMR detection of cationic-polymer–based gene delivery systems using saturation transfer via proton exchange. *J Am Chem Soc* 2001;123:8628–8629.

36. Zhou J, Lal B, Wilson DA, Laterra J, van Zijl PC. Amide proton transfer (APT) contrast for imaging of brain tumors. *Magn Reson Med* 2003;50:1120–1126.

37. Jones CK, Schlosser MJ, van Zijl PC, Pomper MG, Golay X, Zhou J. Amide proton transfer imaging of human brain tumors at 3T. *Magn Reson Med* 2006;56:585–592.

38. Salhotra A, Lal B, Laterra J, Sun PZ, van Zijl PC, Zhou J. Amide proton transfer imaging of 9L gliosarcoma and human glioblastoma xenografts. *NMR Biomed* 2008;21:489–497.

39. Zhou J, Blakeley JO, Hua J, Kim M, Laterra J, Pomper MG, van Zijl PC. Practical data acquisition method for human brain tumor amide proton transfer (APT) imaging. *Magn Reson Med* 2008;60:842–849.

40. Wen Z, Hu S, Huang F, Wang X, Guo L, Quan X, Wang S, Zhou J. MR imaging of high-grade brain tumors using endogenous protein and peptide-based contrast. *Neuroimage* 2010;51:616–622.

41. Zhou J, Payen JF, Wilson DA, Traystman RJ, van Zijl PC. Using the amide proton signals of intracellular proteins and peptides to detect pH effects in MRI. *Nat Med* 2003;9:1085–1090.

42. Sun PZ, Zhou J, Huang J, van Zijl P. Simplified quantitative description of amide proton transfer (APT) imaging during acute ischemia. *Magn Reson Med* 2007;57:405–410.

43. Jokivarsi KT, Grohn HI, Grohn OH, Kauppinen RA. Proton transfer ratio, lactate, and intracellular pH in acute cerebral ischemia. *Magn Reson Med* 2007;57:647–653.

44. Sun PZ, Murata Y, Lu J, Wang X, Lo EH, Sorensen AG. Relaxation-compensated fast multislice amide proton transfer (APT) imaging of acute ischemic stroke. *Magn Reson Med* 2008;59:1175–1182.

45. Ward KM, Balaban RS. Determination of pH using water protons and chemical exchange dependent saturation transfer (CEST). *Magn Reson Med* 2000;44:799–802.

46. Sun PZ, Sorensen AG. Imaging pH using the chemical exchange saturation transfer (CEST) MRI: Correction of concomitant RF irradiation effects to quantify CEST MRI for chemical exchange rate and pH. *Magn Reson Med* 2008;60:390–397.

47. Sun PZ, Benner T, Kumar A, Sorensen AG. Investigation of optimizing and translating pH-sensitive pulsed-chemical exchange saturation transfer (CEST) imaging to a 3T clinical scanner. *Magn Reson Med* 2008;60:834–841.

48. Ward KM, Aletras AH, Balaban RS. A new class of contrast agents for MRI based on proton chemical exchange dependent saturation transfer (CEST). *J Magn Reson* 2000; 143:79–87.

49. Zhou JY, van Zijl PCM. Chemical exchange saturation transfer imaging and spectroscopy. *Progr Nucl Magn Reson Spectrosc* 2006;48:109–136.

50. Kim M, Gillen J, Landman BA, Zhou JY, van Zijl PCM. Water saturation shift referencing (WASSR) for chemical exchange saturation transfer (CEST) experiments. *Magn Reson Med* 2009;61:1441–1450.

51. Hua J, Jones CK, Blakeley J, Smith SA, van Zijl PCM, Zhou JY. Quantitative description of the asymmetry in magnetization transfer effects around the water resonance in the human brain. *Magn Reson Med* 2007;58:786–793.

52. Ali MM, Liu GS, Shah T, Flask CA, Pagel MD. Using two chemical exchange saturation transfer magnetic resonance imaging contrast agents for molecular imaging studies. *Acc Chem Res* 2009;42:915–924.

53. Mougin OE, Coxon RC, Pitiot A, Gowland PA. Magnetization transfer phenomenon in the human brain at 7 T. *Neuroimage* 2010;49:272–281.

54. McMahon MT, Gilad AA, Zhou JY, Sun PZ, Bulte JWM, van Zijl PCM. Quantifying exchange rates in chemical exchange saturation transfer agents using the saturation

time and saturation power dependencies of the magnetization transfer effect on the magnetic resonance imaging signal (QUEST and QUESP): pH calibration for poly-L-lysine and a starburst dendrimer. *Magn Reson Med* 2006;55: 836–847.

55. Sun PZ, Zhou JY, Sun WY, Huang J, Van Zijl PCM. Detection of the ischemic penumbra using pH-weighted MRI. *J Cereb Blood Flow Metabol* 2007;27:1129–1136.

56. Gatenby RA, Gillies RJ. Why do cancers have high aerobic glycolysis? *Nat Rev Cancer* 2004;4:891–899.

57. Zhang S, Merritt M, Woessner DE, Lenkinski RE, Sherry AD. PARACEST agents: Modulating MRI contrast via water proton exchange. *Acc Chem Res* 2003;36:783–790.

58. Aime S, Carrera C, Delli Castelli D, Geninatti Crich S, Terreno E. Tunable imaging of cells labeled with MRI-PARACEST agents. *Angew Chem Int Ed Engl* 2005;44: 1813–1815.

59. Yoo B, Pagel MD. A PARACEST MRI contrast agent to detect enzyme activity. *J Am Chem Soc* 2006;128:14032–14033.

60. Woods M, Woessner DE, Sherry AD. Paramagnetic lanthanide complexes as PARACEST agents for medical imaging. *Chem Soc Rev* 2006;35:500–511.

61. Yoo B, Pagel MD. An overview of responsive MRI contrast agents for molecular imaging. *Front Biosci* 2008;13: 1733–1752.

62. Liu G, Li Y, Pagel MD. Design and characterization of a new irreversible responsive PARACEST MRI contrast agent that detects nitric oxide. *Magn Reson Med* 2007;58:1249–1256.

63. Aime S, Delli Castelli D, Fedeli F, Terreno E. A paramagnetic MRI-CEST agent responsive to lactate concentration. *J Am Chem Soc* 2002;124:9364–9365.

64. Aime S, Delli Castelli D, Lawson D, Terreno E. Gd-loaded liposomes as T1, susceptibility, and CEST agents, all in one. *J Am Chem Soc* 2007;129:2430–2431.

MR Spectroscopy and Spectroscopic Imaging of Tumor Physiology and Metabolism

Marie-France Penet
The Johns Hopkins University

Dmitri Artemov
The Johns Hopkins University

Noriko Mori
The Johns Hopkins University

Zaver M. Bhujwalla
The Johns Hopkins University

9.1 Introduction

Magnetic resonance spectroscopy (MRS) is becoming increasingly important in providing biomarkers for detecting cancer, and managing cancer treatment. As a tool for basic discovery, MRS methods provide valuable information on tumor physiology and metabolism. In this era of targeted molecular medicine, MRS methods provide an understanding of the effects of down-regulation of specific targets on downstream changes in physiology and metabolism that are a rich source to mine for noninvasive biomarkers associated with molecular targets. In understanding and treating a complex disease such as cancer, the multiparametric capabilities of MRS, its ability to transition from bench to bedside, and the feasibility of combining MRS with other imaging modalities make it valuable for molecular and functional imaging of cancer. In this chapter, we begin by providing qualitative and quantitative descriptions of the technique, before reviewing recent *in vivo* applications of MRS and MRS imaging (MRSI) in preclinical models, and the current clinical uses of MRS techniques for diagnosis and in the assessment of therapy efficacy.

9.2 Qualitative Introduction to MRS Techniques

Nuclei with nonzero magnetic moments precess with a magnetic resonance frequency ω_0 in the presence of an external static magnetic field B_0 [1]. By applying a radiofrequency (RF) pulse at the magnetic resonance frequency it is possible to detect RF signals generated by magnetic nuclear spins precessing in the external magnetic field. This forms the basis of the MR signal. The magnetic resonance frequency ω_0 is linearly dependent on B_0 and on the gyromagnetic ratio of the nucleus γ, as $\omega_0 = \gamma |B_0|$. Generally, the MR signal intensity is dependent on the concentration and the gyromagnetic ratio γ of the observed nucleus; two rate constants, the spin-lattice (or longitudinal relaxation time, T_1) and the spin-spin (or transverse relaxation time, T_2) govern the time dependence of the magnetization (please refer to Chapter 2 for a detailed description). After the acquisition of the free induction decay (FID), the signal is converted to an interpretable form, typically a resonance-frequency spectrum [1]. Postprocessing of the FID data is necessary to maximize the information content. First, the digital frequency resolution is improved with zero filling, a data processing technique where zero points are appended to the FID before Fourier transformation to increase the digital resolution in the spectrum. To increase the spectral signal-to-noise ratio (SNR), the FID is apodized by multiplication with a Lorenzian or Gaussian filter. Finally, the FID is Fourier-transformed, and phase and baseline corrections are performed to obtain an interpretable resonance-frequency spectrum [1]. In an MR spectrum, each peak has a characteristic chemical shift defined in parts per million (ppm) that is dependent upon the chemical structure of the compound. The chemical shift is determined from the difference between the frequency of the peak of interest and the frequency of a standard compound, divided by the resonance frequency of the spectrometer. Since the numerator is typically in hertz and the denominator in megahertz, the chemical shift is expressed in parts per million, and is independent of the spectrometer magnetic field strength. By applying MRS techniques, it is possible to distinguish a particular nucleus with respect to its environment in the molecule, since the resonance frequency of a particular nucleus is dependent upon its molecular structure. Moreover, each peak is characterized by an area under the resonance signal, which is proportional to the number of nuclei. This information can be used to derive metabolite concentration. The nuclei commonly used for *in vivo* MRS, their sensitivity, and some of the applications in cancer discovery and treatment are listed in Table 9.1.

The line width of each peak is dependent not only on its intrinsic T_2 but also on the magnetic field inhomogeneity as a consequence of the susceptibility of the tissue. Field homogeneity can be improved and optimized by shimming the local field, using shim coils. High-resolution MRS of tissue extracts, either water- or lipid-soluble, benefit from sample field homogeneity. Shimming typically allows the sample to be shimmed to narrow line widths of ~1–2 Hz, which results in improved spectral resolution. *In vivo*, it is difficult to achieve a comparable spectral resolution due to tissue microscopic (due to paramagnetic components) and macroscopic (e.g., due to air–tissue or bone–tissue interface) susceptibility effects (please see Figure 9.1 for comparison between *in vivo* and extract ^1H, ^{31}P, and ^{13}C spectra).

MR spectra can be acquired with or without spatial localization. Without localization, the signal is acquired from the entire sensitive region of the coil that detects the RF signal. This can also be exploited to provide RF localization using a surface coil.

In the case of localized spectroscopy the localization can either be confined to a single volume element (voxel) or it can provide spectra from multiple voxels (multivoxel). In single-voxel techniques, three orthogonal intersecting slices are used to generate a voxel to minimize signal from regions outside of the volume of interest. In the STEAM technique (stimulated echo acquisition mode), signal from a voxel is obtained by using three orthogonal slice selective 90° pulses, which result in a well-delineated voxel within the region of interest [2,3]. The use of 90° pulses results in the refocusing of only 50% of magnetization, and therefore, SNR is reduced by a factor of 2. In the PRESS technique (point resolved spectroscopy), the voxel is generated by acquiring signal generated by orthogonal slice selective pulses, that consist of a leading 90° pulse followed by two refocusing 180° pulses [4]. The advantage of PRESS is the twofold increase of SNR compared to STEAM because of the 180° refocusing pulses. However, because of the longer echo time, the technique is not suitable to detect metabolites with short T_2 values, or complex J-coupling network patterns. Also, the use of 180° pulses results in higher power deposition. The image-selected *in vivo* spectroscopy (ISIS) technique [5] is based on the selective inversion of magnetization from the volume of interest. The slice can be selected by subtracting one data set collected without inversion from one data set collected with inversion of the selected slice. Since eight acquisitions are required to accomplish the spatial localization, the sequence cannot be used for localized shimming. Variation between these data collections, for instance, due to movement, will degrade the quality of spectra. To minimize motion artifacts, outer volume suppression techniques such as OSIRIS (outer volume suppressed image related *in vivo* spectroscopy) can be used in combination with ISIS [6]. The sequence allows detection of metabolites with very short T_2.

To summarize, STEAM provides good localization, with short echo time, and low RF power deposition, but the SNR is reduced compared with PRESS. PRESS provides high SNR, but the 180° pulses require higher power deposition and longer echo time, making it unsuitable for metabolites with short T_2. ISIS is useful to detect metabolites with very short T_2, but requires several scans and is sensitive to motion.

Chemical shift imaging (CSI) incorporates phase-encoding gradients to generate images of signals obtained at different chemical shifts [7–9] by spatially encoding chemical shift information. This

TABLE 9.1 Nuclei Commonly Used in MRS Studies and Some of Their *In Vivo* Applications in Cancer

Nucleus	^1H	^{19}F	^{31}P	^{13}C
γ (MHz/T)	42.58	40.08	17.25	10.71
Applications	• Metabolism (total choline, lactate, lipids, NAA, citrate) • extracellular pH • pO$_2$ • Treatment efficacy • Identification of metastasis • Apoptosis detection	• Enzyme activity • Drug pharmacokinetics • extracellular pH • pO$_2$	• Energy metabolism (ATP, Pi, PCr) • intracellular pH/extracellular pH • Phospholipid metabolism • Treatment efficacy	• Drug pharmacokinetics and metabolic pathways evaluated after injection of labeled substrate • pH
Sensitivity of detection				

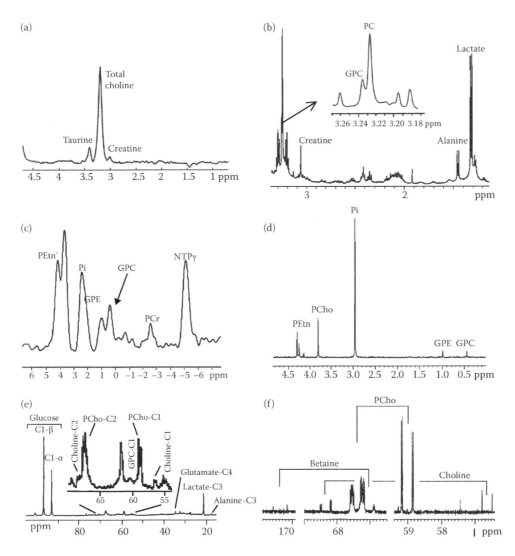

FIGURE 9.1 (a) *In vivo* MR spectrum acquired at 7 T from an MCF-7 tumor showing the signal of taurine, total choline, and creatine. (b) High resolution ¹H MR spectrum of water-soluble metabolites from an MCF-7 tumor extract. Insert: Display of the area between 3.17 and 3.27 ppm. (c) *In vivo* single voxel ³¹P MR spectrum of an MCF-7 tumor xenograft. Resonances were identified as composite phosphoethanolamine/nucleotide monophosphate/sugar phosphate (PEtn), inorganic phosphate (Pi), glycerophosphoethanolamine (GPE), glycerophosphocholine (GPC), phosphocreatine (PCr), and nucleotide triphosphate (NTP). (d) *Ex vivo* ³¹P MR spectrum of MCF-7 tumor centered around the phosphomonoester (PME) and phosphodiester (PDE) region, phosphoethanolamine (PEtn), phosphocholine (PCho), inorganic phosphate (Pi), glycerophosphoethanolamine (GPE), and glycerophosphocholine (GPC). (e) ¹³C spectrum of MCF-7 cells on microcarrier beads. Cells were perfused with medium containing [1-¹³C]glucose (5.6 mM) and [1,2-¹³C]choline (28 μM). Insert: Display of the area between 57 and 65 ppm. (f) ¹³C spectra of intracellular water-soluble metabolites of MCF-7 cells after incubation with [1,2-¹³C]choline. The spectral regions demonstrate signals from [1,2-¹³C]-labeled choline, PCho, and betaine. (a, Adapted from Jensen, L. R. et al., *NMR Biomed*, 23, 56–65, 2010, with permission; c and d, Morse, D. L. et al., *NMR Biomed*, 22, 114–127, 2009, with permission; e, Bogin, L. et al., *Biochim Biophys Acta*, 1392, 217–232, 1998, with permission; f, Katz-Brull, R. et al., *Cancer Res*, 62, 1966–1970, 2002, with permission.)

is particularly useful in heterogeneous tissues such as tumors. Phase encoding can be applied in one, two, or three spatial dimensions. With a single phase-encoding gradient, spectra are acquired from localized slices in the direction of the gradient (1D CSI). When two orthogonal gradients in combination with a slice selective excitation pulse are used, spectra are obtained from an array of voxels within the localized slice (2D CSI). These localized spectra can be processed to obtain images from individual metabolites. The typical spatial resolution of 2D CSI is about 2 × 2 × 2 mm at high-field (4.7 T and higher) preclinical settings and up to 20 × 20 × 20 mm at clinical settings of 1.5–3 T. In 3D CSI, three phase encoding gradients are used to obtain a 3D array of spectra from a volume, and provide a 3D display of individual metabolites. However, this comes with a cost of significantly longer acquisition times. As an example, a 2D CSI data set acquired at 9.4 T from a slice with a thickness of 2 mm and an in-plane resolution of 1 mm would require approximately 17 min (field of view = 16 mm, slice thickness = 2 mm, 16 × 16 phase encode steps, repetition time (TR) = 1 s, number of scans per phase encode step = 4). A 3D data set with a TR = 1 s, 4 scans per phase encode step, and with a spatial resolution of 1 × 1 × 2 mm would require approximately 136 min. However, since the SNR of a 3D experiment is inherently higher, the number of scans could be reduced to one resulting in an acquisition time of approximately 34 min with a comparable SNR. Optimized spherical acquisition weighted *k*-space sampling can be used to further reduce imaging time.

With CSI, there is a possibility of volume averaging effects, where signal, which is only partially within the voxel, is averaged with the other signal in that voxel. This can lead to a distortion in the spatial origin of the metabolite.

9.3 Quantitative Introduction to MRS Techniques

9.3.1 ^{1}H Magnetic Resonance Spectroscopy

Due to its high sensitivity of detection, ^{1}H MRS is widely used in preclinical and clinical studies. Metabolites such as lactate, total creatine, and total choline that have concentrations in the millimolar range are readily detectable. The total choline signal *in vivo* consists of overlapping phosphocholine (PC), glycerophosphocholine (GPC), and free choline (Cho) signals. Similarly, the total creatine signal consists of overlapping phosphocreatine (PCr) and creatine (Cr) signals. At shorter echo times, additional signals from short T_2 metabolites, such as mobile lipids, glutamine, and glutamate are detected. Short echo time single-voxel MRS is usually performed with a STEAM sequence, and CSI is used to acquire 2D MRSI data. Because of their high concentrations, the signals from water (~100 M) and lipid triglycerides (as high as ~10 M in some tissues such as breast) dominate unedited proton spectra. These dominant signals mask signals from other metabolites, and therefore water and lipid suppression methods are required. Examples of proton spectra from *in vivo* tissue and from extracts are provided in Figures 9.1a and b.

In one commonly used technique, chemically selective saturation (CHESS), frequency selective pulses are applied along with a dephasing gradient to suppress the water signal [10]. Variations to this approach include the water suppression enhanced through T_1 effects (WET) technique [11] and the variable power RF pulses and optimized relaxation delays (VAPOR) technique [12]. Another approach for water suppression is using band selective refocusing, where the signal from the water remains defocused resulting in loss of water signal [13,14]. The latter approach has the advantage of retaining signals of exchangeable protons in metabolites that can be suppressed by standard water suppression techniques. Although suppressing the water signal is required to detect metabolite signals, the water signal is useful as an internal concentration reference for absolute quantification of metabolites, as described in more details in paragraph 3.3 [15].

Techniques have also been developed to obtain ^{1}H MR spectra *in vivo* without water suppression [16]. Prior to a standard localization sequence, additional chemical shift selective inversion pulses are either switched off or on to invert the metabolites signals upfield and downfield from water. To obtain the metabolite or the water signals, the two data sets are subtracted or added, allowing the simultaneous detection of signal from water and metabolites [16].

Mobile lipids, the other dominant signal in ^{1}H MRS, can provide useful information on necrosis, apoptosis, or lipid droplet formation [17–20]. ^{1}H MR visible lipids are frequently observed in malignant tissue. Polyunsaturated lipids accumulate in tumors undergoing apoptosis [21]. A positive correlation between ^{1}H MR visible lipids and percentage of necrosis has also been reported [22]. The origins of these lipids are still a matter of debate, and their role as biomarkers in tumor diagnosis and treatment monitoring is still being evaluated. Because these lipid signals dominate the proton spectrum, lipid suppression is required to detect signal from metabolites with chemical shifts close to the lipid signals such as lactate (1.33 ppm) and alanine (1.48 ppm) that occur in the chemical shift range of 0.9 to 1.5 ppm. Because of their short T_2, mobile lipid signals decay faster than metabolite signals. Therefore, by using longer echo times that allow the lipid signal to decay, it is possible to reduce the lipid signal in proton spectra. If the lipids are localized in specific regions (e.g., the subcutaneous lipids in the skull and periprostatic fat around the prostate gland), volume-localized presaturation methods can be used to suppress the lipid signal [13,14].

Several spectral editing methods are available to resolve the lactate signal from lipids [23–25]. Most of these exploit the spin coupling between the lactate-methyl (CH_3) protons at 1.3 ppm and methine (CH) protons at 4.1 ppm that allows either filtering of the lipid signal or subtraction of the lipid signal to obtain the lactate signal [23,26].

9.3.2 ^{31}P MRS, ^{19}F MRS, and ^{13}C MRS

In ^{31}P MR spectra, metabolites such as nucleoside triphosphates (NTP), phosphocreatine (PCr), inorganic phosphate (Pi), phosphodiesters (PDE), and phosphomonoesters (PME) can be detected. Typical *in vivo* ^{31}P MR spectra have three NTP peaks (from β, α, and γ-NTP), that predominantly consist of adenosine triphosphate (ATP), a major source of cellular energy (see Figures 9.1 and 9.2). PCr is a short-term energy reserve that contains a high-energy phosphate bond existing in steady-state exchange with ATP. PCr transfers its phosphate to adenosine diphosphate (ADP) to form new ATP. The reaction is catalyzed by the enzyme creatine kinase [27]. Pi is a product of the breakdown of ATP to ADP and is often elevated in tumors. The phosphomonoester (PME) resonance consists of signal from low-molecular-weight phosphorylated intermediates, including sugar phosphates, such as glucose-6-phosphate and glycerol-3-phosphate, AMP (adenosine monophosphate), and the membrane intermediates phosphoethanolamine (PE) and phosphocholine (PC). PE is produced by hydrolysis of the membrane phospholipid phosphatidylethanolamine

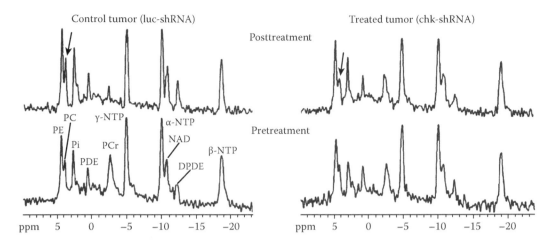

FIGURE 9.2 *In vivo* single-voxel ^{31}P MRS of MDA-MB-231 tumor xenografts. Representative pretreatment and posttreatment spectra of a tumor following systemic delivery of lentivirus expressing luciferase-short hairpin RNA (luc-shRNA) or choline kinase short hairpin RNA (chk-shRNA). Injection of chk-shRNA resulted in a significant reduction of PC (arrow) compared with luc-shRNA. DPDE, diphosphodiester; NAD, nicotinamide adenine dinucleotide; PDE, phosphodiester; PCr, phosphocreatine; PE, phosphoethanolamine. (Adapted from Krishnamachary, B. et al., *Cancer Res*, 69, 3464–3471, 2009. With permission.)

(PtdEtn), and its primary role is in the resynthesis of PtdEtn [27]. PC is synthesized from free choline by the enzyme choline kinase. Like PE, it is an intermediate in the synthesis of the membrane phospholipid phosphatidylcholine (PtdCho) [27]. The PDE signal consists of GPC and glycerophosphoethanolamine (GPE), which are catabolic byproducts, produced by sequential hydrolysis of membrane phospholipids by phospholipases A1 and A2 [27]. Elevated PME (from phospholipid precursors) and PDE (from phospholipid catabolites) can be observed in ^{31}P MR tumor spectra (see Figures 9.1 and 9.2).

Despite its low sensitivity, ^{31}P MRS is a valuable method for evaluating tumor energy metabolism, pH, and choline phospholipid metabolism (Table 9.1) [28–30]. Examples of metabolites that can be detected *in vivo* and with high-resolution spectroscopy of extracts are shown in Figures 9.1c and d.

^{19}F MRS has the advantage of a relatively high sensitivity (84% sensitivity in comparison with ^{1}H), absence of naturally occurring fluorinated compounds resulting in no background signal, and a relatively wide chemical shift range (greater than 200 ppm) [31]. In tumors, it has been used to determine the uptake and pharmacokinetics of ^{19}F-labeled drugs *in vivo* such as the pyrimidine analog 5-fluorouracil (5-FU), which is routinely used as an anticancer agent to treat several cancers [32]. ^{19}F MRS of molecules such as nitroimidazoles, which accumulate in hypoxic cells, has been used to obtain qualitative measurements of tumor hypoxia [33,34].

^{13}C MRS of labeled substrates can be used to study glycolysis and other metabolic pathways, as well as detect drug pharmacokinetics. Examples of *in vivo* and high-resolution ^{13}C spectra are shown in Figures 9.1e and f. The flux of metabolites through various pathways can be determined through the incorporation of the labeled substrate into metabolites. The sensitivity of ^{13}C MRS is relatively low, but indirect detection methods permit the

detection of the ^{13}C label with a sensitivity approaching that of the proton nucleus [35,36]. As discussed subsequently, the use of dynamic nuclear polarization (DNP) of ^{13}C substrates can theoretically increase sensitivity by ~10,000, allowing the *in vivo* acquisition of MRSI of hyperpolarized ^{13}C-labeled substrates. Hyperpolarized MRI is described in detail in Chapter 21. The enriched substrate should be water-soluble and an endogenous or exogenous metabolite with a long T_1 (tens of seconds) relaxation time in the liquid state to enable preparation, transfer, injection, and metabolism of the tracer in the body.

9.3.3 Quantification

In vivo MR spectra are challenging to quantify because of the absence of an internal concentration reference. As a result, metabolite ratios are often used to detect differences in spectra. The water peak, and other metabolites such as creatine (Cr) for the brain or citrate for the prostate [37], have been used as an internal reference, to derive ratios. However, even the use of ratios requires a stable internal marker [1]. As described in Bolan et al., the two components required for MRS quantification are the referencing strategy and the spectral fitting [38]. It is possible to use external referencing, in which the signal from the tissue is compared to an external phantom of known concentration, detected at the same time and under identical conditions [37]. The water signal can also be used as an internal quantitative reference. Total choline levels in breast tissues have been quantitatively measured by using the water signal as an internal reference [38]. Quantification techniques based on the water signal have some limitations, as the water content of tumors is not constant, and can vary due to the presence of edema and differences in cell morphology and content [39].

9.4 Preclinical Applications in Cancer

Multinuclear *in vivo* MRS/MRSI techniques have been widely used in preclinical studies of cancer. As summarized in Table 9.1, with their versatility, they can provide information on tumor metabolism, pH, hypoxia, drug delivery, treatment efficacy, or apoptosis. Examples of some of these applications are described here.

9.4.1 Choline Metabolism

Altered choline metabolism is consistently observed in most cancers and provides unique biomarkers to detect cancer and monitor response to treatment. In addition, enzymes in the choline metabolism pathway provide novel targets for cancer treatment. The aberrant choline metabolism observed in most tumors results in the detection of high total choline in ^{1}H MR spectra of tumors. This high total choline signal is primarily due to increased PC levels in cancer cells, as confirmed by high-resolution ^{1}H MRS studies of cell and tumor extracts [40–43]. Molecular alterations that result in these increased PC levels include a higher rate of choline transport [44], an increased expression and activity of choline kinase [45–47], and an increased activity of phospholipases C and D [48,49]. ^{1}H MRS and ^{31}P MRS have been used to investigate the effects of targeting choline kinase, the enzyme that converts free choline to PC, in preclinical studies of breast cancer [50,51]. Inhibition of choline kinase by small interfering RNA (siRNA) induced a significant reduction in PC and was associated with differentiation and reduced proliferation in MDA-MB-231 cells [52]. Transient transfection of cells with siRNA down-regulating choline kinase resulted in a significant decrease in PC levels in the malignant MDA-MB-231 cells, but did not alter the already low PC concentration of nonmalignant MCF-12A cells [52]. MN58b, a pharmacological inhibitor of choline kinase, was observed to decrease total choline in two human carcinoma models, the HT29 colon cancer xenograft, and the MDA-MB-231 breast cancer xenograft [50]. The decrease was detected noninvasively with ^{1}H MRS and demonstrates the potential utility of using ^{1}H MRS in clinical trials to assess tumor response to choline kinase inhibition [50]. ^{31}P MRS can also be used to detect changes in PC that is part of the PME signal *in vivo*, to follow treatment efficacy. In a recent study, lentiviral vector-mediated down-regulation of choline kinase using short hairpin RNA (shRNA) was achieved in the MDA-MB-231 human breast cancer xenograft model (Figure 9.2) [51]. Concentrated lentivirus encoding shRNA against choline kinase was injected intravenously in the tail vein of MDA-MB-231 tumor-bearing female severe combined immunodeficient (SCID) mice. Noninvasive ^{31}P MR spectra acquired *in vivo* detected lower PC and PME levels in the choline kinase-shRNA treated mice (Figure 9.2). The decreases in PC and PME were associated with reduced tumor growth and proliferation. The study demonstrated the potential use of noninvasive ^{31}P MRS to detect the effect of targeting choline kinase in a human breast cancer xenograft model.

9.4.2 pH Measurements

One of the major contributions of MRS methods in cancer is in the characterization of tumor pH. The high glycolytic activity of tumors and the chaotic tumor vasculature, together with the resulting hypoxia, combine to create an acidic extracellular microenvironment in tumors [53–55]. This acidic extracellular pH (pHe) can influence tumor progression, metastasis, and drug resistance [56,57]. The pH-sensitive chemical shift of the Pi peak coupled with the fact that Pi is primarily intracellular in tumors [58] makes ^{31}P MRS ideal for measuring tumor intracellular pH (pHi). Several pHe-sensitive probes have been developed for ^{31}P and ^{1}H MRS to measure tumor pHe *in vivo*. Acidic tumor pHe values have been reported by using an exogenous ^{31}P-detectable pH marker, 3-aminopropylphosphonate (3-APP) [59]. However, the low sensitivity of ^{31}P MRS results in low spatial resolution. pHe imaging with improved spatial resolution can be performed in tumors *in vivo* with ^{1}H MRSI by acquiring the signal from pH sensitive extracellular compounds such as (imidazol-1-yl)3-ethyoxycarbonylpropionic acid (IEPA) [60] or 2-(imidazol-1-yl)succinic acid (ISUCA) [61].

9.4.3 Hypoxia Reporters

Tumor hypoxia, induced by the abnormal vasculature of tumors, increases the aggressive phenotype of tumors [62] and is a major cause of radiation and chemoresistance [63]. Because of its importance in cancer progression and treatment, the development of noninvasive methods to detect hypoxia is of major importance in cancer discovery and treatment [64]. Perfluorocarbons (PFCs) have been used to measure pO$_2$ using ^{19}F MRS because their relaxivity is dependent on the concentration of dissolved molecular oxygen [65]. More recently, an analogous approach with ^{1}H MRS was described based on the oxygen-dependent relaxivity of hexamethyldisoloxane [66].

Another approach is to use nitroimidazole-based probes, which accumulate in hypoxic cells [67] through biochemical reduction. The reduction products bind to endogenous cellular molecules, and as a result the accumulation of these reporters can be used to estimate hypoxia [68]. Some potential problems are that oxygen-independent binding can occur due to variations in nitro-reductase activity and drug metabolism, and the adducts formed are short-lived [69].

2-Nitro-α-[(2,2,2-trifluoroethoxy)methyl]-imidazole-1-ethanol (TF-MISO) is one such probe that has been used to investigate hypoxia *in vivo* with ^{19}F MRS [64]. Hypoxic regions can be identified by monitoring the uptake, distribution, and retention of the marker using 2D ^{19}F CSI, as shown in Figure 9.3 [64]. The heterogeneity of TF-MISO retention observed in the 2D ^{19}F CSI data set is typical of the heterogeneous distribution of oxygen in tumors.

9.4.4 Glycolysis

One of the characteristic features of cancers is aerobic glycolysis, known after its discoverer, Otto Warburg, as the Warburg effect. The Warburg effect describes the preferred conversion of glucose

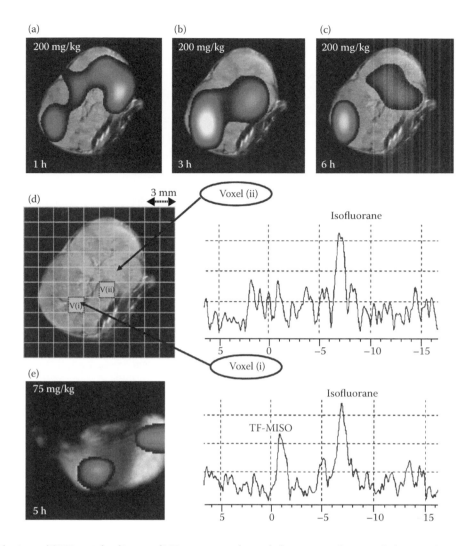

FIGURE 9.3 **(See color insert.)** ^{1}H T$_2$-weighted image of MCa tumor together with the corresponding metabolic map of TF-MISO following a 200-mg/kg dose at (a) 1, (b) 3, and (c) 6 h after injection. (d) Anatomic T$_2$-weighted image with superimposed grid showing the in-plane voxel size and distribution around the tumor. Two representative localized spectra from two voxels showing (i) TF-MISO and isofluorane peaks and (ii) only the isofluorane peak. (e) TF-MISO metabolite map acquired 5 h after an injection of 75-mg/kg dose overlaid on its anatomic reference. (Adapted from Procissi, D. et al., *Clin Cancer Res*, 13, 3738–3747, 2007. With permission.)

to lactate by tumor cells, even in the presence of adequate oxygen. Under oxygenated conditions, most normal cells use glucose for oxidative phosphorylation [70]. Since ^{1}H MRS can be used to detect lactate and ^{13}C MRS can be used to follow the utilization of labeled glucose through metabolic products such as glutamine or lactate, ^{1}H and ^{13}C MRS are useful to study tumor glycolysis in preclinical models. Following an infusion of [1-^{13}C] glucose in mice bearing MCF-7 human breast cancer xenografts, Rivenzon-Segal et al. detected ^{13}C-1 glucose and ^{13}C-3 lactate in ^{13}C MR spectra [71]. The kinetics of ^{13}C glucose and ^{13}C lactate were measured *in vivo* to quantitatively assess tumor glycolysis noninvasively [71].

9.4.5 Drug Delivery and Treatment Efficacy

Tumors typically exhibit heterogeneous perfusion, and perfusion-limited drug delivery is a major barrier to achieving effective treatment [72]. A useful application of MRS is determining drug delivery. *In vivo* detection and localization of anticancer drugs can be performed with ^{13}C MRS, for drugs that can be labeled with ^{13}C [35], and delivered at high concentrations (~100 µM) to overcome sensitivity limitations. Spectroscopic imaging of the intratumoral distribution of the ^{13}C-labeled anticancer agent temozolomide by ^{1}H/^{13}C MRS demonstrated the heterogeneous delivery of this drug, which is used clinically to treat glioblastomas and anaplastic astrocytomas [35].

There is also a continuing search to identify spectroscopic markers of response to treatment. This is especially important with molecular targeting since traditional methods of evaluating tumor response such as reduction of tumor volume may not be useful in determining response. In a ^{1}H MRS study, targeting of the heat shock protein hsp90 with a drug, 17-allylamino,17-demethoxygeldanamycin (17-AAG), in a human prostate cancer model resulted in a significant decrease of total choline within the tumor [73]. Treatment-induced tumor growth inhibition was detected after 14 days, whereas differences in total choline were observed as early as 4 days after treatment. The early changes in total choline levels suggest that the metabolic response may be more sensitive than changes in tumor volume. Morse et al. have shown with ^{31}P MRS that docetaxel, an antimicrotubule drug, induced a reduction in PC levels in a human breast cancer xenograft model [74]. Tumor growth delay was also observed together with an increase in cell death and cell-cycle arrest [74].

^{19}F MRS can also be used to assess the efficacy of treatment. The ideal therapy would target cancer cells while sparing normal tissue. With prodrug therapy, where a drug-activating enzyme is delivered to the tumor followed by the administration of a nontoxic prodrug systemic administration, side effects can be minimized. The ability to image the delivery of the prodrug enzyme can be exploited to time prodrug administration to minimize damage to normal tissue. A prototype agent consisting of the prodrug enzyme cytosine deaminase (CD) labeled with multimodal MR and optical imaging reporters was recently synthesized [75]. The prodrug enzyme, CD, converts a nontoxic prodrug 5-fluorocytosine (5-FC) to 5-FU. Both prodrug (5-FC) and the formation of the cytotoxic drug (5-FU) by the enzyme were detected by ^{19}F MRS.

9.4.6 Lipids and Apoptosis

Several molecular targeted agents induce apoptosis, and therefore MRS methods have been evaluated for the ability to detect apoptosis following such treatments [76]. It has been shown that lipids accumulate in tumor tissues as apoptotic death occurs [20,21,77]. ^{1}H MRSI has been used to follow metabolic response to ganciclovir treatment in a rat glioma model [21]. The treatment induced a decrease in total choline but also an increase of lipids content observable at 8 days posttreatment. The increase of lipids was attributed to membrane lipid catabolism that could be used as a surrogate marker for apoptotic cell death [21]. In a murine lymphoma model, MRS visible lipid levels increased during tumor cell apoptosis induced by treatment with cyclophosphamide and etoposide [20], reinforcing the potential role of analyzing the lipid signal detected by ^{1}H MRS to detect tumor apoptosis *in vivo*.

9.4.7 Hyperpolarization, A New Research Area

Among recent developments in MRS, the use of hyperpolarized ^{13}C is one of the most promising [78]. The hyperpolarization of spins allows an improvement in the sensitivity of detection of the MR signal of ^{13}C-labeled substrates by >10,000 [78]. Hyperpolarized ^{13}C-labeled substrates generated through the process of dynamic nuclear polarization (DNP) are injected systemically, allowing real-time metabolic mapping [78–80]. Metabolic maps acquired after a systemic injection of hyperpolarized ^{13}C-labeled pyruvate, demonstrated that tumor tissue was characterized by a high level of lactate and a lack of alanine [81,82]. Since poorly differentiated aggressive tumors are usually more glycolytic [70], the use of hyperpolarized ^{13}C-labeled pyruvate may, in the future, provide noninvasive determination of tumor grade [83]. Indeed, by injecting hyperpolarized ^{13}C-labeled pyruvate in transgenic TRAMP mice with low- or high-grade prostate cancers, Albers et al. observed that hyperpolarized lactate levels were closely associated with tumor grade [83].

Hyperpolarized ^{13}C substrates are also being used to detect treatment response, as shown by Gallagher et al. [84]. The production of $[1,4\text{-}^{13}C_2]$malate was detected *in vivo* in a murine lymphoma tumor, following the injection of hyperpolarized $[1,4\text{-}^{13}C_2]$fumarate. The conversion of $[1,4\text{-}^{13}C_2]$malate to $[1,4\text{-}^{13}C_2]$ fumarate was attributed to the formation of necrosis following treatment with etoposide, suggesting that this conversion may be used to identify the formation of necrosis following treatment [84]. Another emerging application of hyperpolarization is to measure tumor pHe *in vivo* using hyperpolarized ^{13}C-labeled bicarbonate [85]. Using this technique, pHe maps were obtained from a subcutaneous EL4 lymphoma tumor model (image displayed in Figure 9.4a) as shown in Figure 9.4b. The spatial distribution of $^{13}CO_2$ and $H^{13}CO_3^-$ can be imaged using a gradient echo pulse sequence (Figures 9.4c and d). Images of tumor pHe are generated from the ratio of the signal intensities of hyperpolarized $H^{13}CO_3^-$ and $^{13}CO_2$, with a spatial resolution of $2 \times 2 \times 6$ mm^3. Since bicarbonate is relatively nontoxic, this approach may, in the future, be extended to clinical applications. For more

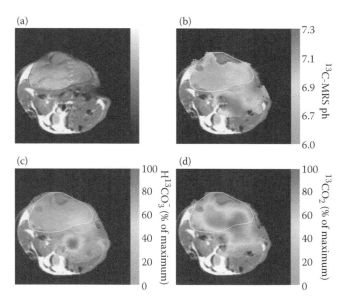

FIGURE 9.4 **(See color insert.)** *In vivo* tumor pH. (a) Transverse ^{1}H MR image of a mouse with a subcutaneous EL4 tumor (outlined in red). (b) pH map of the same animal calculated from the ratio of the $H^{13}CO_3^-$ (shown in c) and $^{13}CO_2$ (shown in d). Voxel intensities in the ^{13}C MRS images were acquired 10 s after intravenous injection of hyperpolarized $H^{13}CO_3^2$. The spatial distribution of (c) $H^{13}CO_3^2$ and (d) $^{13}CO_2$ are displayed as voxel intensities relative to their respective maxima. The tumor margin in b, c, and d is outlined in white. (Adapted from Gallagher, F. A. et al., *Nature*, 453, 940–943, 2008. With permission.)

details on hyperpolarization theory and techniques, please see Chapter 22.

9.5 Applications in Cancer Diagnosis

The abnormal glucose, choline, and lipid metabolism of tumors have provided several potential noninvasive MRS-based biomarkers that are being evaluated not only for differential diagnosis between tumor and benign tissue but also for grading tumor aggressiveness. ¹H MRS has been extensively used in brain tumor classification and diagnosis, and the use of MRS in diagnosis is currently being evaluated in brain, breast, prostate, and other cancers.

9.5.1 Brain Cancer

In the brain, the principal metabolites that can be visualized and analyzed with ¹H MRS are N-acetyl-aspartate (NAA), total choline, total creatine (tCr), glutamine + glutamate (Glx), *myo*-inositol, lactate, and lipids [39]. NAA is a free amino acid, found in neurons, and is present in the brain at relatively high concentrations. Among other functions, NAA is a precursor of NAAG (N-acetylaspartatylglutamate) that is involved in excitatory neurotransmission [86]. NAA is thought to be involved in neuronal and glial signaling [87]. The total creatine signal consists of phosphocreatine and creatine, which provide an energy buffer system used to maintain cellular ATP levels [86]. Glutamate is the most abundant amino acid in the brain and acts as an excitatory neurotransmitter. Glutamine, its precursor, and storage form can be found in astrocytes. At field strengths below 3 T, the separation of glutamate and glutamine signals is difficult, and the signals are detected as a single peak [86]. *myo*-Inositol is thought to play an essential role in cell growth, as an osmolyte, and as a storage form of glucose [86]. Since cancer cells develop, invade, and destroy normal brain tissue, spectra from brain tumors are typically characterized by a decrease in NAA, an appearance of lactate and lipids, a decrease in total creatine, and frequently an increase in total choline [88]. Lactate is normally present at low concentrations in brain tissue, but as the end product of anaerobic glycolysis, its level increases rapidly following impaired blood flow and hypoxia [86]. Howe et al. compared ¹H MR spectra acquired from normal white matter, meningiomas, grade II astrocytomas, anaplastic astrocytomas, glioblastomas, and metastases [88]. A significant decrease of NAA was observed in all tumor spectra. In addition, distinct metabolic characteristics were identified in specific tumors, such as elevated *myo*-inositol in grade II astrocytomas, or decreased *myo*-inositol and creatine in meningiomas. Total choline levels were higher in tumor tissue compared to normal white matter [88]. Metabolic profile identified by ¹H MRS may improve the characterization, grading, and staging of the tumor. Single-voxel ¹H MRS acquired from different brain regions of a patient with a grade IV glioma are shown in Figure 9.5 [89]. These spectra demonstrate how MRS can be used to better characterize a lesion visible on T_2-weighted images by distinguishing malignant tissue

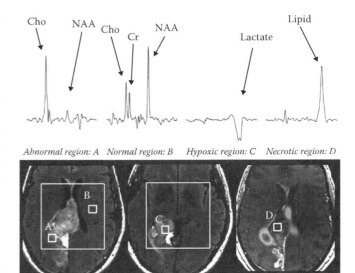

Abnormal region: A   Normal region: B   Hypoxic region: C   Necrotic region: D

FIGURE 9.5   (See color insert.) ¹H MRSI data from a grade IV glioma patient demonstrating the spatial heterogeneity of the tumor shown within the representative voxels. T_2-FLAIR images with the PRESS-selected volume superimposed are shown for two different axial slices. The T_1-weighted image acquired after injection of gadolinium is shown with a superimposed total choline/NAA ratio color map. (Adapted from Osorio, J. A. et al., *J Magn Reson Imaging*, 26, 23–30, 2007. With permission.)

from normal tissue and determining the presence of necrosis within the tumor [89]. Brain metastases can also be identified by ¹H MRS from an increase in the lipid signal [90]. In a majority of brain metastases, increased lipids and total choline were detected at short echo times [90].

9.5.2 Breast Cancer

Several studies of the breast have identified increased total choline in ¹H MR spectra as a marker commonly associated with malignant lesions [91–93]. Katz-Brull et al. have reported an overall sensitivity of 83% and a specificity of 85% for using the detection of total choline as a marker of malignancy in breast cancer [94]. This qualitative approach is limited, since total choline can also be detected in benign lesions and normal tissue at a higher field strength of 4 T, due to increased sensitivity [38]. It is, therefore, necessary to quantify total choline to achieve better accuracy for distinguishing between malignant and benign lesions. Moreover, quantification is important to assess response to treatment. Since water is typically used as an internal concentration standard, this is more challenging in the breast because of differences in water content of glandular and adipose tissues in the breast. By using single-voxel ¹H MRS at 4 T, Bolan et al. have shown quantitatively that the levels of total choline were higher in malignant compared to benign lesions [38]. While most MRS studies of breast cancer have been performed with single-voxel acquisition, the few MRSI studies that have been performed revealed a heterogeneous but elevated distribution of total choline in the malignant lesion [95]. An example of a total choline image derived from ¹H MRSI acquired at 1.5 T is shown in Figure 9.6 and reveals a heterogeneous distribution of total choline within the lesion [96]. In this study the lesion was identified from the enhancement observed in GdDTPA-based dynamic contrast-enhanced (DCE)-MRI [96].

9.5.3 Prostate Cancer

Detection and characterization of prostate cancer can be significantly improved by ¹H MRSI in combination with the anatomical information provided by MRI [97,98]. Normal prostate tissue is characterized by a high level of citrate. Highly specialized secretory epithelial cells produce and secrete citrate that is released by the prostate gland into prostatic fluid. Prostatic citrate provides an energy substrate for sperm [99]. Malignant prostatic epithelial cells are poorly differentiated and have low or undetectable citrate levels [100], and high total choline [97,101]. As a result, healthy prostatic tissue can be distinguished from malignant tissue from the ratio of (choline + creatine)/citrate [97]. At 1.5 T, voxels localized in the peripheral zone where the ratio is at least two standard deviations above the average value in the normal peripheral zone are considered likely to be malignant [102]. However, the lower sensitivity of detection at a field strength of 1.5 T and the resulting decrease of spatial resolution

at 1.5 T limit the ability to detect small lesions [98]. The degree of elevation of total choline has also been found to roughly correlate with Gleason score [103].

9.5.4 Other Cancers

While MRS and MRSI techniques have been applied extensively in brain, prostate, and breast cancer, studies have also been performed with other cancers, such as thyroid carcinoma, head and neck cancer, and ovarian carcinoma. ¹H MRS studies of patients with thyroid carcinoma have revealed that elevated total choline could be detected in all solid carcinomas whereas no choline signal was detected in normal thyroid tissue [104]. An increase of total choline signal has also been shown to be a potential marker for extracranial head and neck cancer. ¹H MRS performed on patients with squamous cell carcinoma revealed a high level of total choline/Cr in the tumors compared to spectra obtained from healthy volunteers [105]. In a recent study characterizing metabolites in both primary and metastatic lesions of ovarian carcinoma, elevated total choline was detected in most but not all tumors. Failure to detect choline was more common in metastatic disease than in primary tumors [106], partly due to technical difficulties such as smaller size resulting in limited SNR and proximity to the bowel that increases susceptibility and motion artifacts [106].

9.6 Clinical Applications in Cancer Therapy

In this age of molecular targeting and personalized medicine, MRS/MRSI methods have the potential to cover a range of applications in cancer therapy such as the detection of drug delivery, radiation therapy treatment planning, and predicting and detecting response to treatment.

9.6.1 Drug Delivery Visualization

Because it is a fluorinated drug, ¹⁹F MRS has been used to visualize the delivery of 5-FU in primary and metastatic cancers [107,108]. Delivery and conversion of a prodrug form of 5-FU, such as capecitabine, can also be detected with ¹⁹F MRS, as shown by Klomp et al., in the liver of patients with advanced colorectal cancer. The study revealed heterogeneity in the spatial distribution of the prodrug and its metabolites. This heterogeneity may be due to a nonhomogeneous distribution of capecitabine, an inhomogeneity in liver morphology induced by the treatment with cytotoxic drugs, or the presence of micrometastases in the liver [107].

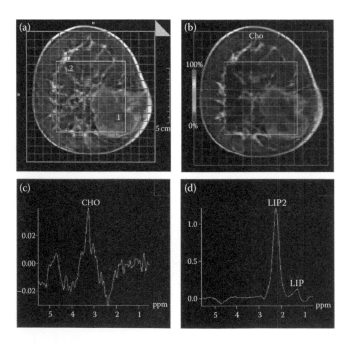

FIGURE 9.6 (See color insert.) Illustration of a "hot" area in a breast cancer lesion. (a) Coronal DCE-MRI of a patient with invasive breast cancer. (b) Color-coded image of total choline superimposed on the corresponding MRI. (c) ¹H MR spectrum from tumor area, labeled as 1 in (a). (d) ¹H MR spectrum from control area, labeled as 2 in (a). (Adapted from Hu, J. et al., *Magn Reson Imaging*, 26, 360–366, 2008. With permission.)

9.6.2 Planning Treatment and Monitoring Treatment Response

To improve the efficacy of radiotherapy treatment, it is necessary to accurately delineate the extent of tumor tissue. MRS techniques have the potential to assist in identifying target volumes in radiotherapy treatment planning [109,110]. Studies with glioma patients revealed differences in the spatial extent of the lesion detected by MRI or MRSI [110]. Combining MRSI with traditional MRI techniques to define the target for focal radiotherapy may improve the effectiveness of the treatment.

In addition to treatment planning, the uses of MRS/MRSI in predicting therapeutic response and providing information on the time course and mechanism of therapeutic response have also been evaluated. MRI/MRSI techniques have been applied to evaluate markers that can be used to predict survival for patients with glioblastoma [111]. A high level of lactate measured before radiotherapy and chemotherapy was associated with poor outcome. This study showed that pretreatment MR data could be used to adapt the treatment protocol and focal therapy for each patient [111].

In low-grade gliomas, treatment with temozolomide induced a reduction in the tumor total choline to water ratio that paralleled tumor volume decrease [112]. A posttreatment mass is frequently found at the site of the primary cancer after a course of chemoradiotherapy or radiotherapy, and it is important to determine whether the mass is benign or is residual cancer. Since an elevated total choline signal has been observed in head and neck squamous cell carcinoma, King et al. have explored the potential use of total choline detected by ^{1}H MRS to identify residual disease after chemoradiotherapy [113]. The presence of total choline was a significant marker for residual cancer [113]. In a multivoxel 3D ^{1}H MRS study designed to identify recurrent glioma from radiation injury, Zeng et al. [114] observed that recurrent tumors were characterized by higher total choline/NAA and total choline/Cr ratio and lower NAA/Cr ratio compared to radiation injury.

To assess the potential use of ^{1}H MRS to predict early response to neoadjuvant chemotherapy, changes in total choline before and 24 h after systemic therapy were measured in patients with locally advanced breast cancer [115]. Not only did the changes in total choline correlate with the change in lesion size but these changes were also significantly different between patients with good response to treatment and those with no response. MRS has also been applied to detect local prostate cancer recurrence after external-beam radiation therapy [116], although the study concluded that additional diagnostic tests were necessary to complement the MRS results, since radiation induced a decrease of citrate and an increase of total choline in benign tissue as well [116].

9.7 Extensions to Fields >3 T

High magnetic field strengths of 9.4 and 11.7 T are routinely used in preclinical small animal studies because of the major advantage of increased sensitivity of detection that translates to improved spatial resolution and higher spectral dispersion for resolving peaks.

The application of high-field MRS for clinical studies is significantly more challenging because of the larger volumes and associated issues with RF power deposition and field homogeneity.

As with preclinical studies, the advantages offered by performing clinical MRS at higher field strengths are increased SNR ratio and a higher spatial and spectral resolution. Typically, the SNR is calculated from the peak height of the metabolite of interest and the noise intensity in an off-resonance region of the MR spectrum [37]. An increase of SNR has been demonstrated at 7 T compared with 4 T, although the RF power required for a 90° pulse at 7 T is twice that of 4 T and the B_1 inhomogeneity is higher at 7 T compared with 4 T [117]. Because of increased spectral dispersion at higher magnetic field strengths, it is possible to resolve signals that would overlap at lower field strengths. Direct detection of glucose is possible at 7 T, without using glucose infusion; inositol and taurine peaks are discernable in the spectrum, and creatine can be resolved from phosphocreatine [118]. The resolution improves significantly for metabolites with J-coupled spin systems, such as glutamine and glutamate, since the overall line widths of the overlapping multiplets are mainly determined by homonuclear J-coupling constants, which are independent of B_0 [118]. *In vivo* ^{1}H MRS at 7 T allows signal assignment with increased reliability, and metabolites such as ascorbate have been detected and quantified in the human brain at 7 T [119]. Feasibility studies of ^{1}H MRS of the prostate at 7 T have been performed with a transmit/receive endorectal coil [120]. To optimize the sensitivity of detection, adiabatic slice-selective refocusing pulses were used [120].

Some of the major technical issues associated with high-field systems are the chemical shift localization error, increased B_1

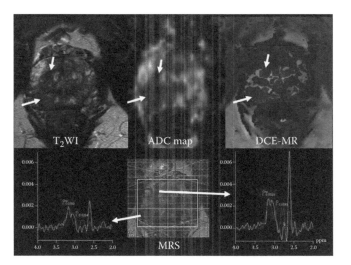

FIGURE 9.7 (See color insert.) T_2-weighted image, DWI/ADC map, MRS, and DCE-MRI in a 61-year-old man with prostate cancer (Gleason score = 6). Multiparametric MRI of the prostate gland revealed focal areas of dark T_2 signal with corresponding decreased ADC and increased total choline along with decreased citrate. Positive DCE-MRI kinetics with increased permeability in the right midgland can be observed, which correlated to the site of prostate cancer. In the normal region (anterior to the lesion) bright on T_2-weighted image, there is normal citrate with a normal ADC map and DCE parameters. (Adapted from Jacobs, M. A. et al., *Top Magn Reson Imaging*, 19, 261–272, 2008. With permission.)

and B_0 inhomogeneities, and increased magnetic susceptibility artifacts. Shimming to optimize the magnetic field homogeneity is crucial in MRS studies, but it is considerably more challenging at higher field strengths since susceptibility effects increase linearly with field strength [37,121]. Spectral resolution can be improved by implementing rapid and efficient shimming methods [122] to get a better field homogeneity in the region of interest within an acceptable time for patient studies [37]. The increase of B_0 field strength also induces MR imaging artifacts, such as lipid sidebands and respiratory-induced frequency shifts [37]. Relaxation rate constants are affected at higher fields; generally, T_1 becomes longer and T_2 shorter, requiring longer repetition time, and resulting in a reduction of signal intensities for a given echo time [37].

Since efficient high-frequency RF coils are critically important for high-field MRI and MRS, there is an increasing need for their development for use with high magnetic field strengths [123]. Due to increased RF power deposition, the use of body RF coils is not feasible at high magnetic field. The development of endorectal surface coils has enabled localized 3D ^{1}H MRSI of the human prostate, improving its diagnostic ability in prostate cancer [97]. Microstrip RF coils have been developed for very high-field MRI/MRS [123]. The use of phased-array coils provides a higher SNR ratio than standard volume coils [89]. The improvement of SNR can be used either to reduce the acquisition time or to improve the spatial resolution, allowing shorter scans times or focusing on smaller volumes of interest.

In conclusion, the major strengths of MRS are its ability to provide a wide range of metabolic and functional information, and its ability to integrate with complementary MRI applications. An example of the integration of MRS with MRI is shown in Figure 9.7 to demonstrate the improvement in specificity and sensitivity that can be achieved [102]. With the increasing emphasis on personalized and targeted molecular medicine, MRS can fulfill important requirements in disease discovery and treatment, especially in multifaceted diseases such as cancer.

Acknowledgment

The authors gratefully acknowledge support from P50 CA103175, P30 CA006973, R01 CA73850, R01 CA82337, R01 CA136576, R01 CA138515, R01 CA138264, R21 CA140904, and R21 CA133600.

References

1. Cousins JP. Clinical MR spectroscopy: Fundamentals, current applications, and future potential. *AJR Am J Roentgenol* 1995;164:1337–1347.
2. Gyngell ML, Ellermann J, Michaelis T, Hanicke W, Merboldt KD, Bruhn H, Frahm J. Non-invasive 1H NMR spectroscopy of the rat brain in vivo using a short echo time STEAM localization sequence. *NMR Biomed* 1991;4:150–156.
3. Frahm J, Bruhn H, Hanicke W, Merboldt KD, Mursch K, Markakis E. Localized proton NMR spectroscopy of brain tumors using short-echo time STEAM sequences. *J Comput Assist Tomogr* 1991;15:915–922.
4. Bottomley PA. Spatial localization in NMR spectroscopy in vivo. *Ann N Y Acad Sci* 1987;508:333–348.
5. Ordidge RJ, Bowley RM, McHale G. A general approach to selection of multiple cubic volume elements using the ISIS technique. *Magn Reson Med* 1988;8:323–331.
6. Connelly A, Counsell C, Lohman JAB, Ordidge RJ. Outer volume suppressed image related in vivo spectroscopy (OSIRIS), a high-sensitivity localization technique. *J Magn Reson* 1988;78:519–525.
7. Maudsley AA, Hilal SK, Simon HE, Wittekoek S. In vivo MR spectroscopic imaging with P-31. Work in progress. *Radiology* 1984;153:745–750.
8. Fernandez EJ, Maudsley AA, Higuchi T, Weiner MW. Three-dimensional 1H spectroscopic imaging of cerebral metabolites in the rat using surface coils. *Magn Reson Imaging* 1992;10:965–974.
9. Buchthal SD, Thoma WJ, Taylor JS, Nelson SJ, Brown TR. In vivo T1 values of phosphorus metabolites in human liver and muscle determined at 1.5 T by chemical shift imaging. *NMR Biomed* 1989;2:298–304.
10. Frahm J, Bruhn H, Gyngell ML, Merboldt KD, Hanicke W, Sauter R. Localized high-resolution proton NMR spectroscopy using stimulated echoes: Initial applications to human brain in vivo. *Magn Reson Med* 1989;9:79–93.
11. Ogg RJ, Kingsley PB, Taylor JS. WET, a T1- and B1-insensitive water-suppression method for in vivo localized 1H NMR spectroscopy. *J Magn Reson B* 1994;104:1–10.
12. Tkac I, Starcuk Z, Choi IY, Gruetter R. In vivo 1H NMR spectroscopy of rat brain at 1 ms echo time. *Magn Reson Med* 1999;41:649–656.
13. Star-Lack J, Nelson SJ, Kurhanewicz J, Huang LR, Vigneron DB. Improved water and lipid suppression for 3D PRESS CSI using RF band selective inversion with gradient dephasing (BASING). *Magn Reson Med* 1997;38:311–321.
14. Shungu DC, Glickson JD. Band-selective spin echoes for in vivo localized 1H NMR spectroscopy. *Magn Reson Med* 1994;32:277–284.
15. Thulborn KR, Ackerman JJ. Absolute molar concentrations by NMR in inhomogeneous B1. A scheme for analysis of in vivo metabolites. *J Magn Reson* 1983;55:357–371.
16. Dreher W, Leibfritz D. New method for the simultaneous detection of metabolites and water in localized in vivo 1H nuclear magnetic resonance spectroscopy. *Magn Reson Med* 2005;54:190–195.
17. Barba I, Cabanas ME, Arus C. The relationship between nuclear magnetic resonance–visible lipids, lipid droplets, and cell proliferation in cultured C6 cells. *Cancer Res* 1999;59:1861–1868.
18. Callies R, Sri-Pathmanathan RM, Ferguson DY, Brindle KM. The appearance of neutral lipid signals in the 1H NMR spectra of a myeloma cell line correlates with the induced formation of cytoplasmic lipid droplets. *Magn Reson Med* 1993;29:546–550.

19. Al-Saffar NM, Titley JC, Robertson D, Clarke PA, Jackson LE, Leach MO, Ronen SM. Apoptosis is associated with triacylglycerol accumulation in Jurkat T-cells. *Br J Cancer* 2002;86:963–970.

20. Schmitz JE, Kettunen MI, Hu DE, Brindle KM. 1H MRS-visible lipids accumulate during apoptosis of lymphoma cells in vitro and in vivo. *Magn Reson Med* 2005;54:43–50.

21. Liimatainen TJ, Erkkila AT, Valonen P, Vidgren H, Lakso M, Wong G, Grohn OH, Yla-Herttuala S, Hakumaki JM. 1H MR spectroscopic imaging of phospholipase-mediated membrane lipid release in apoptotic rat glioma in vivo. *Magn Reson Med* 2008;59:1232–1238.

22. Kuesel AC, Briere KM, Halliday WC, Sutherland GR, Donnelly SM, Smith IC. Mobile lipid accumulation in necrotic tissue of high grade astrocytomas. *Anticancer Res* 1996;16:1485–1489.

23. He Q, Shungu DC, van Zijl PC, Bhujwalla ZM, Glickson JD. Single-scan in vivo lactate editing with complete lipid and water suppression by selective multiple-quantum-coherence transfer (Sel-MQC) with application to tumors. *J Magn Reson B* 1995;106:203–211.

24. Serrai H, Senhadji L, Wang G, Akoka S, Stroman P. Lactate doublet quantification and lipid signal suppression using a new biexponential decay filter: Application to simulated and 1H MRS brain tumor time-domain data. *Magn Reson Med* 2003;50:623–626.

25. Sotak CH. A volume-localized, two-dimensional NMR method for the determination of lactate using zero-quantum coherence created in a stimulated echo pulse sequence. *Magn Reson Med* 1988;7:364–370.

26. Williams SR, Gadian DG, Proctor E. A method for lactate detection in vivo by spectral editing without the need for double irradiation. *J Magn Reson* 1986;66:562–567.

27. Gillies RJ, Morse DL. In vivo magnetic resonance spectroscopy in cancer. *Annu Rev Biomed Eng* 2005;7:287–326.

28. Ronen SM, Leach MO. Imaging biochemistry: Applications to breast cancer. *Breast Cancer Res* 2001;3:36–40.

29. Podo F. Tumour phospholipid metabolism. *NMR Biomed* 1999;12:413–439.

30. de Certaines JD, Larsen VA, Podo F, Carpinelli G, Briot O, Henriksen O. In vivo 31P MRS of experimental tumours. *NMR Biomed* 1993;6:345–365.

31. Wolf W, Presant CA, Waluch V. 19F-MRS studies of fluorinated drugs in humans. *Adv Drug Deliv Rev* 2000;41:55–74.

32. Gade TP, Buchanan IM, Motley MW, Mazaheri Y, Spees WM, Koutcher JA. Imaging intratumoral convection: Pressure-dependent enhancement in chemotherapeutic delivery to solid tumors. *Clin Cancer Res* 2009;15:247–255.

33. Aboagye EO, Lewis AD, Johnson A, Workman P, Tracy M, Huxham IM. The novel fluorinated 2-nitroimidazole hypoxia probe SR-4554: Reductive metabolism and semiquantitative localisation in human ovarian cancer multicellular spheroids as measured by electron energy loss spectroscopic analysis. *Br J Cancer* 1995;72:312–318.

34. Workman P. Pharmacokinetics of hypoxic cell radiosensitizers: A review. *Cancer Clin Trials* 1980;3:237–251.

35. Kato Y, Okollie B, Artemov D. Noninvasive 1H/13C magnetic resonance spectroscopic imaging of the intratumoral distribution of temozolomide. *Magn Reson Med* 2006;55:755–761.

36. van Zijl PC, Chesnick AS, DesPres D, Moonen CT, Ruiz-Cabello J, van Gelderen P. In vivo proton spectroscopy and spectroscopic imaging of [1-13C]-glucose and its metabolic products. *Magn Reson Med* 1993;30:544–551.

37. Mountford C, Ramadan S, Stanwell P, Malycha P. Proton MRS of the breast in the clinical setting. *NMR Biomed* 2009;22:54–64.

38. Bolan PJ, Meisamy S, Baker EH, Lin J, Emory T, Nelson M, Everson LI, Yee D, Garwood M. In vivo quantification of choline compounds in the breast with 1H MR spectroscopy. *Magn Reson Med* 2003;50:1134–1143.

39. Callot V, Galanaud D, Le Fur Y, Confort-Gouny S, Ranjeva JP, Cozzone PJ. (1)H MR spectroscopy of human brain tumours: A practical approach. *Eur J Radiol* 2008;67:268–274.

40. Aboagye EO, Bhujwalla ZM. Malignant transformation alters membrane choline phospholipid metabolism of human mammary epithelial cells. *Cancer Res* 1999;59:80–84.

41. Ackerstaff E, Pflug BR, Nelson JB, Bhujwalla ZM. Detection of increased choline compounds with proton nuclear magnetic resonance spectroscopy subsequent to malignant transformation of human prostatic epithelial cells. *Cancer Res* 2001;61:3599–3603.

42. Bhakoo KK, Williams SR, Florian CL, Land H, Noble MD. Immortalization and transformation are associated with specific alterations in choline metabolism. *Cancer Res* 1996;56:4630–4635.

43. Eliyahu G, Kreizman T, Degani H. Phosphocholine as a biomarker of breast cancer: Molecular and biochemical studies. *Int J Cancer* 2007;120:1721–1730.

44. Katz-Brull R, Degani H. Kinetics of choline transport and phosphorylation in human breast cancer cells; NMR application of the zero trans method. *Anticancer Res* 1996;16:1375–1380.

45. Ramirez de Molina A, Gutierrez R, Ramos MA, Silva JM, Silva J, Bonilla F, Sanchez JJ, Lacal JC. Increased choline kinase activity in human breast carcinomas: Clinical evidence for a potential novel antitumor strategy. *Oncogene* 2002;21:4317–4322.

46. Ramirez de Molina A, Penalva V, Lucas L, Lacal JC. Regulation of choline kinase activity by Ras proteins involves Ral-GDS and PI3K. *Oncogene* 2002;21:937–946.

47. Ramirez de Molina A, Rodriguez-Gonzalez A, Gutierrez R, Martinez-Pineiro L, Sanchez J, Bonilla F, Rosell R, Lacal J. Overexpression of choline kinase is a frequent feature in human tumor-derived cell lines and in lung, prostate, and colorectal human cancers. *Biochem Biophys Res Commun* 2002;296:580–583.

48. Noh DY, Ahn SJ, Lee RA, Park IA, Kim JH, Suh PG, Ryu SH, Lee KH, Han JS. Overexpression of phospholipase D1 in human breast cancer tissues. *Cancer Lett* 2000;161:207–214.

49. Iorio E, Mezzanzanica D, Alberti P, Spadaro F, Ramoni C, D'Ascenzo S, Millimaggi D, Pavan A, Dolo V, Canevari S, Podo F. Alterations of choline phospholipid metabolism in ovarian tumor progression. *Cancer Res* 2005;65: 9369–9376.

50. Al-Saffar NM, Troy H, Ramirez de Molina A, Jackson LE, Madhu B, Griffiths JR, Leach MO, Workman P, Lacal JC, Judson IR, Chung YL. Noninvasive magnetic resonance spectroscopic pharmacodynamic markers of the choline kinase inhibitor MN58b in human carcinoma models. *Cancer Res* 2006;66:427–434.

51. Krishnamachary B, Glunde K, Wildes F, Mori N, Takagi T, Raman V, Bhujwalla ZM. Noninvasive detection of lentiviral-mediated choline kinase targeting in a human breast cancer xenograft. *Cancer Res* 2009;69:3464–3471.

52. Glunde K, Raman V, Mori N, Bhujwalla ZM. RNA interference-mediated choline kinase suppression in breast cancer cells induces differentiation and reduces proliferation. *Cancer Res* 2005;65:11034–11043.

53. Vaupel P, Okunieff P, Kluge M. Response of tumour red blood cell flux to hyperthermia and/or hyperglycaemia. *Int J Hyperthermia* 1989;5:199–210.

54. Gillies RJ, Robey I, Gatenby RA. Causes and consequences of increased glucose metabolism of cancers. *J Nucl Med* 2008;49 Suppl 2:24S–42S.

55. Fukumura D, Jain RK. Tumor microenvironment abnormalities: Causes, consequences, and strategies to normalize. *J Cell Biochem* 2007;101:937–949.

56. Gillies RJ, Gatenby RA. Hypoxia and adaptive landscapes in the evolution of carcinogenesis. *Cancer Metastasis Rev* 2007;26:311–317.

57. Robey IF, Baggett BK, Kirkpatrick ND, Roe DJ, Dosescu J, Sloane BF, Hashim AI, Morse DL, Raghunand N, Gatenby RA, Gillies RJ. Bicarbonate increases tumor pH and inhibits spontaneous metastases. *Cancer Res* 2009;69:2260–2268.

58. Stubbs M, Bhujwalla ZM, Tozer GM, Rodrigues LM, Maxwell RJ, Morgan R, Howe FA, Griffiths JR. An assessment of 31P MRS as a method of measuring pH in rat tumours. *NMR Biomed* 1992;5:351–359.

59. Gillies RJ, Liu Z, Bhujwalla Z. 31P-MRS measurements of extracellular pH of tumors using 3-aminopropylphosphonate. *Am J Physiol* 1994;267:C195–203.

60. van Sluis R, Bhujwalla ZM, Raghunand N, Ballesteros P, Alvarez J, Cerdan S, Galons JP, Gillies RJ. In vivo imaging of extracellular pH using 1H MRSI. *Magn Reson Med* 1999;41:743–750.

61. Provent P, Benito M, Hiba B, Farion R, Lopez-Larrubia P, Ballesteros P, Remy C, Segebarth C, Cerdan S, Coles JA, Garcia-Martin ML. Serial in vivo spectroscopic nuclear magnetic resonance imaging of lactate and extracellular pH in rat gliomas shows redistribution of protons away from sites of glycolysis. *Cancer Res* 2007;67:7638–7645.

62. Chan DA, Giaccia AJ. Hypoxia, gene expression, and metastasis. *Cancer Metastasis Rev* 2007;26:333–339.

63. Rockwell S, Dobrucki IT, Kim EY, Marrison ST, Vu VT. Hypoxia and radiation therapy: Past history, ongoing research, and future promise. *Curr Mol Med* 2009;9:442–458.

64. Procissi D, Claus F, Burgman P, Koziorowski J, Chapman JD, Thakur SB, Matei C, Ling CC, Koutcher JA. In vivo 19F magnetic resonance spectroscopy and chemical shift imaging of tri-fluoro-nitroimidazole as a potential hypoxia reporter in solid tumors. *Clin Cancer Res* 2007;13:3738–3747.

65. Mason RP, Shukla H, Antich PP. In vivo oxygen tension and temperature: Simultaneous determination using 19F NMR spectroscopy of perfluorocarbon. *Magn Reson Med* 1993;29:296–302.

66. Kodibagkar VD, Cui W, Merritt ME, Mason RP. Novel 1H NMR approach to quantitative tissue oximetry using hexamethyldisiloxane. *Magn Reson Med* 2006;55:743–748.

67. Chapman JD, Franko AJ, Sharplin J. A marker for hypoxic cells in tumours with potential clinical applicability. *Br J Cancer* 1981;43:546–550.

68. Chapman JD. The detection and measurement of hypoxic cells in solid tumors. *Cancer* 1984;54:2441–2449.

69. Workman P, Maxwell RJ, Griffiths JR. Non-invasive MRS in new anticancer drug development. *NMR Biomed* 1992;5: 270–272.

70. Kim JW, Dang CV. Cancer's molecular sweet tooth and the Warburg effect. *Cancer Res* 2006;66:8927–8930.

71. Rivenzon-Segal D, Margalit R, Degani H. Glycolysis as a metabolic marker in orthotopic breast cancer, monitored by in vivo (13)C MRS. *Am J Physiol Endocrinol Metab* 2002;283:E623–630.

72. Jain RK. Tumor angiogenesis and accessibility: Role of vascular endothelial growth factor. *Semin Oncol* 2002;29:3–9.

73. Le HC, Lupu M, Kotedia K, Rosen N, Solit D, Koutcher JA. Proton MRS detects metabolic changes in hormone sensitive and resistant human prostate cancer models CWR22 and CWR22r. *Magn Reson Med* 2009;62:1112–1119.

74. Morse DL, Raghunand N, Sadarangani P, Murthi S, Job C, Day S, Howison C, Gillies RJ. Response of choline metabolites to docetaxel therapy is quantified in vivo by localized (31)P MRS of human breast cancer xenografts and in vitro by high-resolution (31)P NMR spectroscopy of cell extracts. *Magn Reson Med* 2007;58:270–280.

75. Li C, Penet MF, Winnard P, Jr., Artemov D, Bhujwalla ZM. Image-guided enzyme/prodrug cancer therapy. *Clin Cancer Res* 2008;14:515–522.

76. Thompson CB. Apoptosis in the pathogenesis and treatment of disease. *Science* 1995;267:1456–1462.

77. Hakumaki JM, Poptani H, Sandmair AM, Yla-Herttuala S, Kauppinen RA. 1H MRS detects polyunsaturated fatty acid accumulation during gene therapy of glioma: Implications for the in vivo detection of apoptosis. *Nat Med* 1999;5:1323–1327.

78. Ardenkjaer-Larsen JH, Fridlund B, Gram A, Hansson G, Hansson L, Lerche MH, Servin R, Thaning M, Golman K. Increase in signal-to-noise ratio of >10,000 times in liquid-state NMR. *Proc Natl Acad Sci U S A* 2003;100:10158–10163.

79. Golman K, Ardenkjaer-Larsen JH, Petersson JS, Mansson S, Leunbach I. Molecular imaging with endogenous substances. *Proc Natl Acad Sci U S A* 2003;100:10435–10439.

80. Golman K, in 't Zandt R, Thaning M. Real-time metabolic imaging. *Proc Natl Acad Sci U S A* 2006;103:11270–11275.

81. Kohler SJ, Yen Y, Wolber J, Chen AP, Albers MJ, Bok R, Zhang V, Tropp J, Nelson S, Vigneron DB, Kurhanewicz J, Hurd RE. In vivo 13 carbon metabolic imaging at 3T with hyperpolarized 13C-1-pyruvate. *Magn Reson Med* 2007; 58:65–69.

82. Golman K, Zandt RI, Lerche M, Pehrson R, Ardenkjaer-Larsen JH. Metabolic imaging by hyperpolarized 13C magnetic resonance imaging for in vivo tumor diagnosis. *Cancer Res* 2006;66:10855–10860.

83. Albers MJ, Bok R, Chen AP, Cunningham CH, Zierhut ML, Zhang VY, Kohler SJ, Tropp J, Hurd RE, Yen YF, Nelson SJ, Vigneron DB, Kurhanewicz J. Hyperpolarized 13C lactate, pyruvate, and alanine: Noninvasive biomarkers for prostate cancer detection and grading. *Cancer Res* 2008;68:8607–8615.

84. Gallagher FA, Kettunen MI, Hu DE, Jensen PR, Zandt RI, Karlsson M, Gisselsson A, Nelson SK, Witney TH, Bohndiek SE, Hansson G, Peitersen T, Lerche MH, Brindle KM. Production of hyperpolarized [1,4–13C2]malate from [1,4–13C2] fumarate is a marker of cell necrosis and treatment response in tumors. *Proc Natl Acad Sci U S A* 2009;106:19801–19806.

85. Gallagher FA, Kettunen MI, Day SE, Hu DE, Ardenkjaer-Larsen JH, Zandt R, Jensen PR, Karlsson M, Golman K, Lerche MH, Brindle KM. Magnetic resonance imaging of pH in vivo using hyperpolarized 13C-labelled bicarbonate. *Nature* 2008;453:940–943.

86. Govindaraju V, Young K, Maudsley AA. Proton NMR chemical shifts and coupling constants for brain metabolites. *NMR Biomed* 2000;13:129–153.

87. Baslow MH. Functions of N-acetyl-L-aspartate and N-acetyl-L-aspartylglutamate in the vertebrate brain: Role in glial cell-specific signaling. *J Neurochem* 2000;75:453–459.

88. Howe FA, Barton SJ, Cudlip SA, Stubbs M, Saunders DE, Murphy M, Wilkins P, Opstad KS, Doyle VL, McLean MA, Bell BA, Griffiths JR. Metabolic profiles of human brain tumors using quantitative in vivo 1H magnetic resonance spectroscopy. *Magn Reson Med* 2003;49:223–232.

89. Osorio JA, Ozturk-Isik E, Xu D, Cha S, Chang S, Berger MS, Vigneron DB, Nelson SJ. 3D 1H MRSI of brain tumors at 3.0 Tesla using an eight-channel phased-array head coil. *J Magn Reson Imaging* 2007;26:23–30.

90. Sjobakk TE, Johansen R, Bathen TF, Sonnewald U, Kvistad KA, Lundgren S, Gribbestad IS. Metabolic profiling of human brain metastases using in vivo proton MR spectroscopy at 3T. *BMC Cancer* 2007;7:141.

91. Jagannathan NR, Singh M, Govindaraju V, Raghunathan P, Coshic O, Julka PK, Rath GK. Volume localized in vivo proton MR spectroscopy of breast carcinoma: Variation of water–fat ratio in patients receiving chemotherapy. *NMR Biomed* 1998;11:414–422.

92. Roebuck JR, Cecil KM, Schnall MD, Lenkinski RE. Human breast lesions: Characterization with proton MR spectroscopy. *Radiology* 1998;209:269–275.

93. Gribbestad IS, Singstad TE, Nilsen G, Fjosne HE, Engan T, Haugen OA, Rinck PA. In vivo 1H MRS of normal breast and breast tumors using a dedicated double breast coil. *J Magn Reson Imaging* 1998;8:1191–1197.

94. Katz-Brull R, Seger D, Rivenson-Segal D, Rushkin E, Degani H. Metabolic markers of breast cancer: Enhanced choline metabolism and reduced choline-ether-phospholipid synthesis. *Cancer Res* 2002;62:1966–1970.

95. Jacobs MA, Barker PB, Bottomley PA, Bhujwalla Z, Bluemke DA. Proton magnetic resonance spectroscopic imaging of human breast cancer: A preliminary study. *J Magn Reson Imaging* 2004;19:68–75.

96. Hu J, Yu Y, Kou Z, Huang W, Jiang Q, Xuan Y, Li T, Sehgal V, Blake C, Haacke EM, Soulen RL. A high spatial resolution 1H magnetic resonance spectroscopic imaging technique for breast cancer with a short echo time. *Magn Reson Imaging* 2008;26:360–366.

97. Kurhanewicz J, Vigneron DB, Hricak H, Narayan P, Carroll P, Nelson SJ. Three-dimensional H-1 MR spectroscopic imaging of the in situ human prostate with high (0.24–0.7-cm³) spatial resolution. *Radiology* 1996;198:795–805.

98. Kurhanewicz J, Vigneron DB. Advances in MR spectroscopy of the prostate. *Magn Reson Imaging Clin North Am* 2008;16:697–710, ix–x.

99. Mazurek MP, Prasad PD, Gopal E, Fraser SP, Bolt L, Rizaner N, Palmer CP, Foster CS, Palmieri F, Ganapathy V, Stuhmer W, Djamgoz MB, Mycielska ME. Molecular origin of plasma membrane citrate transporter in human prostate epithelial cells. *EMBO Rep* 2010;11:431–437.

100. Singh KK, Desouki MM, Franklin RB, Costello LC. Mitochondrial aconitase and citrate metabolism in malignant and nonmalignant human prostate tissues. *Mol Cancer* 2006;5:14.

101. Mueller-Lisse UG, Scherr MK. Proton MR spectroscopy of the prostate. *Eur J Radiol* 2007;63:351–360.

102. Jacobs MA, Ouwerkerk R, Petrowski K, Macura KJ. Diffusion-weighted imaging with apparent diffusion coefficient mapping and spectroscopy in prostate cancer. *Top Magn Reson Imaging* 2008;19:261–272.

103. Kurhanewicz J, Swanson MG, Nelson SJ, Vigneron DB. Combined magnetic resonance imaging and spectroscopic imaging approach to molecular imaging of prostate cancer. *J Magn Reson Imaging* 2002;16:451–463.

104. King AD, Yeung DK, Ahuja AT, Tse GM, Chan AB, Lam SS, van Hasselt AC. In vivo 1H MR spectroscopy of thyroid carcinoma. *Eur J Radiol* 2005;54:112–117.

105. Mukherji SK, Schiro S, Castillo M, Kwock L, Muller KE, Blackstock W. Proton MR spectroscopy of squamous cell carcinoma of the extracranial head and neck: In vitro and in vivo studies. *AJNR Am J Neuroradiol* 1997;18:1057–1072.

106. McLean MA, Priest AN, Joubert I, Lomas DJ, Kataoka MY, Earl H, Crawford R, Brenton JD, Griffiths JR, Sala E.

Metabolic characterization of primary and metastatic ovarian cancer by 1H-MRS in vivo at 3T. *Magn Reson Med* 2009;62:855–861.

107. Klomp D, van Laarhoven H, Scheenen T, Kamm Y, Heerschap A. Quantitative 19F MR spectroscopy at 3 T to detect heterogeneous capecitabine metabolism in human liver. *NMR Biomed* 2007;20:485–492.

108. Kamm YJ, Heerschap A, van den Bergh EJ, Wagener DJ. 19F-magnetic resonance spectroscopy in patients with liver metastases of colorectal cancer treated with 5-fluorouracil. *Anticancer Drugs* 2004;15:229–233.

109. Payne GS, Leach MO. Applications of magnetic resonance spectroscopy in radiotherapy treatment planning. *Br J Radiol* 2006;79 Spec No 1:S16–26.

110. Nelson SJ, Graves E, Pirzkall A, Li X, Antiniw Chan A, Vigneron DB, McKnight TR. In vivo molecular imaging for planning radiation therapy of gliomas: An application of 1H MRSI. *J Magn Reson Imaging* 2002;16:464–476.

111. Saraswathy S, Crawford FW, Lamborn KR, Pirzkall A, Chang S, Cha S, Nelson SJ. Evaluation of MR markers that predict survival in patients with newly diagnosed GBM prior to adjuvant therapy. *J Neurooncol* 2009;91:69–81.

112. Murphy PS, Viviers L, Abson C, Rowland IJ, Brada M, Leach MO, Dzik-Jurasz AS. Monitoring temozolomide treatment of low-grade glioma with proton magnetic resonance spectroscopy. *Br J Cancer* 2004;90:781–786.

113. King AD, Yeung DK, Yu KH, Mo FK, Hu CW, Bhatia KS, Tse GM, Vlantis AC, Wong JK, Ahuja AT. Monitoring of treatment response after chemoradiotherapy for head and neck cancer using in vivo (1)H MR spectroscopy. *Eur Radiol* 2010;20:165–172.

114. Zeng QS, Li CF, Zhang K, Liu H, Kang XS, Zhen JH. Multivoxel 3D proton MR spectroscopy in the distinction of recurrent glioma from radiation injury. *J Neurooncol* 2007;84:63–69.

115. Meisamy S, Bolan PJ, Baker EH, Bliss RL, Gulbahce E, Everson LI, Nelson MT, Emory TH, Tuttle TM, Yee D, Garwood M. Neoadjuvant chemotherapy of locally advanced breast cancer: Predicting response with in vivo (1)H MR spectroscopy—A pilot study at 4 T. *Radiology* 2004;233:424–431.

116. Pucar D, Shukla-Dave A, Hricak H, Moskowitz CS, Kuroiwa K, Olgac S, Ebora LE, Scardino PT, Koutcher JA, Zakian KL. Prostate cancer: Correlation of MR imaging and MR spectroscopy with pathologic findings after radiation therapy—Initial experience. *Radiology* 2005;236:545–553.

117. Vaughan JT, Garwood M, Collins CM, Liu W, DelaBarre L, Adriany G, Andersen P, Merkle H, Goebel R, Smith MB, Ugurbil K. 7T vs. 4T: RF power, homogeneity, and signal-to-noise comparison in head images. *Magn Reson Med* 2001;46:24–30.

118. Tkac I, Andersen P, Adriany G, Merkle H, Ugurbil K, Gruetter R. In vivo 1H NMR spectroscopy of the human brain at 7 T. *Magn Reson Med* 2001;46:451–456.

119. Terpstra M, Ugurbil K, Tkac I. Noninvasive quantification of human brain ascorbate concentration using (1)H NMR spectroscopy at 7 T. *NMR Biomed* 2010;23:227–232.

120. Klomp DW, Bitz AK, Heerschap A, Scheenen TW. Proton spectroscopic imaging of the human prostate at 7 T. *NMR Biomed* 2009;22:495–501.

121. Mountford C, Lean C, Malycha P, Russell P. Proton spectroscopy provides accurate pathology on biopsy and in vivo. *J Magn Reson Imaging* 2006;24:459–477.

122. Gruetter R, Weisdorf SA, Rajanayagan V, Terpstra M, Merkle H, Truwit CL, Garwood M, Nyberg SL, Ugurbil K. Resolution improvements in in vivo 1H NMR spectra with increased magnetic field strength. *J Magn Reson* 1998;135:260–264.

123. Zhang X, Ugurbil K, Chen W. Microstrip RF surface coil design for extremely high-field MRI and spectroscopy. *Magn Reson Med* 2001;46:443–450.

124. Jensen LR, Huuse EM, Bathen TF, Goa PE, Bofin AM, Pedersen TB, Lundgren S, Gribbestad IS. Assessment of early docetaxel response in an experimental model of human breast cancer using DCE-MRI, ex vivo HR MAS, and in vivo 1H MRS. *NMR Biomed* 2010;23:56–65.

125. Morse DL, Carroll D, Day S, Gray H, Sadarangani P, Murthi S, Job C, Baggett B, Raghunand N, Gillies RJ. Characterization of breast cancers and therapy response by MRS and quantitative gene expression profiling in the choline pathway. *NMR Biomed* 2009;22:114–127.

126. Bogin L, Papa MZ, Polak-Charcon S, Degani H. TNF-induced modulations of phospholipid metabolism in human breast cancer cells. *Biochim Biophys Acta* 1998;1392:217–232.

Characterizing
Tissue Properties
with Exogenous
Contrast Agents

10

Contrast Agents for T_1-Weighted MRI

Jason R. Buck
Vanderbilt University

Matthew R. Hight
Vanderbilt University

Dewei Tang
Vanderbilt University

H. Charles Manning
Vanderbilt University

10.1 Introduction

Magnetic resonance imaging (MRI) is a powerful noninvasive technique that provides high-quality 3D images of soft tissues, including information on anatomy, function, and metabolism of tissue *in vivo* [1]. An important attribute of magnetic resonance imaging is that numerous clinically important applications do not require the use of exogenous materials to generate sufficient contrast for routine diagnostic purposes. In the context of medical imaging, contrast is the difference in visual properties that renders structures or fluids within a body (or representation image thereof) distinguishable from other objects and the background. This is possible because adjacent tissues of varying structure and chemical composition tend to possess inherently unique nuclear magnetic resonance (NMR) properties such as relaxation times and proton density. As described in greater detail in Chapter 5, appropriate selection of pulse sequences allows one to exploit these unique properties and thus, in many cases, generate satisfactory contrast between neighboring tissues of interest. The lack of a need for contrast agents in many MRI applications represents a fundamental advantage over other clinical imaging modalities such as nuclear imaging methods (PET/SPECT) and certain X-ray techniques. Despite this advantage, non–contrast-enhanced MRI has important limitations, and a growing number of clinically relevant MRI applications require the application of exogenous contrast agents. These applications include, but are not limited to, differentiation of tissue phenotypes that otherwise cannot be differentiated with non–contrast-enhanced MRI, providing specificity to regions of abnormal signal and/or depiction of tissue vascularity and perfusion [2].

As described in detail in Chapters 2, 4, and 5, the observed signal intensity in MRI depends upon a number of determinants that include proton density, T_1 and T_2 relaxation times, magnetic susceptibility, and field gradient. In typical spin-echo imaging, contrast primarily arises from the local value of the longitudinal relaxation rate of water protons ($1/T_1$) and the transverse relaxation rate ($1/T_2$). In spin-echo imaging, signal intensity tends to increase with increasing $1/T_1$ and decrease with increasing $1/T_2$. From this, an obvious goal of MRI contrast agents is T_1 or T_2 modulation. Several MRI contrast agents have been developed and explored *in vitro* and/or in preclinical models. A small number of agents have been translated into the clinic for application in humans, while an even smaller number is employed routinely within the clinic. As of this writing, the list of clinical contrast agents is short (Table 10.1), but with the development of new chemistries and compounds of increasing utility and safety, this number stands to grow significantly. Furthermore, new imaging applications are likely to drive contrast agent development, such as hyperpolarization MRI (see Chapter 22). As with other modalities, MRI contrast agents must satisfy a number of constraints that extend beyond producing a large effect in an imaging study. As with all exogenous materials administered in excess of tracer levels, the materials must be of minimal toxicity, be of a desirable formulation, possess a desirable and timely excretion pathway, and target the appropriate tissue.

In this chapter, we will discuss the physical principles that affect the mechanisms of action of MRI contrast agents. Excellent reviews of this topic already exist and are recommended for further reading [2,3]. The subsequent effects of these physical principles can directly impact the *in vivo* efficacy of the contrast

TABLE 10.1 Clinical Gadolinium-Based Contrast Agents

Chemical Name	Generic Name	Brand Name	Company	Classification
$[Gd(DTPA)(H_2O)]^{2-}$	Gadopentetate dimeglumine	Magnevist®	Bayer (Germany)	Extracellular
$[Gd(DOTA)(H_2O)]^{-}$	Gadoterate meglumine	Dotarem®	Guerbet (France)	Extracellular
$[Gd(DTPA-BMA)(H_2O)]$	Gadodiamide	Omniscan®	GE Healthcare (UK)	Extracellular
$[Gd(HP-DO3A)(H_2O)]$	Gadoteridol	ProHance®	Bracco (Italy)	Extracellular
$[Gd(DO3A-butrol)(H_2O)]$	Gadobutrol	Gadovist®	Bayer (Germany)	Extracellular
$[Gd(DTPA-BMEA)(H_2O)]$	Gadoversetamide	OptiMARK®	Mallinckrodt (United States)	Extracellular
$[Gd(BOPTA)(H_2O)]^{2-}$	Gadobenate dimeglumine	MultiHance®	Bracco (Italy)	Hepatobiliary/extracellular
$[Gd(EOB-DTPA)(H_2O)]^{2-}$	Gadoxetic acid disodium	Eovist®/Primovist®	Bayer (Germany)	Hepatobiliary
$[Gd(MS-325-L)(H_2O)]^{3-}$	Gadofosveset trisodium	Ablavar®/Vasovist®	Lantheus Medical Imaging (United States)/Bayer (Germany)	Blood pool

agent. This is a unique feature of MRI contrast agents and specifically distinguishes them from X-ray or nuclear agents. By surveying a portion of the more prominent compounds known, we will discuss how the design and structure of contrast agents, both small molecule (<800 Da) and macromolecular (>800 Da), can affect relaxation efficacy in ways unique to this modality. We will describe the rational design of "smart" contrast agents in which the agent's relaxation efficiency is dependent upon tissue microenvironment, a characteristic employed to profile localized biochemistry in tissues and fluids of interest. Rather than discuss specific applications or potential new agents, we will focus our discussion on the design and reasons for the selection of the materials that are in clinical use today, with special emphasis on explaining and rationalizing the choices of gadolinium chelates and macromolecular complexes containing primarily gadolinium. Finally, we will briefly introduce molecularly targeted MRI contrast agents and describe many important limitations inherent to their usage.

10.2 Complexes of Gadolinium

Exogenous agents that reduce T_1 or T_2 in tissue do so by affecting the amplitude and time scale of the magnetic field variation experienced by water molecules. Paramagnetic agents, such as transition or rare earth metals, are used in the majority of MRI contrast agents. Paramagnetic agents contain unpaired electrons that are unbalanced within a chemical bond and therefore generate large, residual magnetic dipoles. Accordingly, these ions form strong, local magnetic fields. Protons of water molecules near these ions experience a reduction in the relaxation time of the water protons. Some of the earliest contrast enhancement MR studies used the transition element manganese (Mn). Lauterbur et al. [4] demonstrated the use of manganese in a canine model of myocardial infarction, showing regional T_1 effects in areas of the excised heart *ex vivo*. The first contrast enhancement MRI *in vivo* studies were conducted by investigators from Hammersmith Hospital, London, in the Central Research laboratories of EMI Ltd. In this work, Doyle et al. [5] measured the time course of enhancement by manganese and other paramagnetic ions in rabbit tissues and demonstrated the feasibility of assessing organ

uptake and excretion. The same group also performed the first human contrast studies by demonstrating the effects of oral ferrous ions on the signal from the gastrointestinal tract [6].

Today, most MRI contrast agents employed clinically contain the lanthanide metal gadolinium (Gd), a silvery white malleable and ductile rare earth metal. As previously noted, unpaired electrons lead to large dipole moments. These corresponding large dipole moments imbue the gadolinium with the ability to locally increase the proton relaxation rate of the water within the targeted tissue; this is known as the measurable property of relaxivity (mM^{-1} s^{-1}). The term *relaxation* describes the numerous processes by which nuclear magnetization, prepared in a nonequilibrium state, returns to the equilibrium distribution. In brief, it is how fast the proton spins of water "forget" the direction in which they are oriented. The rates of this spin relaxation can be measured by both spectroscopy and imaging applications. The most common oxidation state of gadolinium for use in MRI contrast agents is 3+ (Gd(III)/Gd^{3+}). Since Gd^{3+} possesses seven unpaired electrons, exactly a half filled d-shell, it potentially has the largest dipole based upon the number of unpaired electrons alone. The Gd^{3+} ion improves imaging contrast by increasing the longitudinal relaxation time (T_1) of proximal water protons, which appear brighter in T_1-weighted images [7]. Nevertheless, unpaired d-shell electrons are not the sole factor that contributes to the observed relaxivity imparted by Gd^{3+}. Interestingly, two other lanthanide ions, dysprosium (Dy^{3+}) and holmium (Ho^{3+}), possess magnetic moments larger than Gd^{3+}. However, the asymmetry of their electronic states leads to rapid electron spin relaxation [1], while the symmetric S-state of Gd^{3+} leads to slower electronic relaxation rates, subsequently improving gadolinium's ability to enhance the relaxation of water.

As with many transition and rare earth metals, Gd^{3+} is a toxic material. This undesirable attribute is complicated by limitations imposed by the inherent, limited sensitivity of MRI. Therefore, unlike the more sensitive nuclear imaging methods, where picogram quantities of a labeled material are typically used in humans, significant quantities of Gd^{3+} (>1.0 g) must be administered to the patient. To avoid toxicity that would certainly arise when administering such high doses, the metal must be stably complexed/bound to an organic chelator. The chelator is a molecule that forms two or more separate bonds with a single, central atom, in this case, Gd^{3+}. By their very nature, Gd^{3+} salts possess low solubility, and chelation thus helps prevent precipitation at tissue pH. Chelation also effectively alters the excretion route and organ distribution relative to the free Gd^{3+} ion, substantially reducing toxicity [3,8,9].

The word *chelate* is derived from the Greek word for claw, which correctly connotes the structure of the complex. Chelation is the formation or presence of two or more separate bonds between a multiple bonded ligand (chelating agent) and a single central atom (Gd^{3+}) (Figure 10.1). Chelating agents/chelators are routinely used to deliver paramagnetic materials such as Gd^{3+} because they reduce metal toxicity, improve solubility compared to the lone Gd^{3+} ion, and improve biodistribution of the material. Unfortunately, there is a trade-off when utilizing chelators, as they tend to exert substantial effects upon the relaxation properties of the metal ion to which they are bound. Typical chelators, such as ethylenediaminetetraacetic acid (EDTA) and diethylene triamine pentaacetic acid (DTPA), form cage-like structures around the metal ions, with high association constants. The association constant is a measure of the bonding affinity of two molecules at equilibrium. This subsequently restricts access of the water molecules to the metal ion (*vide infra*). Accordingly, chelate design and optimization is an important area for chemists interested in developing MRI contrast agents.

Chelates are generally classified according to their geometry (macrocyclic or linear) and whether they are ionic or not, with cyclic ionic chelates considered the least likely to release the ion and hence, typically the most safe [10]. Chelates safety is often measured by stability constants (thermodynamic and conditional) and percentage of free chelating agent (unbound from its metal ion) in solution [10]. The latter attribute is particularly vital in the design of contrast agents, for upon dissociation from the

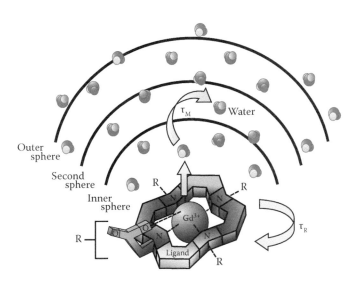

FIGURE 10.1 Relationship between a coordinated metal ion (Gd^{3+}), its chelate, and surrounding (interacting) water molecules that are comprised within multiple spheres (i.e., inner, second, and outer).

chelator, Gd^{3+} can promptly bind to serum proteins and come to reside in the bone in an irreversibly bound state. Chemical factors of the chelate that contribute to stable complexes include high thermodynamic stability, kinetic inertness, and denticity of the ligand systems (the number of atoms of the ligand that bind the metal atom within the complex). Once bound, the final structure is often referred to as the metal–chelate complex, or simply complex.

Structural variation of the chelate imparts different activity and kinetics to the final complex that are manifested through absorption, distribution, metabolism, and excretion (ADME). Within this pharmacological context, the final complex/contrast agent can functionally be viewed as a drug, "a chemical substance used in the treatment, cure, prevention, or diagnosis of disease or used to otherwise enhance physical or mental well-being" [11]. The biological effects and implications of these structural variations are known as the structure–activity relationship (SAR), the relationship between the chemical or 3D structure of a molecule and its biological activity. Analysis of SAR enables the elucidation of key chemical features (macrocyclic or linear) associated with the chelate responsible for evoking a specific biological effect within the organism. This subsequently allows for modification of the effects or potencies of the bioactive compound (contrast agent) by changing its chemical structure (Figure 10.1, through variation of the R groups). Medicinal chemists will often use chemical synthesis techniques to change key chemical features (macrocyclic or linear) or insert new chemical groups (ketone, double bond) into the contrast agent and test the modifications for the biological effects.

Contrast agents currently approved for use in the United States (Table 10.1) fall into one of two categories based upon scaffold (in drug design, scaffold refers to the fixed part of a molecule on which functional groups are substituted or exchanged) geometry: linear, such as DTPA; cyclic, such as 1,4,7,10-tetraazacyclododecane-1,4,7,10-tetraacetic acid (DOTA). Of the eight listed contrast agents currently approved for use by the FDA, seven are based upon the DTPA scaffold (Magnevist®, Omniscan®, MultiHance®, OptiMARK®, Eovist®/Primovist®, Ablavar®/Vasovist®) and one upon the DOTA scaffold (ProHance®) (Figure 10.2).

DTPA is a polyamino carboxylic acid consisting of a diethylenetriamine backbone with five carboxymethyl groups. It is particularly suitable for chelation with Gd^{3+}, forming a complex that is stable within a living organism. The LD_{50} (median lethal dose required to kill half the members of a test population after a specified duration) of the gadolinium complex (Magnevist®/Gd-DTPA) (Figure 10.2), the first reported MRI contrast agent for imaging tumors, inflammation, and vascular lesions, is approximately 10 mmol/kg in rats, nearly half that of common iodinated X-ray agents and 100 times the dose commonly administered in a typical MRI contrast study. Other DTPA analogs include the aforementioned: DTPA-BMA (Omniscan®, for intracranial and spinal lesions in the presence of abnormal vascularity or a compromised blood–brain barrier); BOPTA (MultiHance®, for liver and central nervous system); DTPA-BMEA (OptiMARK®, for lesions of the brain, spine, and liver,

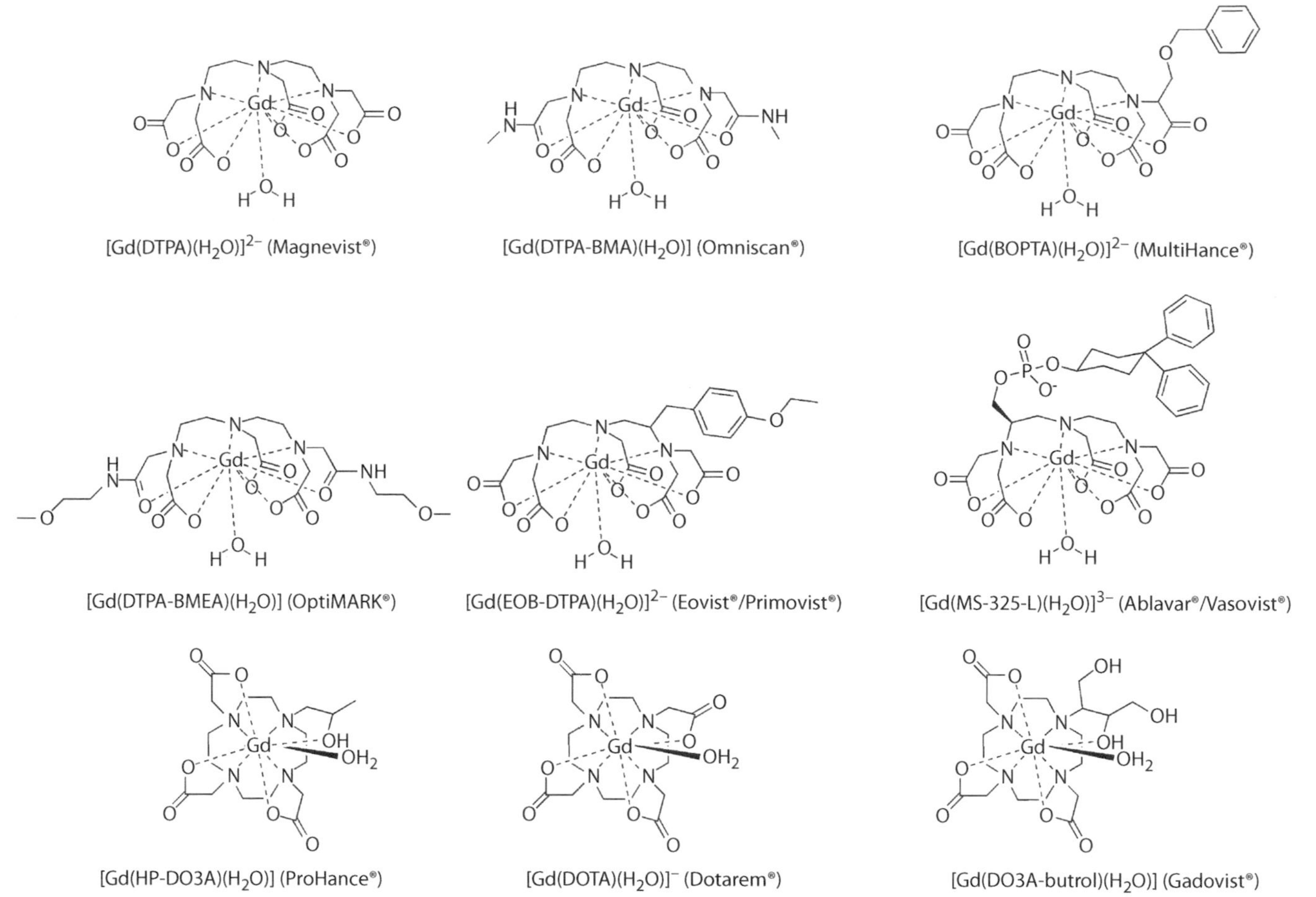

FIGURE 10.2 Clinical gadolinium-based contrast agents.

including tumors); EOB-DTPA (Eovist®/Primovist®, for lesions of the liver); MS-325-L (Ablavar®/Vasovist®, for magnetic resonance angiography/MRA, see Chapter 13) (Figure 10.2).

Cyclic chelators are also commonly employed and result in agents that provide the desired contrast needed for clinical applications. 1,4,7,10-Tetraazacyclododecane-1,4,7,10-tetraacetic acid (DOTA) is a 12-membered cyclen macrocycle modified to form a polyamino carboxylic acid (Figure 10.1, ligand). DOTA, like DTPA, forms simple, nontoxic, and nontargeted imaging agents, safe for human use. Dotarem® exemplifies use of the DOTA scaffold as an effective gadolinium chelator for visualization of intracranial and spinal lesions with an abnormal blood–brain barrier or abnormal vascularity and in whole-body imaging. Other contrast agents in the clinic utilizing DOTA analogues include HP-DO3A (ProHance®) for visualization of intracranial and spinal lesions in the presence of abnormal vascularity or a compromised blood–brain barrier, as well as whole-body contrast-enhanced MRI, including the head, neck, liver, breast, musculoskeletal system, and soft tissue pathologies; DO3A-butrol (Gadovist®) for central nervous system indications and currently in the FDA drug development and approval process for magnetic resonance angiography (MRA, see Chapter 14).

While the majority of these contrast agents are generally safe for use in the clinic, as with any drug, issues and unwanted side effects are possible, such as nephrogenic systemic fibrosis (NSF). Also known as nephrogenic fibrosing dermopathy (NFD), NSF is a rare, relatively recent diagnosis (1997) [12], whose natural history is not well understood. Recent reports have strongly correlated the development of NSF with exposure to gadolinium-based contrast agents (GBCAs) [13]. NSF may occur after exposure to GBCAs, but only in patients either with acute or chronic severe renal insufficiency or with acute renal insufficiency of any severity due to hepatorenal syndrome or in the perioperative liver transplantation period [10]. To date, there is no evidence that other patient groups are at risk [13]. The condition appears to affect male and female populations in approximately equal numbers and has been identified in patients from a variety of ethnic backgrounds (North America, Europe, Asia) [14]. Although confirmed in children and the elderly, it tends to most commonly affect the middle-aged [14]. Currently, the majority of NSF cases have been reported in association with the administration of Omniscan® (gadodiamide, GE Healthcare), with the second highest number of cases associated with the administration of Magnevist® (gadopentetate dimeglumine, Bayer Healthcare)

[13]. Causative association of the advent of NSF is confounded by the fact that GBCAs were already widely in use in renally impaired patients well before identification of the disease and that a few cases of NSF have occurred in patients never exposed to GBCAs [15,16]. Nevertheless, it is appropriate to assume that a potential association may exist, which can be potentially averted if necessary preventative precautions are taken by the healthcare provider(s) [10].

10.3 Effect of Chelators on Metal–Water Interactions

Previously, we noted that complexation of Gd^{3+} with an organic chelator results in a complex that may be relatively nontoxic, with improved biodistribution, but usually with a diminished ability of the Gd^{3+} to affect the relaxation rate of the water protons. This arises because the distance between the paramagnetic center of the complex (Gd^{3+}) and the water molecules to be affected is very important. In fact, the ability of Gd^{3+} to enhance the relaxation rate of neighboring protons is dependent upon the sixth power inverse of this distance [17,18]. Thus, water molecules must be close to the paramagnetic center (inner sphere) for the contrast agent to have a measurable effect and ideally bind directly to the Gd^{3+} (Figure 10.1). However, several major parameters influence relaxivity, including the number of water molecules that coordinate the metal atom (q), rotational correlation time of the metal–chelate complex (τ_R), and mean water residence time (τ_m). The Solomon Bloembergen–Morgan theory further predicts that an increase in τ_R, by slowing down the molecular tumbling of an agent in solution, will result in an increase in relaxivity [19].

Most chelated paramagnetic ions show lower relaxivities than do free ions because of the effects of restricted water access, although these can be offset by positive enhancement effects due to the slower tumbling of a larger complex. A common approach to reducing molecular tumbling is by attaching small Gd^{3+} complexes to a macromolecule through rigid linkages [20–22]. Importantly, the formation of a complex with a multidentate ligand displaces a number of water molecules from the coordination sphere of the metal ion. Alsaadi [23] showed the systematic decrease in relaxation rates for lanthanide complexes with ligands of increasing denticity. An important parameter then becomes the reduction in the number of water molecules bound directly to the metal (q), which is approximately the number of donor atoms of the ligand. Of critical importance is the number of sites for water to access the inner sphere of the ion, but in general, the stability of the complex and the number of sites are inversely related.

Gadolinium ions themselves have $q = 9$ *in aquo*. Both DOTA and DTPA are octadentate ligands, and as such, the complex removes 8 of the 9 water sites, leaving $q = 1$. Depending upon the structure of the ligand/chelate (Figure 10.1, ligand and R, referred as R groups), the association of the water (τ_m) and the rotation of the metal–chelate complex (τ_R) can either be deterred or promoted. The retention of the metal-bound water molecule imparts the performance of Gd^{3+} complexes as MRI contrast agents. This tenet represents a paradoxical tradeoff between denticity and contrast agent performance. To overcome this limitation imposed by chelates on contrast agent performance, one may look to innovative chelate design to improve water access, increase the overall mass of the complex to reduce agent tumbling, and/or design probes that deliver larger payloads of Gd^{3+} to the tissue of interest.

10.4 Macromolecules

There are many advantages to developing higher molecular weight Gd-containing agents. The most common approach to prepare macromolecular structures containing Gd^{3+} chelates involves conjugation of the functionalized chelate to one of three macromolecules: linear polymers; dendrimers; biological molecules. Linear polymers are large molecules composed of a repeated structural unit, typically connected by covalent chemical bonds to form a single backbone with no branches. Dendrimers are large molecules composed of a repeated structural unit (dendron), typically symmetric around a core and often possessing a spherical 3D morphology. Biological molecules are any organic molecule produced by a living organism, including, but not limited to, large polymeric molecules such as nucleic acids, proteins, and polysaccharides. Conjugation of low-molecular-weight chelates (GdDOTA or GdDTPA) to macromolecular species alters both the pharmacological and biophysical properties of the low-molecular-weight agents [24]. For example, conjugation of Gd^{3+} complexes to polymeric material tend to increase the rotational correlation time and may somewhat improve the relaxivity of the gadolinium atom. Furthermore, macromolecules can be decorated with multiple Gd^{3+} complexes, facilitating significant relaxivity enhancement that tends to be essentially additive on a per gadolinium basis. If the macromolecule is a tissue-specific targeting moiety, such as a monoclonal antibody, the conjugate may also have sufficient relaxivity to enable imaging of relatively low-concentration receptors. By virtue of molecular size, high molecular weight conjugates tend to be maintained in circulation and vascular spaces for longer times than monomeric species; therefore, and without surprise, most macromolecular agents tend to be useful for blood pool imaging.

10.4.1 Macromolecules: Linear Species

Gadolinium chelates have been conjugated to polylysine in a variety of molecular weight ranges [25–29], with reported conjugates accommodating nearly 70 Gd^{3+} complexes. Although somewhat lower than anticipated for a truly immobilized monoamide DTPA complex, the relaxivity of polylysine derivatives tend to range from 15 to 20 mM^{-1} s^{-1} at 20 MHz. Lower than predicted relaxivities appear to stem from the flexible nature of the linear polymeric backbone [30,31]. A series of bisamide chelate linear copolymers linked by α,ω-alkyldiamides with polyethylene glycol (PEG) diamines (6.13 mM^{-1} s^{-1}) [32] or a varying number [4–12] of methylenes (9–17 mM^{-1} s^{-1}) [33] have been prepared. These compounds differ from those described previously as the chelates are directly incorporated into the polymeric chain. It has since been proposed that select linear polymers can form intramolecular aggregates, making their structure less rodlike and more globular, thereby increasing the rotational correlation time [3].

10.4.2 Macromolecules: Nonlinear Species

Analogous to linear species, Gd^{3+} complexes have also been conjugated to dendrimers. Dendrimers are prepared by the repeated reaction of the initial, smaller "core" molecules [3]. This highly branched structure has led to a number of interesting molecular attributes for this class of macromolecule [34]. Preeminent among the common dendrimers is the starburst polyamidoamine (PAMAM) series. Depending on the nature of the chelate and the dendrimer structure, the reported relaxivity values of dendrimer conjugates range from ~14 to 36 mM^{-1} s^{-1} at 25 MHz, 37°C. However, some species exhibiting relaxivities approaching (fullerene [35]) and exceeding (nanotube, [36]) 100 mM^{-1} s^{-1} have been reported. These values tend to be higher than those observed for linear analogues because linear polymers based upon polyaminocarboxylate derivatives are known to possess relaxivities limited by slow water exchange, and dendrimers are intrinsically more rigid than their linear counterparts [3,37].

10.4.3 Macromolecules: Protein/ Saccharide Carriers

Select Gd-DTPA derivatives of human or bovine serum albumin (HSA or BSA) have been prepared and explored [38]. Like dendrimers, the intravascular retention of such macromolecules makes them effective blood pool agents for magnetic resonance angiography [39]. There are important practical concerns with using simple albumin-based agents that have precluded their adoption in clinical applications. Aside from possible immunoreactivity of native albumin, these agents are not amenable to heat, limiting conjugation chemistry and sterilization potential. However, a related alternative approach that has been met with greater success employs a paramagnetic metal-containing compound that binds *in vivo* with serum proteins, notably blood albumin, and thus becomes more effective (5- to 10-fold higher relaxivity per ion than Gd-DTPA). This is known as receptor induced magnetization enhancement (RIME), an effective MRI signal amplification strategy whereby receptor binding increases relaxation rates. Targeted MRI contrast agents that utilize the RIME effect are typically bifunctional molecules in which a Gd^{3+} chelate complex is appended to a targeting moiety (HSA) through a tether [40]. The binding is not permanent so that the agent resides intravascularly two to three hours before excretion [41]. Such an approach promises to provide excellent visualization of blood vessels for angiographic imaging [42]. As a scaffold for the attachment of various chelates, dextran has also been explored [43,44]. Dextran is a complex, branched glucan (polysaccharide composed of numerous glucose molecules) composed of chains of varying lengths ranging from 10 to 150 kDa. As

with polylysine, dextrans are available in a variety of molecular weights and can be modified to include chelating agents [45–49]. Dextran affords the advantages of high solubility, limited polydispersity, and robust chemistry with respect to the variability of Gd-chelate linking agents; however, degradation *in vivo* can be a concern if not properly functionalized.

10.4.4 Macromolecules: "Smart" Agents

"Smart" contrast agents can be synthesized to respond to a variety of relevant tissue-based parameters, including pH, enzyme activity, and metal ion gradients. Of these, Gd^{3+}-based complexes aimed at pH sensitivity have been among the best characterized and can be prepared in a number of ways. Typically, pH-sensitive chemistries incorporate at least one structural component into their design that facilitates pH-responsive relaxivity, usually through modulation of the complex hydration spheres (Figure 10.1, inner, second, outer). As an illustrative example of inner sphere modulation, Aime et al. have recently reported a macrocyclic Gd^{3+} complex bearing a pendant β-arylsulfonamide functionality that is highly pH sensitive across a wide range [50]. It has been shown that the pH sensitivity of this probe stems from the change in coordination ability of the β-arylsulfonamide pendant arm at variable pHs; this in turn modulates the number of water molecules directly coordinated to the Gd(III) ion. Alternatively, the Sherry group has shown pH-dependent relaxivity by designing agents that facilitate modulation of outer hydration spheres [51]. Utilizing a tetraamide derivative of DOTA (DOTA-4AmP), the group has shown variable pH sensitivity that is thought to stem primarily from alterations in the hydrogen bond network between coordinated water and pendant phosphonate groups. Contrast agents that enable tissue-based pH measurements will continue to be of particular interest, especially in light of the emerging importance of characterizing the pH gradients known to exist in diseased states such as cancer.

Probes responsive to enzyme activity are becoming more prevalently reported. Several prototypical compounds have been reported where portions of the Gd^{3+} complex serve as substrates for enzyme activity, and the resultant processing of the probe alters the relaxation efficiency. The most useful imaging probes demonstrate increased relaxivity upon enzymatic processing that is usually accomplished by improving water access to the paramagnetic center. An alternative approach was recently suggested where probes were prepared such that following specific enzymatic hydrolysis, the agent was able to bind serum proteins, resulting in an *in situ* macromolecular agent with increased relaxivity [52].

Much consideration has also been devoted toward the development of "smart" contrast agents aimed at reporting concentration gradients of physiologically important metal ions, including calcium [53], and more recently, copper [54]. As with pH-sensitive, Gd^{3+}-based probes, these "smart" agents incorporate complexing ligands that alter their coordination environment in the presence of the metal of interest, which typically results in improved water access and hydration of the paramagnetic center. The *in vivo* utility of metal sensing "smart" agents has not been widely demonstrated, but theoretically these agents could be particularly useful in studies of signal transduction as well as fundamental study of many disease processes.

10.5 Targeted Agents

In theory, small molecules, peptides, and antibodies can be combined with paramagnetic centers to generate disease-specific MRI agents [55]. However, due to the relative insensitivity of MRI, the primary obstacle with delivering such agents remains the quantities of the paramagnetic label delivered [56]. Nevertheless, in select cases, success has been noted. For example, conjugation of polylysine-gadolinium chelates to anticarcinoembryonic antigen monoclonal antibodies has been illustrated [57], whereby paramagnetic loading as high as 24 to 28 metal ions per antibody was obtained without sacrificing immunoreactivity (80–85%) for conjugates PL-Gd-DOTA$_{24-28}$F(ab') and PL-Gd-DTPA$_{24-28}$F(ab)$_2$. *In vivo* studies using these agents on nude mice bearing LS 174T human colorectal carcinoma xenografts showed tumor uptake of the Gd^{3+}-labeled immunoconjugate to be 10–15%. Metal–chelate–dendrimer–antibody constructs for imaging and radioimmunotherapy have also been reported, with the potential to increase overall stability [58]. Pursuant to targeted agents employing small molecules is the work of Manning et al. [59], whereby a known small molecule ligand (PK11195) of the translocator protein (TSPO), conjugated to a trifunctional lanthanide DOTA chelate, proved capable of multimodal imaging of C6 glioblastoma (brain cancer) cells. Density of the expressed target is very critical, especially since many are insufficiently expressed; thus, a significant challenge for future chemists is balancing the need for higher relaxivity with acceptable stability.

10.6 Summary

While the volume of contrast agents typically used in MRI procedures is generally less than that used for contrast-enhanced X-ray procedures. It nevertheless requires tissue concentrations in the millimolar range, unlike agents used in nuclear medicine imaging (positron emission tomography [PET] and single-proton emission computer tomography [SPECT]), where agents are delivered in micromolar concentrations. The future of MRI contrast agents resides firmly in continued innovation of chemical design, from small molecule to macromolecular [60,61]. Ideally, optimization of structure begets optimization of behavior, whereby kinetic and thermodynamic stability, pharmacokinetics/pharmacodynamics, increased relaxivity, and the subsequent cost of such agents will produce contrast agents that are tissue/organ-specific, afford high contrast enhancement at low *in vivo* doses, and are applicable in the clinic. While this chapter has focused exclusively upon Gd^{3+}-based, T_1 contrast agents, there exist other contrast agents that utilize transition metals such as manganese (Teslascan®), which also share in this future. Moreover, there is contrast enhancement through hyperpolarization (see Chapter 22), nuclear spin polarization of a material far beyond thermal equilibrium conditions. Nevertheless, the goal of such research ultimately lies with clinical application. With MRI and the aid of contrast agents, invasive and potentially painful procedures can be precluded and essential diagnostic and possible prognostic information acquired.

References

1. Lauffer RB. Paramagnetic metal-complexes as water proton relaxation agents for NMR imaging—Theory and design. *Chem Rev* 1987;87:901–927.
2. Gore JC, Joers JM, Manning HC, Kennean RP. Contrast agents and relaxation effects. In: Atlas SW, ed. *Magnetic Resonance Imaging of the Brain and Spine*. 4th ed. Philadelphia: Lippincott, Williams & Wilkins; 2008.
3. Caravan P, Ellison JJ, McMurry TJ, Lauffer RB. Gadolinium(III) chelates as MRI contrast agents: Structure, dynamics, and applications. *Chem Rev* 1999;99:2293–2352.
4. Dutton PL, Leigh JS, Scarpa A. *Frontiers of Biological Energetics: Electrons to Tissues*. New York: Academic Press; 1978.
5. Doyle FH, Gore JC, Pennock JM. Relaxation rate enhancement observed invivo by NMR imaging. *J Comput Assist Tomogr* 1981;5:295–296.
6. Witcofski RL, Karstaedt N, Partain CL, *National Cancer Institute (US), Bowman Gray School of Medicine, Vanderbilt University. School of Medicine. NMR Imaging: Proceedings of an International Symposium on Nuclear Magnetic Resonance Imaging*, Bowman Gray School of Medicine of Wake Forest University, Winston-Salem, NC, October 1–3, 1981. Winston-Salem, NC: Bowman Gray School of Medicine, Wake Forest University. Available from the Department of Radiology, Bowman Gray School of Medicine; 1982.
7. Caravan P. Strategies for increasing the sensitivity of gadolinium based MRI contrast agents. *Chem Soc Rev* 2006;35:512–523.
8. Kumar K, Jin TZ, Wang XY, Desreux JF, Tweedle MF. Effect of ligand basicity on the formation and dissociation equilibria and kinetics of Gd3+ complexes of macrocyclic polyamino carboxylates. *Inorganic Chem* 1994;33:3823–3829.
9. Tweedle MF, Wedeking P, Kumar K. Biodistribution of radiolabeled, formulated gadopentetate, gadoteridol, gadoterate, and gadodiamide in mice and rats. *Invest Radiol* 1995;30:372–380.
10. Shellock FG, Spinazzi A. MRI safety update 2008: Part 1, MRI contrast agents and nephrogenic systemic fibrosis. *AJR Am J Roentgenol* 2008;191:1129–1139.
11. Dictionary.com. Drug. 2010.
12. Galan A, Cowper SE, Bucala R. Nephrogenic systemic fibrosis (nephrogenic fibrosing dermopathy). *Curr Opin Rheumatol* 2006;18:614–617.
13. Shellock FG. MRIsafety.com. *MRI Contrast Agents and Nephrogenic Systemic Fibrosis (NSF)* 2010. Accessed July 25, 2010, 2010.

14. Yerram P, Saab G, Karuparthi PR, Hayden MR, Khanna R. Nephrogenic systemic fibrosis: A mysterious disease in patients with renal failure—Role of gadolinium-based contrast media in causation and the beneficial effect of intravenous sodium thiosulfate. *Clin J Am Soc Nephrol* 2007;2:258–263.

15. Anavekar NS, Chong AH, Norris R, Dowling J, Goodman D. Nephrogenic systemic fibrosis in a gadolinium-naive renal transplant recipient. *Australas J Dermatol* 2008;49:44–47.

16. Wahba IM, Simpson EL, White K. Gadolinium is not the only trigger for nephrogenic systemic fibrosis: Insights from two cases and review of the recent literature. *Am J Transplant* 2007;7:2425–2432.

17. Bloembergen L, Morgan LO. Proton relaxation times in paramagnetic solutions. effects of electron spin relaxation. *J Chem Phys* 1961;34:842–850.

18. Solomon I. Relaxation processes in a system of two spins. *Phys Rev* 1955 1955;99:559–565.

19. Toth E, Helm L, Merbach AE. Relaxivity of MRI contrast agents. *Contrast Agents I.* 2002;221:61–101.

20. Song Y, Kohlmeir EK, Meade TJ. Synthesis of multimeric MR contrast agents for cellular imaging. *J Am Chem Soc* 2008;130:6662–6663.

21. Prasuhn DE, Jr., Yeh RM, Obenaus A, Manchester M, Finn MG. Viral MRI contrast agents: Coordination of Gd by native virions and attachment of Gd complexes by azide–alkyne cycloaddition. *Chem Commun (Camb)* 2007;12:1269–1271.

22. Datta A, Hooker JM, Botta M, Francis MB, Aime S, Raymond KN. High relaxivity gadolinium hydroxypyridonate–viral capsid conjugates: Nanosized MRI contrast agents. *J Am Chem Soc* 2008;130:2546–2552.

23. Alsaadi BM, Rossotti FJC, Williams RJP. Hydration of complexone complexes of lanthanide cations. *J Chem Soc Dalton Trans* 1980:2151–2154.

24. Brasch RC. Rationale and applications for macromolecular Gd-based contrast agents. *Magn Reson Med* 1991;22:282–287.

25. Schuhmanngiampieri G, Schmittwillich H, Frenzel T, Press WR, Weinmann HJ. Invivo and invitro evaluation of Gd-DTPA-polylysine as a macromolecular contrast agent for magnetic resonance imaging. *Invest Radiol* 1991;26:969–974.

26. Sieving PF, Watson AD, Rocklage SM. Preparation and characterization of paramagnetic polychelates and their protein conjugates. *Bioconjugate Chem* 1990;1:65–71.

27. Spanoghe M, Lanens D, Dommisse R, Vanderlinden A, Alderweireldt F. Proton relaxation enhancement by means of serum-albumin and poly-L-lysine labeled with DTPA-Gd3+—Relaxivities as a function of molecular-weight and conjugation efficiency. *Magn Reson Imaging* 1992;10:913–917.

28. Desser TS, Rubin DL, Muller HH et al. Dynamics of tumor imaging with Gd-DTPA-polyethylene glycol polymers: Dependence on molecular weight. *J Magn Reson Imaging* 1994;4:467–472.

29. Frank H, Weissleder R, Bogdanov AA, Brady TJ. Detection of pulmonary emboli by using MR-angiography with MPEG-PL-GdDTPA—An experimental study in rabbits. *AJR Am J Roentgenol* 1994;162:1041–1046.

30. Bogdanov AA, Weissleder R, Frank HW et al. A new macromolecule as a contrast agent for MR-angiography—Preparation, properties, and animal studies. *Radiology* 1993;187:701–706.

31. Bogdanov A, Weissleder R, Brady TJ. Long-circulating blood-pool imaging agents. *Adv Drug Delivery Rev* 1995;16:335–348.

32. Toth E, van Uffelen I, Helm L et al. Gadolinium-based linear polymer with temperature-independent proton relaxivities: A unique interplay between the water exchange and rotational contributions. *Magn Reson Chem* 1998;36:S125–S134.

33. Kellar KE, Henrichs PM, Hollister R, Koenig SH, Eck T, Wei D. High relaxivity linear Gd(DTPA)–polymer conjugates: The role of hydrophobic interactions. *Magn Reson Med* 1997;38:712–716.

34. Menjoge AR, Kannan RM, Tomalia DA. Dendrimer-based drug and imaging conjugates: Design considerations for nanomedical applications. *Drug Discovery Today* 2010;15:171–185.

35. Mikawa M, Kato H, Okumura M et al. Paramagnetic water-soluble metallofullerenes having the highest relaxivity for MRI contrast agents. *Bioconjugate Chem* 2001;12:510–514.

36. Hartman KB, Laus S, Bolskar RD et al. Gadonanotubes as ultrasensitive pH-smart probes for magnetic resonance imaging. *Nano Lett* 2008;8:415–419.

37. Toth E, Pubanz D, Vauthey S, Helm L, Merbach AE. The role of water exchange in attaining maximum relaxivities for dendrimeric MRI contrast agents. *Chemistry* 1996;2:1607–1615.

38. Lauffer RB, Brady TJ. Preparation and water relaxation properties of proteins labeled with paramagnetic metal chelates. *Magn Reson Imaging* 1985;3:11–16.

39. Schmiedl U, Ogan M, Paajanen H et al. Albumin labeled with Gd-DTPA as an intravascular, blood pool enhancing agent for MR imaging—Biodistribution and imaging studies. *Radiology* 1987;162:205–210.

40. Caravan P, Greenwood JM, Welch JT, Franklin SJ. Gadolinium-binding helix–turn–helix peptides: DNA-dependent MRI contrast agents. *Chem Commun (Camb)* 2003:2574–2575.

41. Parmelee DJ, Walovitch RC, Ouellet HS, Lauffer RB. Preclinical evaluation of the pharmacokinetics, biodistribution, and elimination of MS-325, a blood pool agent for magnetic resonance imaging. *Invest Radiol* 1997;32:741–747.

42. Li D, Dolan RP, Walovitch RC, Lauffer RB. Three-dimensional MRI of coronary arteries using an intravascular contrast agent. *Magn Reson Med* 1998;39:1014–1018.

43. Casali C, Janier M, Canet E et al. Evaluation of Gd-DOTA-labeled dextran polymer as an intravascular MR contrast agent for myocardial perfusion. *Acad Radiol* 1998;5:S214–S218.

44. Meyer D, Schaefer M, Bouillot A, Beaute S, Chambon C. Paramagnetic dextrans as magnetic-resonance contrast agents. *Invest Radiol* 1991;26:S50–S52.

45. Armitage FE, Richardson DE, Li KCP. Polymeric contrast agents for magnetic resonance imaging: Synthesis and characterization of gadolinium diethylonetriaminepentaacetic acid conjugated to polysaccharides. *Bioconjugate Chem* 1990; 1:365–374.

46. Rebizak R, Schaefer M, Dellacherie E. Polymeric conjugates of Gd3+-diethylenetriaminepentaacetic acid and dextran. 1. Synthesis, characterization, and paramagnetic properties. *Bioconjugate Chem* 1997;8:605–610.

47. Corot C, Schaefer M, Beaute S et al. Physical, chemical and biological evaluations of CMD-A2-Gd-DOTA—A new paramagnetic dextran polymer. *Acta Radiol* 1997;38:91–99.

48. Siauve N, Clement O, Cuenod CA, Benderbous S, Frija G. Capillary leakage of a macromolecular MRI agent, carboxymethyldextran-Gd-DTPA, in the liver: Pharmacokinetics and imaging implications. *Magn Reson Imaging* 1996;14: 381–390.

49. Rebizak R, Schaefer M, Dellacherie E. Polymeric conjugates of Gd3+-diethylenetriaminepentaacetic acid and dextran. 2. Influence of spacer arm length and conjugate molecular mass on the paramagnetic properties and some biological parameters. *Bioconjugate Chem* 1998;9:94–99.

50. Lowe MP, Parker D, Reany O et al. pH-dependent modulation of relaxivity and luminescence in macrocyclic gadolinium and europium complexes based on reversible intramolecular sulfonamide ligation. *J Am Chem Soc* 2001;123:7601–7609.

51. Zhang S, Wu K, Sherry AD. A Novel pH-sensitive MRI contrast agent. *Angew Chem Int Ed Engl* 1999;38:3192–3194.

52. Nivorozhkin AL, Kolodziej AF, Caravan P, Greenfield MT, Lauffer RB, McMurry TJ. Enzyme-activated Gd(3+) magnetic resonance imaging contrast agents with a prominent receptor-induced magnetization enhancement We thank Dr. Shrikumar Nair for helpful discussions. *Angew Chem Int Ed Engl* 2001;40:2903–2906.

53. Li W-J, Fraser SE, Meade TJ. A calcium-sensitive magnetic resonance imaging contrast agent. *J Am Chem Soc* 1999;121: 1413–1414.

54. Que EL, Chang CJ. A smart magnetic resonance contrast agent for selective copper sensing. *J Am Chem Soc* 2006; 128:15942–15943.

55. Accardo A, Tesauro D, Aloj L, Pedone C, Morelli G. Supramolecular aggregates containing lipophilic Gd(III) complexes as contrast agents in MRI. *Coord Chem Rev* 2009; 253:2193–2213.

56. Nunn AD, Linder KE, Tweedle MF. Can receptors be imaged with MRI agents? *Q J Nucl Med* 1997;41:155–162.

57. Curtet C, Maton F, Havet T et al. Polylysine-Gd-DTPA(n) and polylysine-Gd-DOTA(n) coupled to anti-CEA F(ab') (2) fragments as potential immunocontrast agents—Relaxometry, biodistribution, and magnetic resonance imaging in nude mice grafted with human colorectal carcinoma. *Invest Radiol* 1998;33:752–761.

58. Wu C, Brechbeil MW, Kozak RW, Gansow OA. Preparation and characterization of dendrimer-based poly-metal chelates and monoclonal-antibodies conjugates. *J Nucl Med.* May 1994;35:P62–P62.

59. Manning HC, Goebel T, Thompson RC, Price RR, Lee H, Bornhop DJ. Targeted molecular imaging agents for cellular-scale bimodal imaging. *Bioconjug Chem* 2004;15: 1488–1495.

60. Yan G-P, Robinson L, Hogg P. Magnetic resonance imaging contrast agents: Overview and perspectives. *Radiography* 2007;13:e5–e19.

61. Chan KWY, Wong WT. Small molecular gadolinium(III) complexes as MRI contrast agents for diagnostic imaging. *Coord Chem Rev* 2007;251:2428–2451.

Nanoparticles for T_2 and T_2^*-Weighted MRI

Michael Nickels
Vanderbilt University

Wellington Pham
Vanderbilt University

11.1 Introduction

As described in detail in Chapters 2 and 5, the overwhelming majority of magnetic resonance imaging is based on detecting the relaxation and local density characteristics of water protons. Since different tissues have different relaxation rates, they give rise to substantial contrast in a typical MR image. The native contrast between tissues can be enhanced using MRI contrast agents. In the previous chapter, contrast agents were discussed in terms of their ability to impart a direct effect on the surrounding bulk water, thereby providing greater water relaxation in their proximity. This class, known as T_1 contrast agents due to the enhanced longitudinal relaxation they afford, generally possesses a lanthanide metal chelated to a suitable targeting or biocompatible molecule. An alternative approach to increasing the contrast between two tissues is to exploit an enhancement in the transverse relaxation process, T_2, through the introduction of magnetic nanoparticles.

Recall from general chemistry that electrons fall into distinct energy levels around nuclei, and they can either be paired with opposite spins (the Pauli exclusion principle) or remain unpaired. This lack of electron pairing enables elements or compounds to exhibit a net magnetization when placed in an external magnetic field. This intrinsic property of magnetic materials can be divided into six basic categories: diamagnetism, paramagnetism, superparamagnetism, ferrimagnetism, ferromagnetism, and antiferromagnetism [1]. Diamagnetism refers to atoms, ions, or molecules that show either very weak or no net magnetization when placed in an external magnetic field. In contrast, paramagnetism refers to atoms, ions, or molecules that exhibit the ability to accept magnetization when placed in an external field. Ferrimagnetism, ferromagnetism, and antiferromagnetism all refer to the ability of atoms and molecules within a crystal, such as those composed of iron, nickel, cobalt, or Fe_3O_4, to either show a magnetization (ferrimagnetic or ferromagnetic) or to not (antiferromagnetic) when placed in an external magnetic field. The last magnetic classification, which is the most important in the context of this discussion, is that of superparamagnetism.

Superparamagnetic materials are prepared by reducing the overall size of a ferromagnetic or ferrimagnetic crystal until a single magnetic domain particle is formed. In this condition, the particle is composed of thousands of magnetically ordered metal ions. Here, magnetization of the particle represents a single, giant magnetic moment, essentially the sum of all individual magnetic moments. While the size requirement for these substances depends on the material, it has been estimated that the diameter of a spherical sample composed of ferromagnetic material must be less than 300 Å in order to possess superparamagnetic properties [2]. The beneficial characteristic of superparamagnetic materials is that they can achieve complete magnetization even when placed in a weak external magnetic field. Due to thermal agitation, however, these particles cannot retain any of the induced magnetic properties once the field is removed. From a biological perspective, superparamagnetic particles offer many advantages over their paramagnetic counterparts since they can enhance relaxation rates in specific organs, at significantly lower doses, and with no registered toxicity. This is because of their larger magnetic moment, which is generated by thousands of magnetically ordered ions [3]. Furthermore, because of their small size and high magnetic susceptibility, these materials have demonstrated tremendous potential for contrast-enhanced MR imaging.

11.2 Design and Use of T_2 Contrast Agents

Designing stable colloidal materials of uniform size and crystallographic order are key criteria for the *in vivo* application of nanoparticle-based T_2 contrast agents. Specifically, the achievement of high relaxation rates ($R_2 = 1/T_2$) is dependent upon a strong magnetic moment formed by localized electrons; hence, this is strongly dependent on the degree of crystallographic order [4]. Furthermore, the development of nanoparticles that have a high degree of homogeneity would facilitate characterization, reduced toxicity, and enhanced tissue distribution. The efficiency of the particle to be used as an MR contrast agent, its circulation, and its ability to distribute to the targeted tissue depends on the uniform diameter of the particles. Heterogeneous-sized particles (1.5–100 nm) tend to be dramatically pro-inflammatory [5]. This can be attributed to the activation of various types of immune cells that exhibit varying preferences for particles of different sizes. In particular, extremely small nanoparticles are known for being more toxic than larger particles of the same chemical composition since they can be taken up by macrophages that induce the release of cytotoxic cytokines. Another problem associated with heterogeneous particles of colloidal materials is that they may preclude the detection of atypical subpopulations.

Surface modification of the colloid particles for bioconjugation with active biopharmaceutical ligands for targeted imaging is an active area of investigation within nanotechnology. For these reasons, nanoparticles are often designed such that they are composed of three to four distinct sections or layers (Figure 11.1). The innermost layer, referred to as the core, is composed of a paramagnetic metal nanoparticle of Fe, Co, Ni, Mn, Cr, Gd, FeCo, FePt, Fe_3Pt, Fe_3O_4 (magnetite), or γ-Fe_2O_3 (maghemite) [6,7]. Of all the various metals that have been used as cores for nanoparticle imaging, iron has received the most

attention due to its low toxicity, which enables a high *in vivo* concentration of the agent [8]. In particular, superparamagnetic iron oxide nanoparticles (SPIO) have been studied and employed so frequently that they are now considered the representative T_2 MRI contrast agent [9].

The preparation of SPIO nanoparticles can be accomplished by several methods, including microemulsions, flow injection synthesis, hydrothermal reaction, hydrolysis, and thermolysis [10–13]. Each of these methods uses different starting materials and reaction conditions. Thus far, the water-in-oil microemulsion system has proven effective at generating rather small iron nanoparticles from 4 to 8 nm. The chemical reagents used in this synthesis process were ferrous chloride hexahydrate and hexadecyltrimethyl ammonium bromide (HTAB), along with other emulsifying components such as 1-butanol and *n*-octane. The microemulsion is prepared by dissolving HTAB in *n*-octane, followed by the addition of 1-butanol and ferrous chloride aqueous salt solution. Flow injection synthesis also generated small diameter SPIOs of 4–6 nm, albeit through the use of ferrous chloride and ferric chloride salts. The technique involves the controlled injection of the reagents into a carrier stream followed by a chemical reaction that occurs as the reaction mixture travels through a capillary reactor. In this type of synthesis, the hydrodynamic parameters, such as flow rate, have limited influence on particle size and size distribution compared to chemical effects. For instance, sodium hydroxide has been used effectively during the flow injection process to tune the homogeneity of small particles. In the presence of sodium hydroxide, a large number of nuclei were generated, thus reducing the particle size, and maintaining a narrower particle size distribution. Different from other techniques, the hydrothermal synthesis of SPIO nanoparticles generates large particles of approximately 50 nm. Its key reactants are $FeSO_4 \cdot 7H_2O$, ferric chloride salt, and ethylenediamine, which serve both as a source of base and a soft template during extended heating syntheses. The hydrolysis synthesis of SPIOs is useful for generating particles with large form factors from several hundred nanometers to nearly one micrometer. The key chemical reaction process involved with the hydrolysis of iron tri-*n*-butoxide in an octanol/acetonitrile solution. However, the most common method for generating SPIO nanoparticles is the chemical coprecipitation of ferrous and ferric salts in an aqueous environment [14,15]. While a detailed description of these synthetic approaches is beyond the scope of this book, it is important to note that typically, the formation of magnetite Fe_3O_4 under the inert condition shown in thin formula 11.1 is not usually easy to achieve since the ferrous ion is easily oxygenated by oxygen:

$$Fe^{2+} + 2Fe^{3+} + 8OH^- \rightarrow Fe_3O_4 + 4H_2O \tag{11.1}$$

$$4Fe_3O_{4+} O_2 \rightarrow 6\gamma - Fe_2O_3 \tag{11.2}$$

If oxidation occurs, magnetite can be converted into maghemite (γFe_2O_3). This conversion process is characterized by a topotactic transformation in which the iron cations diffuse to the surface,

Surface modification

Coating polymer

Biocompatible material

FIGURE 11.1 Symbolic description of an iron oxide nanoparticle for molecular imaging applications. Various types of polymers have been used to coat the iron core such as dextran, PEG, polyethylene oxide (PEO), or 2-(methacryloyloxy)ethyl phosphorycholine (MPC). The surface modification of iron nanoparticles is a subject of study aimed at using such particles for multimodal and targeted imaging. Depending on the work, surface fabrication can include the attachment of fluorescent dyes, delivery peptides such as Tat or polyarginine peptides, and RNA.

where they are oxidized upon exposure to oxygen (formula 11.2). It is quite difficult to distinguish the two forms since the magnetic properties and the size and shape of maghemite closely resemble the characteristics of magnetite.

The desired physical properties of iron oxide nanoparticles, such as size and shape, can be tuned readily by varying the pH, temperature, or the ratio of Fe^{2+}/Fe^{3+} salts. Additionally, nanoparticle size can also be adjusted by coating the nanoparticles with polymers. Depending on the material used, coating nanoparticles can increase or decrease their overall size. For instance, Massart et al. demonstrated the ability to decrease the diameter of citrate-coated nanoparticles from 8 to 3 nm by increasing the concentration of citrate ions to prevent nucleation [16]. One of the key requirements for the preparation of these agents for *in vivo* imaging is that the particles should be of roughly equal size. Monodispersity is the condition in which all of the nanoparticles share the same size and shape profile. This is essential for high quality molecular imaging since clearance rates are related to the size of the nanoparticles. Ideal particles are required to have a prolonged blood circulation time to maximize their penetration into the small capillaries of targeted tissue. Currently, no "golden rule" exists that specifies the relationship between the sizes of the particles relative to the clearance rate. Nevertheless, observations suggest that SPIOs larger than 30 nm usually tend to be cleared more rapidly through extravasations and renal routes (several minutes) compared to the 1–2 h needed for ultra small SPIO particles (USPIOs), which are smaller than 30 nm [17,18].

Besides tuning particle size to leverage biodistribution, the fabrication of surface characteristics on the SPIO nanoparticles is another critical design factor that has been useful for extending the particles' circulation time. Furthermore, coating the surface of SPIOs also enhances their hydrophilicity and colloidal stability, thus allowing for greater biocompatibility [19]. A variety of coating polymer types, such as polyethylene glycol, dextran, and specialized dendrons, to name just a few, have been used to enhance plasma half life and biocompatibility for SPIOs; specifically, coating SPIO nanoparticles tends to avoid quick opsonization compared to the bare counterparts [20,21]. Colloidal stability refers to the particles' ability to remain isolated in solution and avoid aggregating to form large groups of nanoparticles in a process known as flocculation. For SPIOs, colloidal stability has been shown to be pH dependent, with an optimal pH of around 7 for nonfunctionalized particles. Since physiological pH is approximately 7.4, nonfunctionalized nanoparticles are thus unsuitable for longitudinal *in vivo* applications. However, coating the particles with a suitable polymer or simple compounds capable of interacting and remaining on the iron core enables the nanoparticles to remain in the colloidal state, even across a broad pH range. A specific example of this concept involves coating maghemite (γ-Fe_2O_3) nanoparticles with a mixture of gluconic and citric acid to yield a particle that does not undergo flocculation throughout the pH range of 3.5 to 11 [22]. If gluconic acid was used as the sole coating material, then the ferrofluid in such an environment is considered

acidic, given the pK of carboxylate from gluconic acid is about 3.56. Consequently, the particles are stabilized by the charges of the surface hydroxyl groups. The adsorption of gluconic acid on the surface of the maghemite does not guarantee the stability of the particles in a physiological environment (pH 7.4). To overcome this, citric acid is used as a second coating material to provide a suitable ionizable function to enhance biocompatibility.

While physiological stability enables the nanoparticle imaging agent to be useful for *in vivo* studies, it is desirable that the agents either have a means of accumulating in the intended area of an organism or the ability to target specific disease states. To accomplish these goals, the surface of the nanoparticles must be functionalized with an appropriate targeting agent, ideally, one that is specific to the type of tissue being imaged. Otherwise, the nanoparticles must be delivered to the target tissue by means of a suitable delivery agent. Such agents can include something as seemingly simple as the cells themselves, as in the representative example of using human neural precursor cells, or they can be extremely complex to synthesize and employ, as in the example of polymeric vesicles designed to target folate-positive tumors [23,24]. An alternative to encapsulating the nanoparticles within a delivery agent is the attachment of the nanoparticles to the surface of a larger agent such as mesoporous silica nanoparticles [25]. Examples of this delivery approach have become abundant in the primary literature and have diversified to include the use of monocytes, bone marrow stem cells, embryonic stem cells, mesenchymal stem cells, T cells, macrophages, smooth muscles, and bacteria, with additional examples appearing quite frequently [26–31].

11.3 Targeted Nanoparticles for T_2-Weighted Imaging

Functionalized nanoparticles designed to accumulate at specific biological targets represent one of the largest and most diverse groups being used as imaging agents. Examples of the molecular processes imaged include carbohydrate receptor function, hematopoietic cells, neural progenitor cells, lung cancer, HER2/neu receptor positive breast cancer, transferrin receptor, hepatic asialoglycoprotein receptors, and many others have been studied as well (Table 11.1) [32–43]. As shown in Figure 11.1, the attachment of a targeting agent is accomplished by appending it to a suitable linker or a biocompatible component on the surface of the nanoparticle shell. The principle behind this approach is that if the SPIO surface is coated with a molecule or molecular species that offers an affinity high enough for an appropriate biological target, they will interact with that target and remain in the region of interest. The SPIOs remain in the bloodstream and are ultimately cleared by the liver or kidneys. A large body of research has demonstrated that this approach has been successful in a variety of different biological processes and with an even larger range of functionalized SPIOs and targeting agents or target specific ligands [42,44,45]. However, preparing and employing these agents successfully must first begin with the appropriate surface moiety, which is capable of either direct

TABLE 11.1   Targeted Nanoparticles for T_2-Weighted Imaging

Labeled Ligand	Nanoparticle Type	Molecular Target	Reference
Oligonucleotides	Aminated SPIO	Molecular interactions	[38]
E-selectin abs	CLIO	Proinflammatory marker	[39]
Peptides	CLIO	Cathepsin B	[40]
CD8+ T cells	CLIO	Cancer	[41]
siRNA	SPIO	Cancer	[42]
Folic acid	SPIO	Cancer	[43]
Carbohydrate	SPIO	Cancer	[32]
Tat peptide	CLIO	Stem cells	[33]
Peptide	SPIO	Cancer	[34]
Herceptin abs	CLIO	Cancer	[35]

conjugation to a target-specific ligand or can easily be modified via chemical means for attachment to the ligand.

A major problem with dextran- or PEG-coated SPIOs is that the surface functional groups are essentially alcohol groups (–OH), which are considered weak nucleophiles relative to labeling nanoparticles with other biological active ligands. Converting the alcohol functionality into a more reactive group is a task that can be accomplished using normal solution phase chemistry; however, SPIOs do not follow the same chemical and physical rules specified for solution phase chemistry. This means that converting them into a compound akin to an amine (–NH$_2$) or carboxylic acid (–COOH) is not a straightforward operation. Luckily, a variety of small molecular "linkers" has been developed, which allows for a more simplistic transition from a nonreactive to a reactive functionality [46–49]. One such example uses a new class of linker molecule that possesses a highly reactive epoxide on one end of the molecule and a primary amine on the other, which are separated by several carbon atoms. This linker was shown to efficiently convert a large number of surface alcohols to amines on dextran-coated SPIO nanoparticles. The nanoparticles were further functionalized with folate and shown to have a high affinity for folate receptor-positive cells [43].

As mentioned previously, the use of a suitable linker molecule with a proper targeting agent has been shown in multiple studies to enhance contrast within the target tissue; however, obtaining a suitable level of nanoparticle accumulation can be accomplished in a variety of ways. The ideal method is to place a very high affinity ligand on the surface of a nanoparticle that has a very high affinity for the target tissue or disease state compared with the surrounding tissue or nondiseased tissue. Of note, the targeting of Her-2/neu receptor-positive breast cancer cell lines was shown to be a promising example of this approach [50]. The contrast agent was prepared with polysaccharide-coated SPIOs. These were then conjugated into streptavidin molecules, which allowed for direct and selective attachment of biotinylated Herceptin, a specific monoclonal antibody that binds selectively to Her-2 protein. This approach represents an example of functionalizing

a nanoparticle with a high affinity species that exhibits a high affinity for the targeting ligand, which in turn has a high affinity for the target molecular species. An alternative approach is to functionalize the nanoparticle with a molecule that has limited affinity for anything *in vivo*, but can be "switched on" by applying some form of external stimulus or molecular target. An example of this concept is the use of spiropyran-functionalized nanoparticles. Spiropyrans are a class of molecules that shift conformation between hydrophobic and hydrophilic in response to light. Under specific physiological conditions, the hydrophilic form of the spiropyran molecule is well solvated and remains in solution in contrast to the hydrophobic form, which tends to aggregate under the same conditions. Using this concept, a spiropyran derivative was conjugated to dextran-coated SPIO and shown to aggregate when irradiated at 563 nm [51]. Nevertheless, the use of this type of imaging agent remains limited due to truncated light penetration through skin and tissue at this wavelength. In an effort to avoid the use of external stimuli, a variety of probes target internal species to either aggregate or accumulate as a result of protein–protein or protein–nucleic acid interactions. In this magnetic relaxation switch approach, the formation or interactions of the individual entities prompted a change in the aggregation state of SPIO particles that were labeled to the interacting moieties. An example of this is the targeting of calcium ions via the conjugation of SPIOs with either the calcium-binding protein calmodulin (CaM) with short peptide sequences, such as M13 or RS20, which are derived from rabbit skeletal muscle myosin light chain kinase and smooth muscle myosin light chain kinase, respectively [52]. These peptides are known to associate and dissociate reversibly as a function of calcium concentration. In that way, if CaM-SPIO were exposed to either M13- or RS20-SPIO in the presence of calcium, it would then lead to a selective aggregation and subsequently large decrease in T_2 relaxivity. In a similar approach, Perez et al. developed an innovative relaxation assay where cross-linked iron oxide nanoparticles are conjugated with telomeric-specific oligonucleoties and tested for their ability to report the concentration and enzymatic activity of telomerase in crude tissue samples [53].

11.4 Nanomedicine and Treatment Applications

In addition to imaging of physiological processes, SPIOs can be manipulated in a manner that allows for the treatment of disease or the selective targeting of conditions, such as cancer, which results in a form of treatment that offers pronounced advantages over conventional anticancer treatments such as surgery, radiotherapy, chemotherapy, hormones, and immunotherapy [54]. As discussed previously, cancer targeting can be accomplished via surface functionalization of the SPIO. One of the major shortcomings associated with placing a targeting agent on the surface of the SPIO is the synthetic difficulty associated with preparing the agent and validating that it has been done correctly. An imaging approach that circumvents this problem is found in the preparation of SPIOs with biocompatible coatings and no targeting agent, allowing for an increased systemic circulation time. With an increased circulation time, nanoparticles have a greater likelihood of accumulating within tumors by taking advantage of the enhanced permeability and retention effect present in many types of tumors [55]. This greater accumulation permits differentiation of the tumor from the surrounding tissue, which will generally reflect a lower amount of imaging agent. However, while this approach is both straightforward and simplistic, it does not hold significant therapeutic potential. Functionalization of the polymer-coated particles with therapeutic agents does not guarantee accumulation of the particle within tumors. Moreover, it can actually prevent such accumulation due to decreased circulation time caused by the very presence of the agent on the surface. With that in mind, one must envision agents of increasing sophistication. One potential approach is take advantage of the multivalency offered from nanotechnology: the use of multilayered nanoparticles that can be imaged as well as employed as selective therapeutic agents. This can be accomplished by starting with an amino-modified silica nanoparticle core, which is subsequently functionalized with magnetite nanoparticles, followed by the attachment of gold seed nanoparticles directly onto the silica core. A gold shell can then be grown around both the silica nanoparticle and the magnetite particles. The surface of the gold can then be further functionalized with polyethylene glycol in order to increase the agent's biocompatibility. This type of agent was shown to be effective at shortening the T_2 times of water and was shown to undergo a temperature increase when subjected to femtosecond laser bursts at a wavelength of 800 nm [56]. The temperature increase experienced by the nanoparticle was noted as being sufficient to destroy any cells within its proximity. These characteristics can be useful for targeted therapy once the agent has been functionalized, which can also be accomplished by using the polyethylene glycol surface polymers already present. Moreover, this probe has demonstrated selectivity for HER2/neu receptor-positive breast cancer cell lines when functionalized with the antibody anti-HER2/neu (Ab$_{HER2/neu}$).

11.5 T_2^*-Weighted Imaging

As described in Chapters 2 and 5, T_2^* relaxation refers to the decay of transverse magnetization generated by a combination of magnetic field inhomogeneities and spin–spin transverse relaxation [57]. Further, T_2^*-based imaging has demonstrated great promise when using SPIO contrast agents to image liver cancer. In this approach, a tumor is visualized as an area with a higher signal than the surrounding liver parenchyma, without decreasing signal reflecting decreased uptake of SPIO due to decreased reticuloendothelial function in the tumor [58]. In contrast to the agents that have been discussed thus far, the SPIOs used in these types of studies are not typically functionalized with any targeting agent or attached to any type of directing agent. Rather, they are simply coated with a biocompatible polymer and injected into the study subject. Studies have shown that accumulation of SPIOs within the liver tissue is sufficiently high within just 10 minutes of injection to obtain usable images [59].

This class of imaging has demonstrated utility in a variety of studies that examined hepatocellular carcinoma and related conditions [60–62]. Results have indicated that T_2^*-weighted imaging allows for the detection of both well-differentiated carcinomas and hepatic metastases. For instance, acute hepatitis or liver cirrhosis are associated with a decrease in both the number of Kupffer cells and their phagocytotic functions, which results in hyperintensity lesion detection on SPIO-enhanced T_2^*-weighted images [63]. It has also been shown that there is an advantage to using this approach over traditional X-ray computed tomography (CT) scanning when considering well-differentiated carcinomas. While this finding does not necessarily imply that a CT scan should not be done if a contrast-enhanced MRI is available, it still encourages clinicians to consider both studies when diagnosing potential hepatocellular carcinomas [58]. Studies have also been conducted to determine if combining SPIO-enhanced T_2^*-weighted gradient recalled echo (GRE) and T_2-weighted turbo-spin echo (TSE) sequences in sequentially acquired gadolinium- and SPIO-enhanced MRI exhibit synergistic value compared to each sequence alone, for the detection of hepatocellular carcinomas. The results demonstrated that a combination of the two contrast agents affords better diagnostic capability than each sequence individually for the detection of hepatic tumor [64].

11.6 Concluding Remarks

In the past decade, nanotechnology has contributed significantly to molecular imaging. For example, iron oxide nanoparticles stand out as being among of the most commonly used T_2- and T_2^*-based imaging probes for several imaging applications, from basic to clinical studies. Aside from being used in the form of a nonspecific contrasting agent that extravasates into cancer tissue due to the abundance of the vasculature network, several other innovative chemical modifications enable the particles to be functionalized with the intended functional groups or ligands for targeted imaging studies. While it is increasingly recognized that iron oxide nanoparticles are appreciated for their remarkable stability, long shelf-life, low toxicity, and biocompatibility, it is undoubtedly true that the multivalency afforded by surface modifications make iron nanoparticles versatile probes that can be used for delivery, targeting, and multimodal imaging. One of the current limitations in the development of SPIO imaging probes is the difficulty encountered in characterizing the syntheses. Clearly, issues such as these should be investigated thoroughly, since success in this aspect is an element that could facilitate the translation of targeted SPIO-based imaging to future clinical studies and applications.

References

1. Saini S, Frankel RB, Stark DD, Ferrucci JT, Jr. Magnetism: A primer and review. *AJR Am J Roentgenol* 1988;150:735–743.
2. Bean CP, Livingston JD. Superparamagnetism. *J Appl Phys* 1959;30:120S–129S.

3. Merbach AE, Toth E, editors. *The Chemistry of Contrast Agents in Medical Magnetic Resonance Imaging*; 2001. 471 pp.

4. Morales MP, Bomati-Miguel O, Perez de Alejo R, Ruiz-Cabello J, Veintemillas-Verdaguer S, O'Grady K. Contrast agents for MRI based on iron oxide nanoparticles prepared by laser pyrolysis. *Journal of Magnetism and Magnetic Materials* 2003;266:102–109.

5. Venier C, Guthmann MD, Fernandez LE, Fainboim L. Innate-immunity cytokines induced by very small size proteoliposomes, a *Neisseria*-derived immunological adjuvant. *Clinical and Experimental Immunology* 2007;147:379–388.

6. Fang C, Zhang M. Multifunctional magnetic nanoparticles for medical imaging applications. *J Mater Chem* 2009;19: 6258–6266.

7. Zeng H, Li J, Liu JP, Wang ZL, Sun S. Exchange-coupled nanocomposite magnets by nanoparticle self-assembly. *Nature* 2002;420:395–398.

8. Weissleder R, Stark DD, Engelstad BL, Bacon BR, Compton CC, White DL, Jacobs P, Lewis J. Superparamagnetic iron oxide: Pharmacokinetics and toxicity. *AJR Am J Roentgenol* 1989;152:167–173.

9. Na HB, Hyeon T. Nanostructured T1 MRI contrast agents. *J Mater Chem* 2009;19:6267–6273.

10. Chin AB, Yaacob II. Synthesis and characterization of magnetic iron oxide nanoparticles via w/o microemulsion and Massart's procedure. *J Mater Process Technol* 2007;191:235–237.

11. Salazar-Alvarez G, Muhammed M, Zagorodni AA. Novel flow injection synthesis of iron oxide nanoparticles with narrow size distribution. *Chem Eng Sci* 2006;61:4625–4633.

12. Wan J, Chen X, Wang Z, Yang X, Qian Y. A soft-template-assisted hydrothermal approach to single-crystal Fe3O4 nanorods. *J Crystal Growth* 2005;276:571–576.

13. Kimata M, Nakagawa D, Hasegawa M. Preparation of monodisperse magnetic particles by hydrolysis of iron alkoxide. *Powder Technol* 2003;132:112–118.

14. Jeong U, Teng X, Wang Y, Yang H, Xia Y. Superparamagnetic colloids: Controlled synthesis and niche applications. *Adv Mater* 2007;19:33–60.

15. Kobukai S, Baheza R, Cobb Jared G, Virostko J, Xie J, Gillman A, Koktysh D, Kerns D, Does M, Gore John C, Pham W. Magnetic nanoparticles for imaging dendritic cells. *Magn Reson Med* 2010;63:1383–1390.

16. Bee A, Massart R, Neveu S. Synthesis of very fine maghemite particles. *J Magn Magn Mater* 1995;149:6–9.

17. Yang X, Pilla S, Grailer JJ, Steeber DA, Gong S, Chen Y, Chen G. Tumor-targeting, superparamagnetic polymeric vesicles as highly efficient MRI contrast probes. *J Mater Chem* 2009;19:5812–5817.

18. Lawaczeck R, Menzel M, Pietsch H. Superparamagnetic iron oxide particles: Contrast media for magnetic resonance imaging. *Appl Organometal Chem* 2004;18:506–513.

19. Weissleder R, Bogdanov A, Neuwelt EA, Papisov M. Long-circulating iron oxides for MR imaging. *Adv Drug Deliv Rev* 1995;16:321–333.

20. Duanmu C, Saha I, Zheng Y, Goodson BM, Gao Y. Dendron-functionalized superparamagnetic nanoparticles with switchable solubility in organic and aqueous media: Matrices for homogeneous catalysis and potential MRI contrast agents. *Chem Mater* 2006;18:5973–5981.

21. Mornet S, Vasseur S, Grasset F, Duguet E. Magnetic nanoparticle design for medical diagnosis and therapy. *J Mater Chem* 2004;14:2161–2175.

22. Fauconnier N, Bee A, Roger J, Pons JN. Adsorption of gluconic and citric acids on maghemite particles in aqueous medium. *Progr Coll Polymer Science* 1996;100:212–216.

23. Focke A, Schwarz S, Foerschler A, Scheibe J, Milosevic J, Zimmer C, Schwarz J. Labeling of human neural precursor cells using ferromagnetic nanoparticles. *Magn Reson Med* 2008;60:1321–1328.

24. Song H-T, Choi J-S, Huh Y-M, Kim S, Jun Y-W, Suh J-S, Cheon J. Surface modulation of magnetic nanocrystals in the development of highly efficient magnetic resonance probes for intracellular labeling. *J Am Chem Soc* 2005;127:9992–9993.

25. Lee JE, Lee N, Kim H, Kim J, Choi SH, Kim JH, Kim T, Song IC, Park SP, Moon WK, Hyeon T. Uniform mesoporous dye-doped silica nanoparticles decorated with multiple magnetite nanocrystals for simultaneous enhanced magnetic resonance imaging, fluorescence imaging, and drug delivery. *J Am Chem Soc*;132:552–557.

26. Bulte Jeff WM, Duncan Ian D, Frank Joseph A. In vivo magnetic resonance tracking of magnetically labeled cells after transplantation. *J Cereb Blood Flow Metab* 2002;22:899–907.

27. Zelivyanskaya ML, Nelson JA, Poluektova L, Uberti M, Mellon M, Gendelman HE, Boska MD. Tracking superparamagnetic iron oxide labeled monocytes in brain by high-field magnetic resonance imaging. *J Neurosci Res* 2003;73:284–295.

28. Hinds KA, Hill JM, Shapiro EM, Laukkanen MO, Silva AC, Combs CA, Varney TR, Balaban RS, Koretsky AP, Dunbar CE. Highly efficient endosomal labeling of progenitor and stem cells with large magnetic particles allows magnetic resonance imaging of single cells. *Blood* 2003;102: 867–872.

29. Bos C, Delmas Y, Desmouliere A, Solanilla A, Hauger O, Grosset C, Dubus I, Ivanovic Z, Rosenbaum J, Charbord P, Combe C, Bulte Jeff WM, Moonen Chrit TW, Ripoche J, Grenier N. In vivo MR imaging of intravascularly injected magnetically labeled mesenchymal stem cells in rat kidney and liver. *Radiology* 2004;233:781–789.

30. Kraitchman DL, Heldman AW, Atalar E, Amado LC, Martin BJ, Pittenger MF, Hare JM, Bulte JW. In vivo magnetic resonance imaging of mesenchymal stem cells in myocardial infarction. *Circulation* 2003;107:2290–2293.

31. Riviere C, Boudghene FP, Gazeau F, Roger J, Pons J-N, Laissy J-P, Allaire E, Michel J-B, Letourneur D, Deux J-F. Iron oxide nanoparticle-labeled rat smooth muscle cells: Cardiac MR imaging for cell graft monitoring and quantitation. *Radiology* 2005;235:959–967.

32. El-Boubbou K, Zhu DC, Vasileiou C, Borhan B, Prosperi D, Li W, Huang X. Magnetic Glyco-nanoparticles: A tool to detect, differentiate, and unlock the glyco-codes of cancer via magnetic resonance imaging. *J Am Chem Soc*;132:4490–4499.

33. Lewin M, Carlesso N, Tung C-H, Tang X-W, Cory D, Scadden DT, Weissleder R. Tat peptide-derivatized magnetic nanoparticles allow in vivo tracking and recovery of progenitor cells. *Nat Biotechnol* 2000;18:410–414.

34. Huang G, Zhang C, Li S, Khemtong C, Yang S-G, Tian R, Minna JD, Brown KC, Gao J. A novel strategy for surface modification of superparamagnetic iron oxide nanoparticles for lung cancer imaging. *J Mater Chem* 2009;19:6367–6372.

35. Huh Y-M, Jun Y-W, Song H-T, Kim S, Choi J-S, Lee J-H, Yoon S, Kim K-S, Shin J-S, Suh J-S, Cheon J. In vivo magnetic resonance detection of cancer by using multifunctional magnetic nanocrystals. *J Am Chem Soc* 2005;127:12387–12391.

36. Bulte JWM, Zhang SC, Van Gelderen P, Herynek V, Jordan EK, Duncan ID, Frank JA. Neurotransplantation of magnetically labeled oligodendrocyte progenitors: Magnetic resonance tracking of cell migration and myelination. *Proc Natl Acad Sci U S A* 1999;96:15256–15261.

37. Schaffer BK, Linker C, Papisov M, Tsai E, Nossiff N, Shibata T, Bogdanov A, Jr., Brady TJ, Weissleder R. MION-ASF: Biokinetics of an MR receptor agent. *Magn Reson Imaging* 1993;11:411–417.

38. Perez JM, Josephson L, O'Loughlin T, Hoegemann D, Weissleder R. Magnetic relaxation switches capable of sensing molecular interactions. *Nat Biotechnol* 2002;20:816–820.

39. Kang HW, Josephson L, Petrovsky A, Weissleder R, Bogdanov A, Jr. Magnetic resonance imaging of inducible E-selectin expression in human endothelial cell culture. *Bioconjugate Chem* 2002;13:122–127.

40. Kircher MF, Josephson L, Weissleder R. Ratio imaging of enzyme activity using dual wavelength optical reporters. *Mol Imaging* 2002;1:89–95.

41. Kircher MF, Allport JR, Graves EE, Love V, Josephson L, Lichtman AH, Weissleder R. In vivo high resolution three-dimensional imaging of antigen-specific cytotoxic T-lymphocyte trafficking to tumors. *Cancer Res* 2003;63:6838–6846.

42. Medarova Z, Pham W, Farrar C, Petkova V, Moore A. In vivo imaging of siRNA delivery and silencing in tumors. *Nat Med* 2006;13:372–377.

43. Nickels M, Xie J, Cobb J, Gore JC, Pham W. Functionalization of iron oxide nanoparticles with a versatile epoxy amine linker. *J Mater Chem*;20:4776–4780.

44. Park J, Yu MK, Jeong YY, Kim JW, Lee K, Phan VN, Jon S. Antibiofouling amphiphilic polymer-coated superparamagnetic iron oxide nanoparticles: Synthesis, characterization, and use in cancer imaging in vivo. *J Mater Chem* 2009;19:6412–6417.

45. Medarova Z, Pham W, Kim Y, Dai G, Moore A. In vivo imaging of tumor response to therapy using a dual-modality imaging strategy. *Int J Cancer* 2006;118:2796–2802.

46. Young KL, Xu C, Xie J, Sun S. Conjugating Methotrexate to magnetite (Fe3O4) nanoparticles via trichloro-*s*-triazine. *J Mater Chem* 2009;19:6400–6406.

47. Hwang H-Y, Kim I-S, Kwon IC, Kim Y-H. Tumor targetability and antitumor effect of docetaxel-loaded hydrophobically modified glycol chitosan nanoparticles. *J Controlled Rel* 2008;128:23–31.

48. Ke J-H, Lin J-J, Carey JR, Chen J-S, Chen C-Y, Wang L-F. A specific tumor-targeting magnetofluorescent nanoprobe for dual-modality molecular imaging. *Biomaterials* 2010;31:1707–1715.

49. Liu S, Jia B, Qiao R, Yang Z, Yu Z, Liu Z, Liu K, Shi J, Ouyang H, Wang F, Gao M. A novel type of dual-modality molecular probe for MR and nuclear imaging of tumor: Preparation, characterization and in vivo application. *Mol Pharmaceutics* 2009;6:1074–1082.

50. Artemov D, Mori N, Okollie B, Bhujwalla ZM. MR molecular imaging of the Her-2/neu receptor in breast cancer cells using targeted iron oxide nanoparticles. *Magn Reson Med* 2003;49:403–408.

51. Osborne EA, Jarrett BR, Tu C, Louie AY. Modulation of T2 relaxation time by light-induced, reversible aggregation of magnetic nanoparticles. *J Am Chem Soc* 2010;132:5934–5935.

52. Atanasijevic T, Shusteff M, Fam P, Jasanoff A. Calcium-sensitive MRI contrast agents based on superparamagnetic iron oxide nanoparticles and calmodulin. *Proc Natl Acad Sci U S A* 2006;103:14707–14712.

53. Perez JM, Grimm J, Josephson L, Weissleder R. Integrated nanosensors to determine levels and functional activity of human telomerase. *Neoplasia* 2008;10:1066–1072.

54. Kim DK, Dobson J. Nanomedicine for targeted drug delivery. *J Mater Chem* 2009;19:6294–6307.

55. Lee H, Lee E, Kim DK, Jang NK, Jeong YY, Jon S. Antibiofouling polymer-coated superparamagnetic iron oxide nanoparticles as potential magnetic resonance contrast agents for in vivo cancer imaging. *J Am Chem Soc* 2006;128:7383–7389.

56. Kim J, Park S, Lee JE, Jin SM, Lee JH, Lee IS, Yang I, Kim J-S, Kim SK, Cho M-H, Hyeon T. Designed fabrication of multifunctional magnetic gold nanoshells and their application to magnetic resonance imaging and photothermal therapy. *Angew Chem Int Ed* 2006;45:7754–7758, S7754/7751.

57. Chavhan GB, Babyn PS, Thomas B, Shroff MM, Haacke EM. Principles, techniques, and applications of T_2^*-based MR imaging and its special applications. *Radiographics* 2009;29:1433–1449.

58. Hori M, Murakami T, Kim T, Tomoda K, Nakamura H. CT scan and MRI in the differentiation of liver tumors. *Dig Dis* 2004;22:39–55.

59. Reimer P, Muller M, Marx C, Balzer T. Evaluation of the time window for Resovist-enhanced T2-weighted MRI of the liver. *Acad Radiol* 2002;9 Suppl 2:S336–338.

60. Kim Seong H, Lee Won J, Lim Hyo K, Park Cheol K. SPIO-enhanced MRI findings of well-differentiated hepatocellular carcinomas: Correlation with MDCT findings. *Korean J Radiol* 2009;10:112–120.

61. Araki T. SPIO-MRI in the detection of hepatocellular carcinoma. *J Gastroenterol* 2000;35:874–876.

62. Nasu K, Kuroki Y, Nawano S, Kuroki S, Tsukamoto T, Yamamoto S, Motoori K, Ueda T. Hepatic metastases: Diffusion-weighted sensitivity-encoding versus SPIO-enhanced MR imaging. *Radiology* 2006;239:122–130.

63. Hirohashi S, Hirohashi R, Kitano S, Morimoto K, Yamamoto K, Marugami N, Ueda K, Kichikawa K. SPIO-enhanced MR imaging for liver tumors with pathologic correlation. *Jpn–Deutche Med Ber* 2004;49:15–22.

64. Kim YK, Lee YH, Kwak HS, Kim CS, Han YM. Small malignant hepatic tumor detection in gadolinium- and ferucarbotran-enhanced magnetic resonance imaging: Does combining ferucarbotran-enhanced T_2^*-weighted gradient echo and T2-weighted turbo spin echo images have additive efficacy? *Korean J Radiol* 2008;9:510–519.

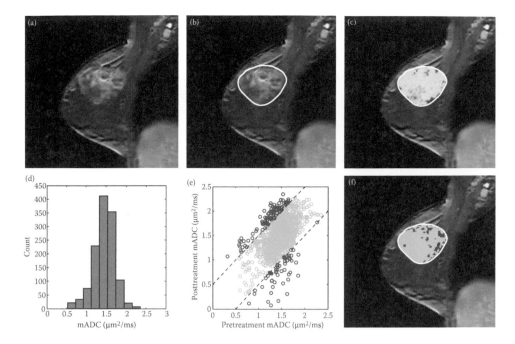

FIGURE 7.6 Examples of common analysis methods in ADC of breast cancer. (a–b) Before performing any quantitative comparisons, the ROI must be defined, either manually (b) or with an automated algorithm. Typically, an anatomical image, such as a contrast enhanced T_1-weighted image (a), is used to define the ROI. (c) For a ROI analysis, the ROI is applied to the ADC map, and the statistical properties (e.g., mean value) of the ADC values within the ROI are then compared either to another ROI of healthy tissue within the same patient, across time points, or between subject populations. (d) In a histogram analysis, the ADC values within the ROI are grouped into bins, and the number of voxels is plotted as a function of the ADC range for the bin. The shape and statistical characteristics of the histograms are then compared. (e–f) fDM provides a means of comparing ADC values across time points with spatial specificity. ADC maps acquired at two different time points (e.g., pretreatment, post-treatment) first have to be aligned. Then, ADC values within the ROI at the second time point are plotted against the corresponding values at the first time point. 95% confidence intervals are then used to determine voxels within the ROI demonstrating significant increases (red) and decreases (blue) in ADC value between the two time points. This information can then be translated back to the parametric map to display where the changes occurred within the ROI.

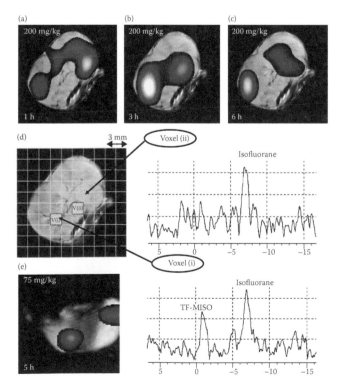

FIGURE 9.3 ^{1}H T_2-weighted image of MCa tumor together with the corresponding metabolic map of TF-MISO following a 200-mg/kg dose at (a) 1, (b) 3, and (c) 6 h after injection. (d) Anatomic T_2-weighted image with superimposed grid showing the in-plane voxel size and distribution around the tumor. Two representative localized spectra from two voxels showing (i) TF-MISO and isofluorane peaks and (ii) only the isofluorane peak. (e) TF-MISO metabolite map acquired 5 h after an injection of 75-mg/kg dose overlaid on its anatomic reference. (Adapted from Procissi, D. et al., *Clin Cancer Res*, 13, 3738–3747, 2007. With permission.)

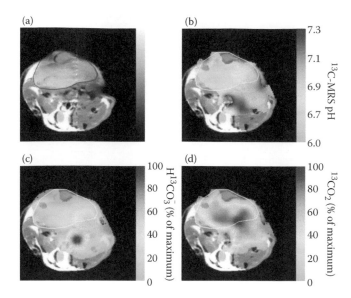

FIGURE 9.4 *In vivo* tumor pH. (a) Transverse ^{1}H MR image of a mouse with a subcutaneous EL4 tumor (outlined in red). (b) pH map of the same animal calculated from the ratio of the $H^{13}CO_3^-$ (shown in c) and $^{13}CO_2$ (shown in d). Voxel intensities in the ^{13}C MRS images were acquired 10 s after intravenous injection of hyperpolarized $H^{13}CO_3^2$. The spatial distribution of (c) $H^{13}CO_3^2$ and (d) $^{13}CO_2$ are displayed as voxel intensities relative to their respective maxima. The tumor margin in b, c, and d is outlined in white. (Adapted from Gallagher, F. A. et al., *Nature*, 453, 940–943, 2008. With permission.)

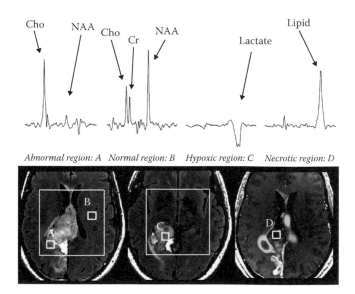

FIGURE 9.5 ^{1}H MRSI data from a grade IV glioma patient demonstrating the spatial heterogeneity of the tumor shown within the representative voxels. T_2-FLAIR images with the PRESS-selected volume superimposed are shown for two different axial slices. The T_1-weighted image acquired after injection of gadolinium is shown with a superimposed total choline/NAA ratio color map. (Adapted from Osorio, J. A. et al., *J Magn Reson Imaging*, 26, 23–30, 2007. With permission.)

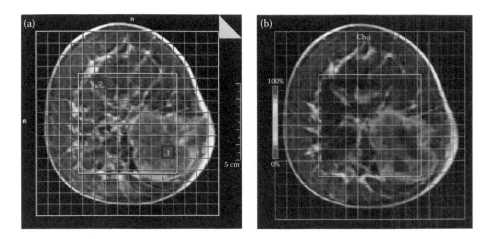

FIGURE 9.6 Illustration of a "hot" area in a breast cancer lesion. (a) Coronal DCE-MRI of a patient with invasive breast cancer. (b) Color-coded image of total choline superimposed on the corresponding MRI. (Adapted from Hu, J. et al., *Magn Reson Imaging*, 26, 360–366, 2008. With permission.)

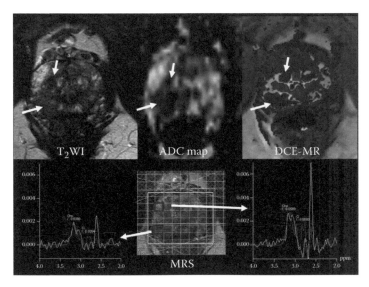

FIGURE 9.7 T_2-weighted image, DWI/ADC map, MRS, and DCE-MRI in a 61-year-old man with prostate cancer (Gleason score = 6). Multiparametric MRI of the prostate gland revealed focal areas of dark T_2 signal with corresponding decreased ADC and increased total choline along with decreased citrate. Positive DCE-MRI kinetics with increased permeability in the right midgland can be observed, which correlated to the site of prostate cancer. In the normal region (anterior to the lesion) bright on T_2-weighted image, there is normal citrate with a normal ADC map and DCE parameters. (Adapted from Jacobs, M. A. et al., *Top Magn Reson Imaging*, 19, 261–272, 2008. With permission.)

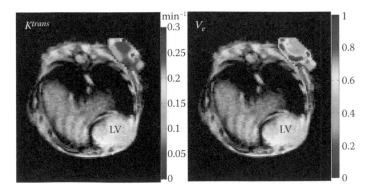

FIGURE 12.8 Axial cross section through the chest of a mouse. Within the field of view is the left ventricle (LV), the liver, and lungs. The left panel shows a parametric K^{trans} map overlain on the anatomical image, while the right panel displays a parametric v_e map.

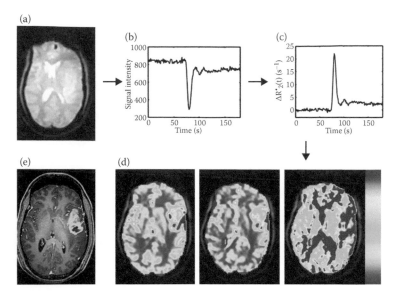

FIGURE 13.1 An *in vivo* example of a typical DSC-MRI experiment in a patient with a grade IV glioblastoma multiforme (GBM). (a) Example echo planar image that is used to acquire serial images with a 1-s temporal resolution. (b) Following CA injection the MRI signal reduces in regions containing functional vessels. (c) MRI signals are converted to the change in the transverse relaxation rate, a parameter that is proportional to the tissue CA concentration time curves. (d) Kinetic analysis of the CA concentration time curves is used to derive maps of BF, BV, and MTT. (e) A high resolution T_1-weighted image acquired after the contrast agent injection is shown to highlight the enhancing tumor volume in the brain.

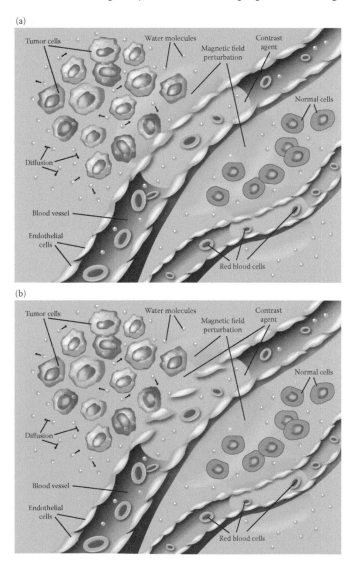

FIGURE 13.2 (a) The compartmentalization of a paramagnetic contrast agent inside normal brain blood vessels can induce magnetic field perturbations in the surrounding tissues. (b) In brain tumor tissues, contrast agent can extravasate, potentially leading to additional field perturbations. As water protons diffuse through these perturbations, they lose phase coherence, resulting in shorter T_2 and T_2^* times and a decrease in the MRI signal intensity.

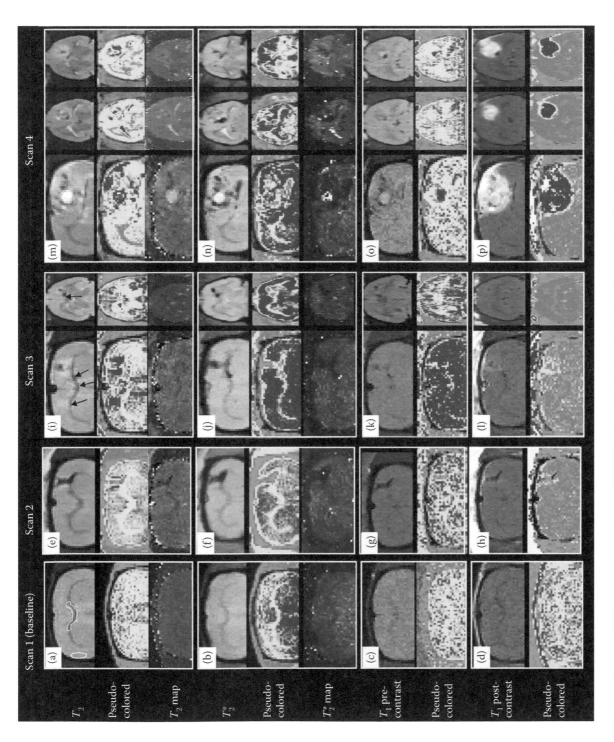

FIGURE 17.9 MR images of a tumor/GRID-labeled neuronal stem cell (NSC) implanted animal acquired at baseline (scan 1, a–d), at 6 days (scan 2, e–h), at 10 days (scan 3, i–l), and at 17 days (scan 4, m–p). Axial orientations were obtained from multiplanar reconstructions using the nICE software. To provide a better visualization of the different signal intensities, pseudocolor-coded images were generated. Cold and warm colors indicate low and high signal intensities, respectively. Arrows point to regions with a signal attenuation in the corpus callosum caused by migrating GRID-labeled NSCs. At the last time point (scan 4, m–p), the tumor appeared very heterogeneous on all image sequences. (Reprinted from Brekke, C. et al., *NeuroImage*, 37 (3), 769–782, 2007. With permission.)

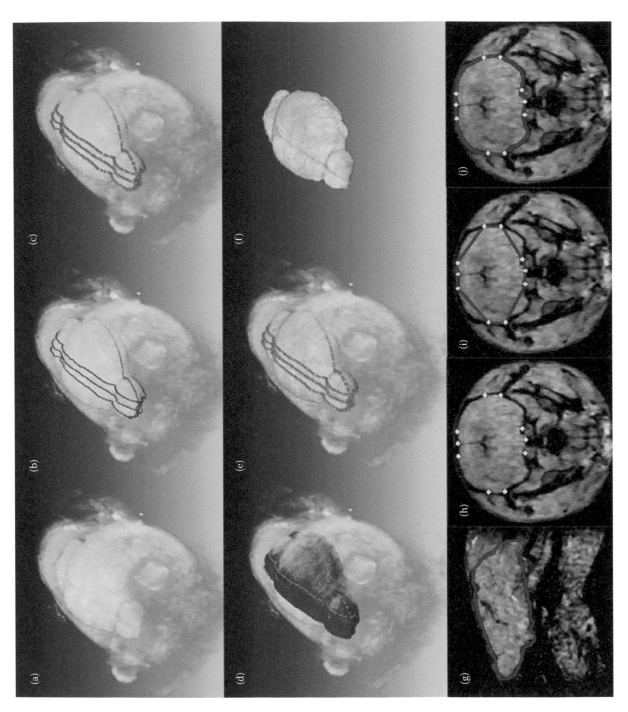

FIGURE 17.10 Results of the brain extraction using 3D constraint level set (CLS) on the T_1 image of a mouse. (a) Mouse head MRI. (b) Brain contours were drawn on sagittal and axial slices. (c) Constraint points were extracted from the contours. (d) Initial zero level surface was constructed using the constraint points as vertices. (e) Final surface after 60 iterations. (f) Segmented brain using CLS method. (g) Contour drawn on the sagittal slice. (h) Constraint points extracted from contours on a coronal slice. (i) Initial surface contour on the coronal slice. (j) Final surface contour on the coronal slice. (Reprinted from Uberti, M. G. et al., *J. Neurosci. Methods*, 179 (2), 338–344, 2009. With permission.)

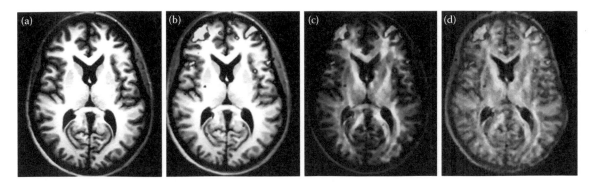

FIGURE 19.1 Simple overlay of four MRI parameter maps: T_1-weighted (gray), fMRI activation map (yellow/red), FA map (green), and MTR map (blue).

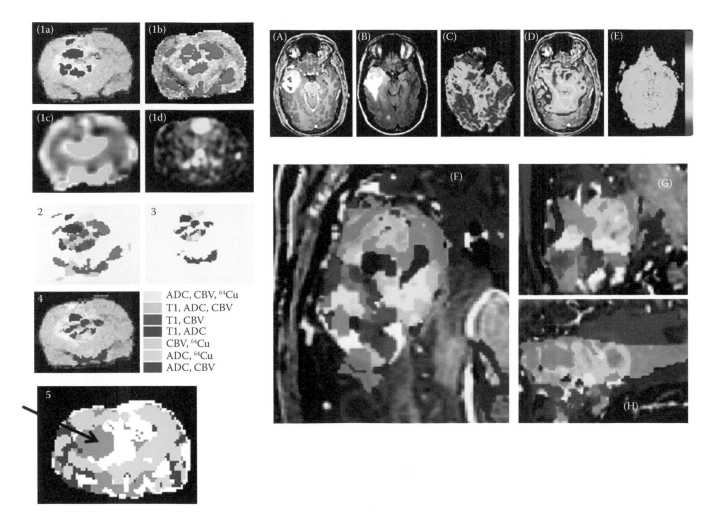

ADC, CBV, ^{64}Cu
T1, ADC, CBV
T1, CBV
T1, ADC
CBV, ^{64}Cu
ADC, ^{64}Cu
ADC, CBV

FIGURE 19.5 Composite map analysis of a preclinical animal study (rat). Input images include postcontrast T_1 (1a), ADC (1b), CBV (1c), and PET image with ^{64}Cu-ATSM (1d). Threshold images are shown in panels 2 and 3. Thresholds are applied and overlaid on the postcontrast T_1-weighted image (panel 4). Colors correspond to the legend on the bottom right and show the most concurrent parameters in the lower medial aspect of the contrast-enhancing lesion in yellow and orange. (Panel 5) Using the same input images, analysis was then performed with the HMRF segmentation method. While these computationally intensive, automated methods are more sophisticated, interpretation of the final maps is nontrivial. However, four image classes were identified, and the tumor is clearly seen (arrow) as a separate tissue class. (A–F). Multiparametric imaging performed preoperatively on a glioma patient. (A) Postcontrast T_1. (B) Postcontrast T_2-FLAIR. (C) Cerebral blood volume. (D) Choline/creatine. (E) Apparent diffusion coefficient. The color images use the continuous color bar on the far right. Composite map (F–H), magnified to show tumor area. Axial (F), coronal (G), and sagittal (H) views of the tumor region. The map shows a highly active, heterogeneous environment, but areas in yellow show regions where all four parameters overlap. While surgery to debulk the tumor was done, maps such as these can give treating physicians increased insight into the functional environment in addition to the anatomical structure.

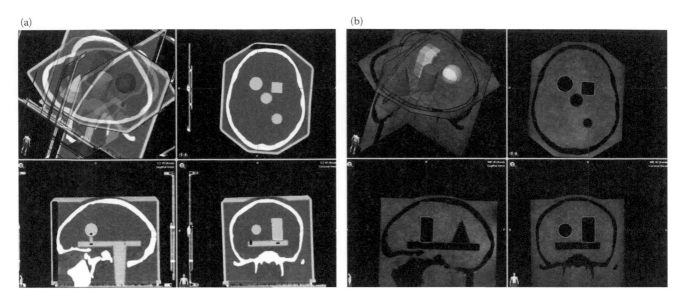

FIGURE 20.3 Comparison of 3D, axial, sagittal, and coronal views of the Radionics Brain Phantom, with (a) CT reconstructed from 2-mm slices and (b) MRI reconstructed from 1.5-mm slices. The four geometric objects were contoured based on CT (a) and MRI (b) images, respectively.

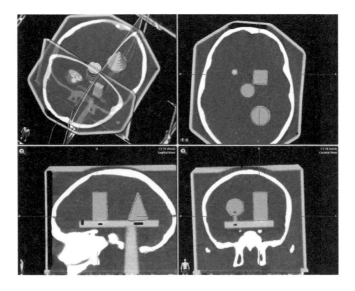

FIGURE 20.4 Contours of four different objects delineated based on CT and MR images mapped onto CT images after image registration are shown in red and blue, respectively, on 3D, axial, sagittal, and coronal views of the Radionics Brain Phantom.

12

Dynamic Contrast-Enhanced MRI: Data Acquisition and Analysis

Mary E. Loveless
Vanderbilt University

Thomas E. Yankeelov
Vanderbilt University

12.1 Qualitative Introduction

12.1.1 Biological Motivation

One of the fundamental phenomena to observe in assessing tumor growth and treatment response is vascularity. In order to grow beyond approximately 1–2 mm³, a tumor must recruit and form new vasculature, as it can no longer rely on the passive diffusion of nutrients [1]. During the process known as angiogenesis, several growth factors such as platelet-derived growth factor (PGF), fibroblast growth factor (FGF), and vascular endothelial growth factor (VEGF) are secreted to promote the migration of endothelial cells to the site. However, the vessels produced from tumor-associated angiogenesis differ from those of normal vasculature. Normal vasculature is arranged as a progression from large arteries feeding into smaller arteries/arterioles which terminate at the capillary bed. From the capillary bed, venules carry the deoxygenated blood to small veins, which empty into larger primary veins. During tumor angiogenesis, the imbalance to antiangiogenic and angiogenic factors can lead to abnormal vasculature formation. Tumor vasculature has been reported to be leaky and poorly constructed, with tortuous topology containing incomplete vessels [1–3].

The importance of tumor angiogenesis has led to the development of many drugs designed to essentially cut off the tumor's blood supply, thereby stunting tumor growth indefinitely. Thus, noninvasive imaging of tumor vasculature or vascular characteristics has become essential for tumor treatment response and therapeutic assessment of drugs that target this pathway [4].

12.1.2 Overview of Dynamic Contrast-Enhanced MRI

One technique that has been shown to assess physiological characteristics of tumor vasculature, including perfusion, blood vessel permeability, blood volume, and the extravascular extracellular volume fraction, is dynamic contrast-enhanced MRI (DCE-MRI) [5,6]. This technique characterizes the pharmacokinetics of an injected contrast agent (CA) as it enters and exits a region or tissue of interest (ROI/TOI). Typical, clinically employed, MRI contrast agents are based on gadolinium (Gd) chelates. As the CA passes through the vasculature, the unpaired electrons in the Gd interact with water hydrogen nuclei, shortening the tissue's native the T_1, T_2, and T_2^*. This shortening of relaxation times will lead to a signal intensity increase on a T_1-weighted image and a signal intensity decrease on a T_2- or T_2^*-weighted image. (This chapter is exclusively concerned with T_1 effects; see Chapter 11 for a discussion of T_2 effects.) By acquiring T_1-weighted images of the MRI signal over time, the kinetics of the CA can be modeled, and biologically relevant parameters can be extracted that represent changes in, for example, blood flow, permeability, and tissue volume fractions. For this type of quantification, three main components, which will be discussed in further detail below, are required: (1) a T_1 map prior to the injection of the contrast agent, (2) serial T_1-weighted images acquired during the injection, and (3) the arterial input function (AIF). Figure 12.1 demonstrates an example time intensity curve (TIC) selected from a voxel within the tumor.

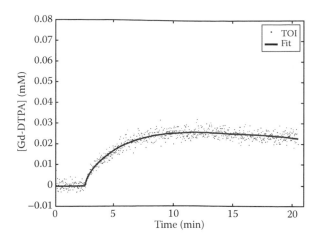

FIGURE 12.1 An example of the change in concentration within the tissue of interest (TOI) over time. The filled circles represent actual data while the solid line represents the model fit to this data.

Each image acquisition represents one time point, and the resulting TIC from each voxel can be analyzed with a mathematical model. Since the kinetics of the CA depend on the physiological characteristics of the tissue, pathological conditions such as cancer show distinct differences from that of normal tissue. This observation can be exploited in order to diagnose cancer and monitor treatment.

12.2 Quantitative Introduction

12.2.1 T_1 Mapping Techniques

As mentioned above, conversion from signal intensity to CA concentration is not as straightforward as CT or PET. The time course of CA concentration is required because these are the data that are used in pharmacokinetic analysis. In a typical DCE-MRI acquisition, there is not enough time to capture a T_1 map at each sample point during a bolus injection. So, in order to calibrate the changes in signal intensity during the injection time course, images to generate a T_1 map must be collected prior to injection to establish the native tissue relaxation rate, T_{10}. Recall from Chapters 2–5 that once magnetization has been tipped into the transverse plane, it will seek to re-establish thermal equilibrium at a rate of R_1 ($\equiv 1/T_1$). How quickly it relaxes back is based on the interaction of spins with their surrounding; hence, T_1 relaxation is also called spin-lattice relaxation (see Chapter 2). Contrast agents work by shortening this relaxation time, thereby changing the R_1 local to the CA. The relaxivity of a contrast agent describes how the R_1 relaxation changes with respect to concentration, and this is dependent on the type of agent used as well as the magnetic field strength at which the experiment is performed. By using the baseline relaxation measurement, R_{10}, and the known relaxivity of the agent (r), the concentration of contrast agent [CA] can be determined by

$$R_1 = r \times [CA] + R_{10} \tag{12.1}$$

Several T_1 precontrast estimation techniques are discussed for accuracy, scan time, and sensitivity to field inhomogeneities below. (It is important to note that Equation 12.1 is a "fast exchange limit" relation; see below.)

For clinical DCE-MRI experiments, a rapid method of determining T_1 is required, and frequently, a spoiled gradient echo recalled (SPGR) sequence is used. The signal intensity for an SPGR sequence at steady state is

$$S = S_0 \frac{[\sin\alpha \times (1 - e^{-TR/T_1})] \times e^{-TE/T_2^*}}{(1 - e^{-TR/T_1} \times \cos\alpha)} \tag{12.2}$$

where the repetition time (TR), echo time (TE), and flip angle (α) are prescribed by the investigator, and S_0 is a constant describing proton density and scanner gain settings. T_1 is the longitudinal relaxation time, and T_2^* is the apparent transverse relaxation time of the tissue or voxel being interrogated. By acquiring images when $TE \ll T_2^*$ (to eliminate T_2^* weighting) at either multiple flip angles or multiple TR values, the images will have different T_1 weighting values. A least-squares fit of the data to Equation 12.2 can provide an estimated T_1. Since varying TR can take significant time, it is often not appropriate for clinical work, and therefore, variable flip angle (or variable nutation) approaches are frequently used, as shown in Figure 12.2a [12,13]. Similarly, by capturing a proton-weighted image and a T_1-weighted image, the ratios of the two have been used to provide an estimate of T_1 values as well [14].

The Look–Locker (LL) method provides T_1 estimations in relatively short acquisition time. In general, the LL method measures magnetization after a series of applied low flip-angle pulses spaced τ time apart after an initial inversion or saturation pulse; thus, the T_1 recovery curve is sampled continuously by tipping a small portion into the transverse plane by the RF pulse.

Another technique for more robust estimations of T_1 is an inversion recovery sequence in which the signal is sampled at various inversion times, TI, after a 180° inversion pulse, as described below by Equation 12.3

$$S = S_0 \times |1 - 2e^{-TI/T_1} + e^{-TR/T_1}| \tag{12.3}$$

where S_0 and S are the signal intensities at baseline and at the inversion times, respectively. While inversion recovery allows for a large dynamic range (sensitivity) for T_1, homogeneous inversion pulses are often difficult to achieve, requiring time and power for complete inversion, but result in more accurate T_1 estimations. An example of inversion recovery data fit to Equation 12.3 is shown in Figure 12.2b. Comparing this to LL, this method yields a total acquisition of approximately $5 \cdot T_1 \cdot$number of samples, whereas LL reduces this time to approximately $5 \cdot T_1$ [15–17].

The dependence on nominal flip angle for methods such as variable flip angle and LL makes these techniques susceptible to RF or B_1 inhomogeneities, particularly at higher fields. However, these methods offer fast alternatives to determine T_1 baseline

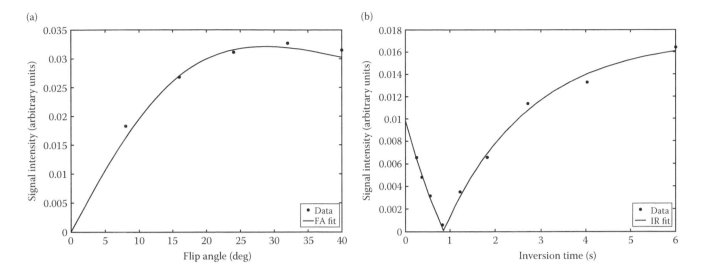

FIGURE 12.2 (a) To obtain a T_1 measurement, signal can be acquired at varied flip angles (FA) (•) and the signal can be fit (solid line) to Equation 12.2. (b) Using an inversion recovery (IR) technique, the signal can be sampled at multiple inversion times (•) and fit (solid line) to Equation 12.3 to obtain T_1 measurements.

measurements for DCE-MRI protocols. Other T_1 measurement techniques are discussed in detail elsewhere [18]; also, please see Chapter 5.

12.2.2 Dynamic Acquisition

In order to observe the T_1 changes induced by the injected CA, heavily T_1-weighted images must be acquired before, during, and after the injection. As with any MRI protocol, trade-offs exist between temporal resolution, spatial resolution, and signal-to-noise ratio. In oncologic applications, it is important to cover as much of the lesion volume as possible; in addition, the acquisition must be rapid enough to characterize the CA kinetics in a heterogeneous tumor region. Typical temporal resolutions can range from 1 to 30 s, depending on the application. If modeling tumor CA kinetics on a voxel-by-voxel basis to map tumor heterogeneity is required, then spatial resolution must be high enough to probe details of lesion. However, increasing the spatial resolution necessarily limits the temporal resolution and signal-to-noise ratio of the acquired data. Thus, the relative importance of temporal resolution, spatial resolution, and signal-to-noise ratio is dependent on the goals of the study; considerations for these trade-offs will be discussed below.

12.2.3 AIF

In order to perform a quantitative analysis of DCE-MRI data, knowledge of the kinetics of the CA in the plasma blood (arterial input function, or AIF) is required. While this is not a unique problem to DCE-MRI, technical demands make this acquisition difficult. Typical characteristics of the AIF from a bolus injection include a rapid wash-in of high CA concentration (A), followed by a short-lived peak concentration value (B), and an

exponentially decaying washout period (C), as shown in the AIF in Figure 12.3. Capturing the peak is one of the most critical and most difficult tasks, due to the need for rapid sampling (1–2 s). Several techniques on obtaining the AIF are discussed.

The current "gold standard" for measuring the AIF is taking arterial blood samples during the imaging acquisition [19,20]. Very accurate AIFs can be obtained with this method in that CA concentration can be determined directly from these samples. However, the invasive nature of this method is a substantial drawback. To characterize the rapid rise and washout of the CA in the blood, obtaining blood samples at a high enough temporal resolution is also very difficult. For preclinical studies, the typical blood volume of a mouse is ~2 ml; thus, the number of blood samples that can be obtained is quite limited.

Another method of estimating the AIF is derived from the image data itself. By strategically including a large vessel or left ventricle of the heart, the signal from the blood can be converted to describe the changing concentration of the CA during the injection. While this technique is certainly less invasive than blood sampling, it requires the presence of a blood pool in the field of view, which is not always conveniently located near the lesion of interest. Additionally, care must be taken to avoid partial volume or flow effects when measuring the AIF in this manner. As mentioned above, rapid scanning is required to capture the AIF peak (seen in Figure 12.3), and this results in lower spatial resolution and/or signal-to-noise ratio. Also, in high concentrations such as those seen during the first pass of the CA bolus, T_2^* effects can diminish the measured MR signal in the image, providing an inaccurate estimation of the peak CA concentration in the blood. Dosage and imaging parameters (*TE*, gradient spoiler amplitude, etc.) should be optimized to eliminate T_2^* effects. An example of this effect is shown in Figure 12.4. If arterial input functions are measured from the left ventricle of a

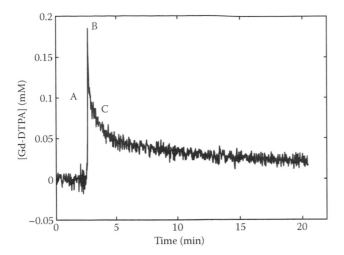

FIGURE 12.3 An example of a typical vascular/arterial input function obtained from the left ventricle of the mouse. Point A illustrates the rapid rise during a bolus injection of contrast agent, while point B depicts the peak amplitude of measured contrast agent concentration, and point C shows the washout of the CA as it diffuses to the tissue.

heart (in this case, from a mouse), dose optimization is critical. If the concentration is too high, as seen with a 0.3 mmol/kg injection of Gd-DTPA, the signal will be severely diminished, but if the dose is too low, the signal-to-noise ratio will not be enough to differentiate the signal from background noise. Some sequences have been designed to measure both T_1 and T_2^* relaxation during a dynamic scan such that the effect of T_2^* can be quantified and corrected at the expense of temporal resolution [21].

Alternatively, reference region (RR) approaches have been applied to DCE-MRI techniques if a large vessel or blood pool is not present in the field of view [22–24]. Specifically, if a sample of tissue is well characterized (i.e., muscle), the time course from that tissue can be used to calibrate the signal from the region of interest, and the need for estimating the AIF is eliminated with this technique. The high temporal resolution required to accurately capture the AIF can now be used to increase spatial resolution, which improves the ability to probe tumor heterogeneity. Studies with an RR model have reported both good correlation with blood sampling analysis [25] and reasonable repeatability [26] and reproducibility [27]. However, if the reference region shows much variability or is poorly characterized, the accuracy of the approach suffers. The mathematics of the reference region method is discussed in the next section.

Because of the difficulty associated with measuring the AIF, many studies have used a cohort of similar subjects to obtain a population average AIF [14,28]. The population average AIF is then applied to other similar subjects in additional studies. An example of this approach is shown in Figure 12.5, where five mice were imaged on separate days. The group averaged, or population, AIF is displayed in black.

Since changes in tissue are much slower than changes in the blood, the temporal resolution required to capture the kinetics of the AIF can now be "traded" for additional spatial resolution and/or signal-to-noise ratio for the region of interest. Thus, during subsequent studies, increased spatial resolution data can be used, and the population AIF can be used to drive the kinetic modeling. However, the obvious drawback to this technique is the inter- and intrasubject variability that may induce systematic errors in the pharmacokinetic parameters extracted from the models. When pursuing this technique, physical variation between subjects must be minimal, and DCE-MRI protocol setups must be identical to that used in obtaining the population-derived AIF.

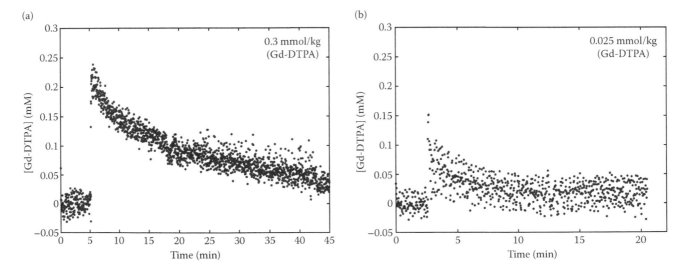

FIGURE 12.4 Example time courses using two different doses: (a) 0.3 and (b) 0.025 mmol Gd-DTPA/kg (b). At high doses, T_2^* effects can diminish the apparent peak concentration while doses that are too low cannot be distinguished from the noise floor.

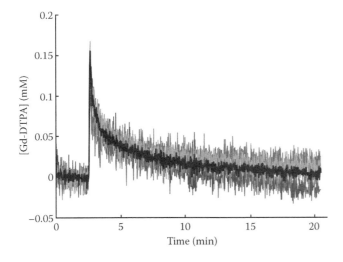

FIGURE 12.5 Multiple vascular input functions measured from the left ventricle of five mice as indicated by the gray-shaded time courses. The black line depicts the population average of this cohort of mice.

12.2.4 Modeling

12.2.4.1 Kety/Tofts and Extended Tofts

Mathematical models are frequently employed to understand and quantify underlying physiology. Models for contrast agents have been developed that describe the introduction, distribution, and clearance of the agent within the body; specifically for DCE-MRI, extravasation of the CA in tissue can indicate physiologic parameters such as blood flow, vessel wall permeability, and tissue volume fractions of the tissue [5]. The most commonly used model, developed by Kety in 1951, divides the body into two compartments: the blood/plasma space (denoted by C_p) and the tissue space (denoted by C_t), as shown in Figure 12.6 [29].

Using the notation standardized for DCE-MRI by Tofts et al. in 1999, K^{trans} represents the transfer constant from the plasma space to the tissue space [5]. This transfer constant has different physiologic interpretations depending on factors such as permeability and blood flow for the tissue of interest. This process can be described in four ways: (1) flow limited (areas with high permeability), (2) permeability-vessel surface area (PS) limited (areas with low permeability), (3) mixed flow and PS, and (4) clearance. Under each of these conditions, the definition of the transfer constant K^{trans} changes. For example, in highly permeable areas, K^{trans} is governed by the flow (F) of the CA in the plasma of the blood (1-hematocrit, or Hct) and can be defined as $K^{trans} = F \cdot \rho \cdot (1-Hct)$, where ρ is tissue density. Similarly, if permeability limits extravasation, K^{trans} is then defined as the product of vessel permeability and surface area for the given tissue ($PS \cdot \rho$). In the mixed case, K^{trans} is defined as the extraction fraction, $E\ (= 1-e^{(-PS/F(1-Hct))})$ times $F \cdot \rho \cdot (1-Hct)$. For tumors, PS limited or mixed models most likely best describe CA extravasation from the characteristic leaky vasculature [5]. Contrast agents of different molecular size can also force specific weighting on K^{trans}. Smaller agents are more permeable so the predominant

component affecting K^{trans} is flow while larger agents are slower to extravasate, allowing K^{trans} to reflect vessel permeability. An example of the use of a larger agent is presented below.

If we assume a homogeneous distribution of CA in both compartments and that there exists no back flux to either compartment, then any of the above cases can be generalized into a linear, first-order ordinary differential equation describing CA kinetics into the tissue:

$$\frac{d}{dt} C_t(t) = K^{trans} \times C_p(t) - (K^{trans}/v_e) \times C_t(t) \qquad (12.4)$$

This equation can then be solved, resulting in

$$C_t(t) = K^{trans} \times \int_0^t C_p(u) \times e^{-(K^{trans}/v_e)(t-u)} du \qquad (12.5)$$

It is important to note that the term (1-*Hct*) has been applied to the blood pool to indicate the portion of the blood, $C_b(t)$, that is well mixed with the CA; $C_p(t)$ represents the CA in the plasma space (1-*Hct*)·$C_b(t)$; thereby, the K^{trans} interpretations above have been simplified.

The aforementioned mathematical model neglects the fraction of tissue that may contain vascular space. Investigators have shown that in some tissues such as muscle, the fraction of vascular space (v_p) is so small that it can be considered negligible [30]. However, some investigators argue that tumor tissue may contain a nontrivial fraction of vascular space due to the increased angiogenic activity. If v_p is ignored for a tissue that contains significant vascular space, the pharmacokinetic modeling will overestimate K^{trans}, leading to erroneous results. An illustration of this extension has been adapted by Daldrup et al., shown in Figure 12.7, and has been used by other investigators [31,32].

An extension of the two-compartmental model has been formalized to improve accuracy in highly vascularized tissues such that the tissue space is composed of not only extravascular extracellular space but also plasma space, described as

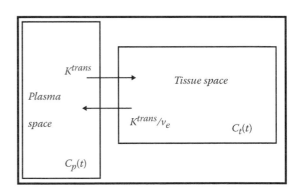

FIGURE 12.6 Two-compartment model showing one compartment representing the plasma space while the other compartment is the tissue space. Now, however, the tissue space is composed of both plasma and interstitial components.

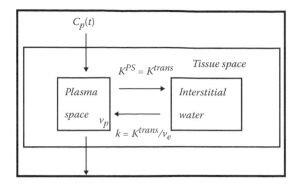

FIGURE 12.7 Two-compartment model showing one compartment representing the plasma space, while the other compartment is the tissue space. The contrast agent leaves the plasma space at a rate represented by K^{trans} and returns by K^{trans}/v_e.

$$C_t(t) = K^{trans} \times \int_0^t C_p(u) \times e^{-(K^{trans}/v_e)(t-u)} du + v_p \times C_p(t) \qquad (12.6)$$

In these forms, if $C_p(t)$ is measured, K^{trans}, v_e, and v_p can be determined using a nonlinear least-squares fit to the collected data, $C_t(t)$. This analysis can be done on a voxel-by-voxel basis in order to produce parametric maps, such as those depicted in Figure 12.8.

12.2.4.2 Reference Region Models

For reference region models that do not require the measurement of $C_p(t)$, two linear, first-order, ordinary differential equations are used to describe the kinetics from the plasma compartment to the tissue of interest and the reference region, respectively [23]:

$$\frac{d}{dt}C_{ROI}(t) = K^{trans,ROI} \times C_p(t) - (K^{trans,ROI}/v_{e,ROI})$$
$$\times C_{ROI}(t)$$
$$\qquad (12.7)$$
$$\frac{d}{dt}C_{RR}(t) = K^{trans,RR} \times C_p(t) - (K^{trans,RR}/v_{e,RR})$$
$$\times C_{RR}(t)$$

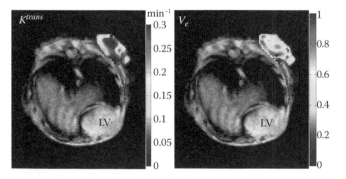

FIGURE 12.8 (**See color insert**) Axial cross section through the chest of a mouse. Within the field of view is the left ventricle (LV), the liver, and lungs. The left panel shows a parametric K^{trans} map overlain on the anatomical image, while the right panel displays a parametric v_e map.

While details are presented elsewhere [23], the final result eliminates the need to explicitly measuring an AIF:

$$C_{ROI}(T) = \frac{K^{trans,ROI}}{K^{trans,RR}} \times C_{RR}(T) + \frac{K^{trans,ROI}}{K^{trans,RR}}$$
$$\times \left[\frac{K^{trans,RR}}{v_{e,RR}} - \frac{K^{trans,ROI}}{v_{e,TOI}} \right] \qquad (12.8)$$
$$\times \int_0^T C_{RR}(t) \exp(-K^{trans,ROI}/v_{e,ROI})(T-t)dt$$

An extension of this model has been developed as well to incorporate v_p [33].

As mentioned previously, a priori information about the reference region (typically muscle) is required, and constant values ($K^{trans,RR}$, $v_{e,RR}$, and $v_{p,RR}$) can be substituted in accordingly. This is also one of the drawbacks of the reference region analysis technique; if there exists variability in the assumed reference region tissue, parameterization of $C_{ROI}(t)$ will not be accurate.

12.2.4.3 Water Exchange Regimes

As previously discussed, the relationship between MRI signal and concentration of CA is not straightforward. The relationship between the T_1 changes and concentration was described in Equation 12.3 as linear. This relationship, known as the fast exchange limit, or FXL, assumes that water exchange between the extravascular–extracellular space and the extravascular–intracellular space (trancytolemmal water exchange) is fast compared to the difference in relaxation rates of these compartments [5]. In other words, the mean intracellular water molecule lifetime (τ) is effectively zero.

Some investigators have emphasized the importance of probing the effects of the trancytolemmal water exchange by incorporating it into the Kety analysis discussed above [34–37]. It has been shown that during extravasation of the CA, the dependence on CA concentration causes the system to transiently drift from the FXL state into a fast exchange regime (FXR). In this scenario, Bloch–McConnell equations can be substituted into the typical SPGR equation to obtain parameters that indicate the intra- and extracellular water lifetime, τ_i and τ_e, respectively. The magnetization from two compartments is described below:

$$\frac{dM_{zi}(t)}{dt} = \frac{1}{T_1}[M_0 - M_{zi}(t)]$$
$$\qquad (12.9)$$
$$\frac{dM_{ze}(t)}{dt} = \frac{1}{T_1}[M_0 - M_{ze}(t)]$$

where M_{zi} and M_{ze} represent the z component of magnetization in the intracellular and extracellular compartments, respectively, and M_0 is the equilibrium magnetization. These equations were modified by McConnell [38] to denote the exchange between these two compartments by incorporating the lifetime, τ_i and τ_e, spent in each compartment, denoted by

$$\frac{dM_{zi}(t)}{dt} = \frac{[M_0 - M_{zi}(t)]}{T_1} - \frac{M_{zi}}{\tau_i} + \frac{M_{ze}}{\tau_e}$$
$$\frac{dM_{ze}(t)}{dt} = \frac{[M_0 - M_{ze}(t)]}{T_1} - \frac{M_{ze}}{\tau_e} + \frac{M_{zi}}{\tau_i} \quad (12.10)$$

Thus, longitudinal relaxation rates change as water transfers from one compartment to another. These linear homogeneous differential equations can be solved, taking the form derived by Woessner [39]:

$$\frac{1}{T_{1i,1e}} = \frac{1}{2}\left(\frac{1}{T_{1i}} + \frac{1}{\tau_i} + \frac{1}{T_{1e}} + \frac{1}{\tau_e}\right) \pm \left[\frac{1}{4}\left(\frac{1}{T_{1i}} + \frac{1}{\tau_i} + \frac{1}{T_{1e}} + \frac{1}{\tau_e}\right)^2\right.$$
$$\left. - \left(\left(\frac{1}{T_{1i}} + \frac{1}{\tau_i}\right)\left(\frac{1}{T_{1e}} + \frac{1}{\tau_e}\right) - \frac{1}{\tau_i \tau_e}\right)\right]^{1/2} \quad (12.11)$$

By combining this equation with Equation 12.5, K^{trans}, v_e, and τ_i can now be determined, providing a more rigorous modeling of DCE-MRI data. This analysis has been applied to several cancer applications, including diagnosing malignant versus benign breast tumors [37,40].

12.2.4.4 *iAUC*

As an alternative to rigorous mathematical modeling of DCE-MRI, semiquantitative measures of the data have also been studied. With these semiquantitative approaches, the signal time courses for each voxel is examined for changes in metrics such as wash-in slope, total amount of CA in the tissue, and initial area under the curve (*iAUC*) [41]. These metrics are typically easier to calculate and do not require the measurement of the AIF while still maintaining correlation with physiologic function such as blood flow and permeability. It is emphasized that these signal curves should be calibrated to precontrast T_1 maps because the raw signal change is dependent upon the native longitudinal relaxation [42]. While these metrics, specifically the *iAUC*, correlate with modeled DCE-MRI parameters, the physiologic definition is still unclear. In fact, studies have shown that *iAUC* measures maintain an intractable relationship with K^{trans}, v_e, and v_p and is therefore considered a mixed metric [27]. Despite these drawbacks, semiquantitative analysis of DCE-MRI data has still provided correlation with tumor regression rate, staging, and discrimination of cancerous tissue [43–45].

12.2.4.5 Empirical Models

Empirical models have also been developed to analyze data collected from DCE-MRI. In these models, the tissue time course is fit to a model such as

$$C(t) = A \times (1 - e^{-\alpha t})^q \times e^{-\beta t} \frac{(1 + e^{-\gamma t})}{2} \quad (12.12)$$

where A represents the maximum of tracer concentration, α is the contrast uptake rate (1/min), β is the overall contrast washout rate (1/min), γ is the initial contrast washout rate (1/min), and q is a parameter representing the curvature of the $C(t)$ time course. Fan et al. demonstrated that fits using this model were significantly better than the conventional Tofts model described above. In addition, A and β demonstrated statistically significant differences in metastatic versus nonmetastatic tumors. In another study examining malignant versus benign breast lesions, this empirical model was able to improve specificity and sensitivity as compared to other semiquantitative metrics [46–48]. Drawbacks to this method are similar to those of other semiquantitative techniques; there is no clear connection to underlying physiology, and therefore interpretation is cumbersome.

12.2.4.6 Statistical (Bayesian)

A significant drawback to the quantitative models described above is that they require an iterative curve fitting approach to converge to a solution using nonlinear regression. The difficulty of fitting tissue data is well known, and convergence to the correct solution (i.e., a local minimum) is not guaranteed. Statistical methods have been developed to alleviate some of the modeling difficulties with the current compartmental models. While models such as those based on Bayesian inference techniques provide the same physiologically relevant parameters, K^{trans}, v_e, and v_p, they potentially offer more robust results [49].

Another example of a statistical approach to DCE-MRI uses cluster analysis to select kinetic curves with similar characteristics in order to separate benign and malignant breast lesions. Current methods that use kinetic curves in diagnosis involve manually drawing ROIs, which has drawbacks associated with inter- and intraobserve variability. The cluster analysis approach examines the time course at each voxel, obtaining characteristics such as initial enhancement, uptake time, peak time, and washout time. These kinetic parameters are then used to differentiate between malignant and benign lesions [50].

Many of these variations of DCE-MRI analysis have proven valuable in diagnosing and assessing cancer treatment. Methods to improve DCE-MRI modeling are ongoing, but many of these described techniques have been used in both preclinical and clinical settings. Below, preclinical and clinical diagnosis and treatment monitoring examples are discussed.

12.3 Preclinical Applications in Cancer

The ability of DCE-MRI to report on tissue vascular status makes it a promising metric for antivascular and antiangiogenic drug efficacy and treatment studies. Antivascular drugs target the vascular endothelium, causing the collapse of tumor vessels. The results of antivascular therapy include reduced vasculature, reduced perfusion, and blood volume. Antiangiogenic drugs are also used in anticancer therapy; these drugs specifically target angiogenic pathways, ultimately inhibiting tumor neovascularization. Frequently targeted pathways include inhibition of the VEGF receptor, inhibition of receptor tyrosine kinases, and disruption of other angiogenic pathways such as FLT1, PI3 kinase, and HIF. [4]. Numerous studies have employed DCE-MRI to delineate drug efficacy at early time points.

12.3.1 Early Drug Efficacy Studies: *iAUC*

ZD6126 is an antivascular drug that acts by disrupting the microtubule cytoskeleton in proliferating immature endothelial cells. It has been previously shown to induce necrosis and inhibit blood flow to the lesion [51,52]. McIntyre et al. tested the hypothesis that DCE-MRI could be used to interrogate potential regrowth of vasculature after a single dose of ZD6126 [53]. Typical ZD6126 efficacy studies examined endpoints at 24 h, and any regrowth after this time point was assessed histologically. The investigators inoculated eight Wistar rats with GH3 prolactinomas on the hind limb. When the tumors reached approximately 1.71 cm^3, pretreatment DCE-MRI data were collected. All animals were then treated with a 50 mg/kg dose of ZD6126 and imaged at 24 h posttreatment. Then two animals were imaged at each of the following terminal time points: 48, 60, 72, and 96 h. DCE-MRI protocols were performed on a 4.7T, where initial T_1 maps were created by using a T_1-weighted spin echo protocol with $TR = 0.12, 0.5, 2$, and 10 s. Then, multislice dynamic spin echo images were acquired with $TR = 120$ ms, providing a temporal resolution of 15.4 s/volume. Five baseline acquisitions were acquired prior to an injection of 0.1 mmol/kg of gadodiamide (GE Healthcare, UK), followed by 40 dynamic acquisitions. Semiquantitative *iAUC* values were calculated for ROIs within the core and the rim of the tumor as well as paraspinal muscle, for reference. Voxels with a normalized value greater than 1 were defined to be the highly enhancing fraction (*HEF*), which is said to correlated with blood volume [54]. A Student's *t* test showed that pre- and posttreatment yielded no significant change in the reference tissue. However, after 24 h posttreatment, both tumor rim and core *HEF* values decreased by approximately 30% and 60%, respectively. At the later time point, no sign of regrowth was visible (i.e., no increase in *HEF* values); in addition, this study confirms a strong inverse correlation with enhancement and necrosis score seen by other studies. While the core enhancement decreased significantly, the rim enhancement only modestly decreased; this finding confirms work by Galbraith et al., who monitored treatment of another antivascular drug, CA4P, with K^{trans} [55]. Thus, DCE-MRI was able to study longitudinal ZD6126 efficacy in preclinical tumor models.

In many cases, chemotherapeutic drugs can be combined to improve overall efficacy [56]. It has been shown that low-dose cytotoxic agents given in a repeated low-dose regime (metronomic) can produce antiangiogenic activity [57,58]. Zhao et al. examined not only the metronomic effect of cyclophosphamide (CTX), an alkylating drug, but also this dose regime used in conjunction with the antiangiogenic drug thalidomide [59]. Thalidomide inhibits tumor necrosis factor (TNF-α), which is a known angiogenic

growth factor [60]. Zhao et al. studied a cohort of Copenhagen 2331 rats inoculated with a slow-growing Dunning R3327 ATI rat prostate tumor line to investigate two hypotheses: (1) the efficacy of the metronomic dose regime of CTX in conjunction with thalidomide and (2) the ability of DCE-MRI to detect early treatment response. The rats were subdivided into five groups: untreated ($n = 6$), thalidomide-treated at 60 mg/kg twice per week ($n = 6$), CTX-treated at 150 mg/kg twice per week ($n = 6$), metronomically dosed CTX (M-CTX) at 30 mg/kg dissolved in water ($n = 12$), and M-CTX with Tha ($n = 13$). Three controls and rats from groups 4 and 5 were selected for DCE-MRI measurements. Imaging was performed at 4.7T, with no initial T_1 map acquired. The dynamic protocol included T_1-weighted spin echo (SE) images with $TR/TE = 220/15$ ms, field of view (FOV) = 40^2 mm, and acquisition matrix = 128^2, yielding a temporal resolution of 59 s for three slices. Three baseline acquisitions were taken before a 0.1 mmol/kg injection of Gd-DTPA-BMA, followed by 10 min of continuous acquisition. Signal *iAUC* was determined over 120 s postinjection. In addition to DCE-MRI, stains for hemotoxylin and eosin (H&E), hypoxia, VEGF, and CD-31 were collected postmortem. The standard CTX dose given to group 3 proved to be incredibly toxic, resulting in three deaths after the third dose. Significant growth delays were found in groups 4 and 5 compared with those in the untreated group. DCE-MRI results showed that *iAUC* values significantly decreased in the center of the tumor for those treated with M-CTX + thalidomide as early as day 3. By day 7, rim and core *iAUC* significantly decreased for this group. In addition, tumor size and low *iAUC* values correlated significantly by day 18 ($r = .85$, $p < .02$).

12.3.2 Early Drug Efficacy Studies: Tofts

By employing the Tofts two-compartmental model described above, quantitative measures of K^{trans} and v_e can provide valuable information regarding the efficacy of anticancer drugs. For example, Checkley et al. examined both drug efficacy and dose-dependent response using DCE-MRI parameters as primary metrics [61]. ZD6474 is a VEGF receptor 2 (KDR) tyrosine kinase inhibitor that prevents endothelial migration and proliferation [62,63]. In this study, athymic mice were inoculated with PC-3 human prostate adenocarcinoma xenografts on the hind limb. After reaching approximately 1.0 cm^3, the mice were divided into controls that received a vehicle treatment and mice that received a dose of ZD6474; this study was repeated for doses ranging from 12.5, 25, 50, to 100 mg/kg, administered in two doses at time 0 h and 22 h. DCE-MRI data were collected pretreatment and 24 h posttreatment. Data were acquired at 4.7T; a saturation recovery experiment with TR values ranging from 120, 500, 2000, to 10,000 ms ($TE = 16$ ms) was used to calibrate signal to precontrast T_1. The dynamic acquisition employed a spin echo sequence with $TR/TE = 120/10$ ms, an acquisition matrix of $512 \times 256 \times 4$ slices, resulting in a spatial resolution of $0.625 \times 0.312 \times 2$ mm. At a temporal resolution of approximately 16 s, five baseline images were acquired, followed by a 0.01 ml/g dose of Gd-DTPA over a 3-s injection. Imaging continued for

11 more minutes, for a total of 40 image sets. Relaxivity for Gd-DTPA was measured using mouse blood samples dosed with known concentrations of Gd-DTPA, and the AIF was obtained from another set of mice (identical physiologically) and fit to a biexponential decay. A reproducibility study was also done using 11 vehicle controls. It was found that there was no significant difference between K^{trans} ($p = .4$) and v_e ($p = .1$), as measured on the same animal during two different imaging sessions. No change in growth was reported during the 24 h treatment study; however, a significant reduction in K^{trans} for doses ≥ 25 mg/kg and in v_e for doses > 50 mg/kg was identified. Chronic dose yielded significant volumetric difference at day 11. Since K^{trans} was not significantly different at 12.5 mg/kg and significant changes in K^{trans} between doses were demonstrated, the authors suggest that K^{trans} is dose dependent. This study illustrates the utility of DCE-MRI in antiangiogenic treatment studies at early response time.

In another study, the two-compartmental model was used to examine the efficacy of a high and low dose of KRN951, an antiangiogenic drug. KRN951 is a quinoline-urea derivative that acts as a VEGF tyrosine kinase inhibitor [64]. Among several *in vitro* and assay studies to characterize this novel drug, Nakamura et al. performed DCE-MRI on a cohort of athymic rats with Calu-6 lung cancer tumor fragments implanted in the hind limb. The rats were randomized, and half of the cohort received a dose of 0.2 mg/kg, while the other received a dose of 1 mg/kg on days 2 and 13. Imaging was performed pretreatment and on posttreatment days 2, 13, and 21. DCE-MRI protocol was performed at 1.5T; to calibrate signal intensity to native T_1 values, T_1 maps were generated using a T_1-weighted gradient echo sequence with varying flip angles from 10° to 90°. Following the T_1 mapping, dynamic images were acquired at a temporal resolution of 15 s/image with the following imaging parameters: $TR/TE = 200/6$ ms, acquisition matrix = 64 × 128, FOV = 40 mm × 80 mm, and slice thickness of 3 mm. Four precontrast images were acquired, followed by rapid bolus injection of 0.15 mmol/kg of Gd-DTPA, with 80 postinjection images acquired. The AIF was acquired from the vena cava and fit to a biexponential decay model. This comprehensive study included results form *in vitro* studies with human umbilical vein endothelial cells (HUVEC) to test the selectivity and inhibition of VEGF-induced proliferation at low concentration of KRN951. In addition, KRN951 was tested on 14 different tumor models, including breast, colon, hepatic, lung, ovarian, pancreatic, and prostate cancer. Ten out of the 14 models showed significant decrease in growth after 2 weeks of 1 mg/kg treatment. K^{trans} estimates obtained by modeling the kinetic data in the rim of the tumor were significantly decreased by ~45% in the low (0.2 mg/kg) dose treatment after day 13, while K^{trans} values in the rim were decreased by 55% and 61% at days 2 and 13 for the high (1 mg/kg) dose. However, interestingly enough, K^{trans} values at day 21 were not significantly different from baseline levels. The author suggests that this could be due to reversible vascularization because treatment stopped at day 13. The author also concludes that DCE-MRI could be an early biomarker for future studies involving KRN951 [64].

12.3.3 Early Drug Efficacy Studies: Extended Tofts

In the extended Tofts model, the fractional volume of plasma within the tissue is an additional parameter (see Equation 12.6). In tumors, the plasma volume fraction is frequently not negligible, so some investigators prefer including this additional term in the analysis. De Lussanet et al. [32] studied anginex and TNP-470. Anginex blocks adhesion and migration of endothelial cells, ultimately causing apoptosis [65], and TNP-470 is an established antiangiogenic agent [66]. A group of 34 C57BL/6 mice were subcutaneously injected with B16F10 melanoma cells on the hind limb. After 6–9 days, 9 were treated with TNP-470 (30 mg/kg IP every other day), 14 were treated with anginex (6 mg/kg every day), and 11 were treated with saline as controls. On day 16, anginex-treated mice ($n = 5$), TNP-470-treated mice ($n = 3$), and controls ($n = 5$) were imaged at 1.5 T. A 3D gradient echo sequence was employed with flip angles of 2°, 5°, 10°, 15°, 25°, and 35°, with $TR/TE = 50/7$ ms, in order to provide the baseline T_1 map. The FOV was 44 × 64 × 32 mm³, with an acquisition matrix of 88 × 128 and 16 slices. This resolution was the same for dynamic acquisition, with the exception of an increased slice thickness to 4 mm. A 3D gradient echo sequence was also used for dynamic acquisition with the TR/TE previously mentioned and a fixed flip angle of 35°. Temporal resolution was approximately 39 s/volume, and 0.1 mmol/kg Gd-DTPA was injected over 30 s after five baseline images had been acquired; subsequently, dynamic imaging continued for approximately 30 min postinjection. The AIF was obtained from the aorta in individual subjects. The two-compartmental model similar to the extended Tofts model was used to produce fit parameters of K^{PS} (K^{trans}), k (K^{trans}/v_e), and f^{PV} (v_p) [31]. Microvessel density maps were created *ex vivo*; these maps were reduced by approximately 50% for both treatment groups. Anginex showed a 68% decrease in growth ($p < .0001$), while TNP-470 reduced tumor volume by 65% ($p < .0005$). However, K^{trans} values were reduced by 64% ($p < .01$) in animals treated with anginex, with 66% ($p < .003$) and 67% ($p < .05$) reductions in the rim and core, respectively. For animals treated with TNP-470 K^{trans}, reductions over the whole tumor, rim, and core were 44% ($p = .17$), 43% ($p = .16$), and 43% ($p = .16$), respectively. Reductions in v_p suggested a decrease in plasma fraction, but changes were not significant. There was also no reported significant change in K^{trans}/v_e. The author concludes by remarking on similarities in the MVD maps and K^{trans} maps; both appeared to be aligned and have similar pattern.

As explained in the previous section, depending on the vascular environment, modeling can be flow-limited, permeability-limited, or mixed. The use of a macromolecular rapid clearance blood flow agent that is larger and therefore crosses the endothelium more slowly can be used to control the weighting of flow and permeability on K^{trans}. In Bradley et al. [67], gadomelitol with a molecular weight of 6.47 kDa (as opposed to the frequently used Gd-DTPA, weighing 0.5 kDa) was used to model components of the tissue curve that represented flow and permeability in order to explore the effects of two antiangiogenic drugs, AZD2171 and vandetanib. AZD2171 and vandetanib both inhibit VEGF receptor type 2 tyrosine kinases and have previously shown a drastic reduction in tumor blood flow and permeability [63,68]. Nineteen male athymic AP rats were inoculated with SW620 colorectal adenocarcinoma tumor cells on the forelimb flank (dorsal to the heart). After 10 days, eight rats were treated with AZD2171 (3 mg/kg), eight rats were treated with vandetanib (50 mg/kg), and three were treated with vehicle controls. All rats were treated again after 22 h and were imaged pretreatment and 23 h posttreatment. One of the emphasized points of this study was the meticulous imaging protocol. Tumors were placed just above the heart in order to capture both the region of interest as well as the left ventricle of the heart to capture the AIF. Imaging was done at 4.7T, and a semikeyhole technique (sampling only a portion of k-space) was incorporated to improve temporal resolution [69]. Gadomelitol gel phantoms of various concentrations were also included in the field of view to calibrate signal intensity to agent concentration. The dynamic acquisition was done with an SPGR sequence with $TR/TE = 13/9/2.3$ s, $\alpha = 15°$, FOV $= 60^2$, and the acquisition matrix $= 128 \times 96$. With these parameters, a temporal resolution of 0.5 s/image was obtained; with this fine a resolution, the rapid enhancing AIF in the left ventricle was adequately sampled for each subject. Baseline images were taken for 15 s, followed by a bolus injection of 0.045 mmol Gd/kg over 2–3 s, with 5 min of additional imaging. The AIF peaked at approximately 1.15 mM of Gd for all animals. A two-compartmental model was used that defined the tumor to be maintaining a vascular component (v_p) and an interstitial component (v_e) while being fed by the vasculature (C_p), as described in the previous section. The tissue curves were analyzed such that F/v_T, v_p, and $PS\cdot\rho$, where F, v_p, P, S, and ρ are as defined in the previous section and v_t is the total volume of the image voxel. Animals that were treated with the angiogenic drug showed significant changes in vascular related parameters, $PS\cdot\rho$ (19.7 $\pm$ 9.5% and 28.9 $\pm$ 14.1%) and v_p (31.2 $\pm$ 19.1% and 54.8 $\pm$ 21%), while no effect was seen in F/v_T. This study explored the use of a larger molecular weight contrast agent to discern effects on permeability and blood flow on antiangiogenic drugs. Furthermore, a diligently devised imaging protocol to help minimize parametric variability was optimized.

12.3.4 Studying Tumor Microenvironment

12.3.4.1 Interstitial Fluid Pressure

There have been recent efforts to extend the utility of DCE-MRI into other aspects of the tumor microenvironment. Below are several examples of contrast-enhanced dynamic MRI data used to probe other properties of cancer. Interstitial fluid pressure (IFP) is normally regulated through interactions between the extracellular matrix and fibroblasts. Several factors contribute to increased tumor IFP, including increased vessel permeability, decreased functional lymphatic vessels, and increased macromolecule accumulation in the tumor stroma [70,71]. Elevated IFP has been shown to impair drug delivery and efficacy [72]. As many current methods of determining IFP are highly invasive,

techniques have been developed to use noninvasive DCE-MRI to map IFP levels on a voxel-by-voxel basis [73,74]. Correlation with DCE-MRI parameters has also been examined, and a weak but negative correlation was found with independent IFP measures and K^{trans} [75]. Uncoupling extravascular extracellular space and pressure components remains difficult and cannot be done with typical DCE-MRI protocols. Another method incorporated using a slow injection of Gd-DTPA in order to reach steady state between plasma and tissue compartments. T_1 maps taken both preinjection and at steady state were used to calculate extravascular extracellular Gd-DTPA concentration in the tissue using FXL assumptions. Further, a linear relationship between the concentration of Gd-DTPA in the extravascular extracellular space at steady state and IFP was assumed based on the fact that flux of Gd-DTPA is dominated by the transcapillary extravasation and that interstitial transport is negligible; since transcapillary extravasation is due to diffusion and the hydrostatic pressure between capillaries and IFP, the authors claim that at steady state, the flux of Gd-DTPA can be assumed to be linearly related to IFP. Maps of IFP were generated by this method that correlated with traditional IFP measurement techniques [73,74]. The dynamic portion of this data was also examined. With an assumed, independently measured v_e for the tumor line, values of K^{trans} and K_{ep} (K^{trans}/v_e) that describe the influx and outflux of contrast agent were determined. The difference between the influx and outflux parameters was termed the *extravasative pressure-gradient–dependent transfer rate*, $K^{\Delta p}$. Correlation was found with $K^{\Delta p}$ and concentration of Gd-DTPA in the extravascular, extracellular space; thus, applying the prior assumption that Gd-DTPA concentration is linearly related to IFP, the spatial distribution of relative IFP measures can be determined [73].

12.3.4.2 Hypoxia

Several studies have attempted to acquire information about hypoxia by using DCE-MRI. Hypoxic tissue ($pO_2 < 10$ mmHg) has been indicative of malignant progression in cancer [76,77]. Patients who have significantly high hypoxia fractions who undergo radiation therapy are typically associated with poor overall survival; therefore, determining hypoxia in a noninvasive manner is appealing such that patients who have significant hypoxia may undergo more aggressive treatment [78]. Hypoxia is based on an imbalance between oxygen supply and consumption. DCE-MRI quantification can provide parameters that indicate blood perfusion or microvascular density with K^{trans} and extravascular extracellular volume fraction (i.e., cellularity) with v_e; it is hypothesized that oxygen supply measured with K^{trans} and oxygen consumption measured with v_e can provide estimates of hypoxia in tumors.

Three studies are mentioned here that have examined different tumor lines and necrosis levels to identify the significance of DCE-MRI metrics for hypoxia. Benjaminsen et al. implanted mice with A-07 human melanoma xenografts [79]. Six mice were used as controls, while 10 mice were irradiated with a single dose of 20 Gy. Both groups were imaged before and 24 h after radiation therapy. Proton density and T_1-weighted images were used to calibrate T_1 changes, and two calibration tubes filled with different concentrations of contrast agent were used to convert T_1 values to concentration. Images were acquired every 14 s for 15 min with the following parameters: $TR/TE = 200/3.2$ ms, $\alpha = 80°$, FOV = 6×3 cm^2, and an acquisition matrix = 256×64. An arterial input function model was used [80] to drive a two-compartment Kety analysis, yielding an estimation of the extraction fraction–perfusion termed $E \cdot F$ ($\approx K^{trans}$) and the partition coefficient λ ($\approx v_e$). Hypoxia fractions were measured using a plastic surface colony assay, where the tumor is resected and minced, and cell-surviving fractions are used to determine hypoxia fractions; the method is detailed elsewhere [81]. For treated animals, the hypoxia fraction was 3.5 ± 1.5 fold higher than for untreated animals. Frequency distributions of $E \cdot F$ and λ yielded no significant difference between treated and untreated animals. Median $E \cdot F$ and λ values, interquartile ranges for $E \cdot F$ and λ, and fraction of voxels with $0.06 < E \cdot F < 0.2$ were determined in addition to a 16-sector analysis of the tumor region. The results of this study indicated no significant difference between any parameters mentioned above in treated or untreated animals, and no correlations were found with the *ex vivo* hypoxia fraction for human melanoma xenografts. The authors state that this could be due to the specific tumor line or necrosis. However, in another study from the same group, two different human cervical carcinoma cell lines were studied with a similar protocol as above. They, in fact, showed a significant inverse correlation between $E \cdot F$ and hypoxia fraction [82].

In a third study, Egeland et al. showed evidence that the level of necrosis confounded DCE-MRI measures of hypoxia [83]. D-12 and U-25 human melanoma xenografts were imaged with an identical protocol as above and *ex vivo* hypoxia fraction was assessed similarly as well. Necrosis fraction was assessed by histological staining of sliding in the central axial plane of the tumor that corresponded with the image slice. K^{trans} was unable to differentiate between regions of necrosis and regions of viable tissue within the tumor; however, v_e values that were between 0.15 and 0.70 were exclusively indicative of viable tissue, while v_e values < 0.15 or > 0.70 correlated with highly necrotic regions. Using this as a method for establishing necrosis fraction ($NF_{DCE-MRI}$), values for each tumor line were plotted from the necrosis fraction derived from histology (NF_{Hist}) and found to be similar. Frequency distributions of K^{trans} and v_e were corrected by thresholding out necrotic regions identified, above yielding values of $K^{trans, corr}$ and $v_{e, corr}$. Uncorrected values of K^{trans} and v_e suggested significant differences among tumors of different sizes for each cell line. However, these differences disappeared after correction, uncovering a significant dependence of these parameters on necrosis. The authors conclude that DCE-MRI may still be useful in providing information on hypoxia, but data must be corrected for regions of necrosis. These studies suggest that determining the utility of DCE-MRI as a surrogate marker for hypoxia is promising, but more studies are necessary.

For a more complete description on the use of MRI for measuring tissue oxygen status, please see Chapter 15.

12.4 Application in Cancer Diagnosis

DCE-MRI has been used frequently in clinical settings and, in particular, to identify and stage various types of cancers including breast, ovarian, and prostate. The next sections illustrate the use of DCE-MRI in this setting compared to other modalities.

12.4.1 Breast

For women with a hereditary predisposition to breast cancer, surveillance and early diagnosis is critical. In particular, women who carry the breast cancer susceptibility gene (BRCA) have higher lifetime risk up to 65–80% [84,85]. Typical screening for women who fall into this category starts at age 30 at the latest. In a comprehensive study by Kuhl et al., surveillance of women with this genetic predisposition was performed with X-ray mammography, ultrasound, and DCE-MRI [46]. In this study, 529 participants underwent constant clinical breast exams and US screening. Upon the presence of an abnormality, patients would have additional X-ray mammogram and DCE-MRI. Of the 529 participants, 43 presented with breast cancer; 34 of these cases were deemed invasive and 9 were reported as DCIS cases. For DCE-MRI studies, a 1.5T magnet was used followed by dynamic acquisition before, during, and after a 0.1 mmol/kg injection of Gd-DTPA with acquisition matrices varying from 256^2 to 512×400. Of the 43 patients that presented with lesions, 40 were diagnosed by imaging. Fourteen were diagnosed with mammography alone, 17 were diagnosed with ultrasound alone, and 21 were diagnosed with mammography and ultrasound combined. Using DCE-MRI, 39 of the 40 cases were diagnosed, and when DCE-MRI and mammography were combined, all 40 cases were accurately diagnosed. DCE-MRI-alone had a sensitivity of 90.7%, surpassing the sensitivities of mammography-alone, US-alone, and mammography–US which were 32.6%, 39.5%, and 48.8%, respectively. DCE-MRI not only had the highest sensitivity but also maintained the highest specificity and positive predictive value for invasive and intraductal cancer. Of this total population, 19 patients were diagnosed with MRI alone compared to only only cancer was diagnosed by mammography alone. This study confirms the value of DCE-MRI when diagnosing and staging breast cancer, particularly for early diagnosis in those women with familial breast cancer.

Microcalcifications are found in both benign and malignant breast lesions. Mammography is highly sensitive to these calcifications but maintains a low specificity to diagnosing severity of this disease. Ductal carcinoma *in situ* (DCIS) presents clustered microcalcifications on mammography, but specificity to this disease is low (10%–35%) [86]. In a study done by Kneeshaw et al., the sensitivity and specificity of DCE-MRI was assessed for patients presenting with clustering microcalcifications specific to DCIS [87,88]. X-ray mammography, ultrasound and DCE-MRI were performed on a group of patients who were classified as "indeterminate," "suspicious of malignancy," and "probably malignant" prior to surgical biopsy. DCE-MRI was performed at 1.5T; after scout images, ~9 slices 5 mm thick (with a 2-mm gap between slices) were selected for dynamic acquisition. Thirty-five 3D volumes were collect using an SPGR sequence with α = 30°, FOV= (30 cm)^2, and an acquisition matrix = 256 × 128. A dose of 0.1 mmol/kg Gd-DTPA was administered over 10 s after the second set of images; the temporal resolution was approximately 13 s/volume. Both pharmacokinetic parameters using the two-compartment modeling technique mentioned above as well as empirical parameters including the enhancement index (*EI*), which is the percentage rise in signal from baseline, and normalized maximum intensity time ratio (*nMITR*) which is defined as

$$nMITR = \frac{S_{max} - S_0}{S_0} \times \frac{1}{T}$$ (12.13)

where S_{max} and S_0 are the signal intensity at maximum and baseline values while *T* is the time for the change in signal intensity to occur. Assessment with these parameters was performed on an ROI including the entire tumor and an ROI including only the nine most enhancing pixels. Receiver operator characteristic curves were used to define cutoff values for each parameter. Both *EI* and *nMITR* were statistically different between benign and malignant patients, but *nMITR* had a greater difference reported between the two classes. *nMITR* achieved the best overall accuracy for the other parameters reported; these parameter accuracies ranged from 79.5% to 81.8%. The overall accuracy of determining benign versus malignant lesions for each modality was also reported; X-ray mammography had a high sensitivity but low specificity, leading to an overall accuracy of 40.9%. Ultrasound had a sensitivity of 55% and a specificity of 81%, producing an overall accuracy of 75%. DCE-MRI parameters yield an overall accuracy of 86.4% with a sensitivity and specificity of 75% and 89.7%, respectively. The authors concluded by suggesting DCE-MRI as a useful addition to X-ray mammography to determine more accurate assessment of DCIS disease.

12.4.2 Ovarian

While there are a multitude of studies examining breast cancer with DCE-MRI, several investigators have examined other types of cancer. Thomassin-Naggara et al. investigated the use of DCE-MRI to classify the malignancy of indeterminate ovarian cancer patients [89]. Typically, ultrasonography is used to image these masses; however, ultrasonography is less accurate when distinguishing between complex masses [90]. Those patients who were deemed indeterminate with ultrasound (*n* = 37) were selected for additional DCE-MRI at 1.5T. T_1 and T_2 weighted scouts were used to locate the tumor, and 3 slices (5 mm thick) were selected for multislice DCE-MRI. An injection of 0.2 ml/kg Gd-DOTA was administered using an automated syringe pump at a rate of 2 ml/min. At the time of injection, T_1-weighted gradient echo images were acquired with the following parameters: *TR/TE* = 38/4.8 ms, α = 70°, FOV = 400–300 mm^2, and an acquisition matrix of 246 × 134. Images were acquired every 5 s for 2 min following injection. Semiquantitative analysis

based of regions of interest in the tumor was used to determine malignancy. First, the shape of the curve was classified into one of three categories. Curve type 1 defined a gradually increasing curve without a defined shoulder, while curve, type 2 was described as a moderately increasing curve with a defined plateau. Curve type 3 described an initial rise that was steeper than that of the reference tissue, which was normal tissue outside of the myometrium. In addition to curve classification, the curves were fit to a sigmoid shape:

$$EI(t) = \frac{A}{1 + (B/t)^C}$$ (12.14)

where *EI* is the normalized enhancement index, *A* is the asymptotic enhancement amplitude (*EA*), *B* is the time of half rising (*THR*), and *C* is the power constant. The maximum slope (*MS*) and *iAUC* for 60 s were also measured (*iAUC$_{60}$*). By taking these parameters and dividing them by the equivalent parameters in the reference tissue, ratios parameters were calculated and defined as *EAr*, *THRr*, and *MSr*. Based on shape classification, curve type 3 seemed to be very specific to invasive ovarian cancers. Curve type 1 seemed statistically specific to benign lesions over malignant lesions. However, there was no significant difference in curve shape reported between benign and borderline lesions. *EAr*, *THRr*, and *MSr* were reported statistically higher in invasive lesions over benign and borderline cases; however, there were, again, no statistically significant differences between borderline and benign lesions. The *iAUC$_{60}$* showed both a statistically significant increase in invasive cases over borderline and benign cases as well as a statistically significant increase in borderline versus benign lesions. By using the *iAUC$_{60}$* ratio < 0.25, this parameters has a 90% sensitivity and 85.2% specificity for predicting benign lesions, while at ratios > 0.39, the sensitivity and specificity for predicting invasive lesions is 93.7% and 81%, respectively. The combination of *MSr* and *iAUC$_{60}$* was able to correctly classify 30 of 37 lesions (81%). The seven lesions that were incorrectly classified were found to be mucinous tumors. The authors concluded by citing DCE-MRI as a valuable tool for accurately classifying complex masses in ovarian cancer.

12.4.3 Prostate

For prostate cancers, one of the primary methods of detection/diagnosis is prostate specific antigen (PSA) levels. PSA has high sensitivity and specificity for this disease but can sometimes be higher even in nonmalignant disease [91,92]. Thus, Hara et al. proposed using DCE-MRI as a method to help visualize and diagnose men with elevated PSA levels. In this study, 90 men who presented with higher than normal PSA levels were clinically staged for prostate cancer with transrectal ultrasound (TRUS), cystourethrography, cystourethroscopy, computerized tomography (CT) scan, isotope-bone-scanning, and DCE-MRI [93]. DCE-MRI was performed at 1.5T with a *TR/TE* of 269.5/4.8 ms, α = 80°, 5 mm thick slices, (300 mm)^2, with an acquisition matrix of 256 × 192. Images were acquired approximately every 30 s for 5 min before, during, and after a bolus injection of 0.1 mmol/kg Gd-DTPA. Diagnoses were classified by curve shape. If the lesion enhanced with an early peak and intermediate washout, it was diagnosed positive for prostate cancer. A slow monotonic increase was deemed no evidence of disease, while a peak within 2 min and a late washout was categorized as "suspicious." Detection of prostate cancer with PSA alone was only 36.8%. DCE-MRI detected prostate cancer with an accuracy of approximately 80% and identified clinically significant prostate cancer with specificity of 96.3%. The authors conclude by highlighting the ability of DCE-MRI to locate early stage prostate cancer with high sensitivity and specificity and emphasize that biopsies should be a last resort depending on PSA and DCE-MRI results.

12.5 Application in Cancer Therapy

While DCE-MRI has a place in clinical diagnosis, the true power of this modality involves longitudinal monitoring of treatment. There exists a multitude of studies using DCE-MRI as a surrogate marker for treatment response; below are a few examples illustrating various techniques in monitoring breast, brain and prostate treatments.

12.5.1 Longitudinal Monitoring of Treatment Response in Breast

Both semiquantitative and quantitative analyses are used for treatment response monitoring. Hayes et al. examined metrics produced by both analysis techniques to compare efficacy reporting [94]. Fifteen patients with confirmed cases of breast cancer were imaged at 1.5 T. A set of T_1 weight images were acquired using a fast gradient echo sequence, while a set of T_2 weighted images were acquired using a fast spin echo sequence to scout the region of interest. A proton weighted saturation recovery turbo flash sequence (SRTF) or fast low angle shot sequence (FLASH) was used to calibrate T_1 values. The dynamic sequence consisted of 42 image sets acquired one of two sequences: (1) FLASH with *TR/TE* = 350/5 ms, α = 20°, averages = 2, acquisition matrix = 256 × 192, FOV = (25 mm)^2 , and 3 slices at 10-mm thickness or (2) SRTF with *TR/TE/recovery time* = 11.7/4.4/10,000 ms, α = 20°, averages = 2, acquisition matrix = 128 × 128, and 5 slices at 8-mm thickness. Each sequence took 9–10 s to acquire a volume and 0.1 mmol/kg Gd-DTPA was injected after the 4th acquisition for both sequences. Signal intensity curves were analyzed using both semiquantitative and quantitative means. The signal intensity curves were first classified into one of five types: (1) no enhancement, (2) slow sustained early and late enhancements, (3) rapid initial enhancement followed by sustained late enhancement, (4) rapid initial enhancement with a stable enhancement, and (5) rapid initial enhancement with rapid washout [95]. Additionally, maximum enhancement (*ME*) and *mean rate of enhancement* (rate from 10% to 90% of maximum enhancement) were calculated. The two-compartmental modeling was used to extract kinetic parameters of K^{trans} and v_e. Using the curve classification method, responders typically exhibited a type 4 pretreatment shape, while nonresponders demonstrated a type 3. There was a trend for responders to decrease to type 3 after treatment while nonresponders increased to type 4; however, there was large variation, and these results were not significant. There existed a significant correlation between curve shape and percent change in K^{trans} for both responders and nonresponders; in cases where K^{trans} increased by > 50%, the patient exhibited a more aggressive curve type (i.e., type 3 to type 4), while K^{trans} decreasing by > 50% showed a step down to a less aggressive curve type. No significant correlation was found between K^{trans} and *ME*, but a significant correlation was found between K^{trans} and *mean rate of enhancement*. The authors also examined using a voxel-by-voxel analysis of the tumor as opposed to a whole tumor ROI approach. They found a linear relationship between the median values of all three parameters (K^{trans}, v_e, and concentration of Gd-DTPA) from the voxel-by-voxel histogram analysis with the average whole tumor ROI values. If a "hot spot" was selected (i.e., highest values of K^{trans}), 88.8% of responders showed a decrease of 25% or more after treatment. This study showed that average curve shape correlated with the K^{trans} in an ROI of the whole tumor in the central slice; thus, semiquantitative and quantitative parameters are clearly related. In addition, they demonstrated that whole tumor average and histogram analysis present very similar results. Finally, the most dynamic results in treatment response were presented with changes in the most extreme K^{trans} measurements.

In another study by Pickles et al., parameters resulting from the two-compartment model, K^{trans} and v_e, were found to have significant differences between responders and nonresponders after treatment [96]. In late stage breast cancer, neoadjuvant chemotherapy is frequently given to downstage or regress the lesion for surgical resection. Methods that can detect efficacy quickly are critical to avoid unnecessary cost or toxicity to the patient. Current methods of mammography and clinical breast exams are insufficient for determining response accurately [97,98]. DCE-MRI was performed on 68 patients to determine if pharmacokinetic parameters from the analysis of the data could predict responders versus nonresponders. On a 1.5T magnet, 3D T_1-weighted and proton-weighted spoiled gradient echo volumes were acquired to determine T_1 values of the native tissue. Thirty-five image sets were acquired before, during, and after an injection of 0.1 mmol/kg Gd-DTPA (<10 s injection) with a temporal resolution of ~11.6 s. *TR/TE/α* were 7.6 ms/4.2 ms/30°, while FOV and the acquisition matrix varied depending on the patient. K^{trans}, K_{ep}, and v_e were extracted by performing two compartment model fits to the lesion data. An ROI was drawn around the whole tumor for ROI$_{whole}$ analysis, and the maximum enhancing 9 voxels (hot spot) were selected for ROI$_{hs}$. If the reduction in tumor size was > 65%, the patient was classified as a responder. Of these 68 patients, 39 had nonspecific invasive carcinomas, 12 had invasive ductal carcinomas, 8 had invasive lobular carcinomas, 1 invasive tubular carcinoma, and 8 had histology that could not be used and were removed from further analysis. Posttreatment DCE-MRI occurred 36–153 days after administration of chemotherapy. Pickles et al.

found that the analysis of percent difference for K^{trans} and v_e in the "hot spot" were most significant between nonresponders and responders. If the absolute difference between pre- and posttreatment parameters were compared, v_e was significant for whole region of interest (ROI$_{whole}$) analysis as well. K_{ep} was borderline significant for ROI$_{hs}$ ($p = .077$ and $.054$ for percent and absolute difference, respectively). This study underlines previous studies showing significant differences in "hot spot" analysis with pharmacokinetic parameters from DCE-MRI [94,99].

12.5.2 Longitudinal Monitoring of Treatment Response in Brain

Optical pathway gliomas (OPGs) are a common pediatric tumor. In many cases, the tumors progress to cause additional problems with visual acuity and require treatment [100,101]. However, in other cases OPGs will spontaneously regress [102]. Individuals with neurofibromatosis 1 (NF1) have a predisposition to developing OPGs [103], but predicting the aggression of the tumors is still unclear. Jost et al. found that tumor volume had no correlation to whether these tumors would progress in children and require extensive treatment, so DCE-MRI and diffusion-weighted MRI were two metrics they hypothesized as biomarkers for tumor classification [104]. In this study, 27 patients with OPGs were imaged at either 1.5 T or 3.0 T using a diffusion weighted protocol and a fast low angle shot (FLASH) DCE-MRI protocol with the following parameters: $TR/TE = 30/6$ ms, $128 \times 128 \times 16$ acquisition matrix yielding a resolution of ($1 \times 1 \times 3$ mm^3), and a temporal resolution of < 1 min for approximately 6 min before, during, and after a bolus injection of 0.1 mmol/kg of Gd-DTPA. DCE-MRI data were processed using a standard Patlak analysis yielding an estimate of K^{trans} values, which were compared between clinically aggressive and clinically stable subgroups. Eleven of the 27 patients were categorized clinically stable while 16 patients were required to undergo treatment by chemotherapy, radiation, or surgery. Jost et al. found that there was a strong correlation between NF1 status and level of aggression, as 14 individuals had NF1. Also, mean K^{trans} values correlated with level of aggression; in fact, K^{trans} was significantly higher in aggressive subtypes than in stable disease. Apparent diffusion coefficient (ADC) values demonstrated no significant predictive value for determining subtypes. (Please see Chapter 7 for a description of diffusion weighted MRI.)

12.5.3 Longitudinal Monitoring of Treatment Response in Prostate

DCE-MRI has also been used to monitor treatment in prostate cancer. In typical cases, prostate cancers are treated by prostatectomy, radiation, or hormone ablation [105]. Hormone ablation is a preferred method and can also be used to reduce the tumor size before resection. Current methods of hormone ablation, including digital rectal exam and PSA levels are invasive [106,107]. It has been shown that prostate VEGF requires continuous stimulation by androgens [108]; thus, Padhani et al. hypothesized that monitoring blood flow parameters with DCE-MRI in response

to androgen treatment might be able to distinguish responders from nonresponders [109]. Fifty-six untreated patients with prostate cancer were imaged before and after (~119 days) treatment with an anti-androgen (luteinizing hormone analog) drug. Imaging took place at 1.5 T and included a proton-weighted sequence to calibrate T_1 values. SRTF or FLASH sequences were then used to acquire dynamic T_1-weighted images during an injection of 0.1 mmol/kg Gd-DTPA. The following parameters were used for each sequence: (1) FLASH with $TR/TE = 35/5$ ms, $\alpha = 70°$, acquisition matrix $= 256 \times 192$, FOV $= (25$ cm$)^2$, and three slices at 10-mm thickness or (2) SRTF with $TR/TE/recovery\ time = 11.7/4.4/10,000$ ms, $\alpha = 20°$, acquisition matrix $= 128 \times 128$, FOV $= (20$ cm$)^2$, and 8-mm thickness. Each sequence had a temporal resolution of ~9–10 s. ROIs were drawn in the tumor, periphery, and central gland; semiquantitative and quantitative analyses were performed on the resulting signal intensity time courses. Four parameters were extracted from the signal intensity curve data: (1) onset time (time between bolus injection and time at >10% enhancement), (2) mean gradient (rate between 10%–90% enhancement), (3) maximum enhancement, and (4) washout score. Washout score was categorized based on curve shape: type A, monophasic increasing curve; type B, peak intensity was reached within 2 min of injection time followed by a sustained or late enhancement; and type C, early initial peak followed by a rapid washout. Pharmacokinetic modeling was also used to extract parameters of K^{trans}, v_e, and maximum tissue gadolinium accumulation. In all patients, PSA was significantly reduced after hormone treatment. The wash patterns before and after treatment also decreased. Before treatment 94% of patients were type B or C, and after treatment, 78% of patients were categorized as type A. K^{trans} and maximum contrast agent accumulation were also found significantly decreased. Within the peripheral zone, K^{trans} was reduced while maximum contrast agent accumulation was significantly increased; v_e showed no significant change. K^{trans} and maximum contrast agent accumulation were found to be significantly different in the central gland posttreatment as well. Overall, this study demonstrates that DCE-MRI parameters are sensitive to hormone therapies related to prostate cancer; 91% of all patients shared a significant decrease in K^{trans} and PSA levels posttreatment.

Similarly, Rouviere et al. examined the efficacy of DCE-MRI techniques to determine reoccurrence of prostate cancer after external beam radiotherapy [110]. The investigators examined the effectiveness of T_2-weighted MRI and DCE-MRI in determining the reoccurrence of cancer as well as providing an intraprostatic map for treatment options. Twenty-two patients with suspected reoccurrence after radiotherapy (as determined by increased PSA levels at three separate visits) were imaged at 1.5 T. The T_2-weighted protocol used was a turbo spin echo (TSE) with $TR/TE_{effective} = 4380/115$ ms, FOV $= (250$ mm$)^2$, acquisition matrix $= 307 \times 512$, and 20 slices at a 4-mm thickness. For the dynamic T_1-weighted scan, a FLASH sequence was utilized with $TR/TE = 148/4.6$ ms, $\alpha = 70°$, FOV $= (250$ mm$)^2$, acquisition matrix $= 184 \times 256$, 14 slices at a 4-mm thickness, and a 20-ml injection of Gd-DOTA at 2 ml/s. The region was divided into 10 sectors, including the sextant in the peripheral zone (PZ), two in the transitional zone (TZ), and two in the seminal vesicles (SV). In the T_2-weighted images, the cancer was considered malignant if there was hypointensity in the PZ, hypointensity in the TZ if it extended to PZ, or hypointensity in the SV lumen indicating invasion. For DCE-MRI, all enhancing regions in the PZ or SV were categorized as malignant. Early enhancement that extended to the PZ or early enhancement in a nodular shape indicated malignancy in the TZ. Three investigators categorized the MRI data; using the T_2-weighted images, the readers found cancerous regions in 15, 15, and 13 out of the known 19 cases of cancer, while readers were able to discriminate all 19 out of 19 cases correctly using DCE-MRI. Sensitivity of DCE-MRI (0.70, 0.74, and 0.74), as opposed to the T_2-weighted images (0.26, 0.42, and 0.44), was significantly higher. While specificity was similar between the two approaches, the accuracy of the DCE-MRI (0.79, 0.73, and 0.74) was better than T_2-weighted image accuracy (0.60, 0.54, and 0.64). Overall, the authors concluded that DCE-MRI may be a useful tool in determining the extent of reoccurrence as well as intraprostatic mapping for salvage surgeries.

12.6 Extensions to Fields > 3 T

DCE-MRI sequences stand to benefit substantially from moving to higher fields. The nature of a dynamic scan requires very rapid sequential images acquired; short *TR* values, small flip angles, and high temporal resolution leave DCE-MRI protocols starved for signal. Since SNR increases proportionally with field strength, it has been shown that the peak SNR for similar DCE-MRI protocols using 4.0 T (not optimized for the higher field strength) was 2.2 times higher than at 1.5 T when using a contrast agent. Contrast-to-noise ratio also increased by a factor of 1.59. Optimization for imaging at field strengths includes Ernst angle modifications due to the change increase in T_1 relaxation at higher fields. Relaxivity of MR contrast agents are affected by factors such as temperature and environment, as well as field strength. The increased SNR allows DCE-MRI protocols to improve temporal resolution so that the integrity of fitting data with pharmacokinetic models may be preserved. Spatial resolution can also be increased, allowing better visualization of local blood flow heterogeneity in lesions [112].

Overall, DCE-MRI is a powerful tool that provides information about blood flow, permeability, tissue volume fractions, and vascularization that can help diagnose or stage various cancers as well as longitudinally monitor treatment. Analysis of the dynamic data can provide both semiquantitative and quantitative information that describes the underlying physiology. Data analysis techniques, new contrast agents, and high field imaging continue to improve DCE-MRI protocols.

References

1. Folkman J. Tumor angiogenesis: therapeutic implications. *N Engl J Med* 1971;285:1182–1186.
2. Ribatti D, Vacca A, Presta M. The discovery of angiogenic factors: a historical review. *Gen Pharmacol* 2000;35:227–231.
3. Jain RK. Determinants of tumor blood flow: a review. *Cancer Res* 1988;48:2641–2658.
4. Leach MO, Brindle KM, Evelhoch JL, Griffiths JR, Horsman MR, Jackson A, Jayson GC, Judson IR, Knopp MV, Maxwell RJ, McIntyre D, Padhani AR, Price P, Rathbone R, Rustin GJ, Tofts PS, Tozer GM, Vennart W, Waterton JC, Williams SR, Workman P. The assessment of antiangiogenic and antivascular therapies in early-stage clinical trials using magnetic resonance imaging: issues and recommendations. *Br J Cancer* 2005;92:1599–1610.
5. Tofts PS, Brix G, Buckley DL, Evelhoch JL, Henderson E, Knopp MV, Larsson HB, Lee TY, Mayr NA, Parker GJ, Port RE, Taylor J, Weisskoff RM. Estimating kinetic parameters from dynamic contrast-enhanced T(1)-weighted MRI of a diffusable tracer: standardized quantities and symbols. *J Magn Reson Imaging* 1999;10:223–232.
6. Yankeelov TE, Gore JC. Contrast enhanced magnetic resonance imaging in oncology: data acquisition, analysis, and examples. *Curr Med Imaging Rev* 2007;3:91–107.
7. Goh V, Padhani AR. Imaging tumor angiogenesis: functional assessment using MDCT or MRI? *Abdom Imaging* 2006;31:194–199.
8. Axel L. Cerebral blood flow determination by rapid-sequence computed tomography: theoretical analysis. *Radiology* 1980;137:679–686.
9. Miles KA, Griffiths MR. Perfusion CT: a worthwhile enhancement? *Br J Radiol* 2003;76:220–231.
10. de Langen AJ, van den Boogaart VE, Marcus JT, Lubberink M. Use of H2(15)O-PET and DCE-MRI to measure tumor blood flow. *Oncologist* 2008;13:631–644.
11. Wei K, Jayaweera AR, Firoozan S, Linka A, Skyba DM, Kaul S. Quantification of myocardial blood flow with ultrasound-induced destruction of microbubbles administered as a constant venous infusion. *Circulation* 1998;97:473–483.
12. Gupta R. New Look at method of variable nutation angle for measurement of spin lattice relaxation times using Fourier transform NMR. *J Magn Reson* 1977;25:231–235.
13. Homer J, Beevers M. Driven-equilibrium single-pulse observation of T1 relaxation—a reevaluation of a rapid new method for determining NMR spin-lattice relaxation times. *J Magn Reson* 1985;63:287–297.
14. Parker GJ, Barker GJ, Tofts PS. Accurate multislice gradient echo T(1) measurement in the presence of non-ideal RF pulse shape and RF field nonuniformity. *Magn Reson Med* 2001;45:838–845.
15. Brix G, Schad L, Deimling M, Lorenz Z. Fast and precise T1 imaging using a TOMROP sequence. *Magn Reson Imag* 1990;8:351–356.
16. Look D, Locker D. Nuclear spin-lattice relaxation measurements by tone-burst modulation. *Phys Rev Lett* 1968;20:987.
17. Kaptein R, Dijkstra K, Rarr C. Single scan Fourier transform method for measuring spin-lattice relaxation times. *J Magn Reson* 1976;24:295–300.

18. Kingsley P. Methods of measuring spin-lattice (T-1) relaxation times: an annotated bibliography. *Concept Magn Res* 1999;11:243–276.

19. Fritz-Hansen T, Rostrup E, Larsson HB, Sondergaard L, Ring P, Henriksen O. Measurement of the arterial concentration of Gd-DTPA using MRI: a step toward quantitative perfusion imaging. *Magn Reson Med* 1996;36:225–231.

20. Larsson HB, Stubgaard M, Frederiksen JL, Jensen M, Henriksen O, Paulson OB. Quantitation of blood-brain barrier defect by magnetic resonance imaging and gadolinium-DTPA in patients with multiple sclerosis and brain tumors. *Magn Reson Med* 1990;16:117–131.

21. Heilmann M, Walczak C, Vautier J, Dimicoli JL, Thomas CD, Lupu M, Mispelter J, Volk A. Simultaneous dynamic T1 and T_2^* measurement for AIF assessment combined with DCE MRI in a mouse tumor model. *Magma* 2007;20:193–203.

22. Kovar DA, Lewis M, Karczmar GS. A new method for imaging perfusion and contrast extraction fraction: input functions derived from reference tissues. *J Magn Reson* Imaging 1998;8:1126–1134.

23. Yankeelov TE, Luci JJ, Lepage M, Li R, Debusk L, Lin PC, Price RR, Gore JC. Quantitative pharmacokinetic analysis of DCE-MRI data without an arterial input function: a reference region model. *Magn Reson Imaging* 2005;23:519–529.

24. Yang C, Karczmar GS, Medved M, Stadler WM. Estimating the arterial input function using two reference tissues in dynamic contrast-enhanced MRI studies: fundamental concepts and simulations. *Magn Reson Med* 2004;52:1110–1117.

25. Yankeelov TE, Cron GO, Addison CL, Wallace JC, Wilkins RC, Pappas BA, Santyr GE, Gore JC. Comparison of a reference region model with direct measurement of an AIF in the analysis of DCE-MRI data. *Magn Reson Med* 2007;57:353–361.

26. Yankeelov TE, DeBusk LM, Billheimer DD, Luci JJ, Lin PC, Price RR, Gore JC. Repeatability of a reference region model for analysis of murine DCE-MRI data at 7T. *J Magn Reson Imaging* 2006;24:1140–1147.

27. Walker-Samuel S, Parker CC, Leach MO, Collins DJ. Reproducibility of reference tissue quantification of dynamic contrast-enhanced data: comparison with a fixed vascular input function. *Phys Med Biol* 2007;52:75–89.

28. Port RE, Knopp MV, Brix G. Dynamic contrast-enhanced MRI using Gd-DTPA: interindividual variability of the arterial input function and consequences for the assessment of kinetics in tumors. *Magn Reson Med* 2001;45:1030–1038.

29. Kety SS. Peripheral blood flow measurements. *Pharmacol Rev* 1951;3:1–41.

30. Shames DM, Kuwatsuru R, Vexler V, Muhler A, Brasch RC. Measurement of capillary permeability to macromolecules by dynamic magnetic resonance imaging: a quantitative noninvasive technique. *Magn Reson Med* 1993;29:616–622.

31. Daldrup H, Shames DM, Wendland M, Okuhata Y, Link TM, Rosenau W, Lu Y, Brasch RC. Correlation of dynamic contrast-enhanced MR imaging with histologic tumor grade: comparison of macromolecular and small-molecular contrast media. *AJR Am J Roentgenol* 1998;171:941–949.

32. de Lussanet QG, Beets-Tan RG, Backes WH, van der Schaft DW, van Engelshoven JM, Mayo KH, Griffioen AW. Dynamic contrast-enhanced magnetic resonance imaging at 1.5 Tesla with gadopentetate dimeglumine to assess the angiostatic effects of anginex in mice. *Eur J Cancer* 2004;40:1262–1268.

33. Faranesh AZ, Yankeelov TE. Incorporating a vascular term into a reference region model for the analysis of DCE-MRI data: a simulation study. *Phys Med Biol* 2008;53:2617–2631.

34. Landis CS, Li X, Telang FW, Molina PE, Palyka I, Vetek G, Springer CS, Jr. Equilibrium transcytolemmal water-exchange kinetics in skeletal muscle in vivo. *Magn Reson Med* 1999;42:467–478.

35. Landis CS, Li X, Telang FW, Coderre JA, Micca PL, Rooney WD, Latour LL, Vetek G, Palyka I, Springer CS, Jr. Determination of the MRI contrast agent concentration time course in vivo following bolus injection: effect of equilibrium transcytolemmal water exchange. *Magn Reson Med* 2000;44:563–574.

36. Yankeelov TE, Rooney WD, Li X, Springer CS, Jr. Variation of the relaxographic "shutter-speed" for transcytolemmal water exchange affects the CR bolus-tracking curve shape. *Magn Reson Med* 2003;50:1151–1169.

37. Li X, Rooney WD, Springer CS, Jr. A unified magnetic resonance imaging pharmacokinetic theory: intravascular and extracellular contrast reagents. *Magn Reson Med* 2005;54:1351–1359.

38. McConnell HM. Reaction rates by nuclear magnetic resonance. *J Chem Phys* 1958;28:430–431.

39. Woessner DE. Nuclear transfer effects in nuclear magnetic resonance pulse experiments. *J Chem Phys* 1961;28:41–48.

40. Yankeelov TE, Rooney WD, Huang W, Dyke JP, Li X, Tudorica A, Lee JH, Koutcher JA, Springer CS, Jr. Evidence for shutter-speed variation in CR bolus-tracking studies of human pathology. *NMR Biomed* 2005;18:173–185.

41. Evelhoch JL. Key factors in the acquisition of contrast kinetic data for oncology. *J Magn Reson Imaging* 1999;10:254–259.

42. Jesberger JA, Rafie N, Duerk JL, Sunshine JL, Mendez M, Remick SC, Lewin JS. Model-free parameters from dynamic contrast-enhanced-MRI: sensitivity to EES volume fraction and bolus timing. *J Magn Reson Imaging* 2006;24:586–594.

43. Gong QY, Brunt JN, Romaniuk CS, Oakley JP, Tan LT, Roberts N, Whitehouse GH, Jones B. Contrast enhanced dynamic MRI of cervical carcinoma during radiotherapy: early prediction of tumour regression rate. *Br J Radiol* 1999; 72:1177–1184.

44. Padhani AR, Gapinski CJ, Macvicar DA, Parker GJ, Suckling J, Revell PB, Leach MO, Dearnaley DP, Husband JE. Dynamic contrast enhanced MRI of prostate cancer: correlation with morphology and tumour stage, histological grade and PSA. *Clin Radiol* 2000;55:99–109.

45. Engelbrecht MR, Huisman HJ, Laheij RJ, Jager GJ, van Leenders GJ, Hulsbergen-Van De Kaa CA, de la Rosette JJ, Blickman JG, Barentsz JO. Discrimination of prostate cancer from normal peripheral zone and central gland tissue by using dynamic contrast-enhanced MR imaging. *Radiology* 2003;229:248–254.

46. Kuhl CK, Mielcareck P, Klaschik S, Leutner C, Wardelmann E, Gieseke J, Schild HH. Dynamic breast MR imaging: are signal intensity time course data useful for differential diagnosis of enhancing lesions? *Radiology* 1999;211: 101–110.

47. Fan X, Medved M, River JN, Zamora M, Corot C, Robert P, Bourrinet P, Lipton M, Culp RM, Karczmar GS. New model for analysis of dynamic contrast-enhanced MRI data distinguishes metastatic from nonmetastatic transplanted rodent prostate tumors. *Magn Reson Med* 2004;51: 487–494.

48. Fan X, Medved M, Karczmar GS, Yang C, Foxley S, Arkani S, Recant W, Zamora MA, Abe H, Newstead GM. Diagnosis of suspicious breast lesions using an empirical mathematical model for dynamic contrast-enhanced MRI. *Magn Reson Imaging* 2007;25:593–603.

49. Schmid VJ, Whitcher BJ, Yang GZ, Taylor NJ, Padhani AR. Statistical analysis of pharmacokinetic models in dynamic contrast-enhanced magnetic resonance imaging. *Med Image Comput Comput Assist Interv Int Conf Med Image Comput Comput Assist Interv* 2005;8(Pt 2):886–893.

50. Chen W, Giger ML, Bick U, Newstead GM. Automatic identification and classification of characteristic kinetic curves of breast lesions on DCE-MRI. *Med Phys* 2006;33: 2878–2887.

51. Blakey DC, Westwood FR, Walker M, Hughes GD, Davis PD, Ashton SE, Ryan AJ. Antitumor activity of the novel vascular targeting agent ZD6126 in a panel of tumor models. *Clin Cancer Res* 2002;8:1974–1983.

52. Davis PD, Dougherty GJ, Blakey DC, Galbraith SM, Tozer GM, Holder AL, Naylor MA, Nolan J, Stratford MR, Chaplin DJ, Hill SA. ZD6126: a novel vascular-targeting agent that causes selective destruction of tumor vasculature. *Cancer Res* 2002;62:7247–7253.

53. McIntyre DJ, Robinson SP, Howe FA, Griffiths JR, Ryan AJ, Blakey DC, Peers IS, Waterton JC. Single dose of the antivascular agent, ZD6126 (N-acetylcolchinol-O-phosphate), reduces perfusion for at least 96 hours in the GH3 prolactinoma rat tumor model. *Neoplasia* 2004;6:150–157.

54. Robinson SP, McIntyre DJ, Checkley D, Tessier JJ, Howe FA, Griffiths JR, Ashton SE, Ryan AJ, Blakey DC, Waterton JC. Tumour dose response to the antivascular agent ZD6126 assessed by magnetic resonance imaging. *Br J Cancer* 2003; 88:1592–1597.

55. Galbraith SM, Maxwell RJ, Lodge MA, Tozer GM, Wilson J, Taylor NJ, Stirling JJ, Sena L, Padhani AR, Rustin GJ. Combretastatin A4 phosphate has tumor antivascular activity in rat and man as demonstrated by dynamic magnetic resonance imaging. *J Clin Oncol* 2003;21:2831–2842.

56. Kerbel RS, Kamen BA. The anti-angiogenic basis of metronomic chemotherapy. *Nat Rev Cancer* 2004;4:423–436.

57. Browder T, Butterfield CE, Kraling BM, Shi B, Marshall B, O'Reilly MS, Folkman J. Antiangiogenic scheduling of chemotherapy improves efficacy against experimental drug-resistant cancer. *Cancer Res* 2000;60:1878–1886.

58. Man S, Bocci G, Francia G, Green SK, Jothy S, Hanahan D, Bohlen P, Hicklin DJ, Bergers G, Kerbel RS. Antitumor effects in mice of low-dose (metronomic) cyclophosphamide administered continuously through the drinking water. *Cancer Res* 2002;62:2731–2735.

59. Zhao D, Jiang L, Hahn EW, Mason RP. Continuous low-dose (metronomic) chemotherapy on rat prostate tumors evaluated using MRI in vivo and comparison with histology. *Neoplasia* 2005;7:678–687.

60. D'Amato RJ, Loughnan MS, Flynn E, Folkman J. Thalidomide is an inhibitor of angiogenesis. *Proc Natl Acad Sci U S A* 1994;91:4082–4085.

61. Checkley D, Tessier JJ, Kendrew J, Waterton JC, Wedge SR. Use of dynamic contrast-enhanced MRI to evaluate acute treatment with ZD6474, a VEGF signalling inhibitor, in PC-3 prostate tumours. *Br J Cancer* 2003;89:1889–1895.

62. Hennequin LF, Stokes ES, Thomas AP, Johnstone C, Ple PA, Ogilvie DJ, Dukes M, Wedge SR, Kendrew J, Curwen JO. Novel 4-anilinoquinazolines with C-7 basic side chains: design and structure activity relationship of a series of potent, orally active, VEGF receptor tyrosine kinase inhibitors. *J Med Chem* 2002;45:1300–1312.

63. Wedge SR, Ogilvie DJ, Dukes M, Kendrew J, Chester R, Jackson JA, Boffey SJ, Valentine PJ, Curwen JO, Musgrove HL, Graham GA, Hughes GD, Thomas AP, Stokes ES, Curry B, Richmond GH, Wadsworth PF, Bigley AL, Hennequin LF. ZD6474 inhibits vascular endothelial growth factor signaling, angiogenesis, and tumor growth following oral administration. *Cancer Res* 2002;62:4645–4655.

64. Nakamura K, Taguchi E, Miura T, Yamamoto A, Takahashi K, Bichat F, Guilbaud N, Hasegawa K, Kubo K, Fujiwara Y, Suzuki R, Kubo K, Shibuya M, Isae T. KRN951, a highly potent inhibitor of vascular endothelial growth factor receptor tyrosine kinases, has antitumor activities and affects functional vascular properties. *Cancer Res* 2006;66:9134–9142.

65. Griffioen AW, van der Schaft DW, Barendsz-Janson AF, Cox A, Struijker Boudier HA, Hillen HF, Mayo KH. Anginex, a designed peptide that inhibits angiogenesis. *Biochem J* 2001;354(Pt 2):233–242.

66. Yamaoka M, Yamamoto T, Ikeyama S, Sudo K, Fujita T. Angiogenesis inhibitor TNP-470 (AGM-1470) potently inhibits the tumor growth of hormone-independent human breast and prostate carcinoma cell lines. *Cancer Res* 1993;53: 5233–5236.

67. Bradley DP, Tessier JL, Checkley D, Kuribayashi H, Waterton JC, Kendrew J, Wedge SR. Effects of AZD2171 and vandetanib (ZD6474, Zactima) on haemodynamic variables in an SW620 human colon tumour model: an investigation using dynamic contrast-enhanced MRI and the rapid clearance blood pool contrast agent, P792 (gadomelitol). *NMR Biomed* 2008;21:42–52.

68. Wedge SR, Kendrew J, Hennequin LF, Valentine PJ, Barry ST, Brave SR, Smith NR, James NH, Dukes M, Curwen JO, Chester R, Jackson JA, Boffey SJ, Kilburn LL, Barnett S, Richmond GH, Wadsworth PF, Walker M, Bigley AL,

Taylor ST, Cooper L, Beck S, Jurgensmeier JM, Ogilvie DJ. AZD2171: a highly potent, orally bioavailable, vascular endothelial growth factor receptor-2 tyrosine kinase inhibitor for the treatment of cancer. *Cancer Res* 2005;65: 4389–4400.

69. Kuribayashi H, Bradley DP, Checkley DR, Worthington PL, Tessier JJ. Averaging keyhole pulse sequence with presaturation pulses and EXORCYCLE phase cycling for dynamic contrast-enhanced MRI. *Magn Reson Med Sci* 2004;3:207–210.

70. Reed RK, Berg A, Gjerde EA, Rubin K. Control of interstitial fluid pressure: role of beta1-integrins. *Semin Nephrol* 2001;21:222–230.

71. Fukumura D, Jain RK. Tumor microenvironment abnormalities: causes, consequences, and strategies to normalize. *J Cell Biochem* 2007;101:937–949.

72. Mimeault M, Hauke R, Batra SK. Recent advances on the molecular mechanisms involved in the drug resistance of cancer cells and novel targeting therapies. *Clin Pharmacol Ther* 2008;83:673–691.

73. Hassid Y, Eyal E, Margalit R, Furman-Haran E, Degani H. Non-invasive imaging of barriers to drug delivery in tumors. *Microvasc Res* 2008;76:94–103.

74. Hassid Y, Furman-Haran E, Margalit R, Eilam R, Degani H. Noninvasive magnetic resonance imaging of transport and interstitial fluid pressure in ectopic human lung tumors. *Cancer Res* 2006;66:4159–4166.

75. Haider MA, Sitartchouk I, Roberts TP, Fyles A, Hashmi AT, Milosevic M. Correlations between dynamic contrast-enhanced magnetic resonance imaging-derived measures of tumor microvasculature and interstitial fluid pressure in patients with cervical cancer. J *Magn Reson Imaging* 2007;25:153–159.

76. Vaupel P, Harrison L. Tumor hypoxia: causative factors, compensatory mechanisms, and cellular response. *Oncologist* 2004; 9(Suppl 5):4–9.

77. Rofstad EK. Microenvironment-induced cancer metastasis. *Int J Radiat Biol* 2000;76:589–605.

78. Rofstad EK, Sundfor K, Lyng H, Trope CG. Hypoxia-induced treatment failure in advanced squamous cell carcinoma of the uterine cervix is primarily due to hypoxia-induced radiation resistance rather than hypoxia-induced metastasis. *Br J Cancer* 2000;83:354–359.

79. Benjaminsen IC, Melas EA, Mathiesen BS, Rofstad EK. Limitations of dynamic contrast-enhanced MRI in monitoring radiation-induced changes in the fraction of radiobiologically hypoxic cells in human melanoma xenografts. *J Magn Reson Imaging* 2008;28:1209–1218.

80. Benjaminsen IC, Graff BA, Brurberg KG, Rofstad EK. Assessment of tumor blood perfusion by high-resolution dynamic contrast-enhanced MRI: a preclinical study of human melanoma xenografts. *Magn Reson Med* 2004;52:269–276.

81. Rofstad EK, Henriksen K, Galappathi K, Mathiesen B. Antiangiogenic treatment with thrombospondin-1 enhances primary tumor radiation response and prevents growth of dormant pulmonary micrometastases after curative radiation therapy in human melanoma xenografts. *Cancer Res* 2003;63:4055–4061.

82. Ellingsen C, Egeland TA, Gulliksrud K, Gaustad JV, Mathiesen B, Rofstad EK. Assessment of hypoxia in human cervical carcinoma xenografts by dynamic contrast-enhanced magnetic resonance imaging. *Int J Radiat Oncol Biol Phys* 2009;73:838–845.

83. Egeland TA, Gaustad JV, Benjaminsen IC, Hedalen K, Mathiesen B, Rofstad EK. Assessment of fraction of hypoxic cells in human tumor xenografts with necrotic regions by dynamic contrast-enhanced MRI. *Radiat Res* 2008;169:689–699.

84. Nunes LW, Schnall MD, Orel SG, Hochman MG, Langlotz CP, Reynolds CA, Torosian MH. Breast MR imaging: interpretation model. *Radiology* 1997;202:833–841.

85. Boetes C, Barentsz JO, Mus RD, van der Sluis RF, van Erning LJ, Hendriks JH, Holland R, Ruys SH. MR characterization of suspicious breast lesions with a gadolinium-enhanced TurboFLASH subtraction technique. *Radiology* 1994;193:777–781.

86. Moon WK, Im JG, Koh YH, Noh DY, Park IA. US of mammographically detected clustered microcalcifications. *Radiology* 2000;217:849–854.

87. Kneeshaw PJ, Lowry M, Manton D, Hubbard A, Drew PJ, Turnbull LW. Differentiation of benign from malignant breast disease associated with screening detected microcalcifications using dynamic contrast enhanced magnetic resonance imaging. *Breast* 2006;15:29–38.

88. Gilles R, Zafrani B, Guinebretiere JM, Meunier M, Lucidarme O, Tardivon AA, Rochard F, Vanel D, Neuenschwander S, Arriagada R. Ductal carcinoma in situ: MR imaging-histopathologic correlation. *Radiology* 1995;196:415–419.

89. Thomassin-Naggara I, Bazot M, Darai E, Callard P, Thomassin J, Cuenod CA. Epithelial ovarian tumors: value of dynamic contrast-enhanced MR imaging and correlation with tumor angiogenesis. *Radiology* 2008;248:148–159.

90. Kinkel K, Hricak H, Lu Y, Tsuda K, Filly RA. US characterization of ovarian masses: a meta-analysis. *Radiology* 2000;217:803–811.

91. Pansadoro V, Emiliozzi P, Defidio L, Scarpone P, Sabatini G, Brisciani A, Lauretti S. Prostate-specific antigen and prostatitis in men under fifty. *Eur Urol* 1996;30:24–27.

92. Ornstein DK, Smith DS, Humphrey PA, Catalona WJ. The effect of prostate volume, age, total prostate specific antigen level and acute inflammation on the percentage of free serum prostate specific antigen levels in men without clinically detectable prostate cancer. *J Urol* 1998;159:1234–1237.

93. Hara N, Okuizumi M, Koike H, Kawaguchi M, Bilim V. Dynamic contrast-enhanced magnetic resonance imaging (DCE-MRI) is a useful modality for the precise detection and staging of early prostate cancer. *Prostate* 2005;62: 140–147.

94. Hayes C, Padhani AR, Leach MO. Assessing changes in tumour vascular function using dynamic contrast-enhanced magnetic resonance imaging. *NMR Biomed* 2002;15:154–163.

95. Daniel BL, Yen YF, Glover GH, Ikeda DM, Birdwell RL, Sawyer-Glover AM, Black JW, Plevritis SK, Jeffrey SS, Herfkens RJ. Breast disease: dynamic spiral MR imaging. *Radiology* 1998;209:499–509.

96. Pickles MD, Manton DJ, Lowry M, Turnbull LW. Prognostic value of pre-treatment DCE-MRI parameters in predicting disease free and overall survival for breast cancer patients undergoing neoadjuvant chemotherapy. *Eur J Radiol* 2009; 71:498–505.

97. Gilles R, Guinebretiere JM, Toussaint C, Spielman M, Rietjens M, Petit JY, Contesso G, Masselot J, Vanel D. Locally advanced breast cancer: contrast-enhanced subtraction MR imaging of response to preoperative chemotherapy. *Radiology* 1994;191:633–638.

98. Trecate G, Ceglia E, Stabile F, Tesoro-Tess JD, Mariani G, Zambetti M, Musumeci R. Locally advanced breast cancer treated with primary chemotherapy: comparison between magnetic resonance imaging and pathologic evaluation of residual disease. *Tumori* 1999;85:220–228.

99. Delille JP, Slanetz PJ, Yeh ED, Halpern EF, Kopans DB, Garrido L. Invasive ductal breast carcinoma response to neoadjuvant chemotherapy: noninvasive monitoring with functional MR imaging pilot study. *Radiology* 2003;228:63–69.

100. Adams C, Fletcher WA, Myles ST. Chiasmal glioma in neurofibromatosis type 1 with severe visual loss regained with radiation. *Pediatr Neurol* 1997;17:80–82.

101. Champion MP, Robinson RO. Screening for optic gliomas in neurofibromatosis type 1: the role of neuroimaging. *J Pediatr* 1995;127:507–508.

102. Brzowski AE, Bazan C, 3rd, Mumma JV, Ryan SG. Spontaneous regression of optic glioma in a patient with neurofibromatosis. *Neurology* 1992;42(3 Pt 1):679–681.

103. Friedman JM. Epidemiology of neurofibromatosis type 1. *Am J Med Genet* 1999;89:1–6.

104. Jost SC, Ackerman JW, Garbow JR, Manwaring LP, Gutmann DH, McKinstry RC. Diffusion-weighted and dynamic contrast-enhanced imaging as markers of clinical behavior in children with optic pathway glioma. *Pediatr Radiol* 2008;38:1293–1299.

105. Chodak GW, Thisted RA, Gerber GS, Johansson JE, Adolfsson J, Jones GW, Chisholm GD, Moskovitz B, Livne PM, Warner J. Results of conservative management of clinically localized prostate cancer. *N Engl J Med* 1994;330:242–248.

106. Shearer RJ, Davies JH, Gelister JS, Dearnaley DP. Hormonal cytoreduction and radiotherapy for carcinoma of the prostate. *Br J Urol* 1992;69:521–524.

107. Pinault S, Tetu B, Gagnon J, Monfette G, Dupont A, Labrie F. Transrectal ultrasound evaluation of local prostate cancer in patients treated with LHRH agonist and in combination with flutamide. *Urology* 1992;39:254–261.

108. Haggstrom S, Lissbrant IF, Bergh A, Damber JE. Testosterone induces vascular endothelial growth factor synthesis in the ventral prostate in castrated rats. *J Urol* 1999;161:1620–1625.

109. Padhani AR, MacVicar AD, Gapinski CJ, Dearnaley DP, Parker GJ, Suckling J, Leach MO, Husband JE. Effects of androgen deprivation on prostatic morphology and vascular permeability evaluated with MR imaging. *Radiology* 2001;218:365–374.

110. Rouviere O, Valette O, Grivolat S, Colin-Pangaud C, Bouvier R, Chapelon JY, Gelet A, Lyonnet D. Recurrent prostate cancer after external beam radiotherapy: value of contrast-enhanced dynamic MRI in localizing intraprostatic tumor—correlation with biopsy findings. *Urology* 2004;63:922–927.

111. Redpath TW. Signal-to-noise ratio in MRI. *Br J Radiol* 1998;71:704–707.

112. Takahashi M, Uematsu H, Hatabu H. MR imaging at high magnetic fields. *Eur J Radiol* 2003;46:45–52.

13

Dynamic Susceptibility MRI: Data Acquisition and Analysis

C. Chad Quarles
Vanderbilt University

This chapter will introduce the reader to dynamic susceptibility contrast magnetic resonance imaging (DSC-MRI) methods and describe their use to assess the vascular and hemodynamic status of tumor-bearing tissue. A brief qualitative introduction (Section 13.1) is provided to demonstrate the types of parameters and images that are derived from a DSC-MRI experiment. The next section (Section 13.2) represents the bulk of the chapter and focuses on the derivation of the tracer kinetic models employed in DSC-MRI data analysis, DSC-MRI acquisition methods, the theory underlying DSC-MRI signals in cancer, and the significance of selecting appropriate arterial input functions. In the remaining sections, pre-clinical and clinical applications of DSC-MRI and the advantages and potential limitations of high field DSC-MRI studies are discussed. The goal of this chapter is to provide a thorough description of the theoretical basis and experimental implementation of DSC-MRI, so that the reader is able to comprehend and critically evaluate studies in this field, plan and execute studies, and better appreciate the potential of DSC-MRI to interrogate the tumor microenvironment.

13.1 Qualitative Introduction to Dynamic Susceptibility Contrast MRI

DSC-MRI involves the serial acquisition of MR images of a tissue of interest (e.g., a tumor) before and after an intravenous injection of a contrast agent (CA). As the CA traverses a tissue under investigation, the native relaxation times (T_1, T_2, and T_2^*) of tissue water decrease to an extent that is determined by the concentration of the agent and the geometry of the tissue structures containing the CA. By considering a set of images acquired before, during, and after a CA infusion, the signal intensity time course for individual voxels can be related to changes in CA concentration. By fitting the DSC-MRI data to an appropriate pharmacokinetic model, physiological parameters relating to blood volume (*BV*), blood flow (*BF*), and mean transit time (*MTT*) can be extracted [1–3]. DSC-MRI is most commonly used to assess hemodynamics within normal and/or tumor-bearing brain tissue, but its applicability in noncerebral tumors (e.g., breast and prostate) is also under investigation [4,5]. Abnormal angiogenesis is a hallmark of most cancer types and DSC-MRI is particularly well-suited for the noninvasive evaluation of morphological and functional characteristics of a developing vascular network. Indeed, DSC-MRI derived tumor hemodynamic maps have demonstrated a correlation with brain tumor grade [2,6–8] and treatment response [9–11].

An *in vivo* example of a typical DSC-MRI experiment in a patient with a grade IV glioblastoma multiforme (GBM) is illustrated in Figure 13.1. Using a rapid magnetic resonance (MR) image acquisition pulse sequence (e.g., echo planar or fast low-angle shot), images are collected for several minutes with a temporal resolution of ≤1.5 s per image (Figure 13.1a). A bolus of CA is injected via an antecubital vein, and upon reaching the brain, causes the MRI signal to drop in the arteries and the tissues they are supplying as shown in Fig. 13.1b. Using the formalism described in Section 13.2.2, the signal for each voxel is converted into the change in the transverse relaxation rate,

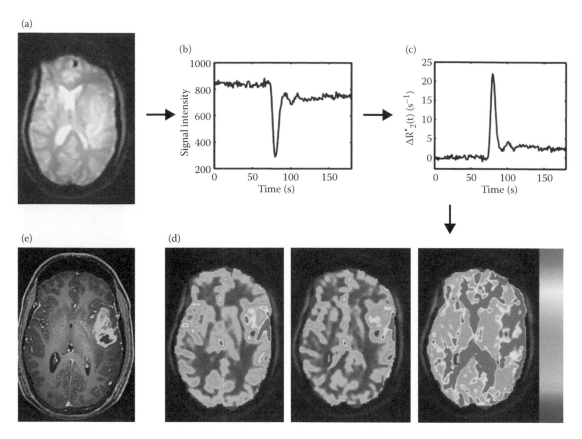

FIGURE 13.1 **(See color insert.)** An *in vivo* example of a typical DSC-MRI experiment in a patient with a grade IV glioblastoma multiforme (GBM). (a) Example echo planar image that is used to acquire serial images with a 1-s temporal resolution. (b) Following CA injection the MRI signal reduces in regions containing functional vessels. (c) MRI signals are converted to the change in the transverse relaxation rate, a parameter that is proportional to the tissue CA concentration time curves. (d) Kinetic analysis of the CA concentration time curves is used to derive maps of BF, BV, and MTT. (e) A high resolution T_1-weighted image acquired after the contrast agent injection is shown to highlight the enhancing tumor volume in the brain.

a parameter that is proportional to the tissue CA concentration time curves (Figure 13.1c). Finally, kinetic analysis of the CA concentration time curves is used to derive maps of *BF*, *BV*, and *MTT* (Figure 13.1d). For reference, Figure 13.1e is a high resolution T_1-weighted image acquired after the contrast agent injection and is shown to highlight the enhancing tumor volume in the brain. Note that tumor hemodynamic parameters are unlike those found in the surrounding healthy-appearing gray and white matter with higher *BF*, *BV*, and *MTT* values that are markedly heterogeneous across the tumor. As a point of reference, Table 13.1 summarizes *BF*, *BV*, and *MTT* values measured in gray and white matter as well as the range of values reported for brain tumors (of all grades).

TABLE 13.1 Typically Reported Values of BF, BV, and MTT in Healthy Brain Tissue and Tumors

Tissue Type	BF (ml/100 g/min)	BV (ml/100 g)	MTT (s)
Gray matter	60	4	4
White matter	25	2	4.8
Brain tumor	0–120	0–16	0–16

13.2 Quantitative Introduction to DSC-MRI

13.2.1 Tracer Kinetic Theory

The basis of perfusion quantification using DSC-MRI is derived from the principles of indicator dilution theory for nondiffusable tracers [12,13]. For completeness, the mathematical description of the DSC-MRI method will be developed from basic principles [12]. In the classic outflow experiment, a known quantity of indicator, q_I, in units of mass, is injected into a system at time zero (an infinitely short bolus) and the concentration at the output, $C_O(t)$, in units of mass per volume, is measured. The quantity of indicator, q_O, that leaves the system during a short time interval, t to $t + dt$, is equal to the product of the concentration of the indicator at the output, C_O, and the volume of fluid that leaves the system during this time interval, $F \cdot dt$, where F is the flow, in units of volume per time, and describes the rate at which fluid travels through the system, that is, $q_O = F \cdot C_O \, dt$.

After all of the indicator leaves the system, the sum of the individual quantities leaving the system during all time intervals must equal the total amount injected, q_I, or

$$q_I = F \cdot \int_0^\infty C_O(t)dt \qquad (13.1)$$

which can be rewritten to describe the flow through the system:

$$F = \frac{q_I}{\int_0^\infty C_O(t)dt} \qquad (13.2)$$

This is referred to as the Stewart-Hamilton equation [14–16].

It is useful to consider the fraction of indicator leaving the system per unit time, $h(t)$, which is the rate at which the indicator leaves the system divided by the total injected amount:

$$h(t) = F \cdot C_O(t)/q_I \qquad (13.3)$$

Summing over all the fractions that leave the system yields the following property of $h(t)$:

$$\int_0^\infty h(t)dt = 1 \qquad (13.4)$$

Physically, the function $h(t)$ describes the distribution of transit times through the system, or the frequency function of traversal times, and can be used to derive an expression for the volume of fluid in the system. Since the rate of fluid entry is F and the fraction of indicator which requires times between t and $t + dt$ to leave is $h(t)dt$, the rate this fraction of indicator enters or leaves the system is $F \cdot h(t)dt$. The volume this fraction of indicator occupies is simply the rate at which it leaves the system multiplied by the time required for it to leave, $dV = t \cdot F \cdot h(t)\ dt$. Thus, the total volume of the system, V, is the sum over all such time intervals:

$$V = F \int_0^\infty th(t)dt \qquad (13.5)$$

Since $h(t)$ is the transit time distribution, the integral portion of Equation 13.5 is, by definition, the mean transit time (MTT):

$$MTT = \int_0^\infty th(t)dt \qquad (13.6)$$

Combining Equations 13.6 and 13.5 yields one of the fundamental equations in quantitative DSC-MRI analysis, termed the central volume theorem:

$$MTT = V/F \qquad (13.7)$$

Thus, by measuring the concentration of tracer at the output of a system, information regarding the underlying kinetics, in terms of the flow, volume and transit times, can be evaluated.

For *in vivo* applications, the bolus that is injected into the circulatory system is non-ideal because it is distributed in time. Such a bolus is typically termed the arterial input function, or $C_A(t)$. Due to these temporal characteristics, the output or venous concentration, $C_V(t)$, depends linearly on the arterial input function and can be described as the convolution of $h(t)$ and $C_A(t)$:

$$C_V(t) = \int_0^t C_A(\tau)h(t-\tau)d\tau \qquad (13.8)$$

In a biological tissue, $h(t)$ is a kernel that characterizes the tissue's impulse response function, or, more practically, the transport of blood in the vasculature. Thus, the convolution describes the tissue's response, in terms of the venous concentration, to a temporally dependent arterial input function as the sum of individual, very short arterial impulses.

Unlike the outflow experiment described above, DSC-MRI relies on measuring a given tissue's, or imaging voxel's, total contrast agent concentration, termed $C_T(t)$, rather than just the venous output concentration. For reasons that will be detailed below, a key assumption for DSC-MRI is that the contrast agent remains compartmentalized within the blood vessels. As such, for a given blood contrast agent concentration, increased quantities of contrast agent within a voxel of interest (or increased tissue concentration) indicates higher vascular density. The vascular density can be described in terms of the fraction it occupies within a voxel according to the following equation:

$$v_B = \frac{V_B}{V_B + V_{EES} + V_{EIS}} \qquad (13.9)$$

where v_B is the blood volume fraction (unitless) and V_B, V_{EES}, and V_{EIS} are the absolute volumes occupied by the vascular space, the extravascular extracellular space (EES) and the extravascular intracellular space, respectively. Thus, a voxel's contrast agent concentration will be lower, by the fraction v_B, than that found in its blood vessels alone as given by $C_T(t) = v_B \cdot C_B(t)$, where $C_B(t)$ is the blood contrast agent concentration. An expression to compute the blood volume fraction can be derived by considering conservation of mass; the total amount of agent delivered to a tissue will equal the total amount leaving the tissue or the total amount within the tissue. Considering Equation 13.1, this can be expressed as

$$F \int_0^\infty C_A(t)dt = F \int_0^\infty C_V(t)dt = F \int_0^\infty C_B(t)dt = \frac{F}{v_B} \int_0^\infty C_T(t)dt \qquad (13.10)$$

For constant flow, the blood volume fraction is then

$$v_B = \frac{\int_0^\infty C_T(t)dt}{\int_0^\infty C_A(t)dt} \qquad (13.11)$$

Thus, if measurements are acquired of each voxel's contrast agent concentration time series, along with an arterial input function, Equation 13.11 can be used to generate parametric maps of the blood volume fraction within the tissue. In practice, the arterial input function can be difficult to identify and accurately measure, and as such, its integral is often assumed to be constant across tissues, which enables the computation of *relative* blood volume maps using

$$v_B \propto \int_0^\infty C_T(t)dt \qquad (13.12)$$

Using a similar conservation of mass argument, the quantity of contrast agent in a voxel, $q_T(t)$, can be described in terms of $C_A(t)$ and $C_V(t)$ if we consider that the mass or quantity of an agent in a voxel at a given time is equal to the difference between the total mass that has entered and exited the voxel during that time:

$$q_T(t) = F\int_0^t C_A(\tau)d\tau - F\int_0^t C_V(\tau)d\tau \qquad (13.13)$$

The concentration of the agent in this tissue, $C_T(t)$, is then equal to the quantity of agent divided by the absolute voxel volume, V_T: $C_T(t) = q_T(t)/V_T$. If Equation 13.8 is substituted into Equation 13.13, the tissue contrast agent concentration can be expressed in terms of the arterial input function and the transit time distribution as

$$C_T(t) = \frac{F}{V_T}\left[\int_0^t C_A(\tau)d\tau - \int_0^t\int_0^\tau C_A(\upsilon)h(\tau-\upsilon)d\upsilon d\tau\right] \qquad (13.14)$$

which can be further simplified into the following form:

$$C_T(t) = \frac{F}{V_T}\left[\int_0^t C_A(\tau)\left(1 - \int_0^{t-\tau} h(\omega)d\omega\right)d\tau\right] \qquad (13.15)$$

where $\omega = \tau - \upsilon$. Recall that the transit time distribution, $h(t)$, describes the fraction of indicator leaving the system per unit time. Since DSC-MRI measures the concentration of contrast agent remaining in the tissue, it is more relevant to describe the fraction of tracer still present in the vasculature at a given time.

This function is termed the residue function, $R(t)$, and is related to $h(t)$ by the following expression:

$$R(t) = 1 - \int_0^t h(\tau)d\tau \qquad (13.16)$$

By definition, $R(0) = 1$ and $R(t)$ is a positive, decreasing function of time. Substitution of Equation 13.16 into Equation 13.15 yields the following:

$$C_T(t) = \frac{F}{V_T}\int_0^t C_A(\tau)R(t-\tau)d\tau \qquad (13.17)$$

By convention, tissue blood flow (*TBF*) is generally expressed in units of milliliters (ml) of blood per 100 grams of tissue per minute, or as ml of blood per 100 ml of tissue per minute, and not in absolute flow units (ml blood/min) as expressed in Equation 13.17. However, by multiplying V_T by 100 and the brain density, ρ, which is 1.04 g/ml, the ratio, F/V_T, is then equivalent to *TBF* (in units of ml of blood per 100 grams of tissue per minute). To keep the units consistent for the tissue contrast agent concentration, the expression must also be scaled by the brain density. Note also that this derivation has neglected to account for the hematocrit difference between large ($H_{LV} = 0.45$) and small ($H_{SV} = 0.25$) vessels, which could potentially alter the determination of the arterial input function. Including the brain density and hematocrit scaling factors in Equation 13.17 yields the central equation for DSC-MRI studies of tissue hemodynamics:

$$C_T(t) = \frac{\rho}{k_H}\cdot TBF\cdot\int_0^t C_A(\tau)R(t-\tau)d\tau \qquad (13.18)$$

where $k_H = (1-H_{LV})/(1 - H_{SV})$. The blood volume fraction described in Equation 13.11 can also be converted into the conventional tissue blood volume units (ml of blood per 100 grams) by simply scaling the values by 100 and taking into account the brain density (such that a blood volume fraction of 0.04 corresponds to a tissue blood volume of 4 ml/100 g tissue):

$$TBV = \frac{k_H}{\rho}\cdot\frac{\int_0^\infty C_T(t)dt}{\int_0^\infty C_A(t)dt} \qquad (13.19)$$

For simplicity, a value of 1 is often assumed for ρ and k_H.

Both clinically and preclinically, the *TBF* and the tissue blood volume are the most widely used DSC-MRI-derived parameters to assess tumor hemodynamics. To derive *TBF* from Equation 13.18, the residue function has to be determined by a process called deconvolution, which, in essence, fits $TBF \cdot R(t)$ from the experimental data ($C_T(t)$ and $C_A(t)$). Since $R(0) = 1$, the *TBF* is determined as the initial height of the *TBF* and *R*(*t*) product. Two primary techniques have been proposed to perform the deconvolution: (1) model-dependent methods, where an analytical form of the residue function is assumed (e.g., decaying exponential function) and (2) model-independent approaches, where the residue function is determined via deconvolution. In practice, the model-independent methods are more readily used, as they make no assumptions about the shape of the residue function, which, in essence, characterizes the underlying tissue vascular network. This is particularly important for cancer imaging, as the tumor vascular architecture is known to be extremely heterogeneous and arbitrary models will likely introduce additional error into the measurements. Numerous deconvolution methods have been proposed [17] to compute the flow residue function product, but the singular value decomposition (SVD) approach [3,18] has shown to be the most robust at signal-to-noise ratio levels typically encountered with clinical echo planar imaging measurements [19].

13.2.2 DSC-MRI Signal Theory

The conversion of DSC-MRI signals into CA concentration time curves is critical for the reliable quantification of a tissue's hemodynamic parameters. Unlike computed tomography, position emission tomography or single photon emission computed tomography, whose signals are a direct measure of the amount of CA within a tissue, DSC-MRI's assessment of CA concentration relies on the indirect effects of CA on the magnetic properties of tissue water protons. As a contrast agent traverses through a voxel, it alters the inherent relaxation times (T_1, T_2 and T_2^*) of the tissue. The CA will reduce longitudinal relaxation times, T_1, due to the dipolar interactions of the paramagnetic ions with the protons of water molecules, termed as a "microscopic" effect, as described in Chapter 9. The strength of these relaxation effects decreases very rapidly with distance, so water protons must approach the immediate hydration sphere of the contrast agent to experience its influence [20].

The tissue transverse relaxation rate is not only dependent on the microscopic interactions between the CA and water molecule protons, but also upon magnetic field perturbations between intravoxel tissue structures (e.g., between vessels and the extravascular space) with differing susceptibilities (termed as a "mesoscopic" effect) and upon larger-sized static magnetic fields introduced by physical characteristics, such as shimming imperfections and sample shape (termed as a "macroscopic" effect) [21]. As illustrated in Figure 13.2a, the compartmentalization of paramagnetic agents inside blood vessels can induce magnetic fields by creating a susceptibility gradient between the vessel and the surrounding tissue [22]. As water protons diffuse through these gradients, they lose phase coherence, resulting in shorter T_2 and T_2^* times and a decrease in the MRI signal intensity. During the course of a DSC-MRI study, only the microscopic and mesoscopic effects dynamically alter the measured signal. Considering the magnitude of these effects, DSC-MRI is predominantly concerned with the mesoscopic changes that influence T_2 and T_2^*.

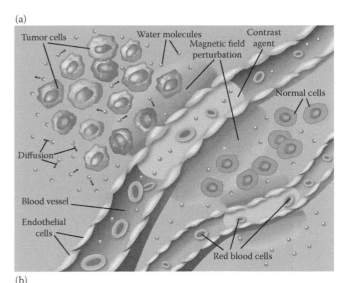

(a)

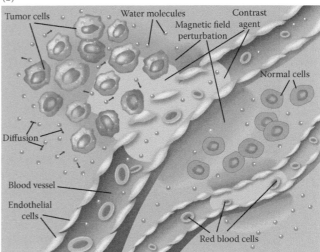

(b)

FIGURE 13.2 (See color insert.) (a) The compartmentalization of a paramagnetic contrast agent inside normal brain blood vessels can induce magnetic field perturbations in the surrounding tissues. (b) In brain tumor tissues, contrast agent can extravasate, potentially leading to additional field perturbations. As water protons diffuse through these perturbations, they lose phase coherence, resulting in shorter T_2 and T_2^* times and a decrease in the MRI signal intensity.

Given the long history of studies on the origin of susceptibility effects in biological nuclear magnetic resonance, it was quickly realized that susceptibility-induced signal losses depend on field strength, the precise nature of the pulse sequence, as well as the geometrical arrangement of the microcirculation [20]. Simulations from this early work demonstrate that increases in the CA containing vascular fraction yield proportional decreases in the MRI signal intensity. It was also found that changes in the relaxation rate are higher at higher concentrations, at higher field strengths, and if the susceptibility gradients overlap when capillaries are closely spaced [20].

A unique and physiologically relevant feature of susceptibility contrast imaging is the dissimilar sensitivity of gradient echo (GE) and spin echo (SE) pulse sequences to blood vessel diameter

[23–25]. Numerical simulations [23,24] and measurements [26] indicate that SE and GE images have different sensitivities to the size scale of the field inhomogeneities, resulting in a differential sensitivity to vessel diameter. The vessel size sensitivity of GE and SE signals is illustrated in Figure 13.3. For a given contrast agent concentration, the SE transverse relaxation rate change (ΔR_2) increases, peaks for microvascular-sized compartments (~5 μm for typical doses of CA), then decreases inversely with radius. Conversely, the GE transverse relaxation rate changes (ΔR_2^*) increase, and then plateaus, and remains independent of size beyond the microvascular-sized compartments. Consequently, the SE changes are assumed maximally sensitive to the microvasculature, while the GE changes are sensitive to the total vasculature. The variable sensitivities of the GE and SE signals can be further exploited, because both computer simulation and experimental studies predicted that the ratio of their relaxation rate changes is proportional to a tissue's mean vessel size, a method commonly termed vessel size imaging [27]. Thus, by simultaneously acquiring GE and SE DSC-MRI data, imaging metrics reflecting mean vessel size, as well as total vascular and microvascular *BV, BF,* and *MTT,* can be computed [6,28]. The preclinical and clinical advantages of such an approach are demonstrated below.

For clinically approved contrast agent doses of gadolinium-based CAs, it is generally assumed, and experimental evidence supports, that an approximate linear relationship exists between the tissue contrast agent concentration and the change in the T_2^* (or T_2):

$$\Delta R_2^*(t) = k \cdot C_T(t) \qquad (13.20)$$

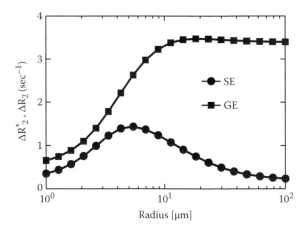

FIGURE 13.3 Sensitivity of SE and GE relaxation rates on perturber (or vessel) size. For a given contrast agent concentration, the SE transverse relaxation rate change (ΔR_2) increases, peaks for microvascular-sized compartments, then decreases inversely with the radius. The GE transverse relaxation rate change (ΔR_2^*) increases, and then plateaus to remain independent of size beyond the microvascular-sized compartments. Consequently, the SE changes are assumed to be maximally sensitive to the microvasculature, while the GE changes are sensitive to the total vasculature.

where $R_2^* \equiv 1/T_2^*$ and k is a proportionality constant and typically termed the susceptibility calibration factor. This factor is discussed in more detail below. Since it is assumed that the CA remains compartmentalized within blood vessels, Equation 13.20 can be rewritten in terms of the fractional vascular volume (v_p) and the contrast agent concentration in the blood, $C_p(t)$:

$$\Delta R_2^*(t) = k \cdot v_p C_p(t) \qquad (13.21)$$

Equation 13.21 assumes that the contrast agent concentration in the EES and the extravascular intracellular space is zero.

As stated above, the changes in the transverse and longitudinal relaxation rates following contrast administration are reflected in changes in the MRI signal. DSC-MRI techniques typically employ gradient echo (GE) or spin echo (SE) pulse sequences. The MRI signal from a GE sequence, before contrast injection (S_{pre}), can be written as

$$S_{pre} = \frac{S_o(1 - e^{-TR/T_{10}})e^{-TE/T_{20}^*} \cdot \sin(\alpha)}{1 - e^{-TR/T_{10}} \cdot \cos(\alpha)} \qquad (13.22)$$

where S_o is a constant describing the scanner gain and proton density, *TR,* is the repetition time, *TE* is the echo time, α is the flip angle, and T_{10} and T_{20}^* are the precontrast longitudinal and transverse relaxation times of tissue water, respectively. Note that a similar expression can be written for the SE sequence, where T_{20}^* would be replaced with T_{20} in Equation 13.22. Following contrast injection, the signal as a function of time can be expressed as

$$S(t) = \frac{S_o(1 - e^{-TR/T_1(t)})e^{-TE/T_2^*(t)} \sin(\alpha)}{1 - e^{-TR/T_{10}} e^{-TR/T_1(t)} \cos(\alpha)} \qquad (13.23)$$

where $T_1(t)$ and $T_2^*(t)$ are the tissue longitudinal and transverse relaxation times, respectively, shortened by the CA. Assuming fast water exchange (as reviewed in Chapter 11), the relationship between $T_1(t)$ and $T_2^*(t)$ and the concentration of CA can be written as follows:

$$\frac{1}{T_1(t)} = r_1 C_T(t) + \frac{1}{T_{10}} \qquad (13.24)$$

and

$$\frac{1}{T_2^*(t)} = k C_T(t) + \frac{1}{T_{20}^*} \qquad (13.25)$$

where r_1 is the CA T_1 relaxivity in units of mM⁻¹s⁻¹. As is conventional for most DSC-MRI descriptions, the microscopic T_2 effects have been ignored in Equation 13.25. This simplification is considered below. Equation 13.22 can then be rewritten in terms of the changes in the longitudinal and transverse relaxation rates:

$$S(t) = \frac{S_o(1 - e^{-TR/T_{10}}e^{-TR\cdot\Delta R_1(t)})e^{-TE/T_{20}^*}e^{-TE\cdot\Delta R_2^*(t)}\sin(\alpha)}{1 - e^{-TR/T_{10}}e^{-TR\cdot\Delta R_1(t)}\cos(\alpha)}, \quad (13.26)$$

where $\Delta R_1(t) = r_1 C_T(t)$ and $\Delta R_2^*(t) = k C_T(t)$.

DSC-MRI methods are most typically applied in normal brain tissue, where the blood-brain barrier is intact. Under such conditions, the contrast agent will change T_1 marginally because only a small fraction, the blood volume fraction, of the tissue water can interact with the contrast agent. The blood volume fraction in normal brain gray and white matter is four and two percent, respectively. A similar argument applies to microscopic T_2 effects. Thus, in normal brain tissue, the mesoscopic effects account for most of the contrast agent-induced signal change. By combining Equations 13.20 and 13.26 and assuming minimal T_1 changes, the ΔR_2^*, and thus, concentration of contrast agent, within a normal brain voxel can be expressed as

$$\Delta R_2^*(t) = k C_T(t) = \frac{-1}{TE} \cdot \log\left(\frac{S(t)}{S_{pre}}\right) \quad (13.27)$$

Typically, the susceptibility calibration factor is assumed to be constant, or simply set to a value of one, so that the computed ΔR_2^* is taken to be equivalent or proportional to the contrast agent concentration. However, theoretical and experimental studies have found that the calibration factor depends on the underlying vascular architecture, the contrast agent properties, and magnetic field strength [20–21,23–26,45]. Nevertheless, the computed concentration time curves for each voxel are then used in Equations 13.18 and 13.19 to derive measures of blood flow and blood volume. Since the CA concentration time curves are scaled by the unknown susceptibility calibration factor, the derived estimates of blood flow and blood volume are expressed as relative, as opposed to absolute measures of these quantities.

13.2.3 DSC-MRI Techniques

Before considering the above theory in the context of DSC-MRI studies for brain tumors, it is useful to review the imaging methods utilized to collect data in normal brain tissue. As mentioned above, the most common method for acquiring DSC-MRI data is to utilize multislice GE or SE echo planar pulse sequences. To adequately sample the first pass kinetics of an injected bolus of contrast agent, the minimal temporal resolution has been shown to be 1.5 s/image. For an echo planar imaging (EPI) acquisition, the following parameters are typically employed for DSC-MRI studies at 1.5 T: TR (1–1.5 s), GE TE (30–50 ms), SE TE (50–80 ms), field of view (~230 × 230 mm²), matrix size (64², 96², or 128²), slice thickness = 3–5 mm, and 5–10 slices. The contrast agent,

diethylenetriaminepentaacetic acid (Gd/DTPA), is injected at a dose of 0.1 to 0.2 mmol/kg and at a rate of 5 ml/s using an MRI compatible power injector. Following the injection of contrast agent, a 10 ml bolus of saline is also injected at the same rate to flush any residual contrast agent into the blood stream and to generate a tight bolus of CA. The precontrast signal, S_{pre}, is computed as the average of the signal acquired for a specified duration, typically 15 to 60 s, prior to the injection of contrast agent. The total image acquisition time ranges between 1 and 3 min.

13.2.4 DSC-MRI in Tumors

It is now well recognized that the signal theory derived above is only applicable for normal brain tissue with an intact blood–brain barrier or when large intravascular CAs are used. A confounding factor with DSC-MRI methods to study tumors results from the fact that all five FDA-approved gadolinium-based contrast agents: gadopentetate dimeglumine (Bayer HealthCare, Germany), gadodiamide (GE, USA), gadoversetamide (Mallinckrodt, USA) gadobenate dimeglumine (Bracco, USA), and gadoteridol (Bracco) are small molecular weight compounds. In brain tumors and in all other tissues lacking a blood–brain barrier, these contrast agents leak out of the vasculature, resulting in additional relaxation effects in the extravascular space, thereby affecting the acquired signal intensity time course. In such cases, the assumption that the microscopic T_1 and T_2 effects are negligible is not necessarily accurate. Furthermore, the spatial redistribution of the contrast agent within the vascular and extravascular space alters the mesoscopic effects, which complicates the relationship between the signal change and the contrast agent concentration. With the commonly used DSC-MRI acquisition and analysis methods described above, contrast agent extravasation has been shown to lead to unreliable estimates of blood volume and blood flow in brain tumors [6,29,30].

For typical gradient or spin echo EPI sequences, the most commonly reported contrast agent leakage effect is the marked reduction of the T_1 of extravascular extracellular water (which accounts for ~20% of the voxel volume). In regions of high permeability, such effects are observed during and following the first pass of the CA through the vasculature, as a substantial increase of the MRI signal above the precontrast baseline as illustrated in Figure 13.4 [6,29,30]. The contrast agent concentration time curves derived from such data, using Equation 13.27, are often underestimated and can result in negative, and thus nonphysiologic, values as well. The integration of such curves, as described in Equation 13.19, results in the underestimation of tumor blood volume. The presence and magnitude of these effects are spatially heterogeneous and depend upon the underlying permeability characteristics of the tumor vessels.

The reduction or elimination of T_1 leakage effects has been investigated over the past two decades. In the traditional clinical setting, the simplest methods to reduce such effects are to employ lower flip angles and longer echo times to reduce the MRI signal's T_1 sensitivity. Such methods can be quite effective but depend on the inherent relaxation properties of the tissue

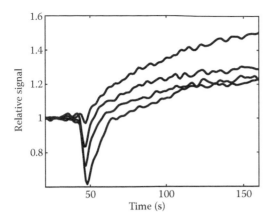

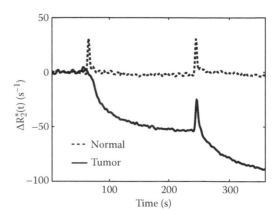

FIGURE 13.4 Example of GE echo planar signals acquired in a 9L rat brain tumor model at 3 T highlighting the impact of T_1 leakage effects in multiple regions of the tumor. In regions with a disrupted blood–brain barrier, these dynamic signal enhancements (above the pre-contrast baseline) are often observed. Such effects reduce the reliability of the computed hemodynamic parameters.

FIGURE 13.5 The GE transverse relaxation rate change after two bolus injections of a contrast agent highlights the effectiveness of the loading dose approach to minimize T_1 leakage effects. Two single doses of Gd-DTPA (0.1 mmol/kg) were administered to a rat bearing a C6 brain tumor and gradient echo EPI images were acquired continuously during both injections. The ΔR_2^* time course for normal brain tissue is similar in appearance and magnitude for both injections. For the tumor tissue, the T_1 effects overwhelm the expected susceptibility-induced changes during the preload as evidenced by completely negative ΔR_2^* values. During the second bolus injection, the tumor has features more consistent with the desired first-pass susceptibility effects but still retains some sensitivity to T_1 leakage effects as evidenced by the further reduction in ΔR_2^* values.

of interest, their blood volume, and the degree of contrast agent extravasation. As an alternative, the use of a loading dose of CA prior to the bolus injection can be given to reduce the EES T_1, such that subsequent T_1 changes are smaller in magnitude. Typically, a single dose (0.1 mmol/kg) of Gd-DTPA will be injected 5–10 min before a single or double dose bolus injection is given for the DSC-MRI study. Figure 13.5 demonstrates the effectiveness of this approach. Two single doses of Gd-DTPA were administered to a rat bearing a C6 brain tumor. Gradient echo EPI images were acquired continuously during the preload and the primary bolus injection. The ΔR_2^* time course, computed using Equation 13.23, for normal brain tissue, is similar in appearance and magnitude for both injections. For the tumor tissue, the T_1 effects overwhelm the expected susceptibility induced changes during the preload as evidenced by completely negative ΔR_2^* values. During the second bolus injection, the tumor has features consistent with the desired first-pass susceptibility effects, but still retains some sensitivity to T_1 leakage effects as evidenced by further reduction in ΔR_2^* values.

The pulse sequence parameter modifications and/or the preload approaches are only partially successful at reducing EES T_1 effects, such that the resulting DSC-MRI signals are still unreliable for measuring tumor hemodynamics. To further reduce the impact of these effects, numerous model-based post-processing correction methods have been proposed. These permeability compensation methods either attempt to estimate and remove the temporal extravasation effects, or others arbitrarily apply a post-bolus baseline correction [6,11,30,31]. Unfortunately, no studies have experimentally validated whether or not these techniques improve the reliability of blood flow and blood volume estimates. However, the use of preloads of CA and the use of post-processing correction methods have been shown to improve the diagnostic potential of the hemodynamic measures. Several studies have demonstrated that uncorrected blood flow

and blood volume values typically do not correlate with brain tumor grade, whereas, corrected ones do [6,7,32].

While the above methods are designed to minimize T_1 leakage effects, an alternative approach is to use a dual echo signal acquisition pulse sequence, which theoretically eliminates any T_1 contribution to the acquired DSC-MRI signal [33]. Equation 13.27 can be rewritten to include the signal acquired at two echo times (TE1 and TE2):

$$\Delta R_2^*(t) = kC_T(t) = \frac{1}{(TE2 - TE1)} \log\left(\frac{S_{TE1}(t)/S_{TE1}(0)}{S_{TE2}(t)/S_{TE2}(0)} \right) \quad (13.28)$$

where S_{TE1} and S_{TE2} are the signals acquired from the first and second echoes, respectively. It is left to the reader to verify that the static and dynamic T_1 contributions to the computed C_T time series are removed in the above expression. An advantage to this approach is that the dual echo data can also be used to derive a signal that is only sensitive to the dynamic T_1 changes, which is, in essence, equivalent to the DCE-MRI approach, described in Chapter 11, in that it primarily reflects contrast agent traversing the EES. As proposed by Kuperman et al, a signal time course sensitive to T_1 changes only, $S_{T1}(t)$, (no T_2^*-dependence) can be derived using the following expression [34]:

$$S_{T1}(t) = S_{TE1}(t) \cdot (S_{TE1}(t)/S_{TE2}(t))^\beta \quad (13.29)$$

where $\beta = TE1/(TE2 - TE1)$. This is equivalent to extrapolating the signal back to TE = 0 at each time point [35]. The separation of the T_1 and T_2^* contributions to the measured MRI signal using this approach was first applied to sort out complex relaxation effects in kidney DCE-MRI data, but was later applied by Vonken et al for DSC-MRI studies of human brain tumors [35]. By substituting Equation 13.22, for two different echo times, into Equation 13.29, an analytical solution for $C_T(t)$ can be derived:

$$C_T(t) = \frac{-1}{r_1 \cdot TR} \log \left[\frac{S_{T1}(t) - S_o \cdot \sin(\alpha)}{S_{T1}(t) \cdot \cos(\alpha) - S_o \cdot \sin(\alpha)} \right] - \frac{1}{T_{10}} \quad (13.30)$$

The precontrast T_1 can be determined using any commonly used method for DCE-MRI studies and, as is typical, the relaxivity, r_1, can be assigned to previously measured values for a given field strength and CA. The contrast agent concentration time curves determined from Equation 13.29 can then be inserted into any of the kinetic models developed for DCE-MRI for the computation of the volume transfer constant, K^{trans}, and the volume fraction of the EES, v_e.

For the low flow angle, long and dual-echo time methods, an unanticipated consequence of their reduced T_1 sensitivity is an enhanced sensitization to extravascular T_2 and/or T_2^* effects that occur as the CA traverses the EES. Such pulse sequences also exhibit enhanced sensitivity to the recirculation of CA within the vasculature. Unlike the T_1 leakage effects, which increase the signal intensity, T_2 and/or T_2^* effects can act to decrease the signal intensity. Deriving tissue concentration time curves from such data can result in an overestimation of the tumor blood volume. Similar to the post-processing methods designed to remove T_1 leakage effects, studies are currently under way to develop models and correction strategies for EES T_2^* leakage effects prior to the computation of BV and BF [31,36,37].

The T_2^* leakage effects in the EES originate from both microscopic and mesoscopic CA interactions, with the latter being the more predominant effect [37]. Field gradients created by CA within the EES can influence both intravascular and extravascular intracellular water as illustrated in Figure 13.2b. As with intravascular CA, the influence of this additional magnetic field perturbation on the measured transverse relaxation rate depends on the underlying geometry of the EES, which implies yet another unknown susceptibility calibration factor specific to this compartment. As there are no geometrical models to describe the extravasated CA induced magnetic field gradients, or their interaction with gradients created by CA within the vasculature, Quarles et al. proposed an analytical linear model that relates the change in the transverse relaxation rate to the susceptibility field gradients that can occur between the blood, EES, and EIS. The following expression was derived to model the mesoscopic effects:

$$R_2^{meso}(t) = K_p v_p \left(v_e \left| C_p(t) - C_e(t) \right| + v_i C_p(t) \right) + K_e v_e v_i C_e(t) , \quad (13.31)$$

where C_e, v_e, C_i, v_i, C_p, v_p denote the CA concentrations and fractional volume fractions of the EES, EIS, and blood plasma, respectively, and K_p and K_e are vascular and extravascular susceptibility calibration factors [37].

Equation 13.31 helps to illustrate a few key points about the influence of contrast agent leakage on R_2^{meso}. First, CA extravasation can reduce the susceptibility difference between blood plasma and the EES as illustrated by the absolute value of the difference in the compartmental CA concentrations. If this were the predominant effect, T_2^*-based contrast agent leakage effects would tend to decrease the magnitude of the signal drop that would have occurred had the CA remained within blood vessels. Second, as seen in the second half of Equation 13.31, the CA compartmentalized within the EES can induce additional susceptibility differences between the EES and the extravascular intracellular space, which increase R_2^{meso} and cause a greater than expected DSC-MRI signal decrease. Such effects are scaled by the tissue-specific extravascular susceptibility calibration factor, K_e. Similar to the vascular susceptibility calibration factor, which depends on the tissue vascular geometry, K_e depends on the spatial distribution or geometry of the CA compartmentalized within the EES. For a wide range of physiologically and physically relevant input parameters, the proposed model predicts that the latter of these two effects dominate the T_2^*-based contrast agent leakage effects. Thus, one of the major conclusions from this study is that when contrast agent leaks out of blood vessels, its predominant effect on tissue T_2^* is to shorten it and thereby cause further signal decreases. This conclusion has significant implications for the development of strategies to correct CA leakage effects and the potential of K_e to be used as a new contrast mechanism to characterize the geometry of the EES.

To date, there is no consensus on which data acquisition or post-processing correction methods yield the most reliable and quantitative estimates of tumor blood flow and blood volume when DSC-MRI data are acquired in the presence of contrast agent extravasation. However, improved understanding of the biophysical basis of DSC-MRI signals has led to the development of more sophisticated acquisition and analysis methods, which should improve leakage correction strategies and enable the extraction of additional physiologically relevant information. In contrast, there is more consensus on which of the many leakage correction methods yield the most robust and clinically relevant diagnostic indices of brain tumor status as described in Paulson et al [38]. A great need exists for studies focused on the validation of existing and novel leakage correction methods using gold standard metrics of tumor blood flow and blood volume. This will be especially critical for the application of DSC-MRI methods outside the brain.

13.2.5 Arterial Input Function

The selection of a suitable arterial input function is particularly important for quantification of hemodynamic indices. Ideally, the arterial input function would be measured for each imaging voxel, but since this is impractical, the most common way to select an arterial input function is to arbitrarily select a common or global concentration time curve that is measured from a voxel(s) located near major arterial vessels and exhibit kinetic patterns consistent with arterial delivery such as early contrast enhancement, high amplitude, and rapid wash-in/out kinetics [39]. The MRI signal from voxels located within major vessels is confounded by signal saturation and partial volume effects [40,41]. To overcome this limitation, phase-image–derived arterial input functions have been explored as phase changes within vessels demonstrate improved linearity with contrast agent concentration [39,42]. More advanced and automated methods are increasingly being explored to produce more robust and accurate arterial input functions, including the use of multiple localized versions that originate from vessels in specific regions and supply a tissue of interest [43,44].

13.3 Preclinical Applications in Cancer

DSC-MRI preclinical studies have primarily focused on the validation of the derived hemodynamic parameters or the assessment of anti-angiogenic treatment response in rodent brain tumor models. In 2003, Pathak et al. investigated the GE susceptibility calibration factor in normal brain tissue, as well as in orthotopically implanted 9L gliosarcoma tumors [45]. Recall that DSC-MRI studies assume a constant calibration factor for all tissue types, and as such, any heterogeneity in this parameter will translate into potentially erroneous hemodynamic estimates. Since *in vivo* contrast agent concentrations are difficult to assess, in this study, tissue ΔR_2^* values were measured for varying amounts of CA. Following the MRI studies, the tissue fractional vascular volume (f_v) was assessed stereologically with a vascular casting technique. Since $\Delta R_2^* = k_G f_v C_p$, variations in the MRI-based slope measurements (ΔR_2^* vs. $C_p(t)$) between tissue types reflect changes in either k_G or f_v. By comparing the ratio of MRI slopes and the ratio of f_v values between two tissue types, changes attributable to k_G variations can be assessed. Specifically, if these two ratios are similar, the k_G values across the two tissue types are also similar. Such was the case for normal brain gray and white matter. In contrast, if these two ratios are unequal, the k_G values are dissimilar, as was the case when tumor and contralateral normal-appearing brain tissue values were compared. Upon further analysis, it was estimated that tumor k_G values were approximately 0.55 times that found in normal tissue. Consequently, it was concluded that DSC-MRI–derived blood volume measurements in this tumor model will be underestimated by a similar amount. This work highlights the impact of susceptibility calibration factors on DSC-MRI signals acquired in heterogeneous tumor tissue. To date, no *in vivo* method exists to estimate these factors or to eliminate their influence on DSC-MRI signals.

As described above, the variable dependence of GE and SE relaxation rates enables the estimation of the mean vessel size (or caliber) within an imaging voxel. Typically, the validation of this approach is achieved through a practically difficult-to-achieve registration of *in vivo* magnetic resonance results to *ex vivo* histological data. However, Farrar et al. recently compared intravital microscopy and magnetic resonance measures of vessel size in a U87 mouse brain tumor model for the first *in vivo* validation of this approach [46]. They demonstrate that the estimation of vessel size using the ratio of GE and SE relaxation rates was in excellent quantitative agreement with that derived from intravital microscopy. Specifically, the mean tumor vessel size, normalized to that measured in normal brain, was 1.67 ± 0.67 and 1.64 ± 0.1 for intravital microscopy and DSC-MRI, respectively.

Another common application of DSC-MRI is for the noninvasive assessment of anti-angiogenic treatment induced changes in tumor hemodynamics. As an example, Quarles et al. investigated the influence of an antiangiogenic agent on brain tumor perfusion using a simultaneously acquired GE and SE DSC-MRI approach to derive perfusion parameters (blood flow, blood volume, and mean transit time), which are sensitive to both the total vasculature (from the GE data) and the microvasculature (from the SE data), as well as providing a measure of the mean vessel diameter [28]. This approach was used to evaluate the response of the 9L rat brain tumor model to 20 mg/kg and 40 mg/kg of the antiangiogenic agent SU11657, which exerts its effects via a potent inhibition of vascular endothelial growth factor receptors in addition to multiple platelet-derived growth factor receptors. The 20 mg/kg dose significantly reduced the mean vessel diameter by 29.9% ($p = 0.02$). The 40 mg/kg dose significantly decreased the mean vessel diameter by 30.4% ($p = 0.0007$), SE blood volume by 31.8% ($p = 0.03$), GE and SE mean transit time by 46.9% ($p = 0.03$) and 62.0% ($p = 0.0005$), and increased GE and SE blood flow by 36.6% ($p = 0.04$) and 52.6% ($p = 0.02$). It was concluded that 40 mg/kg of SU11657 administered from days 10 through 14 post inoculation: (1) inhibits tumor growth, (2) suppresses "total" and "microvascular" tumor blood volume expansion and vessel dilation associated with neoplastic angiogenesis, and (3) maintains or improves tumor blood flow and mean transit time. The last result was unexpected and suggested that this treatment regimen either inhibited the development of high resistance and low flow tortuous vascular networks, which are commonly observed in tumors, or altered (or "normalized") the vascular architecture and function so that the tumor blood perfusion was improved with increases in blood flow and decreases in transit times. These findings highlight the potential of DSC-MRI perfusion methods as a tool for the noninvasive evaluation of morphological and functional changes in tumor vasculature in response to therapy.

13.4 DSC-MRI Applications in Cancer Diagnosis

As described above, a potential diagnostic use of DSC-MRI is the use of blood volume maps to assess brain tumor grade [2,6,7,32,47,48]. While there is no broad consensus on the most reliable method to collect and analyze DSC-MRI data, studies have found that GE-derived blood volume values, which reflect the total tumor vasculature, generally increase, and significantly correlate, with brain tumor grade. In contrast, SE-derived blood volume values, which are maximally sensitive to the microvasculature, do not show the same degree of correlation [6,7]. These findings are consistent with the growth of abnormally large vessels within high-grade aggressive tumors and indicate that GE acquisition methods are best suited for the comprehensive interrogation of brain tumor vasculature, especially as it pertains to the determination of brain tumor grade.

A potential confounding factor in the use of blood volume maps to determine brain tumor grade is the extravasation of contrast agent as described above in Section 13.2.4. In a cohort of patients with WHO grades II ($n = 11$), III ($n = 9$), and IV ($n = 23$) glioma, Boxerman et al. demonstrated that GE-based relative blood volume maps corrected for contrast agent leakage correlate with brain tumor grade, whereas, uncorrected maps do not [32]. The mean uncorrected rBVs, normalized to contralateral normal-appearing brain tissue, were 1.53, 2.51, and 2.14 (grade II, III, and IV), whereas corrected rBVs were 1.52, 2.84, and 3.96. The correlation between corrected rBV values and tumor grade was significant ($r = .60$, $p < .0001$), whereas it was not for uncorrected values ($r = 0.15$, $p < .35$). It was also noted that though leakage correction resulted in a significant correlation between blood volume and tumor grade, there was still extensive intragrade variability of rBV values that increased with tumor grade. Thus, it was concluded that assigning grade to an unknown tumor based on rBV maps alone can be potentially problematic, particularly when differentiating between grades III and IV. As many physiological and molecular factors determine tumor status, there is an increasing interest in the use of multiparametric imaging protocols, including DSC-MRI, to more comprehensively characterize tumor tissue.

13.5 DSC-MRI Applications in Cancer Therapy

While DSC-MRI hemodynamic measures may ultimately exhibit limited diagnostic utility, numerous studies have demonstrated their clinical potential as sensitive biomarkers and prognostic indicators of treatment response. In a study of 35 patients with histologically confirmed low-grade gliomas (including astrocytomas, oligodendromas, and oligoastrocytomas), Law et al. investigated whether blood volume maps could be used to predict clinical response [49]. Pretreatment blood volume maps were compared with four response categories, including complete response, stable disease, progressive disease, and death. Patients received standard care treatments, including surgical resection, radiation, and/or chemotherapy. Wilcoxon statistical tests revealed that blood volume maps could not discriminate patients with nonadverse events (response versus stable disease) or adverse events (progressive disease versus death). However, patients who had an adverse event presented significantly higher pretreatment tumor blood volume values compared to those who did not have an adverse event. A Kaplan-Meier survival analysis revealed that lesions with blood volume values less than 1.75 (relative to normal-appearing white matter) had a median time to progression of 4620 days, whereas those with higher than 1.75 had a median time to progression of 245 days. It was concluded that while DSC-MRI may have low specificity for diagnosing histologically based low-grade gliomas, as described above, it has a much higher specificity for predicting clinical endpoints by identifying those that are likely to rapidly progress and undergo malignant transformation despite standard treatment regimes.

Given the heterogeneous nature of tumor perfusion and its response to therapy, Tsien et al. recently evaluated the use of parametric response mapping, a voxel-wise analysis technique, to distinguish progression from pseudoprogression in high-grade glioma patients receiving chemoradiotherapy (cRT) [50]. Pseudoprogression is a term that describes the frequently observed (15%–30% of patients undergoing cRT) changes in imaging parameters that occur early after treatment initiation and are indistinguishable from tumor progression. For example, radiation therapy can increase local vascular permeability and gadolinium enhancement on traditional T_1-weighted images, thereby confounding the identification of residual tumor mass or subsequent tumor progression. Tsien et al. hypothesized that voxel-wise serial differences in the hemodynamic maps would be a better indicator of progression than changes in the whole tumor mean values. The parametric response maps were computed as the difference between the pre- and 3 weeks post-treatment blood volume maps in 27 high-grade glioma patients who were scheduled for cRT. Eight patients exhibited progressive disease, while six displayed pseudoprogressive disease. The mean tumor blood volume, blood flow, and tumor size were not statistically different between these two cohorts of patients, whereas a significant difference was found in the parametric response blood volume map demonstrating its potential as an early imaging biomarker of treatment response. This study highlights the importance of voxel-wise parametric analysis techniques for DSC-MRI data. DSC-MRI studies that rely on mean tumor values or "hot spot"–based metrics potentially fail to capture the complexity and heterogeneity of the underlying tissue and its response to treatment.

Another widely explored use of DSC-MRI hemodynamic indices is as biomarkers of antiangiogenic treatment response. Batchelor et al. incorporated the simultaneous GE/SE DSC-MRI approach into a multiparametric imaging protocol, which also included DCE-MRI, diffusion weighted MRI, to assess the response of 16 glioblastoma patients to a pan-vascular endothelial growth factor receptor tyrosine kinase inhibitor, AZD2171 (trade name Cediranab) [51]. Multiparametric images were serially acquired on days –5, –1, +1 (after the first dose of AZD2171), +26–28, and +54–56, and patients received daily doses of AZD2171 until the end of the study. One of the expressed goals of this study was to assess whether the imaging protocol could be used to identify the vascular normalization window, which is a phrase used to describe the acute structural and functional normalization of tumor vessels after the initiation of antiangiogenic therapy. Previous studies have demonstrated that the effectiveness of chemotherapy and radiation therapy is enhanced during this period of vascular normalization. In this study, the DSC-MRI parameters revealed that relative tumor vessel size significantly decreased as early as one day and remained reduced up to 56 days after the start of treatment in the majority of patients, which potentially represents the extent of the normalization window. With the vessel-size dependent indices afforded by the GE and SE DSC-MRI approach, it was also found that AZD2171 predominantly affects larger microvessels, as indicated by the GE blood volume metric, compared to the smaller microvessels assessed by the SE blood volume. Blood flow was decreased in both small and large vessels. In addition to the DSC-MRI parameter changes, a significant reduction in the DCE-MRI derived volume transfer constant, K^{trans}, was demonstrated, indicative of reduced vascular permeability, and unlike the vessel size, K^{trans} remained decreased throughout the duration of the study. The diffusion weighted imaging-derived apparent diffusion coefficient, along with the DCE-MRI derived extravascular extracellular volume fraction, also indicated that vasogenic edema was significantly reduced after initiation of therapy. The physiological changes are consistent with those found in preclinical studies where vascular normalization was directly visualized using intravital microscopy, and suggest that such markers could be used to detect the normalization window. It should be noted, however, that the reported changes in the imaging parameters are not unique to the vascular normalization process, as traditional antiangiogenic therapy has frequently been shown to reduce blood volume, blood flow vessel size, and permeability. More unique indicators are often reported in preclinical studies, where vascular normalization is marked by an acute period of decreased vascular density and interstitial fluid pressure along with improved perfusion and oxygenation. Ultimately, the validation of these imaging metrics as biomarkers of the vascular normalization window will depend upon their ability to positively identify a distinct period of time, where the effectiveness of secondary administered therapies is significantly enhanced.

13.6 Extensions to Fields > 3 T

Compared to conventional T_1, T_2, or proton-density-weighted imaging, the signal-to-noise ratio of DSC-MRIs are generally lower due to the rapid imaging requirements (temporal resolution less than 1.5 s) needed to characterize the rapid transit of the contrast agent through the tissue. Since signal-to-noise ratio scales approximately linearly with field strength, transitioning DSC-MRI studies to higher fields will increase the signal-to-noise ratio, which will be partially reduced owing to the decreased T_2^* and T_2 relaxation times at higher fields. A potential confound of using T_2^*- weighted sequences at high field is the increased sensitivity to susceptibility artifacts, which increase exponentially with field strength. However, at high-fields, the contrast-agent-induced signal change will be enhanced due to this increased sensitivity to susceptibility effects, thus, enhancing the contrast-to-noise ratio. By employing parallel imaging techniques, image acquisition speeds could be further increased

and also allow shortening of the TE, which would reduce the influence of susceptibility related artifacts [52]. Ultimately, the combination of improved data acquisition paradigms and increased signal-to-noise ratio and contrast-to-noise ratio available at high fields can be used to gain higher temporal or spatial resolution, enhancing the clinical utility of this approach to assess cancer hemodynamics.

Another potential confound of DSC-MRI studies of cancer at high fields is an increased sensitivity to contrast agent leakage effects. With increasing field strength tissue T_1 times increase, T_2 and T_2^* times decrease, and contrast agent–induced susceptibility effects are more pronounced, all of which enhance the sensitivity of DSC-MRI signals to both T_1 and T_2 microscopic effects, as well as the T_2^* mesoscopic extravasation effects. It is likely that DSC-MRI pulse sequences, as well as postprocessing strategies, will need to be optimized to minimize the influence of such effects. However, through the enhanced sensitivity afforded at high fields and improved biophysical characterization of leakage effects, it is possible that even more physiologically relevant information can be extracted from the acquired DSC-MRI signals (e.g., physical properties of the extravascular space).

13.7 Summary

DSC-MRI can noninvasively characterize many features of a tumor's microenvironment, including its vascular architecture, morphology, function, and potentially, even features of the extravascular space. The focus of ongoing and future research centers on the development, optimization, and validation of imaging methods to reliably acquire DSC-MRI data in tumors outside the brain where leakage effects are more pronounced. Despite the confounding effects of contrast agent leakage, DSC-MRI remains a particularly valuable modality for the assessment of antiangiogenic and antivascular therapy and will be increasingly incorporated into clinical research and practice.

References

1. Aronen HJ, Cohen MS, Belliveau JW, Fordham JA, Rosen BR. Ultrafast imaging of brain tumors. *Top Magn Reson Imaging* 1993;5:14–24.
2. Aronen HJ, Gazit IE, Louis DN, Buchbinder BR, Pardo FS, Weisskoff RM, Harsh GR, Cosgrove GR, Halpern EF, Hochberg FH. Cerebral blood volume maps of gliomas: comparison with tumor grade and histologic findings. *Radiology* 1994;191:41–51.
3. Ostergaard L, Weisskoff RM, Chesler DA, Gyldensted C, Rosen BR. High resolution measurement of cerebral blood flow using intravascular tracer bolus passages. Part I: Mathematical approach and statistical analysis. *Magn Reson in Med* 1996;36:715–725.
4. Delile J, Slanetz P, Yeh E, Kopans D, Garrido L. Breast Cancer: Regional blood flow and volume with magnetic susceptibility-based MR imaging. *Radiology* 2002;223: 558–565.

5. Ludemann L, Prochnow D, Rohlfing T, Franiel T, Warmuth C, Taupitz M, Rehbein H, Beyersdorff D. Simultaneous quantification of perfusion and permeability in the prostate using dynamic contrast-enhanced magnetic resonance imaging with an inversion-prepared dual-contrast sequence. *Ann Biomed Eng* 2009;37:749–762.

6. Donahue KM, Krouwer HG, Rand SD, Pathak AP, Marszalkowski CS, Censky SC, Prost RW. Utility of simultaneously acquired gradient-echo and spin-echo cerebral blood volume and morphology maps in brain tumor patients. *Magn Reson Med* 2000;43:845–853.

7. Schmainda KM, Rand SD, Joseph AM, Lund R, Ward BD, Pathak AP, Ulmer JL, Badruddoja MA, Krouwer HG. Characterization of a first-pass gradient-echo spin-echo method to predict brain tumor grade and angiogenesis. *AJNR Am J Neuroradiol* 2004;25:1524–1532.

8. Hakyemez B, Erdogan C, Ercan I, Ergin N, Uysal S, Atahan S. High-grade and low-grade gliomas: differentiation by using perfusion MR imaging. *Clin Radiol* 2005;60:493–502.

9. Fuss M, Wenz F, Essig M, Muenter M, Debus J, Herman TS, Wannenmacher M. Tumor angiogenesis of low-grade astrocytomas measured by dynamic susceptibility contrast-enhanced MRI (DSC-MRI) is predictive of local tumor control after radiation therapy. *Inte Jo Radiat Oncol, Biol, Phys* 2001;51:478–482.

10. Weber, MA, Thilmann C, Lichy MP, Gunther M, Delorme S, Zuna I, Bongers A et al. Assessment of irradiated brain metastases by means of arterial spin-labeling and dynamic susceptibility-weighted contrast-enhanced perfusion MRI: initial results. *Invest Radiol* 2004;39:277–287.

11. Lee MC, Cha S, Chang SM, Nelson SJ. Dynamic susceptibility contrast perfusion imaging of radiation effects in normal-appearing brain tissue: changes in the first-pass and recirculation phases. *J Magn Reson Imaging* 2005;21:683–693.

12. Zierler KL. Theoritcal basis of indicator-dilution methods for measuring blood flow and volume. *Circ Res* 1965;10:393–407.

13. Zierler KL. Equations for measuring blood flow by external monitoring of radioisotopes. *Circ Res* 1965;16:309–321.

14. Stewart GN. Researches on the Circulation Time and on the Influences which affect it. *J Physiol* 1897;22:159–183.

15. Stewart GN. Pulmonary circulation time: Quantity of blood in the lungs and the output of the heart. *Am J Physiol* 1921;58.

16. Hamilton WF, Moore JW, Kinsman JM, Spurling RG. Studies on the circulation: IV. Further analysis of the injection method, and of changes in hemodynamics uner physiological and pathological conditions. *Am J Physiol* 1932;99:534.

17. Smith AM, Grandin CB, Duprez T, Mataigne F, Cosnard G. Whole brain quantitative CBF and CBV measurements using MRI bolus tracking: comparison of methodologies. *Magn Reson Med* 2000;43:559–564.

18. Ostergaard L, Sorensen AG, Kwong KK, Weisskoff RM, Gyldensted C, Rosen BR. High resolution measurement of cerebral blood flow using intravascular tracer bolus passages. Part II: Experimental comparison and preliminary results. *Magn Reson Med* 1996;36:726–736.

19. Ostergaard L. Principles of cerebral perfusion imaging by bolus tracking. *J Magn Reson Imaging* 2005;22:710–717.

20. Kennan RP, Zhong J, Gore JC. On the relative importance of paramagnetic relaxation and diffusion-mediates susceptibility losses in tissues. *Magn Reson Med* 1991;22:197–203.

21. Kiselev VG. Transverse relaxation effect of MRI contrast agents: a crucial issue for quantitative measurements of cerebral perfusion. *J Magn Reson Imaging* 2005;22:693–696.

22. Villringer A, Rosen BR, Belliveau JW, Ackerman JL, Lauffer RB, Buxton RB, Chao YS, Wedeen VJ, Brady TJ. Dynamic imaging wiht lanthanide chelates in normal brain: Contrast due to magnetic susceptibility effects. *Magn Reson Med* 1988;6:164–174.

23. Fisel CR, Ackerman JL, Buxton RB, Garrido L, Belliveau JW, BRR, Brady TJ. MR contrast due to microscopically heterogeneous magnetic susceptibility: numerical simulations and applications to cerebral physiology. *Magn Reson Med* 1991;17:336–347.

24. Boxerman JL, Hamberg LM, Rosen BR, Weisskoff RM. MR contrast due to intravascular magnetic susceptibility perturbations. *Magn Reson Med* 1995;34:555–566.

25. Weisskoff RM, Zuo CS, Boxerman JL, Rosen BR. Microscopic susceptibility variation and transverse relaxation: theory and experiment. *Magn Reson Med* 1994;31:601–610.

26. Majumdar S, Zoghbi S, Gore JC. The influence of pulse sequence on the relaxation effects of superparamagnetic iron oxide contrast agents. *Magn Reson Med* 1989;10:289–301.

27. Tropes I, Grimault S, Vaeth A, Grillon E, Julien C, Payen J-F, Lamalle L, Decorps M. Vessel Size Imaging. *Magn Reson Med* 2001;45:397–408.

28. Quarles CC, Rand SD, Krouwer HG, Schmainda KM. The anti-angiogenic drug, SU11657, decreases brain tumor size and normalizes perfusion as indicated by DSC-MRI perfusion parameters. 2004; Kyoto, Japan. p 2000.

29. Bruening R, Kwong KK, Vevea MJ, Hochberg FH, Cher L, Harsh GRT, Niemi PT, Weisskoff RM, Rosen BR. Echo-planar MR determination of relative cerebral blood volume in human brain tumors: T1 versus T2 weighting. *AJNR Ame J Neuroradiol* 1996;17:831–840.

30. Quarles CC, Ward BD, Schmainda KM. Improving the reliability of obtaining tumor hemodynamic parameters in the presence of contrast agent extravasation. *Magn Reson Med* 2005;53:1307–1316.

31. Uematsu H, Maeda M, Sadato N, Matsuda T, Ishimori Y, Koshimoto Y, Yamada H et al. Vascular permeability: Quantitative measurement with double-echo dynamic MR imaging—Theory and clinical application. *Radiology* 2000;214:912–917.

32. Boxerman JL, Schmainda KM, Weisskoff RM. Relative cerebral blood volume maps corrected for contrast agent extravasation significantly correlate with glioma tumor grade, whereas uncorrected maps do not. *AJNR Am J Neuroradiol* 2006;27:859–867.

33. Vonken EJ, van Osch MJ, Bakker CJ, Viergever MA. Measurement of cerebral perfusion with dual-echo multi-slice quantitative dynamic susceptibility contrast MRI. *J Magn Reson Imaging* 1999;10:109–117.

34. Kuperman VY, Karczmar GS, Blomley MJ, Lewis MZ, Lubich LM, Lipton MJ. Differentiating between T1 and T_2^* changes caused by gadopentetate dimeglumine in the kidney by using a double-echo dynamic MR imaging sequence. *J Magn Reson Imaging* 1996;6:764–768.

35. Vonken EP, van Osch MJ, Bakker CJ, Viergever MA. Simultaneous quantitative cerebral perfusion and Gd-DTPA extravasation measurement with dual-echo dynamic susceptibility contrast MRI. *Magn Reson Med* 2000;43:820–827.

36. Johnson G, Wetzel SG, Cha S, Babb J, Tofts PS. Measuring blood volume and vascular transfer constant from dynamic, T(2)*-weighted contrast-enhanced MRI. *Magn Reson Med* 2004;51:961–968.

37. Quarles CC, Gochberg DF, Gore JC, Yankeelov TE. A theoretical framework to model DSC-MRI data acquired in the presence of contrast agent extravasation. *Phys Med Biol* 2009;54:5749–5766.

38. Paulson ES, Schmainda KM. Comparison of dynamic susceptibility-weighted contrast-enhanced MR methods: Recommendations for measuring relative cerebral blood volume in brain tumors. *Radiology* 2008;249:601–613.

39. Conturo TE, Akbudak E, Kotys MS, Chen ML, Chun SJ, Hsu RM, Sweeney CC, Markham J. Arterial input functions for dynamic susceptibility contrast MRI: requirements and signal options. *J Magn Reson Imaging* 2005;22:697–703.

40. Duhamel G, Schlaug G, Alsop DC. Measurement of arterial input functions for dynamic susceptibility contrast magnetic resonance imaging using echoplanar images: Comparison of physical simulations with in vivo results. *Magn Reson Med* 2006;55:514–523.

41. Kjolby BF, Mikkelsen IK, Pedersen M, Ostergaard L, Kiselev VG. Analysis of partial volume effects on arterial input functions using gradient echo: A simulation study. *Magn Reson Med* 2009;61:1300–1309.

42. Kotys MS, Akbudak E, Markham J, Conturo TE. Precision, signal-to-noise ratio, and dose optimization of magnitude and phase arterial input functions in dynamic susceptibility contrast MRI. *J Magn Reson Imaging* 2007;25:598–611.

43. Lorenz C, Benner T, Lopez CJ, Ay H, Zhu MW, Aronen H, Karonen J, Liu Y, Nuutinen J, Sorensen AG. Effect of using local arterial input functions on cerebral blood flow estimation. *J Magn Reson Imaging* 2006;24:57–65.

44. Lorenz C, Benner T, Chen PJ, Lopez CJ, Ay H, Zhu MW, Menezes NM et al. Automated perfusion-weighted MRI using localized arterial input functions. *J Magn Reson Imaging* 2006;24:1133–1139.

45. Pathak AP, Rand SD and Schmainda KM. The effect of brain tumor angiogenesis on the in vivo relationship between the gradient-echo relaxation rate change (DeltaR2*) and contrast agent (MION) dose. *J Magn Reson Imaging* 2003;18:397–403.

46. Farrar CT, Kamoun WS, Ley CD, Kim YR, Kwon SJ, Dai G, Rosen BR, di Tomaso E, Jain RK, Sorensen AG. In vivo validation of MRI vessel caliber index measurement methods with intravital optical microscopy in a U87 mouse brain tumor model. *Neuro Oncol* 2010;12:341–350.

47. Law M, Yang S, Wang H, Babb JS, Johnson G, Cha S, Knopp EA, Zagzag D. Glioma grading: Sensitivity, specificity, and predictive values of perfusion MR imaging and proton MR spectroscopic imaging compared with conventional MR imaging. *AJNR: Am J Neuroradiol* 2003;24:1989–1998.

48. Weber MA, Zoubaa S, Schlieter M, Juttler E, Huttner HB, Geletneky K, Ittrich C, Lichy MP, Kroll A, Debus J, Giesel FL, Hartmann M, Essig M. Diagnostic performance of spectroscopic and perfusion MRI for distinction of brain tumors. *Neurology* 2006;66:1899–1906.

49. Law M, Oh S, Babb JS, Wang E, Inglese M, Zagzag D, Knopp EA, Johnson G. Low-grade gliomas: dynamic susceptibility-weighted contrast-enhanced perfusion MR imaging—Prediction of patient clinical response. *Radiology* 2006;238:658–667.

50. Tsien C, Galban CJ, Chenevert TL, Johnson TD, Hamstra DA, Sundgren PC, Junck L et al. Parametric response map as an imaging biomarker to distinguish progression from pseudoprogression in high-grade glioma. *J Clin Oncol* 2010;28:293–9.

51. Batchelor TT, Sorensen AG, di Tomaso E, Zhang WT, Duda DG, Cohen KS, Kozak KR et al. AZD2171, a pan-VEGF receptor tyrosine kinase inhibitor, normalizes tumor vasculature and alleviates edema in glioblastoma patients. *Cancer Cell* 2007;11:83–95.

52. Duyn JH, van Gelderen P, Talagala L, Koretsky A, de Zwart JA. Technological advances in MRI measurement of brain perfusion. *J Magn Reson Imaging* 2005;22: 751–753.

Magnetic Resonance Angiography

Ronald R. Price
Vanderbilt University

Ronald C. Arildsen
Vanderbilt University

Jeffrey L. Creasy
Vanderbilt University

14.1 Introduction

Magnetic resonance angiography (MRA) methods have evolved over the past two decades, resulting in an ever increasing number of clinical applications for assessing blood vessels throughout the body as well as for monitoring tumor vasculature status and response to therapy. MRA offers a less invasive and safer option to conventional catheterization/iodine contrast–based X-ray angiography that typically requires arterial access as well as exposure to ionizing radiation.

Originally, MRA was primarily dependent upon non–contrast-enhanced methods that relied on the inherent motion sensitivity of the MRI pulse sequences. Unfortunately, early non–contrast flow–related methods (NC-MRA) often suffer from artifacts and variability in vessel enhancement. As a result, there was a general movement to the use of exogenous gadolinium-based contrast–enhanced MRA (CE-MRA) methods. CE-MRA generally provided more consistent vascular enhancement, fewer artifacts and greater coverage than NC-MRA. However, because of the association of the use of gadolinium-based contrast agents with nephrogenic systemic sclerosis (or nephrogenic systemic fibrosis) in patients with renal insufficiency, there has been a renewed interest in developing new methods of NC-MRA.

MRA continues to benefit from the development of faster MRI methods. This increased imaging speed has resulted from both hardware (e.g., parallel imaging, faster and stronger imaging gradients, optimized receiver coil geometry, and sensitivity) and software (e.g., improved *k*-space sampling, 3D rendered images, and optimized pulse sequences) developments. In general, the capability to image faster has allowed better artifact-free definition of vasculature structures and improved temporal sampling that has provided better definition of the dynamic patterns of tumor blood flow.

14.2 Basics of MRI and Motion

Historically, nuclear magnetic resonance (NMR) measurements, and later, NMR imaging (MRI), assumed that the spins remained stationary during the measurement process. The effect of spin motion on the NMR signal was first observed by Hahn [1] in 1950. At that time, the intent was specifically to understand the effects of molecular self-diffusion. Those early investigations led to the development of NMR pulse sequences that can be used to quantify diffusion motion and, most recently, to the clinical use of diffusion-weighted MRI for localizing and characterizing malignant lesions. (Diffusion pulse sequences and their clinical applications are covered in Chapter 7). The effect of bulk spin movement due to flowing blood was not considered until 1959. In 1959, Singer [2] described NMR measurements of blood flow in the mouse tail. In 1983, Singer and Crooks [3] described a technique for quantitative measurement of venous (jugular) blood flow in the human head using MRI. It is Singer's work that is generally credited for recognizing that in flowing blood produced a greater signal than stationary blood due to its T_1 relaxation rate being effectively shorter than surrounding tissues. Extensions to Singer's original observations have led to the development of time-of-flight MR angiography (TOF-MRA) that now constitutes one of the more commonly used clinical pulse sequences.

In the 1960s, Hahn [4] demonstrated that spins moving at a constant velocity (v) through a linear gradient field of strength (G) with gradient duration τ, would acquire a shift in the phase

($\Delta\phi \sim Gv\tau^2$) of the transverse magnetization relative to the phase of stationary spins that is directly proportional to the velocity of the blood flow (v). These observations have led to the development of phase-contrast MRA (PC-MRA) methods that constitutes a common clinical tool used most frequently in cardiac related MRI.

The purpose of this chapter is to review the different MRA methods and techniques, the physical basis of each and to present examples with typical acquisition parameters and pulse sequences with specific emphasis on sequences being used for MRA evaluation of cancer. The discussion of the methods will be divided into the two broad categories of non–contrast- and contrast-enhanced methods.

14.3 Non–Contrast-Enhanced MRA

Some of the earliest MRI observations of blood flow effects [5,6] were made using spin echo (SE) pulse sequences. In SE images, flowing blood would most often result in "flow voids." Flow voids resulted from spins within a slice being excited by a 90° pulse but, because of their motion, would not experience the companion refocusing 180° pulse and thus produce no signal (Figure 14.1). This phenomenon is often referred to as the "wash-out" effect. In addition to flow voids, flow enhancement could also occur, especially in multislice SE sequences. In multislice acquisitions, occasional vascular enhancement would depend upon multiple factors, including entry or exit slice, flow pulsatility, the timing of the RF pulses relative to the cardiac cycle, flow direction, and flow velocity. For example, if the flow velocity is very slow, spins within the excited slice could experience both the 90° and 180° pulses in an SE sequence and may appear brighter than the surrounding tissues. These enhancements in SE sequences were often identified as "anomalous" or "paradoxical." Though somewhat unpredictable, SE images certainly contain useful information that can be helpful for characterizing vascular flow. However, because of the complex relationship between

flow and signal intensity, SE sequences are typically not used for bright-blood MRA. Due in large part to a more controllable and predictable vascular signal, gradient echo (GE) imaging sequences have become the dominant imaging pulse sequence used for bright-blood MRA. The increased acquisition speed of GE sequences is also another important aspect of gradient echo MRA (GE-MRA).

GE-MRA may utilize either two-dimensional (2D) (Figure 14.2) or three-dimensional (3D) (Figure 14.3) image sequences [7] and rely primarily on wash-in effects to produce vascular enhancement. Specifically, when fresh, fully relaxed spins (i.e., spins not having experienced prior RF pulses and thus not partially "saturated") are washed into a voxel and then read out with a GE, the MR signal from the flowing spins will thus be larger than the surrounding tissues that have a partially saturated signal strength. Specifically, the spins flowing into the voxel will have their magnetization (M_z) fully aligned with the longitudinal B_0 field, unlike the surrounding stationary spins that retain a degree of alignment along the transverse axis (M_{xy}) as a result of repeated RF pulses. The degree of transverse axis alignment of the stationary spins will depend upon the number of prior

2D sequence with magnetization prepulse

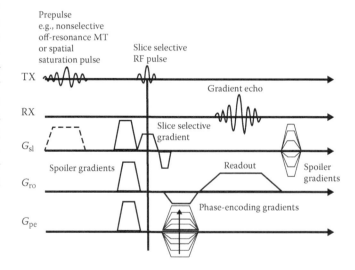

FIGURE 14.2 A typical pulse sequence timing diagram for a generic 2D imaging. The 2D sequence is characterized by a slice selective RF pulse, phase-encoding gradients for spatial encoding in one axis, and a frequency encoding (readout) gradient for spatial encoding in the orthogonal axis. The sequence often contains a prepulse, that is, an RF pulse prior to echo formation and imaging. The prepulse may either be at the water proton resonance frequency (on resonance) or centered at another frequency, such as for fat protons (nominally, 3.5 ppm frequency offset from water protons) or centered significantly off-resonance, as is the case for magnetization transfer (MT) pulses. If the prepulse is applied along with a gradient, the pulse is spatially selective. On-resonance pulses used to saturate the signal from spins flowing into the image volume are typically spatially selective. MT and fat saturation (fat-sat) pulses are typically not spatially selective and thus affect the entire field of view.

2D Slice excitation

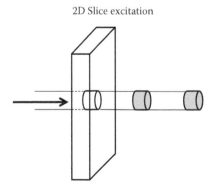

FIGURE 14.1 Flow voids are produced in spin echo images when blood flowing into the image volume is excited by the 90° pulse (dark plug), but when the subsequent refocusing 180° pulse is applied to the same slice that now contains fresh blood with no net transverse magnetization, there will be no observed signal.

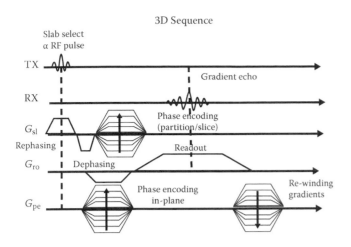

3D Sequence

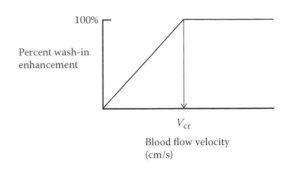

100%

Percent wash-in
enhancement

V_{cr}

Blood flow velocity
(cm/s)

FIGURE 14.3 A generic 3D imaging sequence. The 3D sequence is characterized by either a nonselective RF pulse (no gradient) that excites an entire 3D volume or a slab-selective pulse (weak gradient) that excites a smaller 3D volume. Unlike the 2D sequence, the 3D sequence uses phase encoding for two axes and frequency encoding for the third axis. Strong "spoiler" gradients are typically applied after readout and prior to the next volume-selective RF pulse for the purpose of destroying (spoiling) any residual transverse magnetization.

FIGURE 14.4 In gradient echo pulse sequences, when fresh blood, with fully relaxed spins (i.e., only longitudinal and no transverse magnetization), completely displaces spins that were originally in the slice, maximum signal strength will result. When only a fraction of the original spins are displaced, the signal will be less and the minimum signal will result when no wash-in of new spins occurs. The signal strength will increase as the fraction of original spins replaced with fresh blood increases. The rate of displacement depends upon the flow velocity. The signal will increased with increased flow velocity up to a critical velocity (V_{cr}) at which 100% of the original spins have been replaced by fresh spins; beyond this critical velocity, no further signal increase is observed.

pulses, the flip angle of the pulses, and the repetition time (TR). Since the spins that wash into the image plane are fully aligned, they are said to exhibit an effectively shorter T_1 than the surrounding tissues. The take-away message is that spins within voxels with relatively shorter T_1 will exhibit a higher MR signal and spins within voxels that exhibit less T_1 recovery (i.e., greater transverse magnetization) will produce a lower signal. Thus, enhancement can be maximized by both shortening the T_1 of the blood and saturating the signal from spins located in surrounding background tissue voxels; that is, we can "improve" MRA vessel/tissue contrast by increasing the in-flow enhancement of the blood and/or by increasing the background signal suppression.

14.3.1 Wash-in Effects

For a 2D slice being imaged by a short TR gradient echo sequence, the relative wash-in enhancement [8] is determined by the fraction of the excitation slice thickness (T) that is displaced out of the original slice position during the TR interval by in-flowing spins (Figure 14.4). A simple way to understand the signal intensity in GE images is to recognize that the magnitude of the signal is dependent upon how much T_1 recovery has taken place during the interval between the prior RF pulse and the time of the GE readout. For a constant flow velocity (v), the portion of the slice (d) that will be moved out of the slice is

$$d = v \cdot TR \qquad (14.1)$$

As the velocity increases, a larger fraction of the spins at the original slice position will be replaced, and therefore exhibit a

shorter effective T_1 and higher signal. The signal will increase up to a critical velocity v_{cr} at which the spins in the slice have been completely replaced by fully relaxed spins.

$$V_{CR} = \frac{T}{TR} \qquad (14.2)$$

Once the velocity reaches v_{cr}, there will be no further in-flow enhancement. As an example, for a 3-mm slice thickness and a TR = 30 ms, the critical velocity would be about 10 cm/s. With current MRI systems, it is quite easy to achieve TRs short enough to ensure maximum wash-in effects. As a reference, typical peak (not average) flow velocity in the abdominal aorta would be 100–180 cm/s, velocity in the common carotid artery 80–120 cm/s, and velocity in the intracranial arteries from 40–60 cm/s. In comparison, the average venous flow velocity is 2–5 cm/s.

In addition to the intentional saturation of the surrounding background tissue spins, further reduction in the background signal can be achieved by using an off-resonance, non–spatially selective, magnetization-transfer (MT) prepulse (see Chapter 8). The MT prepulse is used to preferentially saturate the signal from the nonmobile protons located in macromolecules and lipids (Figure 14.2) which in turn further suppress the signal from the mobile or free protons of water and fat.

For 3D scan volumes, much like the stationary tissues, once the flowing blood enters the 3D volume, it will also experience a number of RF pulses from multiple phase-encoding steps. Specifically, the moving spins will experience an additional RF pulse during each TR interval as long as the spins remain within the 3D volume. As a consequence, saturation of the flowing

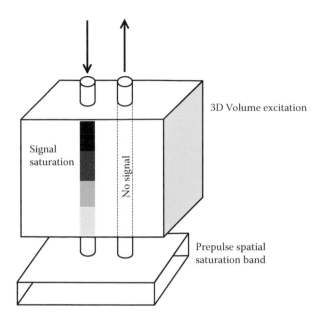

FIGURE 14.5 In 3D TOF-MRA and without the benefit of a contrast agent, the blood signal will become increasingly saturated (reduced signal strength) the longer it resides within the 3D excitation volume. Thus, the vessel will exhibit unequal contrast throughout its length. Selection of smaller, overlapping volumes that are subsequently combined (multiple overlapping thin slab acquisition) is used to minimize the saturation effect. On-resonance, spatially selective prepulses may be used to suppress signal from specific flow directions, allowing separation of venous and arterial flow.

spins will begin to occur, and spin saturation will continue to increase as the number of pulses that the spins experience increases (Figure 14.5). In summary, the saturation effect will be greater as the image volume increases and/or as the velocity decreases.

In order to compensate for this saturation effect, one can reduce the size of the 3D volume by using the so-called multiple overlapping thin slab acquisition technique [9]. An alternative method that may be used to minimize the saturation effect is to sequentially reduce the flip angle of the RF pulses during each phase-encoding cycle. The use of variable flip angles effectively alters the sensitivity profile of the excitation volume and has been referred to as "tilted-plane" excitation. Tilted-plane excitation results in a more balanced enhancement by proportionally saturating the spins that have been within the volume the longest time less than the spins that enter volume later in the acquisition. Further contrast enhancement may also be achieved by using fat suppression pulses and/or by cardiac gating to minimize pulsatile flow effects.

14.3.2 Two-Dimensional Time-of-Flight MRA

For 2D noncontrast MRA (NC-MRA), multiple continuous thin slices gradient echo images are acquired and then reassembled as

a volume set. The volume is then viewed as a series of maximum intensity projection (MIP) images [6] reconstructed at multiple viewing angles typically taken perpendicular to the original imaging planes. When displayed as a continuous movie (cine) the vascular pattern can be appreciated in 3D space.

When the flow direction is oriented perpendicular to the image plane, a companion on-resonance, spatially "traveling" saturation band (see Figure 14.5) can be applied on the inflow side of the slice being imaged to eliminate either arterial or venous flow into the slice. For example, when imaging the carotid arteries, if the traveling saturation band is placed superior to the slice being acquired, venous flow from above will be suppressed. Alternatively, when imaging the jugular veins, the traveling saturation band is placed a few millimeters inferior to the slice to eliminate the arterial flow. The spatial saturation prepulse is illustrated in Figure 14.2 and described in detail in Chapter 4. Typical 2D timing parameters are TR of 20–30 ms, TR of 5–10 ms, and flip angles of 40°–50°. Slice thicknesses are typically 1–2 mm.

The primary advantages of 2D acquisitions are that they are less sensitive to saturation effects on the in-flowing blood and can be made sensitive to very slow flow. The disadvantages are lower signal-to-noise ratio (SNR), relatively thick slices resulting in through-plan resolution, and poor sensitivity to vessels that flow entirely within the plane. If patient motion, vessel pulsations, or uncorrected pulsatile flow occurs, there will often be "venetian blind" artifacts in the maximum intensity projection (MIP) images.

14.3.3 Three-Dimensional Time-of-Flight MRA

Many aspects of 3D TOF-MRA [10–12] are similar to 2D TOF-MRA [13,14], with the obvious difference being that the image data are acquired with spatially nonselective or "relatively" nonselective RF pulses (i.e., volume or slab excitation) rather than slice selective RF pulses. The 3D image acquisition relies upon phase encoding along two axes (i.e., adding phase encoding in the "slice" direction) rather than phase encoding just along a single axis used for 2D image acquisitions. The additional phase encoding allows for thinner slices than are possible with 2D acquisitions and thus, better through-plane spatial resolution. In 3D acquisitions, voxel sizes are typically submillimeter along each axis. The 3D TOF-MRA volume acquisition times are also generally shorter than the typical 2D acquisition needed to cover the same volume. As noted previously, the signal-to-noise per unit time is also higher with 3D sequences.

The 3D TOF pulse sequence can typically use short echo times (~5 ms) because of the shorter nonselective excitation pulses relative to the longer slice-selective pulse required for 2D sequences. The TRs are often longer than used or 2D sequences (TR ~40 ms), and the flip angles are smaller (~20°). The in-plane image matrix size is typically high (256 × 256) and the through-plane, phase-encoding direction will typically be 64 or 128. The through-plane direction may often be referred to as "partitions" rather than slices.

The 3D TOF-MRA images are also viewed as MIPs or, alternatively, as surface-rendered images. The primary advantages of 3D TOF-MRA are the thinner slices, higher spatial resolution (smaller voxels), higher SNR, shorter TEs, and because of smaller slab-selective gradient amplitudes, there are fewer phase artifacts than for the 2D methods. The disadvantages of 3D TOF-MRA are that the blood signal becomes saturated when large acquisition volumes are used or when the vascular flow is slow. Shorter T_1 tissues such as fat or old "bleeds" may appear bright and may occasionally be mistaken for vessels. Similar to 2D methods, spatial presaturation pulses may also be used to distinguish arterial from venous flow.

The requirement of phase encoding in two axes can, of course, lead to relatively long acquisition times (> 10 min). However, with the advent of a number of acceleration methods (see Chapter 4) such as parallel imaging with sensitivity encoding, simultaneous acquisition of spatial harmonics or sparse k-space sampling (e.g., VIPR: Vastly undersampled isotropic projection reconstruction), 3D TOF-MRA sequence can be shortened to 2 min or less, depending upon the field of view and matrix sizes.

14.3.4 Steady-State Free Precession Sequences

Another important approach to noncontrast vascular imaging, sometimes called "fresh blood imaging," is becoming very popular and uses an ultrafast balanced SSFP (bSSFP) sequence [15,16] with prepulse fat suppression. The bSSFP sequence, also known as true FISP, is a free precession GE sequence that has balanced gradients in all three orthogonal directions and has the effect of rephasing and maintaining the transverse magnetization of the flowing blood, thus producing a bright blood signal. The sequence may also be applied using ECG gating to further minimize transverse dephasing and vessel signal loss. A sequence that uses "balanced" gradients means that for each positive gradient pulse, there is an equal and opposite negative gradient. These compensating gradients are sometimes referred to as "re-winding" gradients (Figure 14.6). The sequence can have extremely short TR/TE ~2 ms/1.5 ms and yields an image intensity weighting that is a combination of T_1 and T_2. Because of the balanced gradients, the bSSFP is relatively motion insensitive and produces a bright blood signal and is often for coronary artery imaging [17–20], abdominal vessels [16], as well as breast imaging [21]. The signal in the bSSFP images is dominated by long T_2 or T_2^* tissues and is essentially the sum of both a spin echo and a gradient echo. A short inversion recovery (STIR) prepulse may also be used to reduce the signal from the surrounding fat tissues, and the sequence can also be applied after Gd contrast, as is the case for most breast cancer protocols.

A significant difference between the bSSFP images from images made with other MRA sequences is that the signal from background tissues is retained rather than suppressed. Thus, bSSFP

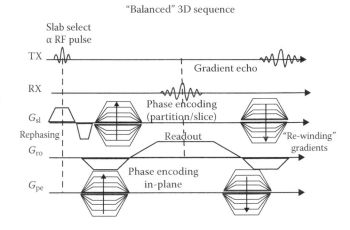

FIGURE 14.6 The balanced steady-state free precession (bSSFP) pulse sequence is designed to create a steady state in the transverse magnetization. Thus, no "spoiling" or "crusher" gradients are used. Since the application of any gradient also creates spin dephasing, the bSSFP attempts to reverse all dephasing effects by "rewinding" (i.e., reverse polarity with equal duration) all gradients used in all axes.

images not only have bright blood but also significant anatomical details (Figure 14. 7). Through the use of half-Fourier acquisition and acceleration from parallel imaging, total acceleration factors of six or more are possible, allowing volume acquisitions on the order of 90 s. Due to recent concerns over renal toxicity of the Gd contrast agents, as well as concerns for cost containment, SSFP sequences are being used more frequently for MRA.

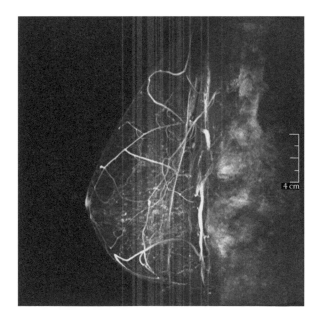

FIGURE 14.7 Breast MRI illustrating the bright blood characteristic of 3D bSSFP pulse sequences. The image illustrated (TE/TR/FA = 3.2 ms/6.5 ms/12°) employed parallel imaging acceleration and was acquired in 1:35 min.

14.3.5 Phase Contrast MRA

As noted previously, PC-MRA [22–24] relies upon dephasing the moving spins submitted to a bipolar gradient. For a bipolar gradient of a given intensity, (G) and time (t), the moving spins will dephase and acquire a shift in the phase of the transverse magnetization (Figure 14.8) relative to stationary spins that is proportional to their velocity (v).

Starting with the Larmor equation

$$\omega(x) = \gamma B(x) \tag{14.3}$$

where, ω is the precessional frequency of protons at location x, and $B(x)$ is the magnetic field strength that the protons experience at that location. The magnetic field strength and thus the related precessional frequency that a proton will experience in a linear gradient field (G) are determined by the position at which the proton resides within the field. For moving protons, the position is dependent upon the velocity and the time over which the observation is made: $x = vt$. Spins moving at a velocity (v) in the presence of a bipolar gradient pulse that consists of a negative lobe of duration (τ) followed by a gradient reversal, positive lobe of the same duration (Figure 14.9), will induce a phase shift ($\Delta\phi$), relative to stationary spins ($\Delta\phi = \gamma GV\tau^2$). This can be shown by rewriting the Larmor equation in terms of the time rate of change of the net magnetization phase angle ($\omega = d\phi/dt$) and B in terms of the linear gradient strength and the time:

$$B(t) = Gx(t) = Gvt$$
$$d\phi/dt = \gamma Gvt \tag{14.4}$$

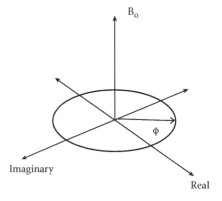

Transverse magnetization (Mxy) phase angle

FIGURE 14.8 The measurement of the phase angle (ϕ) is made possible by quadrature receivers. Quadrature receivers typically utilize two receiver coils or channels, referred to as real and imaginary, that are nominally displaced by 90°, allowing the orientation in space of the net-magnetization vector to be tracked. The reference angle is typically taken to be that of the protons that are located at the isocenter of the magnetic field.

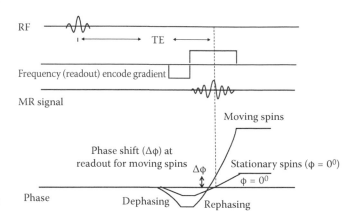

FIGURE 14.9 For moving spins, a bipolar gradient will induce a phase shift ($\Delta\phi$) relative to stationary protons. Illustrated is a generic gradient echo readout showing that at the center of the readout gradient pulse, the stationary spins will be completely rephased ($\phi = 0$), while the phase angle of the transverse magnetization of moving spins will be nonzero.

$$d\phi = \int_0^\tau \gamma(-G)vt\,dt + \int_\tau^{2\tau} \gamma(+G)vt\,dt \tag{14.5}$$

$$\phi = \gamma Gv\tau^2$$

From this result, it is worth noting that (1) the bipolar pulse thus induces a phase shift that is directly proportional to the velocity, (2) the phase shift is quadratic in the gradient duration (τ), and (3) that stationary spins will acquire no net phase shift. Thus, by adding a bipolar gradient [25] in the imaging pulse sequence (Figure 14.10), it is possible to create a shift in

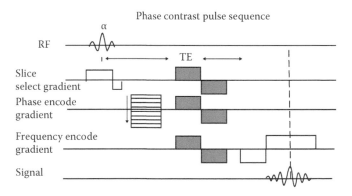

FIGURE 14.10 When bipolar flow-encoding gradient pulses (dark shaded) are added to an imaging sequence, it is necessary that they be applied sequentially in each axis. For 3D flow encoding, it thus requires three separate image acquisitions. The reason for this is that the gradient magnetic fields add as vectors, that is, if two equal gradients are turned on at the same time simply results in a new gradient direction. For example, if equal strength x- and y-gradients are turned on at the same time, the resultant gradient would be oriented at 45° with respect to the original gradient directions.

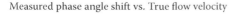

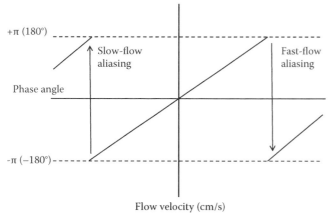

FIGURE 14.11 In PC-MRA, the induced shift in the phase angle will be directly proportional to the flow velocity. However, when the phase shift is greater than 180° (i.e., π radians) it becomes redundant in that it cannot distinguish positive from negative shift since blood flowing in opposite directions along a gradient direction will induce shifts of the opposite directions. This redundant result is referred to as "aliasing" and in some cases can be easily recognized when there is a sudden shift from positive to negative flow. It is most desirable to make sure that the vessel being measured be encoded with a gradient that will produce a phase shift between −180° and +180°.

the phase of the transverse magnetization that is directly proportional to the flow velocity (v). Because of the cyclic nature of the phase, the valid phase values will range from −π to +π. Beyond this range of values, aliasing will occur, meaning that the phase angle has moved through a complete (2π or 360°) cycle and resulted in an aliased value for v that is both in the wrong direction and incorrect magnitude (Figure 14.11). Thus, it is important to pick an appropriate velocity-encoding gradient strength and duration to assure that the phase shift remains between −π and +π. In practice, the operator is required to enter the anticipated maximum flow speed that will be encountered in the image volume. If the entered maximum is too large, the image contrast will be reduced, and if the entered maximum is too small, aliasing will result. Because of the importance of setting an appropriate encoding gradient, it may be necessary to repeat the acquisition to eliminate aliasing or to improve image contrast. It is also important to point out that the velocity-encoding gradient can be applied along any of the three orthogonal gradient directions and thus, actually encodes the component of flow velocity along each respective axis. The implication of this is that in order to determine the velocity vector in 3D space, it will be necessary to make three separate acquisitions, with each acquisition providing the flow component along the respective axes. In application, a fourth acquisition is always acquired without any of the velocity-encoding gradients in order to account for any inadvertent phase shifts that result from other sources, such as nonperfect gradient pulses and field inhomogeneity. PC acquisitions can also be ECG gated to provide quantitative flow images at different phases of the cardiac cycle. Figure 14.12 illustrates PC images through the heart and aorta taken at two different phases of the cardiac cycle. Such "cine" PC images are commonly used clinical sequences for cardiac blood flow evaluation.

In the PC images, the pixel intensity is directly proportional to the flow velocity. The convention is to assign bright intensities (positive number) to voxels in which the flow direction is along the flow-encoding gradient and to assign dark intensities (negative numbers) to voxels in which the flow is in the opposite direction. Stationary voxels with zero phase shift are displayed

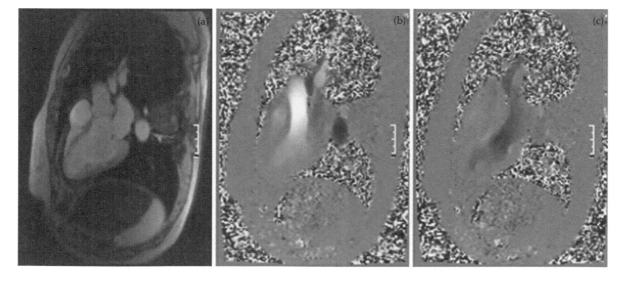

FIGURE 14.12 Sagittal cine PC image of the aorta. Maximum velocity (150 cm/s), in-plane flow encoding. (a) Standard magnitude MRI, (b) PC image at peak systolic velocity, and (c) PC image at low diastolic flow.

as gray. When one is not specifically concerned with the flow direction, absolute magnitude images are computed by combining the phase-shift information from all three orthogonal directions. Much like other MRA images, in the PC velocity–magnitude image, the voxels are displayed as bright against the dark stationary tissues.

14.3.5.1 2D Phase Contrast MRA

For 2D PC-MRA, images are typically acquired using a thick 2D slice with the velocity encoding in the slice direction and as near as possible with the vessel axis perpendicular to the slice plane [5]. Since the method does not rely on T_1 effects, the TR can be very short, resulting in short acquisition times. Advantages of 2D PC-MRA are that it is more quantitative than TOF-MRA, has good background suppression, and can be made sensitive to a large range of flow velocities. A disadvantage is that its flow sensitivity in only one direction actually requires a minimum of two acquisitions (one flow-compensated and one flow-encoded) and can be unreliable with complicated pulsatile flow profiles. When the intent is to acquire 3D PC-MRA, it may be useful to begin with a 2D sequence because the 2D single-slice acquisition is faster and can be useful for testing different encoding speeds that would then be used for the 3D acquisition. This technique can also be employed in vascular flow cine imaging, using pulse or ECG synchronization; thus, a flow velocity profile as a function of the cardiac cycle can be obtained. When the flow velocity (cm/s) is multiplied by the vessel cross-sectional area (cm^2) obtained from the anatomical MRI, one can calculate the vessel's volumetric flow rate.

14.3.5.2 3D Phase Contrast MRA

As with TOF-MRA, 3D PC-MRA can be used to provide higher spatial resolution. As noted previously, 3D PC-MRA does require at least four volume acquisitions [7] and thus a greater time commitment. However, scan times can be significantly reduced with parallel imaging. It is also common to reduce acquisition time by reducing the number of phase encoding steps. The 3D images are also typically viewed as MIP images.

Both 2D and 3D PC-MRA are more sensitive than 3D TOF-MRA for slow flow vessels and thus, can have benefit for 3D cerebral venous imaging. It should also be noted that PC-MRA can be used in conjunction with CE-MRA following the injection of a contrast agent. By combining PC-MRA with contrast-enhanced TOF-MRA, one can achieve high-quality anatomical vascular depiction, as well as quantitative estimates of true blood flow in individual vessels. In summary, the advantages of 3D PC-MRA are the ability to provide accurate quantification of 3D vascular flow velocity and by adding vascular dimensions derived from TOF-MRA, traditional vascular blood flow in units of volume/time (e.g., ml/s) can be obtained. PC-MRA also has excellent background suppression by not being sensitive to short T1 tissues as well as flow sensitivity over a large range of velocities.

14.4 Contrast-Enhanced MRA

The physical basis for CE-MRA [26–31] is essentially the same as for noncontrast TOF-MRA. Specifically, each depends upon the T_1 difference between the blood and the surrounding tissues. However, with TOF-MRA, when the flowing blood within the excitation volume experiences repeated RF pulses, the blood signal becomes more saturated, making the intensity difference (image contrast) between the flowing blood and the surrounding stationary word missing progressively smaller. The result is especially true for large excitation volumes. Specifically, only the blood on the entry side of the volume will have retained its original magnetization and thus bright signal image contrast, whereas blood having flown deeper into the volume will be more saturated, with the result of less image contrast relative to the surrounding tissue.

14.4.1 Three-Dimensional Contrast-Enhanced MRA

The use of the paramagnetic contrast agent (e.g., gadolinium chelates) substantially shortens the T_1 of the blood to the extent that it effectively eliminates the problem of spin saturation in 3D TOF-MRA noted previously (Figure 14.5). In addition to the shorter blood T_1 providing an enhanced in-flow effect (Figure 14.13.), there are a number of other important advantages that have also been recognized. A significant advantage of 3D CE-MRA relative to non–contrast-enhanced methods is that the excitation volume can be much larger. The reason for this is that with the shorter T_1, the flowing blood experiences minimal saturation and thus maintains good contrast between the blood and the surrounding tissues. In addition to the larger field of view, CE-MRA is less subject to flow voids, gives much more consistent image contrast in vessels with pulsatile or turbulent flow, and provides much more accurate vascular dimensions needed for estimating volumetric blood flow.

In order to achieve good image contrast in noncontrast MRA, it is important to orient the vessel perpendicular to the imaging volume to achieve maximum inflow effect. Specifically, the vessel should not be oriented with its flow within the imaging plane. This requirement is generally not necessary with CE-MRA. Finally, sequences used for CE-MRA are generally faster, allowing breath-hold imaging of the pulmonary vasculature.

The larger acquisition volume, without vessel saturation, made possible by CE-MRA, is illustrated in the head and neck in Figure 14.14. In this study, the image volume extends from the aortic arch to the Circle of Willis with essentially constant vascular contrast throughout the entire volume. Because of the approximately isotropic resolution now allowed by the large volume 3D image acquisitions, a variety of 3D display options are available, such as the surface rendered image illustrated in Figure 14.15.

Large vascular coverage also depends strongly upon the volume sensitivity of the coil being used to acquire the image.

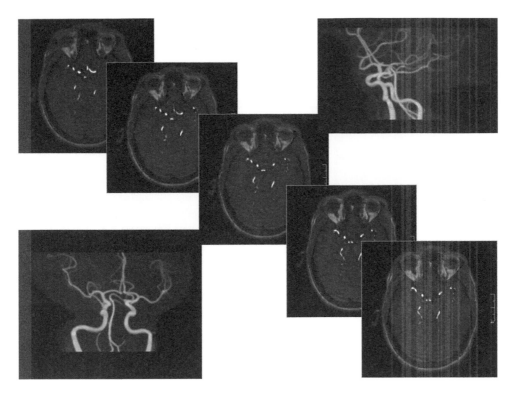

FIGURE 14.13 CE-MRA of the head illustrating equal enhancement throughout all slices. Also illustrated are MIP reconstructions at two different angular orientations.

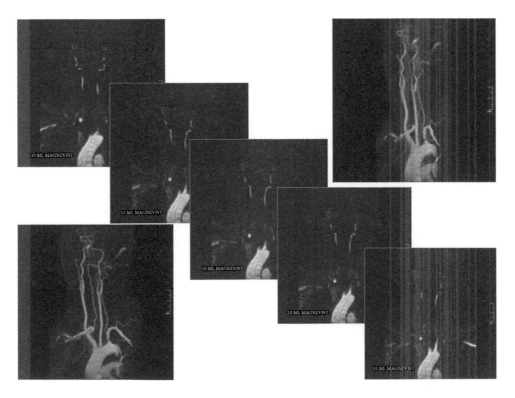

FIGURE 14.14 CE-MRA of the carotids. In this example, the reconstruction FOV = 28 cm. The image was acquired at 3 T using a 16-channel parallel 3D-fast GE imaging sequence (TE/TR/FA = 1.5 ms/4.8 ms/30°). Essentially equal vascular contrast is seen throughout the volume.

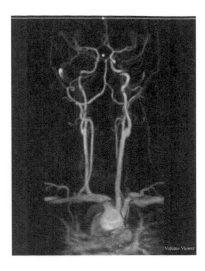

FIGURE 14.15 A variety of 3D image visualization tools are now available to assist in viewing 3D CE-MRA studies. This image illustrates one of a series of surface rendered images.

Fortunately, with the advent of the multichannel parallel imaging capability of current systems, many unique coil designs are now available for MRA. Figure 14.16 illustrates two of the many different multicoil arrays that are optimized for specific MRA applications. The coil in Figure 14.16a is a 16-channel

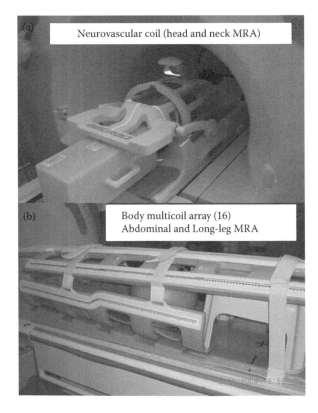

FIGURE 14.16 (a) Sixteen-channel receive-only neurovascular coil array for MRA of the head and neck. (b) Sixteen-channel receive-only body coil array for MRA of the peripheral circulation.

neurovascular array that not only provides uniform signal sensitivity throughout the head, but also includes coils that allow good sensitivity all the way to the aortic arch for carotid MRA. The coil in Figure 14.16b illustrates one of the many coil arrays that can be used for full-body MRA and specifically MRA of the legs.

Even though both 2D and 3D acquisitions are possible, primarily because of speed of acquisition and improved spatial resolution, rapid 3D CE-MRA is currently the most common clinical sequence. CE-MRA sequences are typically T_1 weighted gradient echo sequences with gradient spoiling of the residual transverse magnetization. In order to keep echo time short, the gradients are also not typically flow compensated.

14.4.2 Filling *k*-Space

To achieve the best possible image contrast, the timing of the data acquired at the center of *k*-space should correspond to the moment in time at which the peak intravascular concentration of the contrast agent occurs. *k*-Space filling [32–38] can be optimized (spiral or elliptical centrifugal trajectory, partial refilling, shared data on the *k*-space periphery) to rapidly acquire its center (image contrast) and meet the constraints of the transient passage of the contrast agent (Figure 14.17). With the gain in speed that it entails, parallel imaging is particularly well suited to CE-MRA. Fat suppression may also be added to further reduce the background signal and improve vascular contrast. Timing can also be used to selectively enhance and emphasize either arterial or venous flow by using the appropriate delay between contrast injection and the acquisition of the center of *k*-space to make the center of *k*-space coincide (Figure 14.18) with the peak blood concentration of the contrast agent.

14.5 Preclinical Applications

In mouse models of human cancer, MRA has been used to characterize blood vessels surrounding malignant tumors. The direction of the work has been primarily to characterize vessel geometries (shape and size) and to correlate these parameters with malignancy. Results reported by Brubaker [39] and Bullitt [40] indicated that increased vessel number did not become apparent until the tumor volume had reached approximately 80 mm^3. However, measures of abnormal vessel shape became apparent more quickly. These reports used velocity-compensated non–contrast-enhanced 3D MRA sequences with isotropic voxel sizes of $0.1 \times 0.1 \times 0.1$ mm^3 and were acquired at 3 T. Additional work by the Bullitt group [41] has reported quantitative measures of vessel tortuosity and the relationship of these measures to tumor malignancy.

Other investigators [42] have utilized CE-MRA in mice and have specifically reported whole-body 3D-Micro-MRA. These investigators have used MRA to monitor tumor microvasculature changes in cancerous tissues following antiangiogenesis therapy.

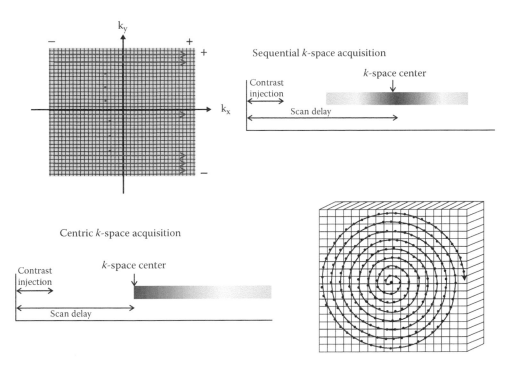

FIGURE 14.17 In order to achieve maximum vascular enhancement, the timing of the contrast injection should be chosen so that the center of *k*-space is acquired when the maximum vascular contrast agent concentration occurs. It is often necessary to use a small-volume test injection of the contrast agent, followed by a rapid imaging sequence to determine the most appropriate delay between injection and imaging. In order to determine the delay it is necessary to know how *k*-space is being filled for a specific imaging sequence. Sequential filling most commonly refers to phase encoding gradients being applied starting first with the highest gradient strengths and then sequentially reducing the phase-encoding gradient strength, passing through zero and then moving to the maximum strength of the opposite polarity. In this case the center of *k*-space is at the center of the acquisition time. Alternatively, centric *k*-space filling sequences start with the zero phase gradient strength and then move toward the maximum value. The specific pattern may be spiral, elliptical, or other. For centric filled sequences, the center of *k*-space is at the beginning of the acquisition.

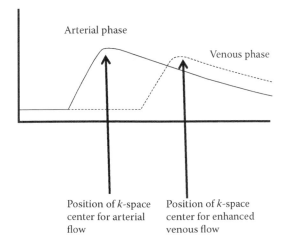

FIGURE 14.18 Timing the center of *k*-space may also be used to selectively enhance flow in either arteries or veins by selecting the timing of the peak concentration of the contrast agent for either arterial or venous flow.

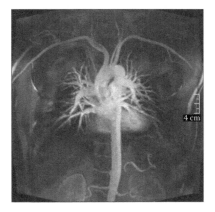

FIGURE 14.19 3D FLASH CE-MRA of the chest and central circulation acquired in coronal axis. ϕ/TE/TR = 15°/0.93/2.51 ms. 400 × 400 mm FOV, 156 × 246 × 384 matrix.

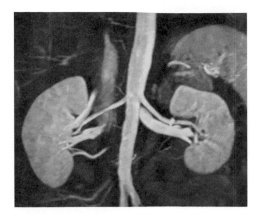

FIGURE 14.20   3D FLASH CE-MRA of the abdominal aorta, vena cava, renal arteries, and renal veins.

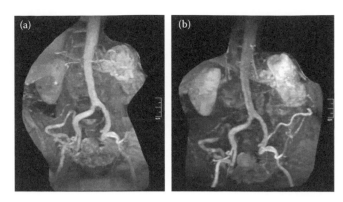

FIGURE 14.22   (a) Coronal view of a maximum intensity projection (MIP) image of the TOF-MRA of the renal vein tumor described in Figure 14.19. (b) Rotated MIP of same 3D volume.

14.6 Clinical Applications

A significant advantage of MRA as a tool for monitoring vascular involvement of cancer or for monitoring vascular response to antiangiogenesis chemotherapy is that it is equally applicable throughout the body with millimeter to submillimeter spatial resolution. 3D MRA of the central vasculature, including pulmonary vessels (Figure 14.19) as well as the abdominal and renal vessels (Figure 14.20), can be acquired with isotropic resolution and reformatted for 3D viewing.

As illustrated in Figure 14.21, MRA is particularly helpful as an easily repeatable and relatively noninvasive tool for assessing altered hemodynamics as a consequence of cancer invasion. Figure 14.21a is a contrast-enhanced axial TOF-MRA taken through the abdomen above the level of the renal veins in a patient with advanced renal cancer. The MRA illustrates an

enlarged left renal vein with a tumor thrombus (arrow) extending from the renal vein into the inferior vena cava (IVC). Figure 14.22 shows MIP images reconstructed at two different viewing angles from the base image shown in Figure 14.21, clearly demonstrating the extent of the cancer involvement.

The images in Figure 14.23 also illustrate how MRA can be a valuable tool for assessing tumor vessels and for determining vascular/tumor anatomy prior to intervention. In Figures 14.23a and 14.23b, by viewing the 3D rendered images, one can easily demonstrate that the tumor does not have significant vascular

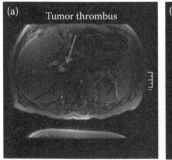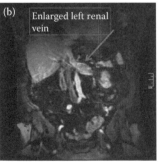

FIGURE 14.21   (a) Axial 2D TOF image through the abdomen above the level of the renal veins, demonstrating high signal intensity in the IVC and low signal from tumor thrombus (arrow). (b) Coronal T1-weighted, gradient echo sequence following gadolinium contrast injection demonstrating heterogeneously enhancing tumor thrombus grossly enlarging the left renal vein (arrow) and extending into the IVC.

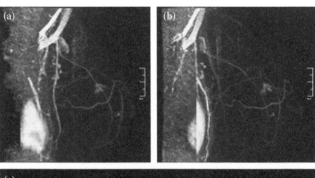

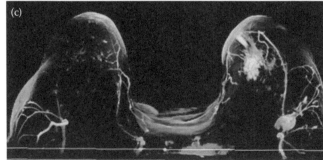

FIGURE 14.23   3D bSSFP images of a breast with suspicious lesion. (a,b) MIP images reconstructed at two different viewing angles to allow determination of vascular involvement. (c) Volumetric rendering of 3D bSSFP image of a breast lesion with significant tumor vasculature.

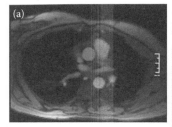

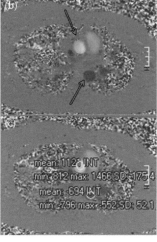

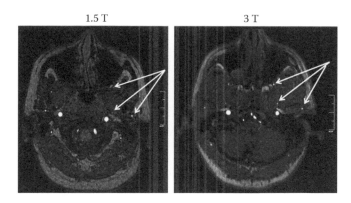

FIGURE 14.25 Basis axial images from CE-MRA studies taken at 1.5 and 3 T. Studies are from different patients; however, the improved small vessel detection (arrows) can be noted in the 3 T images relative to the images acquired at 1.5 T.

FIGURE 14.24 Axial PC image through the aorta. Maximum velocity (150 cm/s, through-plane (slice) flow encoding. (a) Magnitude image, (b) PC image, and (c) region of interest average flow values +112 cm/s in ascending aorta and −69 cm/s in descending aorta.

involvement. The 3D volume rendered image shown in Figure 14.23c, however, illustrates how MRA can clearly provide the definitive information on tumor neovasculature and the invasion of normal tissues (see also Figure 14.24).

14.7 MRA at Field Strengths > 3 T

The linear increase in SNR allows for smaller voxel sizes and, consequently, visualization and detection of smaller vessels (Figure 14.25). In addition, field strengths of 3 T or greater may also offer benefits from the increased T_1 that occurs at higher field strengths as well as from greater enhancement from contrast agents. For non–contrast-enhanced sequences, including bSSFP and true FISP, the longer T_1 allows longer inversion times (TI) and more inflow without compromising background signal suppression. For contrast-enhanced studies, Trattnig et al. [43] recommends that at 3 T, the amount of Gd contrast agent may be decreased by approximately one-half and still maintain enhancement comparable to 1.5 T. This is significant in light of the increased concern for renal toxicity of Gd agents (nephrogenic systemic fibrosis).

14.8 Conclusion

MR angiography offers a high-resolution three-dimensional minimally invasive alternative to intravascular catheter angiography for evaluating tumor blood supply and vascular architecture.

The use of a contrast agent to further decrease the T_1 of the blood produces additional contrast relative to noncontrast MRA that now allows visualization of smaller vessels that can be seen when a contrast agent is not used. Specifically, because of the minimally invasive nature of 3D MRA, it is an excellent tool for assessing arterial and venous anatomy/pathology. For the evaluation of the brain, the use of a contrast agent brings the added benefit of identifying breaches in the blood–brain barrier because, with an intact blood–brain barrier, the agent will remain intravascular.

References

1. Hahn EL. Spin echoes. *Physical Rev*. 1950;80:580–594.
2. Singer JR. Blood flow rates by nuclear magnetic resonance measurements. *Science*. 1959;130:1652–1653.
3. Singer JR, Crooks, LE. Nuclear magnetic resonance blood flow measurements in the human brain. *Science*. 1983;221:654–656.
4. Hahn EL. Detection of sea-water motion by nuclear precession. *J Geophys Res*. 1960;65:776–777.
5. Haacke EM, Lin W, Li D. Whole body magnetic resonance angiography. In: *Methods in Biomedical Magnetic Resonance Imaging and Spectroscopy*. Editor, IR Young. John Wiley & Sons, New York. 2000;488–500.
6. Laub G, Gaa J, Drobnitzky M. Magnetic resonance angiography techniques. *Electromedica*. 1998;66:68–75.
7. Ivancevic MK, Geerts L, Weadock et al. Technical principles of MR angiography methods. *Magn Reso Imaging Clin N Am*. 2009;17:1–11.
8. Laub G. Time-of-flight method of MRA. In: *Methods in Biomedical Magnetic Resonance Imaging and Spectroscopy*. Editor, IR Young. John Wiley & Sons, New York. 2000;500–504.
9. Parker GL, Yuan C, Blatter DD. MR angiography by multiple thin-slab 3D acquisition. *Magn Reson Med*. 1991;17:434–457.
10. Masaryk TJ, Modic MT, Ruggieri PM, et al. Three-dimensional (volume) gradient-echo imaging of the carotid bifurcation: Pulmonary clinical experience. *Radiology*. 1989;171:801–806.
11. Marchal G, Bosmans H, Van Fraeyenhoven L, et al. Intracranial vascular lesions: Optimization and clinical evaluation of three-dimensional time-of-flight MR angiography. *Radiology*. 1990;175:443–448.

12. Ruggieri PM, Laub GA, Masaryk TM, et al. Intracranial circulation pulse-sequence considerations in three-dimensional (volume) MR angiography. *Radiology.* 1989;171:785–791.

13. Keller PJ, Draper BP, From EK, et al. MR Angiography with two-dimensional acquisition and three dimensional display. *Radiology.* 1989;173:527–532.

14. Litt AW, Eidelman FM, Pinto RS, et al. Diagnosis of carotid artery stenosis: Comparison of 2DFT time-of-flight MR angiography with contrast angiography in 50 patients. *AJNR Am J Neuroradiol.* 1991;12:149–154.

15. Scheffler K, Lehnhardt S. Principles and applications of balanced SSFP techniques. *Eur Radiol.* 2003;13:2409–2418.

16. Stafford RB, Sabati M, Haakstad MJ, et al. Unenhanced MR angiography of the renal arteries with balanced steady-state free precession Dixon method. *AJR Am J Roentgenol.* 2008;191:243–246.

17. Finn JP, Naal K, Deshpande V, et al. Cardiac MR imaging: State of the technology. *Radiology.* 2006;241:338–354.

18. Yang Q, Kuncheng L, Liu X, et al. Contrast-enhanced whole heart coronary magnetic resonance angiography at 3T. *J Am Coll Cardiol.* 2009;54:69–76.

19. Dhawan S, Dharmashankar KC, Tak T. Role of magnetic resonance imaging in visualizing coronary arteries. *Clin Med Res.* 2004;3:173–179.

20. Deshpande VS, Shea SM, Laub CT, et al. 3D magnetization-prepared true-FISP: A new technique for imaging coronary arteries. *Magn Reson Med.* 2001;46:494–502.

21. Kaiser WA, Zeitler E. MR imaging of the breast. Fast imaging sequences with and without Gd-DTPA. *Radiology.* 1989;170:681–686.

22. Axel L, Morton D. MR flow imaging by velocity-compensated/uncompensated difference images. *J Comp Assist Tomo.* 1987;11:31–34.

23. Dumoulin CL, Souza SP, Walker MF, et al. Three-dimensional phase contrast angiography. *Mag Reson Med.* 1989;9:139–149.

24. Pelc NJ, Bernstein MA, Shimakawa A, et al. Encoding strategies for three-dimensional phase contrast MR imaging of flow. *J Magn Reson Imaging.* 1991;1:405–413.

25. Dumoulin CL, Turski PA. Phase contrast MRA. In: *Methods in Biomedical Magnetic Resonance Imaging and Spectroscopy.* Editor, IR Young. John Wiley & Sons. New York. 2000;504–517.

26. Creasy JL, Price RR, Presbrey T, et al. Gadolinium-enhanced MR angiography. *Radiology.* 1990;175:280–283.

27. Maki JH, Prince MK, Chenevert T. Optimizing three-dimensional gadolinium enhanced magnetic resonance angiography. *Invest Radiol.* 1998;33:528–537.

28. Zhang H, Maki JH, Prince MR. 3D contrast-enhanced MR angiography. *J Magn Reson Imaging.* 2007;25:13–25.

29. Lu H, Nagae-Poetscher LM, Golay X, et al. Routine clinical brain MRI sequences for use at 3T. *J Magn Reson Imaging.* 2005;22:13–22.

30. Londy FJ, Rohrer S, Kumar S, et al. Comparison of 1.5 Tesla and 3.0 Tesla for skull base lesion enhancement. *J HK Coll Radiol.* 2008;11:19–23.

31. Leiner T, de Vries M, Hoogeveen R, et al. Contrast enhanced peripheral MR angiography at 3.0 Tesla: Initial experience with a whole-body scanner in healthy volunteers. *J Magn Reson Imaging.* 2003;17:609–614.

32. Maki JH, Prince MR, Londy FJ, et al. The effects of time varying intravascular signal intensity and k-space acquisition order on three-dimensional MR angiography image quality. *J Magn Reson Imaging.* 1996;6:642–651.

33. Huston J, Fain SB, Wald JT, et al. Carotid artery elliptic centric contrast-enhanced MR angiography compared with conventional angiography. *Radiology.* 2001;18:138–143.

34. Willinek WA, Gieseke J, Conrad R. et al. Randomly segmented central k-space ordering in high-spatial-resolution contrast-enhanced MR angiography of the supraaortic arteries: Initial experience. *Radiology.* 2003;225:583–588.

35. Sodickson DK, Manning WJ. Simultaneous acquisition of spatial harmonics (SMASH): Fast imaging with radiofrequency coal arrays. *Magn Reson Med.* 1997;38:591–603.

36. Pruessmann KP, Weiger M, Scheidegger MB, et al. SENSE: Sensitivity encoding for fast MRI. *Magn Reson Med.* 1999;42:952–962.

37. Griswold MA, Jakob PM, Heidemann RM, et al. Generalized autocalibrating partially parallel acquisitions (GRAPPA). *Magn Reson Med.* 2002;47:1202–1210.

38. Wilson GJ, Hoogeveen RM, Willinek WA, et al. Parallel imaging in MR angiography. *Top Magn Reson Imaging.* 2004;15:169–185.

39. Brubaker L, Bullitt E, Yin C, et al. Magnetic resonance angiography visualization of abnormal tumor vasculature in genetically engineered mice. *Cancer Res.* 2005;65: 8218–8223.

40. Bullitt E, Wolthusen PA, Brubaker L, et al. Malignancy-associated vessel tortuosity: A computer-assisted, MR angiographic study of choroid plexus carcinoma in genetically engineered mice. *AJNR Am J Neuroradiol.* 2006;27:612–619.

41. Bullitt E, Aylward S, Van Dyke T, et al. Computer-assisted measurement of vessel shape from 3T magnetic resonance angiography of mouse brain. *Methods.* 2007;43:29–34.

42. Kobayashi H, Sato N, Hiraga A, et al. 3D-micro-MR angiography of mice using macromolecular MR contrast agents with polyamidoamine dendrimer core with reference to their pharmacokinetic properties. *Magn Reson Med.* 2001;45:454–460.

43. Trattnig S, Ba-Ssalamah A, Noebauer-Huhmann et al. MR contrast agent at high-field MRI (3 Tesla). *Top Magn Reson Imaging.* 2003;14:365–374.

Image Processing
in Cancer

15

Imaging Tissue Oxygenation Status with MRI

Nilesh Mistry
University of Maryland

C. Chad Quarles
Vanderbilt University

Reduced oxygenation or hypoxia is a significant factor in tumor progression and treatment. In particular, the hypoxic environment in tumors is known to (a) promote angiogenesis, (b) promote malignancy, (c) promote metastases, (d) promote genetic instability, and (e) reduce the effectiveness of radiation- and chemotherapy. Additionally, hypoxia is associated with poor prognosis for patients suffering from several cancers, including that of the head and neck, brain, cervix, and prostate. Hence, it is important to understand the role of hypoxia in tumor growth, progression, and treatment.

The most common cause of hypoxia is believed to be the imbalance between oxygen consumption and delivery. Oxygen consumption is much higher in a rapidly growing tumor mass, due to the high metabolic rate of tumor cells. In healthy tissues, oxygen is supplied through a network of capillaries from which oxygen diffuses into the tissue. However, in tumors, cell proliferation is much faster, relative to healthy tissue and this leads to areas of hypoxia 100–150 μm away from the capillaries [1]. For a tumor to continue growing beyond 1–2 mm, it needs to undergo neovascularization which is often unregulated, leading to severe structural and functional abnormalities [2]. Together, the abnormalities in tumor vasculature, the exaggerated intercapillary spacing and low oxygen tension in regions lead to distinctive microenvironments that do not exist in healthy tissues and into which neither oxygen nor many traditional anticancer agents can penetrate.

Currently, there are no known techniques to noninvasively and repeatedly measure tissue oxygenation with high precision. Eppendorf pO_2 probes [3] and fluorescence optical probes [4] have often been used to measure tissue oxygenation, but these techniques are invasive and often the readings are complicated due to the tissue injury caused by the probes themselves. The techniques are also limited by poor spatial resolution that is a result of selective tissue sampling to measure pO_2.

Imaging-based techniques provide an alternative to studying tissue oxygenation status. The development of appropriate methods to noninvasively characterize tumor hemodynamics and oxygenation can provide new insights into the angiogenic processes, and play an important role in the optimization of individualized treatment protocols. The techniques are noninvasive and may enable repeated measurements in the same tumor over time, either during the same imaging session or over multiple imaging sessions. The development of oxygen-sensitive imaging techniques has spanned multiple imaging modalities, including positron emission tomography (PET), single photon emission computed tomography, magnetic resonance imaging (MRI), and electron paramagnetic resonance.

[18]F-labeled fluoro-misonidazole is a widely used PET imaging agent that provide measures of blood flow (within the first five minutes post ^{18}F-MISO administration) and tissue hypoxia (after one to two hours post administration) [5]. Another radionuclide sensitive to tumor hypoxia is Copper (II) diacetyl-bis(N^4)-methylthiosemi-carbazone, which has been shown to have higher levels of retention in hypoxic cells as compared to ^{18}F-labeled fluoro-misonidazole [6]. While both these methods have a place for imaging tissue oxygenation, the inherent spatial and temporal resolution limitations of PET imaging provide an avenue for other techniques such as MRI to image oxygenation status in tissues at a much higher spatial and temporal resolution.

Several MRI-based techniques have been developed to image tissue oxygenation status, including [19]Fluorine MRI, blood oxygen level–dependent MRI (BOLD MRI), and Overhauser MRI. While [19]Flourine MRI and BOLD MRI are both strictly MRI techniques, Overhauser MRI relies on electron paramagnetic resonance to provide tissue oxygenation contrast while using MRI to provide spatial localization. The technique requires significant modifications to standard MRI systems, and hence, will not be covered in the text below. Interested readers are requested to refer to Matsumoto et al. [7]. In this chapter, we will focus on ^{19}F MRI and BOLD MRI.

15.1 Qualitative Introduction to MR Imaging of Tissue Oxygenation

15.1.1 Fluorine MRI

One of the most reliable techniques for quantification of tissue oxygen tension employs ^{19}F-NMR relaxation rate measurements of exogenously administered perfluorocarbons (PFCs). For several decades, it has been known that the ^{19}F spin lattice relaxation rate ($1/T_1 \equiv R_1$) of PFCs is linearly dependent on oxygen tension [8], and consequently there is a continuing interest in using PFCs for imaging tissue oxygenation [9–13]. With proper calibration, such studies have demonstrated the ability of this approach to quantify tissue oxygen tension (e.g., in units of mmHg) at equilibrium, and repeatedly following vasoactive or metabolic perturbations. However, given the difficulty of delivering sufficient quantities of PFCs to tumor tissue, as many of these agents require intratumoral injection, this method has remained a preclinical tool.

15.1.2 BOLD MRI

Unlike ^{19}F MRI, BOLD imaging relies upon endogenous contrast mechanisms to assess oxygenation. Ogawa et al. first described the basis for the BOLD contrast mechanism in rat studies using NMR at high magnetic fields. They found that the gradient echo (GE) image contrast reflected the venous blood oxygen saturation and more specifically the concentration of the paramagnetic substance deoxyhemoglobin (dHb) [14]. This phenomenon is attributed to the creation of susceptibility-induced magnetic field inhomogeneities surrounding blood vessels containing dHb. The spin echo (SE) and GE transverse relaxation rates, R_2 ($=1/T_2$) and R_2^* ($= 1/T_2^*$), of water in the blood and in the tissue surrounding the blood vessels are enhanced by the presence of dHb [15]. These relaxation rates vary with the concentration of dHb, [dHb], and as such, provide a sensitive measure of changes in blood oxygen saturation. It has since been shown that BOLD contrast correlates with spatial and temporal changes in tumor tissue pO_2 [16]. This approach has been used to monitor phototherapy treatment response [17], assess the tumor response to vasomodulators [18–23], predict radiotherapeutic response [24,25], and characterize tumor vascular architecture, maturation, and function [26,27]. Whereas ^{19}F MRI is used to assess absolute tissue

oxygenation levels (e.g., in terms of mmHg), BOLD is primarily used to assess relative changes in oxygen delivery.

15.2 Quantitative Introduction to MR Imaging of Tissue Oxygenation

15.2.1 Fluorine MRI: PFC Selection and Administration

The selection of the PFC is critical to the success of ^{19}F-NMR studies of tissue oxygenation. An ideal PFC for ^{19}F-NMR should have the following characteristics: a high sensitivity to oxygen tension, a high signal-noise ratio (large number of equivalent ^{19}F nuclei), low dependence on temperature, and suitable biodistribution with the tumor tissue. Some of the PFCs that have been used for *in vivo* tissue oxygenation studies include perfluoro-15-crown-5-ether [28,29], oxypherol [30], and hexafluorobenzene [31,32]. Hexaflourobenzene has been favored by Mason et al. [32] due to its favorable characteristics, such as single resonance NMR peak, high temperature stability, high sensitivity to pO_2, and linear relationship across the entire range from anoxia to hyperbaric levels of O_2. Other researchers have used perfluoro-15-crown-5-ether due to its single NMR peak, higher sensitivity, and longer transverse relaxation time [29].

PFCs can be introduced into the body via two distinct techniques for detection in tumors: (a) systemic intravascular injection and (b) direct intratumoral injection. While a systemic injection is preferable due to ease of administration, there are several factors that go against this choice. First, the tumor uptake of the PFC is very slow, which means that one needs to wait for a long time (typically 24 h post injection) between the injection and imaging session. Second, because a significant amount of PFC is taken up by the reticuloendothelial system, there is a significant collection of the PFC in the liver, spleen, and bone marrow, and this has been reported to result in hepatomegaly in mice [33]. Third, the overall quantity of fluorine that is taken up by the tumor is very small compared to the administered dose, which potentially means low tumor signal relative to background signal. Fourth, the distribution is preferentially limited to well-perfused regions of the tumor, which biases the detection of hypoxia to well-vascularized regions. A direct intratumor injection to enable the visualization and study of nonperfused tumor regions is preferred by some research groups, and they have demonstrated the technique in preclinical settings. However, there are several factors that should be considered when using a direct intratumoral injection technique. The number of PFC injection sites and the distance between those sites may bias the detection. Also, the technique can be carried out effectively only in superficial tumors, which restricts the potential translation of the technique into the clinic, where the tumor site might be difficult to access (e.g., tumors of the liver, kidney, etc.).

15.2.2 Fluorine MRI Methods

The majority of fluorine imaging studies have been performed in small animals using custom built, dual tuned ^{1}H/^{19}F birdcage, or surface coils. The type of imaging coil, along with the strength of the magnet system used for the study, can influence the choice of imaging sequences, both for imaging the distribution of fluorine and the relaxation rate. The choice of the imaging coil influences the uniformity of the B_1 field that can adversely affect accurate measurements of relaxation times. Higher field strengths lead to longer T_1 times resulting in longer acquisition times. In this section, we will cover some of the standard techniques used to work around these issues, particularly focusing on imaging the PFC distribution and their relaxation rates.

A standard SE sequence can be used to image the distribution of ^{19}F. The typical implementation of an SE imaging sequence is shown in Figure 15.1, where RF is the radio-frequency pulse used to create the B_1 field applied in a particular direction at fluorine resonance, where G_x, G_y, and G_z are the gradients used for spatial location. The imaging parameters on a 2T small animal MRI system are typically 128 × 64 pixel resolution, TR = 5000 ms, TE = 25 ms, 64 phase encode steps (gradient magnitude of G_y), number of signal acquisitions = 4, resulting in a total acquisition time of approximately 21 min [29]. This standard implementation results in a very long total acquisition time due to the long T_1 associated with fluorine. The resulting signal from the SE saturation recovery experiment is governed by

$$S = kM_0\{1 - 2e^{-t_2/T_1} + e^{-(t_1+t_2)/T_1}\} \cdot e^{-2t_1/T_2} \qquad (15.1)$$

where S is the resulting transverse signal, k is a proportionality constant, M_0 is the total magnetization, T_1 is the longitudinal relaxation time, T_2 is the transverse relaxation time, and t_1 and t_2 are shown in Figure 15.1.

An alternative technique to reduce the total acquisition time for fluorine MRI was proposed by Mason et al. [30]. The approach is based on driven equilibrium sequences [34] that force T_1

recovery. The technique relies on the application of an additional 180° radio-frequency pulse, following data acquisition combined with the appropriate gradient that refocuses the transverse magnetization, and then applying a 90° radio-frequency pulse to reorient the longitudinal magnetization. The implementation is shown in Figure 15.2, where the times t_1, t_2, t_3, and t_4 can be used to manipulate the signal. The imaging parameters on a 4.7 T MRI system are typically 128 × 32 × 8 pixel resolution, TR = 150 ms, TE = 8 ms, 33 ms 90° excitation pulse, and number of signal acquisitions = 16, resulting in a total acquisition time of approximately 10 min. This particular implementation reduces the total acquisition time of a standard SE sequence from 21 min for a single slice to 10 min for a 3D volume [31]. The signal is not dependent on T_1 or T_2 alone, but is proportional to T_1/T_2. Thus, while there is a significant improvement in the acquisition time over that achieved with a normal SE image, this comes at the cost of mixed contrast. The resulting signal from the SE-driven equilibrium experiment is governed by

$$S = kM_0 \frac{1 - e^{-t_4/T_1}}{1 - e^{-(t_1+t_2+t_3)/T_2}e^{-t_4/T_1}} \cdot e^{-2t_1/T_2} \qquad (15.2)$$

where S is the resulting transverse signal, k is a proportionality constant, M_0 is the total magnetization, T_1 is the longitudinal relaxation time, T_2 is the transverse relaxation time, and t_1, t_2, t_3, and t_4 are shown in Figure 15.2.

The fundamental idea behind using ^{19}F MRI and PFCs to study tissue oxygenation status arises from their linear dependency on the partial pressure of oxygen. A common technique to estimate R_1 (or $1/T_1$) is based on inversion recovery–SE imaging. However, SE imaging is a very slow technique since the sequence needs to acquire a line of k-space (data space) and wait for the

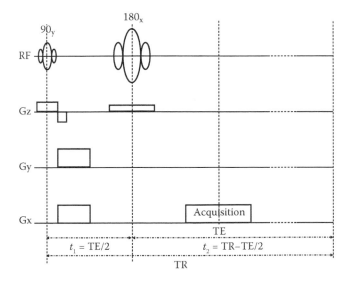

FIGURE 15.1 Spin echo imaging sequence uses a 90–180 RF excitation, but needs to wait for the complete T_1 recovery. This results in a long acquisition time due to the long T_1.

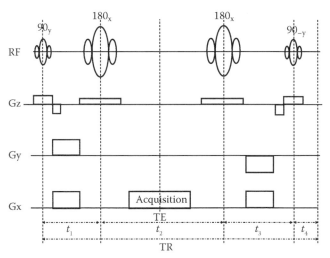

FIGURE 15.2 Driven equilibrium sequence using additional 180–90 RF combinations to improve the time efficiency of the spin echo sequence.

magnetization to completely recover before another line can be acquired. To reduce the imaging time required to measure R_1, a technique was proposed that combined inversion recovery with echo planar imaging (EPI) acquisition [29]. The pulse sequence diagram for the same is shown in Figure 15.3, where an SE-EPI (sawtooth gradient waveform) sequence is preceded by a 180° nonslice selective inversion pulse. The inversion times used for the technique were as follows: TI = 0.08, 0.2, 0.5, 1, 2, 4, and 8 s. The typical parameters for the imaging sequence are as follows: TE = 70 ms, TR = 10,000 ms, number of averages = 8, sampling bandwidth ± 30 kH, at a 64 × 64 pixel resolution. If we assume that the T_1 is 2 or 4 s, we can see the sampling of the longitudinal relaxation curve in Figure 15.4, which shows the T_1 recovery and the samples that are acquired at the inversion times mentioned above.

Estimation of R_1 from the IR-EPI sequence is carried out on a voxel-by-voxel basis using a three parameter nonlinear least square fitting algorithm to the following equation:

$$S(t) = A + B \cdot e^{-t \cdot R_1} \qquad (15.3)$$

where $S(t)$ is the signal intensity of the voxel for each inversion time, while A and B are the magnetization components.

The above technique works well for imaging ^{19}F with birdcage coils with good radiofrequency field uniformity are used. When using surface coils, however, the accuracy of the method may be challenging. It is known that a pulse burst method can help provide accurate T_1 measurements in a minimum amount of time when using surface coils [9]. Combining the pulse burst technique with an SE-EPI acquisition called PBSR, EPI can provide

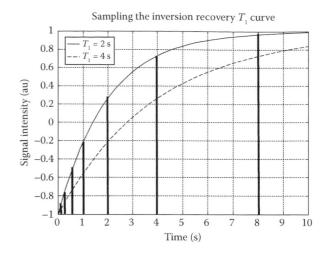

FIGURE 15.4 Sampling of the inversion recovery curve at different inversion times.

further improvement in acquisition time, while maintaining the accuracy of T_1 when using surface coils [11,31,32]. The pulse sequence is shown in Figure 15.5, where a series of nonslice selective 90° hard pulses are applied followed by a variable delay shown by τ, after which a standard SE EPI sequence is executed. The typical implementation of the PBSR–EPI technique uses (a) 20 hard 90° saturating pulses with a 20-ms spacing, (b) 14 different values of τ in the range of 200 ms to 90 s, and (c) a 32 × 32 pixel resolution with a sampling bandwidth of 40 kHz and 4 signal acquisitions per delay. This sequence results in a total

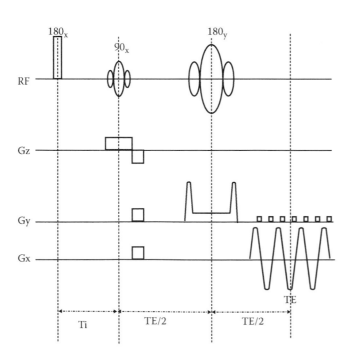

FIGURE 15.3 Pulse sequence depicting a typical inversion recovery–spin echo–EPI sequence. The 180 hard RF excitation pulse prepares the magnetization by inverting the spins.

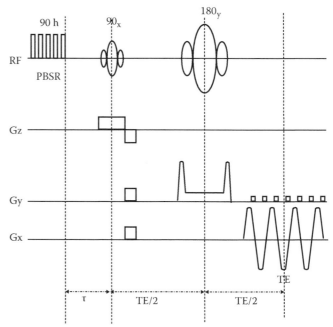

FIGURE 15.5 A typical pulse sequence diagram of a pulse burst saturation recovery–EPI relaxometry sequence. The series of hard 90 RF pulses followed by standard saturation recovery pulses, followed by sampling using an echo planar acquisition.

acquisition time of approximately 20 min. to yield data that can be used to generate an R_1 map. R_1 is calculated on a voxel-by-voxel basis using a three-parameter nonlinear least square fitting algorithm using the following equation:

$$S_i = A(1 - (1 + W) \cdot e^{-\tau_i \cdot R_1}) \tag{15.4}$$

Here, A and W are the magnetization evolution and noise parameters, R_1 is the relaxation rate, and τ_i are the 14 different delay times used for mapping R_1.

Finally, tumor pO_2 maps can be derived from the relaxation maps using the techniques presented above and calibration curves relating the ^{19}F signal to pO_2. The calibration curves are typically derived using *in vitro* studies of ^{19}F NMR T_1 measurements. T_1 measurements are carried out for several mixtures of PFC emulsions with varying amounts of oxygen, nitrogen, and air. The calibration curve can be derived using the following equation [35]:

$$pO_2 = (1/T_1 - A)/S \tag{15.5}$$

where S is the slope of the linear dependency between $1/T_1$, and A is the intercept at anoxia. The parameters of this linear calibration often depend on the temperature and the field strength, among other variables. While most small animal imaging studies do control the core body temperature, there can be fluctuations in the tumor temperature that may not be reflected by the core body temperature.

15.2.3 BOLD MRI Methods

The greatest advantage of BOLD MRI is the simplicity of its data acquisition. Most BOLD studies utilize readily available GE, SE, or EPI imaging methods with an echo time optimized for T_2^* or T_2 contrast. Studies seeking more quantitative measures employ multiple echo sequences for the derivation of absolute R_2^* and/or R_2 values. Such methods have been extensively used to evaluate oxygen perturbations within tissue.

While the acquisition of BOLD data is straightforward, its analysis and interpretation are more complex. Consider a traditional GE BOLD study, which consists of measuring R_2^* before and after an oxygen perturbation (i.e., breathing air versus breathing carbogen). The difference in R_2^* between the two conditions, ΔR_2^*, reflects changes in the blood's deoxyhemoglobin content. However, it has been shown that the measured R_2^* also depends on blood oxygen saturation, blood volume, and hematocrit since these parameters influence the blood's concentration of dHb [36,37]. These confounding factors may act counter to one another, with opposite effects on R_2^* [38,39]. For example, a hypercapnic challenge could increase or decrease R_2^* depending on differences in regional baseline blood volume and whether the change in blood volume delivers more blood highly saturated in oxygen, as may be the case in normal tissue, or poorly saturated in oxygen as could occur in tumors. Other studies

have indicated that within and among tumor types, oxygen perturbations can lead to vasodilation and/or vasoconstriction, potentially masking changes in the blood deoxyhemoglobin content [27,40]. These competing effects, which are generally unmeasured and unknown, have limited the capability of BOLD imaging to providing only qualitative measures of tumor oxygen modulation [20,39].

Analysis of tumor oxygen perturbation BOLD data typically involves comparing the change in T_2^* weighted signal intensity or the change in R_2^*. The disadvantage of these methods, as discussed above, is their inherent sensitivity to the baseline blood volume, changes in blood volume, and in blood oxygen saturation. The recently proposed contrast-referenced BOLD MRI approach, however, allows for the separation and quantification of baseline relative blood volume, vascular reactivity, and oxygen modulation [41,42]. The baseline GE transverse relaxation rate, R_2^*, can be modeled as the sum of the background tissue relaxation and a vascular component that depends on the oxygenation state of the blood:

$$R_2^* = R_{2tissue}^* + kV(X_B - X_T) = R_2 + kV(X_{BT}) \tag{15.6}$$

where $R_{2tissue}^*$ is the background tissue relaxation, and the product kVX_{BT} is the contribution from the vasculature (k is a proportionality constant that depends on field strength, vessel morphology, and hematocrit, V is the blood volume, and X_B and X_T are the susceptibilities of the blood and surrounding tissue. The difference, $X_B - X_T$ or X_{BT}, depends on the blood oxygen saturation. For example, an increase in X_{BT} would reflect lower blood oxygen saturation and higher deoxyhemoglobin content. As the equation indicates, this would increase R_2^* (a decrease in T_2^*). In the presence of an intravascular contrast agent the rate becomes

$$R_{2A}^* = R_{2tissue} + kV(X_{BT} + X_A) \tag{15.7}$$

where X_A is the susceptibility of the contrast agent. By measuring the rates before and after administering the contrast agent, a map of relative blood volume in the tissue can be obtained:

$$R_{2A}^* - R_2^* = kVX_A \tag{15.8}$$

Oxygen perturbation causes (potentially) changes in both the blood volume and oxygenation:

$$R_{2oxy}^* = R_{2tissue}^* + k(V + \Delta V)(X_{BT} + \Delta X_{oxy}) \tag{15.9}$$

where ΔX_{oxy} is the change in the blood susceptibility during the episode of oxygen perturbation. In the presence of the contrast agent, this rate becomes

$$R_{2(oxy+A)}^* = R_{2tissue}^* + k(V + \Delta V)(X_{BT} + \Delta X_{oxy} + X_A) \tag{15.10}$$

The difference between Equations 15.9 and 15.10 is

$$R_{2(oxy+A)}^{*} - R_{2oxy}^{*} = k(V+\Delta V)\mathbf{X}_A = kV\left(1+\frac{\Delta V}{V}\right)\mathbf{X}_A \quad (15.11)$$

from which $\Delta V/V$ (given $kV\mathbf{X}_a$ from Equation 15.8) can be computed. Thus, a map of the fractional change in blood volume, or the vascular reactivity, due to the effect of oxygen perturbation, can be obtained. Furthermore, using this modeling approach, maps of the ratio of the change in blood magnetic susceptibility following perturbation and the susceptibility of the contrast agent, $\Delta\mathbf{X}_{oxy}/\mathbf{X}_A$ can be computed. This ratio should directly correlate with changes in blood oxygen saturation. As a point of reference, if this ratio is negative, it indicates an increase in blood oxygen saturation, or a decrease in the blood susceptibility, following the oxygen perturbation. This simple model requires four measurements of R_2^{*}: baseline, with oxygen modulation, with oxygen modulation in the presence of a contrast agent, and back at baseline with contrast agent on board.

Few studies have evaluated the use of contrast-referenced BOLD MRI in tumors [42]. It could be particularly useful for evaluating oxygen modulators with acute (< 1 hr) effects (e.g., carbogen, allosteric modifiers of hemoglobin). Agents with long-lasting effects are more difficult to assess because of the requirement to measure R_2^{*} in four states. The clinical potential of this approach has yet to be determined, but it is currently limited because it requires the use of non–FDA-approved intravascular iron-oxide based contrast agents. Overcoming this limitation is the subject of ongoing research.

15.3 Preclinical Applications in Cancer

15.3.1 Fluorine MRI

In this section, we provide a brief overview of select preclinical studies reporting on the use of the ^{19}F MRI methods discussed above in studying hypoxia. This is far from a comprehensive review and merely serves to highlight some of the methods described in earlier sections.

^{19}F MRI techniques have been demonstrated primarily in preclinical studies of rodent and murine tumor models [11,13,29–33,43–46]. The studies typically measure the variation of oxygenation within tumors and differences across various tumor lines [30]. The techniques have also been used to monitor changes in tumor pO_2 in response to interventions to increase radiosensitivity. One such radiosensitizing technique that has often been used in small animal studies is the breathing of carbogen (95% O_2 + 5% CO_2). Several studies have demonstrated changes in tumor oxygenation when the gas mixture was changed from air to pure oxygen or carbogen [11,31,45,46]. A study published by Le et al. [11] measured changes in pO_2 in rat tumor models of prostate cancer using PBSR-EPI technique. The study showed skewed pO_2 distributions when rats breathed 33% O_2 along with considerable radiobiological hypoxia. Changing

the gas mixture to pure oxygen shifted the distribution toward increased pO_2. Using the EPI technique allowed the authors to study the changes in oxygenation in these tumors. It was observed that those voxels with baseline $pO_2 > 30$ torr showed a significant change ($p < .05$) in the relaxation rate R_1, as opposed to voxels with $pO_2 < 16$ torr.

Another study by Hees and Sotak [43] studied the effect of nicotinamide, which is known to improve tumor blood flow and, in turn, oxygenation in radiation induced fibrosarcomas in mice. The authors monitored the changes in oxygenation status using ^{19}F relaxation rates of an emulsified PFC. They showed statistically significant improvements in tumor pO_2 for the nicotinamide treated group, as compared to the saline-treated group. The authors concluded that ^{19}F spectroscopic relaxometry can measure therapeutically meaningful changes in tumor oxygenation.

Another study, published by McNab et al., investigated the changes in tumor oxygenation through the tumor growth and regression cycle in an androgen-dependent Shionogi tumor model [47]. The study was performed using PBSR-EPI to monitor the T_1 changes after the intravascular injection of perfluoro-15-crown-5-ether. Nonlocalized, T_1 measurements (converted to pO_2) were made at multiple time points throughout the tumor growth-regression cycle. The authors reported that the values of pO_2 were significantly lower for tumors that were growing, as opposed to the ones that were regressing.

15.3.2 BOLD MRI

In this section, we provide a brief overview of select preclinical BOLD MRI studies reporting on the use of the methods discussed above in small animal models of disease. This is not a comprehensive review, but rather serves to highlight some of the methods described in previous sections.

Building on their own seminal work [19] in the field of MRI of tumor oxygen status was the 1997 paper by Robinson et al. [23]. In this effort, the authors investigated the use of carbogen challenges to affect the signal intensity in GE images across six different rodent models of human cancer. The investigators found that while four of the models displayed increases in T_2^{*} weighted GE images during carbogen challenge, they noted that this observation was consistent with increases in tumor oxygenation (and blood flow), as previously hypothesized. However, in the remaining two models, there was a fall in MR signal intensity in response to the carbogen challenge. Since carbogen breathing should result in a decrease in deoxyhemoglobin, and therefore an increase in a T_2^{*}-weighted MRI, this result was somewhat surprising. The authors hypothesized that this could be the result of a transient decrease in tumor perfusion but the authors also discussed a series of alternate possibilities in their discussion. Whatever the reason for the decrease in T_2^{*}, the variation in changes of the gradient echo signal across tumor models in response to carbogen challenges was one phenomenon, among others, that motivated the execution of validation studies.

In 1998, Al-Hallaq et al. [16] published a validation study on carbogen-induced changes in tumor oxygenation. The authors correlated the noninvasive MR measurement of water linewidth with the invasive, gold-standard data afforded by microelectrodes. The hypothesis that was tested was that a decrease in the water linewidth due to an oxygen challenge is related to increases in extravascular pO_2. T_1-weighed images were acquired of rats bearing R3230AC adenocarcinomas, with and without a carbogen challenge, while microelectrode measurements were made from the same imaging plane. The microelectrodes showed that carbogen breathing significantly ($p < .05$) increased pO_2, while the water linewidth strongly correlated with increased pO_2 levels. This study gave early indications that BOLD-based carbogen measurements could have clinical utility in assessing oxygenation status of tumors. The authors concluded that, given the validation of the method, such an MR-based approach could guide the design and use of tumor oxygenation agents, that could then be employed clinically for optimizing treatments on an individual basis.

A second study aimed at validating the BOLD method of measuring tumor oxygenation status via T_2^* measurements was offered by Baudelet and Gallez in 2002 [39]. This validation study was based on the use of fiberoptic microprobes that were able to measure pO_2 (and erythrocyte flux). In contrast to the study discussed in the previous paragraph, in which the electrodes were positioned after the animal was removed from the scanner, this study performed simultaneous MR and microprobe measurements. Baudelet and Gallez found that while there was a temporal relationship between the BOLD response and changes in pO_2, the magnitude of the BOLD signal was not indicative of the absolute pO_2 value. The authors stated that BOLD contrast could provide a relative marker of tumor pO_2, and could do this at high spatial resolution and noninvasively; however, BOLD did not provide an absolute quantification of tumor pO_2. They therefore concluded that the T_2^* measurements were limited to qualitative descriptions of tumor oxygen status. Note that this study does not necessarily contradict the Al-Hallaq paper, [16] since the MR measurements were different in the two efforts.

Moving from initial observations and validation studies to applications, Rodrigues et al. [25] studied the use of quantitative $R_2^* \left(\equiv 1/T_2^* \right)$ maps before and after carbogen breathing prior to radiotherapy, and correlated these data to tumor growth delay; these two T_2^* maps were employed to construct ΔR_2^* maps. The group showed that animals showing a large ΔR_2^* pretreatment displayed a substantial reduction in tumor volume, whereas those that displayed only a modest ΔR_2^* showed minimal growth inhibition. The authors also derived a simple expression for how carbogen-induced ΔR_2^* can be shown to be directly proportional to the change in tissue deoxyhemoglobin content. They concluded by stating that quantitative measurements of ΔR_2^* provide a noninvasive, prognostic indicator of radiotherapy response, and that the technique may provide an attractive method for clinical use of predicting the response of human tumors to such therapy. This and other data [38,48,49] are examples that paved the way for clinical application.

15.4 Clinical Applications in Cancer

In one of the earlier publications of clinical use of BOLD MRI in imaging tumor hypoxia, Taylor et al. [50] described the use of T_2^* weighted GE imaging to demonstrate improved oxygenation in tumors during carbogen (95% O_2 and 5% CO_2) inhalation. The authors imaged human subjects with a variety of tumors (including prostate, breast, head, and neck) before, during, and after the administration of carbogen, which acts as a radiosensitizer as well as a vasodilator, in order to identify the regions that undergo oxygenation and flow changes. Significant signal increases were observed in 56% of the patients with tumors, suggesting improved tissue oxygenation and blood flow, which in turn would help identify these patients as most likely to benefit from carbogen-based radiosensitization.

A second study that focused on the head and neck tumors was published in 2002 by Rijpkema et al. [51] that aimed to identify the parameters needed to be monitored to improve the tumor treatment response to accelerated radiotherapy with carbogen and nicotinamide. The study included MRI to monitor the blood flow using dynamic contrast-enhanced MRI and blood oxygenation using BOLD MRI during the inhalation of a hyperoxic hypercapnic gas mixture (98% O_2 + 2% CO_2). While the study showed no changes in the dynamic contrast-enhanced MRI tumor vascularity metrics during the inhalation of the gas mixture, it showed significant increases in the absolute T_2^* value measured with a multigradient echo sequence. The same group later reported a similar study in patients with meningiomas [52], where the results indicated a strong correlation between T_2^* changes and vascularity changes measured by the uptake of gadolinium-based contrast agents. The authors concluded that the presence of both vascular and oxygenation effects, along with the heterogeneity of the response to inhalation of the gas mixtures, necessitates the individual assessment of these effects in patients with meningiomas.

Moving from the initial studies of carbogen inhalation, Hoskin et al. [53] explored the correlation between baseline R_2^* values and immunohistochemistry measures of hypoxia in prostate cancer patients. The study aimed to assess the ability of BOLD MRI to depict clinically significant prostate tumor hypoxia. Multiple gradient echo images were acquired in patients prior to undergoing radical prostatectomy to measure tissue oxygenation. Similar images were acquired following the injection of a contrast agent in order to assess relative blood volume. Histological staining was performed with pimonidazole that was administered preoperatively. The sensitivity of R_2^* in depicting tumor hypoxia was reported to be 88% in comparison to the histological staining, which improved to 95% with the inclusion of low blood volume information to the map. However, the specificity of R_2^* values was only 36%. The authors concluded that despite the low specificity, R_2^* maps could serve as a potential noninvasive technique for mapping tumor hypoxia.

A translational study conducted by Alonzi et al. [54] demonstrated the effectiveness of using carbogen to improve oxygenation in prostate cancer. The study tested the effects of breathing carbogen in human prostate xenografts in mice and human patients. A quantitative assessment of R_2^* was carried out in both mice and humans using susceptibility weighted imaging. The authors reported reduction in tumor R_2^* in both animal models and human patients as a result of carbogen inhalation, indicating an increase in tumor oxygen saturation. The inhalation of the gas mixture improved tumor oxygenation, which provided more evidence for the combination of carbogen breathing during radiotherapy in prostate cancer patients.

15.5 Summary

MRI methods to assess tissue oxygenation could serve as valuable clinical tools for identifying hypoxic tumors during the treatment-planning phase, and for evaluating the effects of oxygen enhancing therapeutic strategies. It is clear that further research is needed to develop clinically viable exogenous MRI contrast agents sensitive to tissue oxygen tension. Unlike radiotracers used for nuclear imaging, such MRI agents could be fashioned in such a way as to alter tissue relaxation rates only when activated (via a chemical modification) in the presence of hypoxic tissue. Such strategies have been developed to enhance the detection of molecular imaging agents. With regard to BOLD MRI, studies will continue to explore methods to make this approach more quantitative and specific to tissue oxygenation. Regardless of its complex biophysical basis, its acquisition simplicity renders it a valuable option for evaluating changes in tumor oxygen delivery.

References

1. Gray, L.H., Conger, A.D., Ebert, M., Hornsey, S., Scott, O.C.A. The Concentration of Oxygen Dissolved in Tissues at the Time of Irradiation as a Factor in Radiotherapy. *British Journal of Radiology* 1953;26:638–648.
2. Vaupel, P. The role of hypoxia-induced factors in tumor progression. *Oncologist* 2004;9(Suppl 5):10–17.
3. Kallinowski, F., Zander, R., Hoeckel, M., Vaupel, P. Tumor tissue oxygenation as evaluated by computerized-pO2-histography. 1990; 953–961.
4. Griffiths, J.R., Robinson, S.P. The OxyLite: A fibre-optic oxygen sensor. *Br J Radiol* 1999;72(859):627–630.
5. Martin, G., Caldwell, J., Rasey, J., Grunbaum, Z., Cerqueira, M., Krohn, K. Enhanced binding of the hypoxic cell marker [3H]fluoromisonidazole in ischemic. *J Nucl Med* 1989;30(2):194–201.
6. Lewis, J.S., McCarthy, D.W., McCarthy, T.J., Fujibayashi, Y., Welch, M.J. Evaluation of 64Cu-ATSM in vitro and in vivo in a hypoxic tumor model. *Journal of Nuclear Medicine* 1999;40(1):177–183.
7. Matsumoto, S., Yasui, H., Batra, S., Kinoshita, Y., Bernardo, M., Munasinghe, J., Utsumi, H. et al. Simultaneous imaging of tumor oxygenation and microvascular permeability using Overhauser enhanced MRI. *Proc Natl Acad Sci U S A* 2009;106(42):17898–17903.
8. Joseph, P.M., Fishman, J.E., Mukherji, B., Sloviter, H.A. In vivo 19F NMR imaging of the cardiovascular system. *Journal of computer assisted tomography* 1985;9(6):1012–1019.
9. Evelhoch, J.L., Ackerman, J.J.H. NMR T1 measurements in inhomogeneous B1 with surface coils. *Journal of Magnetic Resonance (1969)* 1983;53(1):52–64.
10. Clark, L.J., Ackerman, J., Thomas, S., Millard, R., Hoffman, R., Pratt, R., Ragle-Cole, H., Kinsey, R., Janakiraman, R. Perfluorinated organic liquids and emulsions as biocompatible NMR imaging agents for 19F and dissolved oxygen. *Adv Exp Med Biol* 1984;180:835–845.
11. Le, D., Mason, R.P., Hunjan, S., Constantinescu, A., Barker, B.R., Antich, P.P. Regional tumor oxygen dynamics: F-19 PBSR EPI of hexafluorobenzene. *Magnetic Resonance Imaging* 1997;15(8):971–981.
12. Aboagye, E.O., Kelson, A.B., Tracy, M., Workman, P. Preclinical development and current status of the fluorinated 2-nitroimidazole hypoxia probe N-(2-hydroxy-3,3,3-trifluoropropyl)-2-(2-nitro-1-imidazolyl) acetamide (SR 4554, CRC 94/17): A non-invasive diagnostic probe for the measurement of tumor hypoxia by magnetic resonance spectroscopy and imaging, and by positron emission tomography. *Anti-Cancer Drug Design* 1998;13(6):703–730.
13. Procissi, D., Claus, F., Burgman, P., Koziorowski, J., Chapman, J.D., Thakur, S.B., Matei, C., Ling, C.C., Koutcher, J.A. In vivo F-19 magnetic resonance spectroscopy and chemical shift imaging of tri-fluoro-nitroimidazole as a potential hypoxia reporter in solid tumors. *Clinical Cancer Research* 2007;13(12):3738–3747.
14. Ogawa, S., Lee, T.M., Nayak, A.S., Glynn, P. Oxygenation-sensitive contrast in magnetic resonance image of rodent brain at high magnetic fields. *Magnetic Resonance in Medicine* 1990;14:68–78.
15. Thulborn, K., Watertown, J., Matthews, P., GK, R. Oxygenation dependence of the transverse relaxation time of water protons in whole blood at high field. *Biochimica et Biophysica Acta* 1982;714:265–270.
16. Al-Hallaq, H.A., River, J.N., Zamora, M., Oikawa, H., Karczmar, G.S. Correlation of magnetic resonance and oxygen microelectrode measurements of carbogen-induced changes in tumor oxygenation. *International Journal of Radiation Oncology, Biology, Physics* 1998;41:151–159.
17. Gross, S., Gilead, A., Scherz, A., Neeman, M., Salomon, Y. Monitoring photodynamic therapy of solid tumors online by BOLD-contrast MRI. *Nature Medicine* 2003;9:1327–1331.
18. Karczmar, G.S., Kuperman, V.Y., River, J.N., Lewis, M.Z., Lipton, M.J. Magnetic resonance measurement of response to hyperoxia differentiates tumors from normal tissue and may be sensitive to oxygen consumption. *Investigative radiology* 1994;29(Suppl 2):S161–S163.
19. Robinson, S.P., Howe, F.A., Griffiths, J.R. Noninvasive monitoring of carbogen-induced changes in tumor blood flow and oxygenation by functional magnetic resonance imaging. *International Journal of Radiation Oncology, Biology, Physics* 1995;33:855–559.

20. Robinson, S.P., Collingridge, D.R., Howe, F.A., Rodrigues, L.M., Chaplin, D.J., Griffiths, J.R. Tumor response to hypercapnia and hyperoxia monitoried by FLOOD magnetic resonance imaging. *NMR in Biomedicine* 1999;12:98–106.

21. Kuperman, V.Y., River, J.N., Lewis, M.Z., Lubich, L.M., Karczmar, G.S. Changes in T_2^*-weighted images during hyperoxia differentiate tumors from normal tissue. *Magnetic Resonance in Medicine* 1995;33:318–325.

22. Tayler, N.J., Baddeley, H., Goodchild, K.A., Powell, M.E., Thoumine, M., Culver, L.A., Stirling, J.J. et al. BOLD MRI of human tumor oxygenation during carbogen breathing. *Journal of Magnetic Resonance Imaging* 2001;14(2):156–163.

23. Robinson, S.P., Rodrigues, L.M., Ojugo, A.S., McSheehy, P.M., Howe, F.A., Griffiths, J.R. The response to carbogen breathing in experimental tumor models monitored by gradient-recalled echo magnetic resonance imaging. *British Journal of Cancer* 1997;75(7):1000–1006.

24. Al-Hallaq, H.A., Zamora, M., Fish, B.L., Farrell, A., Moulder, J.E., Karczmar, G.S. MRI measurements correctly predict the relative effects of tumor oxygenating agents on hypoxic fraction in rodent BA1112 tumors. *International Journal of Radiation Oncology, Biology, Physics* 2000;47(2):481–488.

25. Rodrigues, L.M., Howe, F.A., Griffiths, J.R., Robinson, S.P. Tumor R_2^* is a prognostic indicator of acute radiotherapeutic response in rodent tumors. *Journal of Magnetic Resonance Imaging* 2004;19:482–488.

26. Neeman, M., Dafni, H., Bukhari, O., Braun, R.D., Dewhirst, M.W. In vivo BOLD contrast MRI mapping of subcutaneous vascular function and maturation: validation by intravital microscopy. *Magnetic Resonance in Medicine* 2001;45(5):887–898.

27. Robinson, S.P., Rijken, P.F., Howe, F.A., McSheehy, P.M., van der Sanden, B.P., Heerschap, A., Subbs, M., van der Kogel, A.J., Griffiths, J.R. Tumor vascular architecture and function evaluated by non-invasive susceptibility MRI methods and immunohistochemistry. *Journal of Magnetic Resonance Imaging* 2003;17(4):445–454.

28. van der Sanden, B.P.J., Heerschap, A., Hoofd, L., Simonetti, A.W., Nicolay, K., van der Toorn, A., Colier, W., van der Kogel, A.J. Effect of carbogen breathing on the physiological profile of human glioma xenografts. *Magnetic Resonance in Medicine* 1999;42(3):490–499.

29. Dardzinski, B.J. and Sotak, C.H. Rapid tissue oxygen-tension mapping using F-19 inversion-recovery echo-planar imaging of perfluoro-15-crown-5-ether. *Magnetic Resonance in Medicine* 1994;32(1):88–97.

30. Mason, R.P., Antich, P.P., Babcock, E.E., Constantinescu, A., Peschke, P., Hahn, E.W. Noninvasive determination of tumor oxygen-tension and local variation with growth. *International Journal of Radiation Oncology Biology Physics* 1994;29(1):95–103.

31. Hunjan, S., Zhao, D.W., Constantinescu, A., Hahn, E.W., Antich, P.P., Mason, R.P. Tumor oximetry: Demonstration of an enhanced dynamic mapping procedure using fluorine-19 echo planar magnetic resonance imaging in the Dunning prostate R3327-AT1 rat tumor. *International Journal of Radiation Oncology, Biology, Physics* 2001;49(4):1097–1108.

32. Mason, R.P., Rodbumrung, W., Antich, P.P. Hexafluorobenzene: A sensitive F-19 NMR indicator of tumor oxygenation. *Nmr in Biomedicine* 1996;9(3):125–134.

33. Mason, R., Antich, P., Babcock, E., Gerberich, J., Nunnally, R. Perfluorocarbon imaging in vivo: a 19F MRI study in tumor-bearing mice. *Magn Reson Imaging* 1989;7(5):475–485.

34. Van Uijen, C.M.J. and Den Boef, J.H. Driven-equilibrium radiofrequency pulses in NMR imaging. *Magnetic Resonance in Medicine* 1984;1(4):502–507.

35. Robinson, S.P. and Griffiths, J.R. Current issues in the utility of 19F nuclear magnetic resonance methodologies for the assessment of tumour hypoxia. *Philosophical Transactions: Biological Sciences* 2004;359(1446):987–996.

36. Yablonskiy, D.A. and Haacke, E.M. Theory of NMR signal behavior in magnetically inhomogeneous tissues: the static dephasing regime. *Magnetic Resonance in Medicine* 1994;32:749–763.

37. van Zijl, P.C., Eleff, S.M., Ulatowski, J.A., Oja, J.M., Ulug, A.M., Traystman, R.J., Kauppinen, R.A. Quantitative assessment of blood flow, blood volume and blood oxygenation effects in functional magnetic resonance imaging. *Nature Medicine* 1998;4(2):159–167.

38. Howe, F.A., Robinson, S.P., McIntyre, D.J., Subbs, M., Griffiths, J.R. Issues in flow and oxygenation dependent contrast (FLOOD) imaging in tumors. *NMR in Biomedicine* 2001;14:497–506.

39. Baudelet, C. and Gallez, B. How does blood oxygen level-dependent (BOLD) contrast correlate with oxygen partial pressure (pO(2)) inside tumors? *Magnetic Resonance in Medicine* 2002;48(6):980–986.

40. Dunn, J.F., O'Hara, J.A., Zaim-Wadghiri, Y., Lei, H., Meyerand, M.E., Grinberg, O.Y., Hou, H., Hoopes, P.J., Demidenko, E., Swartz, H.M. Changes in oxygenation of intracranial tumors with carbogen: A BOLD MRI and EPR oxyimetry study. *Journal of Magnetic Resonance Imaging* 2002;15(5):511–521.

41. Kennan, R.P., Scanley, B.E., Gore, J.C. Physiologic basis for BOLD MR signal changes due to hypoxia/hyperoxia: Separation of blood volume and magnetic susceptibility effects. *Magnetic Resonance in Medicine* 1997;37(6):953–956.

42. Quarles, C., Gore, J.C., Price, R.R. Quantitative assessment of vascular maturation and function using contrast referenced BOLD MRI. 2007; Berlin, Germany.

43. Hees, P.S. and Sotak, C.H. Assessment of changes in murine tumor oxygenation in response to nicotinamide using 19F NMR relaxometry of a perfluorocarbon emulsion. *Magn Reson Med* 1993;29(3):303–310. Erratum in: Magn Reson Med 1993 May;1929(1995):1716.

44. Baldwin, N.J. and Ng, T.C. Oxygenation and metabolic status of KHT tumors as measured simultaneously by 19F magnetic resonance imaging and 31P magnetic resonance spectroscopy. *Magnetic Resonance Imaging* 1996;14(5):541–551.

45. Song, Y.L., Worden, K.L., Jiang, X., Zhao, D.W., Constantinescu, A., Liu, H.L, Mason, R.P. Tumor oxygen dynamics: Comparison of F-19 MR EPI and frequency domain NIR spectroscopy. In: Dunn JF, Swartz HM, editors. Oxygen Transport to Tissue xxiv. Volume 530, Advances in Experimental Medicine and Biology. New York: *Kluwer Academic/Plenum Publ*; 2003. p 225–236.

46. Zhao, D., Constantinescu, A., Chang, C-H., Hahn, E.W., Mason, R.P. Correlation of tumor oxygen dynamics with radiation response of the dunning prostate R3327-HI tumor. *Radiation Research* 2003;159(5):621–631.

47. McNab, J.A., Yung, A.C., Kozlowski, P. Tissue oxygen tension measurements in the Shionogi model of prostate cancer using F-19 MRS and MRI. *Magnetic Resonance Materials in Physics Biology and Medicine* 2004;17(3–6):288–295.

48. Howe, F.A., Robinson, S.P., Rodrigues, L.M., Griffiths, J.R. Flow and oxygenation dependent (FLOOD) contrast MR imaging to monitor the response of rat tumors to carbogen breathing. *Magn Reson Imaging* 1999;17(9):1307–1318.

49. Robinson, S.P., Kalber, T.L., Howe, F.A., McIntyre, D.J., Griffiths, J.R., Blakey, D.C., Whittaker, L., Ryan, A.J., Waterton, J.C. Acute tumor response to ZD6126 assessed by intrinsic susceptibility magnetic resonance imaging. *Neoplasia* 2005;7(5):466–474.

50. Taylor, N.J., Baddeley, H., Goodchild, K.A., Powell, M.E.B., Thoumine, M., Culver, L.A., Stirling, J.J. et al. BOLD MRI of human tumor oxygenation during carbogen breathing. *Journal of Magnetic Resonance Imaging* 2001;14(2): 156–163.

51. Rijpkema, M., Kaanders, J., Joosten, F.B.M., van der Kogel, A.J., Heerschap, A. Effects of breathing a hyperoxic hypercapnic gas mixture on blood oxygenation and vascularity of head-and-neck tumors as measured by magnetic resonance imaging. *International Journal of Radiation Oncology, Biology, Physics* 2002;53(5):1185–1191.

52. Rijpkema, M., Schuuring, J., Bernsen, P.L., Bernsen, H.J., Kaanders, J., van der Kogel, A.J., Heerschap, A. BOLD MRI response to hypercapnic hyperoxia in patients with meningiomas: correlation with Gadolinium-DTPA uptake rate. *Magnetic Resonance Imaging* 2004;22(6):761–767.

53. Hoskin, P.J., Carnell, D.M., Taylor, N.J., Smith, R.E., Stirling, J.J., Daley, F.M., Saunders, M.I., et al. Hypoxia in prostate cancer: Correlation of bold-MRI with pimonidazole immunohistochemistry—Initial observations. *International Journal of Radiation Oncology, Biology, Physics* 2007;68(4):1065–1071.

54. Alonzi, R., Padhani, A.R., Maxwell, R.J., Taylor, N.J., Stirling, J.J., Wilson, J.I., d'Arcy, J.A., Collins, D.J., Saunders, M.I., Hoskin, P.J. Carbogen breathing increases prostate cancer oxygenation: a translational MRI study in murine xenografts and humans. *British Journal of Cancer* 2009;100(4):644–648.

16

Clinical Assessment of the Response of Tumors to Treatment with MRI

Mia Levy
Vanderbilt University

Assessment of treatment response is an important clinical and research task in cancer medicine. The goal of treatment response assessment is to categorize the efficacy or toxicity of a treatment for an individual patient or patient cohort. For cancer treatment efficacy, response is typically assessed by evaluating changes in the patient's tumor burden before and after treatment. Tumor burden is an estimate of the amount of tumor in a patient's body at a particular time point and can be described qualitatively or quantitatively. Tumor burden is often estimated with imaging modalities such as magnetic resonance imaging (MRI), CT, and PET, blood tests such as prostate-specific antigen and carcinoembryonic antigen, or by physical exam. Serial imaging studies are used to assess changes in the location, size, and metabolic activity of tumors over time. This chapter describes the general approach to assessing tumor treatment response, formal cancer response criteria, requirements for the future development of MRI-based response criteria, and information systems to support application and development of response criteria.

16.1 Approach to Tumor Treatment Response Assessment

At the individual patient level, tumor treatment response assessment aids clinicians in making decisions regarding continuing or discontinuing the current therapy. In the advanced disease setting, tumors often shrink or remain stable in size with therapy for some period of time but eventually acquire mechanisms of resistance that allow them to grow again. For the individual patient,

response assessment procedures estimate the rate and direction of response (e.g., shrinking, stable, growing) so that effective treatments are continued and ineffective treatments discontinued.

In everyday clinical practice, tumor response assessment is a relatively informal and qualitative process. Imaging modalities (e.g., CT, MRI, or PET) are selected depending upon the primary type of cancer, the location of any metastasis, and patient tolerance to the procedure (e.g., i.v. contrast allergy may be a contraindication for certain CT protocols). MRI is commonly used to evaluate response for tumors located in the breast, brain, extremities, spine, head, and neck. Tumors located in the chest, abdomen, and pelvis are often evaluated with CT or PET-CT. Bone metastases are often evaluated with bone scans, PET, or MRI in case of spinal metastasis. Imaging studies are performed at baseline some time prior to the start of treatment, and then again at interval follow-up periods some (biologically appropriate) time after the start of treatment. Follow-up studies are typically performed every 6–12 weeks, depending upon the pattern of treatment cycles (e.g., three or four week treatment cycles), the expected rate of response to treatment (e.g., cytostatic vs. cytotoxic treatment), the expected rate of progression of the disease (e.g., fast for lung cancer vs. slow for prostate cancer), and the appearance of any new patient symptoms indicative of progression (e.g., new pain).

The radiologist reporting the follow-up case typically compares the imaging study to the most recent study and reports qualitative changes in overall tumor burden as generally increasing, decreasing, or remaining stable. A report of new tumor lesions defines the event of disease progression. Figure 16.1 shows an example of a pretreatment and posttreatment

MRI one minute post contrast of liver tumor

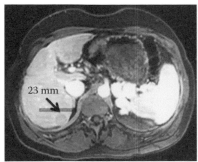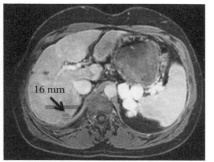

| Pretreatment liver tumor | 4 month posttreatment liver tumor |

FIGURE 16.1 Radiographic imaging study used to measure tumor lesions pretreatment and posttreatment. The MRIs of the abdomen show the pretreatment liver tumor of 23 mm in longest diameter, and the same liver tumor 4 months posttreatment measuring 16 mm in longest diameter.

MRI of the abdomen with a liver mass that has decreased in size with treatment. The oncologist uses the images, radiology reports, and additional clinical features, such as patient toxicity to treatment, to decide if treatment should be continued or discontinued. If the patient's disease is improving (i.e., decreasing in size) or stable, treatment is continued. If the tumor burden is increasing or there are new lesions present, then the treatment is discontinued and new treatment options are considered.

On the other hand, for patients participating in therapeutic clinical trials, the response assessment process is much more formalized and quantitative. Novel cancer therapies are often evaluated first in the metastatic or advanced disease setting for their antitumor activity. The primary therapeutic goal for many antitumor treatments in the advanced disease setting is delay in tumor growth and ideally tumor shrinkage. Delay in tumor growth can correlate with improved quality of life, morbidity, and mortality [1,2]. Historically, tumor shrinkage has been the hallmark of antitumor activity for cytotoxic therapies, which cause tumor cell death, and thus have the potential to shrink tumor masses. Tumor shrinkage in a proportion of patients in phase II studies of cytotoxic drugs has been shown to be predictive of improvement in survival in phase III studies [1,3,4]. Noncytotoxic therapies on the other hand are typically cytostatic, and may not cause tumor shrinkage but rather tumor stability. Several noncytotoxic therapies have also demonstrated improvement in overall survival in randomized trials [5,6]. For such cases, delay in tumor growth can also be used as evidence of antitumor activity [7]. The time to objective tumor growth is referred to as the time to progression and is often a primary or secondary endpoint for phase II and III cancer clinical trials.

The goal of therapeutic clinical trials is, thus, to determine if an experimental therapy is efficacious and safe. In order to compare the efficacy and toxicity of the experimental treatment among the patients within a clinical trial, the trial protocol is developed to standardize treatment and response assessment procedures for all participating subjects. Formal response criteria have been developed to help standardize tumor response assessment across clinical trials so that trial results can be compared.

For each clinical trial, the response assessment protocol specifically defines the imaging modality, image acquisition protocol, the timing of the baseline and follow-up assessments, and the response criteria that should be used to quantify and classify response. Cancer response criteria standardize the approach for estimating tumor burden, defining quantitative and qualitative changes in tumor burden, and classifying tumor response to treatment in clinical trial cohorts. These formal response assessment outcomes are incorporated into the clinical trial decision algorithms that define the conditions when an experimental treatment should be continued or discontinued. In this way, the clinical trial protocol helps to ensure consistent treatment decisions across trial subjects.

The response criteria also enable quantification and classification of each patient's response to treatment so that they can be aggregated and compared across trial arms. Figure 16.2 shows a waterfall plot of each patient's best quantitative response to treatment (response rate). Response rates below zero on the *y*-axis correspond to partial tumor shrinkage with treatment, while those above zero correspond to tumor growth. This type of visualization provides a

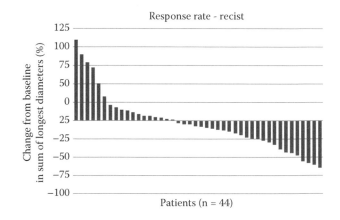

FIGURE 16.2 Aggregation of cohort response using a waterfall plot of best response rate. The best percent change in the tumor burden from baseline (*y*-axis) is plotted for each patient in the cohort (*x*-axis).

concise and informative summary of the entire cohort's response. The mean of the quantitative response rate can also be taken and compared between clinical trial arms as a quantitative estimate of differences in treatment efficacy. Trial arms will also compare the median time to progression and the percentage of subjects with a particular response category. A clinically and statistically significant difference in the time to progression between two trial arms is often used as an intermediate endpoint for novel drug approval by regulatory agencies. This enables more rapid approval of novel therapies than waiting for overall survival endpoints to be reached, which for many cancers can be several years.

16.2 Cancer Response Criteria

16.2.1 History of Cancer Response Criteria

Several generations of solid tumor response criteria have been developed since the first cancer response criteria, called the World Health Organization (WHO) criteria, were published in 1981 [8]. The goal of the WHO criteria was to develop response criteria that could objectively assess tumor response to treatment, as compared to previous approaches that utilized clinical symptoms to assess treatment response. While clinical symptoms of the disease are an important aspect of assessing individual patient responses, they tend to be subjective assessments and not all patients have such symptoms. The WHO criteria thus recommended evaluating tumor lesions with radiographic modalities, taking bidimensional tumor measurements of observed cancer lesions in the images, and taking the sum of the products to generate a quantitative estimate of tumor burden. This was done at baseline and again at follow-up. The quantitative response was calculated as the percent change in the tumor burden (sum of products), taking as reference the baseline assessment.

$$\text{Response Rate} = \frac{(\text{Tumor Burden}_{\text{follow-up}} - \text{Tumor Burden}_{\text{baseline}})}{\text{Tumor Burden}_{\text{baseline}}}$$

The quantitative response was calculated at each follow-up assessment period and the minimum tumor burden achieved, since baseline was used to calculate the best response rate. This quantitative assessment was used to classify the nature of the patient's response into four response categories based on heuristically-defined thresholds. The response categories defined and in use to this day are complete response, partial response, stable disease, and disease progression. With qualitative descriptions, these response categories are forms of ordinal scale. The criteria also specified observations including the appearance of any new lesions as defining the event of disease progression.

The WHO criteria were widely used in cancer clinical trial protocols. However, the designers of clinical trials began to make modifications to the criteria on an ad hoc basis, to incorporate new imaging technologies and address underspecified aspects of the original document. As a result, there was a lack of standardization and it became difficult again to consistently confirm trial results and compare trial outcomes [9,10]. In the mid-1990s an international, multidisciplinary committee called the International Working Party was established to simplify and standardize the criteria. In 2000, the committee published the response evaluation criteria in solid tumors (RECIST) guidelines [11], the next generation of solid tumor response criteria. RECIST had some significant differences from the original WHO criteria. First, it defined more specifically what was considered measurable disease both by the anatomic location of the lesion and by a minimum size requirement. Second, up to five lesions per organ could be selected with a maximum of 10 lesions total to estimate tumor burden. Lastly, the criteria utilized unidimensional measurement of lesions and the sum of the longest diameters to estimate total anatomic tumor burden. As a result of changing from bidimensional to unidimensional measurement, the thresholds for defining response categories were also changed.

Over the last decade, RECIST has been widely utilized in cancer clinical trials and has become a "requirement" for government regulatory approval of drugs in certain contexts [12]. However, just as with the original WHO criteria, after the continued application of RECIST in more settings and the development of new imaging technologies such as metabolic imaging, it became necessary to revise the RECIST criteria. RECIST 1.1 was published in 2009 [12], and is so named because it is more of an evolution of the guideline than a complete departure. RECIST 1.1 further specifies and, in some ways, simplifies the criteria. Major changes include specific definitions for what is considered measurable with respect to lymph nodes, and a decrease in the number of target lesions used to estimate tumor burden from 10 total to five total with a maximum of two lesions per organ. It is anticipated that RECIST 1.1 will become the new standard for response assessment for solid tumor clinical trials, as it has been endorsed by several international cancer centers and government regulatory agencies.

However, RECIST is not an appropriate response criterion for every type of cancer (Figure 16.3). Two generations of response

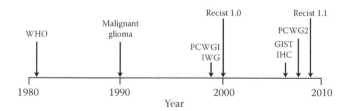

FIGURE 16.3 Timeline of publication of Oncology Response Assessment Criteria. WHO = World Health Organization Criteria (1981) [8], RECIST = Response Evaluation Criteria in Solid Tumors (2000) [11] and (2009) [36], IWG = International Working Group for Lymphoma Response Criteria (1999) [13], IHC = International Harmonization Criteria for Lymphoma Response Assessment (2007) [14], PCWG = Prostate Cancer Clinical Trials Working Group (PCWG1, 1999) [15] and (PCWG2, 2008) [16], Malignant Glioma Response Assessment Criteria (1990) [17], amd GIST = Gastrointestinal Stromal Tumor Response Criteria (2008) [37].

criteria have also been developed for lymphoma [13,14] and prostate cancer [15,16] to accommodate novel assessment modalities and disease characteristics. In addition, the solid tumor criteria have not been found to be sufficient for response assessment for intracranial neoplasms [17], nor for the very rare gastrointestinal stromal tumor [18,19] and mesothelioma [20] neoplasms. As such, response criteria have been developed for these specific diseases [17,21]. Newer approaches that incorporate metabolic imaging with FDG-PET have also recently been proposed [22]. Given historical trends and the continued development and validation of new response biomarkers and criteria (including many of the MRI methods discussed in this text), it is anticipated that cancer response assessment criteria will continue to evolve.

16.2.2 Detailed Example of RECIST 1.1 Criteria

Response criteria such as RECIST 1.1 are sufficiently complex that it is helpful to describe a concrete example of how it is applied to patients and cohorts during the prospective management and analysis of clinical trial data. The RECIST procedure includes the process for estimating tumor burden, estimating qualitative and quantitative changes in tumor burden, classification of response at each follow-up time point and over the entire course of treatment, and aggregate estimates of cohort response in summary statistics.

RECIST recommends a baseline assessment no earlier than four weeks prior to the start of treatment, and follow-up assessments approximately every six to eight weeks, depending upon the type of solid tumor and anticipated kinetics of the experimental treatment. RECIST recommends that solid tumors be assessed with imaging modalities, typically CT scans or MRI, PET, or chest X-ray when appropriate. Once the patient is enrolled in the trial and prior to starting treatment, a baseline imaging study is required to assess the locations and sizes of the tumor lesions. In the current clinical workflow, the oncologist orders an imaging study, the patient has the study performed and the radiologist is the first to review the images. The radiologist summarizes her/his findings in a text report and records detailed measurements as image markups. Figure 16.1 shows a baseline MRI scan of the liver of a patient with metastatic colon cancer. The green markings are an example of the image markups created by the radiologist that measures the longest diameter and its respective perpendicular (short) axis. The report is then sent to the oncologist who independently reviews the report and images. Tumor burden is assessed both quantitatively and qualitatively, depending upon the location of the tumor lesion and the type of imaging modality. At baseline, all identified tumor lesions are recorded in a RECIST flow sheet (Figure 16.4), and given a lesion identifier (Lesion ID) with a corresponding description of the type of lesion and its location. Quantitative measurements are recorded where applicable; otherwise, a Boolean description is made, denoting the persistence or resolution of a lesion. RECIST flow sheets are often managed with paper forms or computer spreadsheets.

Tumor lesions are further classified based on their anatomic location and length measurements into measurable and nonmeasurable lesions. RECIST 1.1 first distinguishes between lymph node and non-lymph node cancer lesions. Generally, cancer lesions assessed by CT and MRI are considered measurable

	Lesion ID	Anatomic location/lesion type	B 7/8/06	F1 9/15/06	F2 11/22/06	F3 1/29/07
Target lesions	1	Liver mass	1.8	1.0	0.9	1.3
	2	Liver mass	3.7	2.0	1.9	2.4
Non-target lesions	3	Liver mass	1.1	0.5	-	-
	4	T4 bone lesion	+	+	NE	++
	5	Right 8th rib bone lesion	+	+	NE	+
	6	Left pleural space effusion	+	+	-	-
New lesion	7	Liver mass				+
Response analysis		Sum of diameters (SD)	5.5	3.0	2.8	3.7
		% Change SD from baseline		−45%	−49%	−33%
		% Change SD from nadir		0	0	+32%
		Target lesion response		PR	PR	PD
		Non-target lesion response		Non-CR/ Non-PD	Not all evaluated	PD
		Response		PR	PR	PD
		Overall best response	Partial response (RR=−49%)			

FIGURE 16.4 RECIST flow sheet: Example RECIST flow sheet for patient LL. B = baseline, F1 = first follow-up, F2 = second follow-up, F3 = third follow-up, NA = not applicable, (+) = present, (−) = absent, SD = sum of diameters, PR = partial response, PD = progressive disease, non-PD/non-CR = nonprogressive disease/noncomplete response.

at baseline if they are at least 10 mm in longest diameter, or at least 15 mm in short axis for lymph nodes. Nonmeasurable disease includes small lesions not meeting size criteria, bone lesions, fluid collections, leptomeningeal disease, and lymphangitic involvement of skin or lung. A subset of the measurable lesions is selected to be included as target lesions for quantitative calculation of tumor burden. This same set of target lesions will be tracked in follow-up for quantitative calculation of the change in tumor burden. For RECIST 1.1, the largest two lesions per organ are selected as target lesions up to a total of five lesions. The remaining measurable and nonmeasurable lesions make up a qualitative estimate of tumor burden. Each cancer lesion is tracked at follow-up and classified according to its persistence or resolution. The quantitative tumor burden is estimated by the sum of the diameters of the target lesions, using the short axis for enlarged lymph nodes and the long axis for all other cancer lesions.

The quantitative and qualitative tumor burden is estimated at baseline and again at each follow-up period. Any new lesions are also noted. Quantitative changes in the tumor burden are estimated through the calculation of the percent change in the sum of diameters from baseline, as well as the percent change from the smallest sum of diameters. Figure 16.5 shows the thresholds that define the classification of the quantitative target lesion response at a follow-up time point using these two parameters. For the sample case in Figure 16.4, the baseline sum of diameters is 5.5 cm. At the first follow-up, the percent change from baseline is −45%, indicating tumor shrinkage and giving a response classification of partial response. The patient would stay on therapy at that time, given demonstration of its efficacy. At the third follow-up assessment, however, the percent change from the smallest sum of diameters has increased by 32%, denoting an increase in tumor burden that meets the threshold for classification of progressive disease.

The nontarget lesion response is classified according to qualitative assessments of nontarget lesion persistence, resolution, and progression. The target lesion and nontarget lesion response are combined with the appearance of any new lesions to give a final response category for a given time point. The patient case in Figure 16.4 meets the criteria for progressive disease at the third

follow-up period based on the presence of new lesions. At that point the patient would be taken off therapy, given the assumption that this particular treatment is no longer efficacious. The time to disease progression is then calculated as the temporal interval between the start of the investigational therapy and the date when progressive disease was first documented. The primary study outcome, best overall response, is calculated from the smallest sum of diameters achieved during the study, in this case, at the second follow-up time point with an objective response rate (RR) of minus 49% and a response classification of partial response.

This procedure of tumor response assessment occurs for each patient in the trial cohort. The cohort's mean response rate and median time to progression are typically compared to a historical control for single arm studies or to the control arm in multi-arm randomized studies. These are some of the intermediate outcomes used as primary endpoints to assess the efficacy of a novel therapy. Other types of aggregated temporal outcomes include the proportion of patients that experience disease progression or recurrence by a certain period of time; for example, progression-free survival at six months. Interpretation of the cohort response outcomes as a positive or negative trial result depends upon the specific disease and primary hypothesis of the trial.

16.3 Requirements for the Development of MRI-Based Cancer Response Criteria

Cancer response criteria continue to evolve as new imaging modalities are introduced into clinical practice. A rational approach to the development of response criteria has recently been proposed as a guide to ongoing research in this area [23]: (1) the assessment modality should be widely available and the technique reproducible, (2) the measurement technique within the modality should be reproducible and accurate, (3) changes in the estimate of the tumor burden should correlate with time to progression and overall survival, (4) new criteria should be evaluated with data from multiple clinical centers, and (5) with multiple disease subtypes, if applicable. RECIST 1.1 has several advantages and disadvantages with respect to these guiding principles. These are discussed in the context of the requirements for the development of new MRI-based imaging response biomarkers and cancer response criteria.

16.3.1 Assessment Modality

For a response criterion to be widely applicable, the imaging modality should be a reproducible technique that is available at most centers internationally, so as not to restrict patient eligibility. For this reason RECIST recommends the use of CT and MRI, which are widely available with reproducible techniques. This has been the main argument for lack of introduction of PET as a quantitative image biomarker for response criteria due to the relative lack of availability of this modality internationally

Percent change from baseline (y)	Percent change from minimum (x)	Follow-up quantitative target lesion response
Any	x ≥ 20	Progressive disease (PD)
Any	0 < x < 20	Stable disease (SD)
0 < y < −30	x ≤ 0	Stable disease (SD)
y ≤ −30	x ≤ 0	Partial response (PR)
y = −100	x ≤ 0	Complete response (CR)

FIGURE 16.5 Classification of the quantitative target lesion response at a follow-up time point t_i using percent change from baseline in the sum of diameters and the percent change from the smallest sum of diameters.

and due to interinstitution variation and reproducibility of the image acquisition techniques.

However, despite the reproducibility of CT and MRI modalities, the protocols for these modalities often vary by clinical indication, leading to different protocols being applied for the baseline and follow-up studies. Ideally, response assessment should be performed using the same image acquisition protocol for the baseline and follow-up study. In particular, the modality, slice thickness, anatomic coverage, and use of contrast agents should be the same from the baseline to follow-up study. However, in the clinical setting, the selection of a particular image acquisition protocol for a given patient is often driven by the billing codes on the requisition order form that describes the indication for the procedure. A baseline assessment where the tumor is first diagnosed may have a different indication code compared to the follow-up study, and thus may result in variations in the image acquisition protocol. This is a common occurrence that makes it difficult to compare the baseline and follow-up studies, both for clinical trial and standard of care patients. As such, new imaging assessment modalities such as diffusion-weighted MRI (DW-MRI, see Chapter 7) and dynamic contrast-enhanced magnetic resonance imaging (DCE-MRI, see Chapter 12) not only need to be reproducible imaging techniques, but also have consistent image acquisition protocols established for cancer treatment response assessment.

16.3.2 Measurement Technique

RECIST has several limitations with respect to the interreader reproducibility and accuracy of its measurement technique for estimating tumor burden. In particular, RECIST utilizes manual human measurement of the lesion's longest diameter and human selection of the target lesions, to include in the sum of diameters. Depending upon the size of the cancer lesion, response criteria classification thresholds may be sensitive to even millimeter differences in manually measured lesions. In addition, when multiple lesions are present for a particular organ, RECIST requires the selection of at most two target lesions per organ, and no more than five target lesions total. The selection of the target lesion in this case is left to the judgment of the reviewer. Two reviewers will often pick different target lesions in these cases, which may result in difference in response classification. In particular, differences have been observed in the interpretation of response outcomes between the local providers managing patients in the trial and central reviewers [24]. The FDA reported a 24–29% rate of discordance in response assessment at the patient level in oncology clinical trials, between local investigator and central reviewer assessment of progression status [25,26]. These differences at the patient level result in an average of 41% decrease in the cohort mean tumor response rate that is reported by central reviewers compared to local investigators in a review of nine clinical trials [24]. Even when the central review process is used, discordance exists between readers. One study in particular reported a rate of discordance for progression status of up to 38% among multiple central reviewers [27]. This lack of concordance is related in part

to the complexity, lack of precision, and lack of specification of response criteria.

However, changes in the longest dimension or volume of tumor are fundamentally anatomic and morphologic changes that are temporally downstream manifestations of underlying pathophysiologic changes, which may occur earlier. Many noncytotoxic therapies do not produce tumor shrinkage, but can produce improvements in time to progression and overall survival. Yet, for many of these noncytotoxic treatments, metabolic changes can be seen prior to detection of tumor shrinkage. Clearly, methods are needed to characterize the underlying pathophysiological changes induced by specific targeting agents. Such methods may be considerably more likely to offer earlier— and more specific—information on response to treatment when compared to changes in longest tumor dimensions. Several aspects of MRI data analysis, acquisition, and synthesis have matured to the point where they can offer quantitative information on breast tumor status and response to therapy. Specifically, DCE-MRI and DW-MRI may offer quantitative approaches for estimating changes in tumor oxygen status, blood flow, and metabolism.

16.3.3 Correlation of Changes in Tumor Burden with Clinical Endpoints

Change detection is necessary, but not sufficient for a useful response biomarker. In order to be an effective intermediate endpoint for clinical trials, the change in the tumor burden biomarker should correlate with overall survival and time to progression. As has been stated earlier, the RECIST percent change in sum of diameters often does not correlate with time to progression or overall survival for some targeted and noncytotoxic therapies [5,6]. In addition, early disease response detection could be used to triage patients away from therapies that are less likely to be efficacious toward potential alternatives. Thus, evaluation methods for novel image response biomarkers should include correlation with the particular therapy, time to progression, and overall survival with the goal of early change detection.

The ongoing multicenter I-SPY II (investigation of serial studies to predict one's therapeutic response with imaging and molecular analysis 2) neoadjuvant breast cancer clinical trial [28] is attempting to address some of these issues by evaluating if breast MRI response predicts treatment outcomes. The National Cancer Institute's cancer Biomedical Informatics Grid program is supporting the I-SPY trial through integration and analysis of diverse data types, including clinical, MRI, and tumor biomarkers, throughout the breast cancer treatment cycle. The integrated platform is designed to enable correlation of molecular data with MRI patterns to identify surrogate markers for early treatment response.

16.3.4 Validation of Response Criteria

Validation of response criteria requires (1) evaluation of the reproducibility of the approach at multiple institutions, and, if applicable, (2) for multiple cancer types. RECIST has been widely used with multiple types of solid tumors and across multiple institutions internationally. Several European and American research centers are involved in the development and testing of new oncology response assessment criteria. For the recent update of the RECIST criteria, from the original version published in 2000 to the RECIST 1.1 version published in early 2009, a large retrospective database of target lesions was developed to test the impact of modifications to the criteria [29,30]. Metadata on 18,000 potential target lesions were obtained from 6512 patients in 16 metastatic cancer clinical trials. These trials represented multiple types of solid tumor malignancies. The database was used to evaluate the impact of changes to RECIST on the classification of patient response to treatment. The RECIST 1.1 criterion was, thus, validated only by comparing it to the previous standard approach, and not by evaluating its correlation with survival endpoints.

However, these databases are not publicly available to enable the development and validation of image processing approaches to quantitative response criteria. In order to facilitate development and validation of novel response criteria, large data sets are needed that contain baseline and follow-up imaging studies and image annotations, along with the corresponding diagnoses, therapies, and clinical outcomes.

16.4 Biomedical Informatics Systems to Support Treatment Response Assessment

The discussion above outlines some of the challenges with consistent application and development of image-based cancer treatment response criteria. At the individual patient level, response criteria are often inconsistently applied due to variations among reviewers in lesion selection, lesion measurement, and interpretation of complex response criteria. Furthermore, new response criteria are difficult to develop and validate in part due to the lack of publicly available image repositories linked to patient level treatments and clinical outcomes. Several biomedical informatics systems have been developed that could support the application and development of response criteria. These include systems to support (1) the creation of image repositories, (2) the acquisition, transfer, and storage of image metadata, and (3) the automated interpretation of response data.

16.4.1 Image Repositories

The development and validation of image-based response criteria requires access to repositories of serial imaging studies, linked to patient-level information regarding treatments and clinical outcomes. Many clinical research institutions and pharmaceutical companies utilize mature clinical trial management system software for recording patient-level information on diagnosis, treatment, and clinical outcomes for their clinical trial participants. However, these systems typically do not contain the primary image data collected as part of response assessment for clinical trials. For single institution clinical trials, the DICOM (digital imaging and communications in medicine) image files are typically maintained on the local institution's PACS (picture archiving and communications system). Only the RECIST flow sheets are stored as part of the clinical trial management system as scanned case report forms. For multi-site clinical trials where a central review is required, the image files are typically stored on CDs that are then shipped to the central review site. However, clinical trial patients may have their baseline studies performed at another institution prior to enrollment, and these outside CDs are often lost. Because of these issues, many large multicenter clinical trials often have incomplete image data sets when it is time to perform a secondary central review.

Several systems have been developed to support the development of image repositories for clinical research. The National Biomedical Imaging Archive (NBIA) provides Web-based access to de-identified DICOM images, markups, and annotations, using role-based security. Publicly available image collections available though the NBIA include the I-SPY trial and the reference image database to evaluate therapy response, which includes serial DCE-MRI studies of the breast. These publicly available image data sets can be accessed via the National Cancer Institute–hosted NBIA, but institutions can adapt the open source NBIA software for data storage by setting up a local instance of NBIA at their institution.

The American College of Radiology Imaging Network has developed the transfer of images and data system, to support the management of large image repositories for clinical research. The transfer of images and data system Web client allows users to search, download, and view available DICOM image series, while the graphical user interface client provides advanced functions for DICOM series routing, image processing, and annotation layer management. Transfer of images and data system is used to manage the image data for all of the American College of Radiology Imaging Network's clinical trials, but remains proprietary software.

16.4.2 Image Annotation

Image metadata regarding tumor lesions for response assessment is typically stored on lesion flow sheets as shown in Figure 16.4- or as case report forms within a clinical trial management system. These documents are often paper-based or recorded on Excel spreadsheets that are scanned into the clinical trial management system as unstructured data. The use of manually transcribed records of response has several disadvantages, including transcription errors, ambiguity of lesion identifiers when multiple lesions are present in the same organ or image, lack of a direct link to the source image data, and limited functionality for calculating response. The lack of a direct link between the recorded lesion values and the raw image data source also makes it difficult for outside reviewers to audit the response assessment interpretation results.

Radiologists and oncologists reviewing imaging studies for treatment response will typically use the caliper tool in their DICOM viewer to measure tumor lesions. The image markup created by these tools, however, is stored in a proprietary format by each vendor and is used only for visualization of the markup on the images. The markup is not typically exported as part of the DICOM files and often cannot be viewed by another DICOM viewer. As such, the markup provides limited utility for the problem of uniquely identifying tumor lesions for lesion tracking. Furthermore, the metadata associated with the markup, such as length measurement, is not stored in a way that would allow for automated interpretation of response.

Recent advances have distinguished between image markup and image annotation, where image markup describes how to visualize a region of interest on an image, and image annotation describes the image metadata that describes the content of images. The annotation and image markup (AIM) [31] standard has been developed as part of the cancer biomedical informatics grid program of the National Cancer Institute. AIM provides the data structure for storing the key semantic lesion information. "Semantics" refers to meaning in images, including image metadata pertinent to quantitative criteria of disease, such as lesion identification, location, size measurements, method of measurement, and other quantitative features. Image annotation tools have been developed [32] that implement the AIM standard for creating image metadata in a structured format that can be transferred via extensible markup language to relational databases for storage. The AIM files maintain information about the image markup, including the source image file, image coordinates, quantitative features such as measurements, and qualitative observations. The image observations are encoded utilizing a structured terminology such as RadLex [33].

These image annotation tools have recently been evaluated for tumor lesion tracking and semantic annotation of image metadata to automatically populate lesion flow sheets for several cancer clinical trials [34]. The use of a semantic image annotation tool to directly populate lesion flow sheets provides several advantages. First, the use of an information model and structured terminology to encode cancer lesion image findings provides a consistent representation for storage and sharing of image metadata needed for cancer clinical trials. The information model also provides a foundation to enable reasoning with and querying over image metadata for response assessment. Image annotation disambiguates lesion identifiers by linking them directly to the source image and image markup. Image annotation tools also directly generate metadata about image markup, such as length calculations, eliminating possible transcription errors for length measurements.

Image annotations could also be generated by image processing algorithms. These approaches could increase the reproducibility of lesion feature extraction, particularly for quantitative image biomarkers such as volumetric assessments. However, these approaches are not yet widely available or validated and, thus, are not recommended by response assessment development committees. Additional research is needed to validate these approaches.

16.4.3 Automated Response Interpretation Methods

Response criteria contain multiple calculation and classification rules that are often inconsistently applied within and across studies. This makes it difficult to compare trial response outcomes between cohort arms within the same study and between different trials within the same clinical domain. Automated interpretation methods are needed to improve the consistency of response classification. However, manual data collection in unstructured data formats does not facilitate automated data interpretation.

Automated response assessment methods have recently been developed to calculate RECIST variables, such as sum of diameters and the percent change from baseline. For example, a semantic reasoning method for response assessment utilizing AIM data was recently developed [35]. The method evaluates both the qualitative and quantitative features of tumor lesion annotations to calculate and classify treatment response according to the RECIST criteria. A vendor solution developed by MEDIAN Technologies (MEDIAN Technologies Inc., Minneapolis, MN) includes systems to measure and track cancer lesions, and calculate quantitative RECIST metrics. Use of such automated approaches could improve the consistency and ease with which response criteria are applied in clinical research and routine care.

These recent informatics advances in image repositories, image processing, image annotation and automated response assessment methods could provide the necessary infrastructure to support the development of MRI-based response criteria. However, these approaches must also be combined with linked clinical data on patient treatments and overall survival in order to correlate novel imaging response biomarkers with clinical outcomes.

Acknowledgment

We thank Dr. Ronald Arildsen for supplying the MRIs in Figure 16.1.

References

1. Buyse M, Thirion P, Carlson RW, Burzykowski T, Molenberghs G, Piedbois P. Relation between tumour response to first-line chemotherapy and survival in advanced colorectal cancer: A meta-analysis. *Meta-Analysis Group in Cancer*. Lancet 2000;356:373–378.

2. Buyse M, Piedbois P. On the relationship between response to treatment and survival time. *Stat Med* 1996;15:2797–2812.

3. El-Maraghi RH, Eisenhauer EA. Review of phase II trial designs used in studies of molecular targeted agents: Outcomes and predictors of success in phase III. *J Clin Oncol* 2008;26:1346–1354.

4. Paesmans M, Sculier JP, Libert P et al. Response to chemotherapy has predictive value for further survival of patients with advanced non-small cell lung cancer: 10 years experience of the European Lung Cancer Working Party. *Eur J Cancer* 1997;33:2326–2332.

5. Escudier B, Eisen T, Stadler WM et al. Sorafenib in advanced clear-cell renal-cell carcinoma. *N Engl J Med* 2007;356:125–134.

6. Ratain MJ, Eisen T, Stadler WM et al. Phase II placebo-controlled randomized discontinuation trial of sorafenib in patients with metastatic renal cell carcinoma. *J Clin Oncol* 2006;24:2505–2512.

7. Goffin J, Baral S, Tu D, Nomikos D, Seymour L. Objective responses in patients with malignant melanoma or renal cell cancer in early clinical studies do not predict regulatory approval. *Clin Cancer Res* 2005;11:5928–5934.

8. Miller AB, Hoogstraten B, Staquet M, Winkler A. Reporting results of cancer treatment. *Cancer* 1981;47:207–214.

9. Baar J, Tannock I. Analyzing the same data in two ways: A demonstration model to illustrate the reporting and mis-reporting of clinical trials. *J Clin Oncol* 1989;7:969–978.

10. Tonkin K, Tritchler D, Tannock I. Criteria of tumor response used in clinical trials of chemotherapy. *J Clin Oncol* 1985;3:870–875.

11. Therasse P, Arbuck SG, Eisenhauer EA et al. New guidelines to evaluate the response to treatment in solid tumors. European Organization for Research and Treatment of Cancer, National Cancer Institute of the United States, National Cancer Institute of Canada. *J Natl Cancer Inst* 2000;92:205–216.

12. Eisenhauer EA, Therasse P, Bogaerts J et al. New response evaluation criteria in solid tumours: Revised RECIST guideline (version 1.1). *Eur J Cancer* 2009;45:228–247.

13. Cheson BD, Horning SJ, Coiffier B et al. Report of an international workshop to standardize response criteria for non-Hodgkin's lymphomas. NCI Sponsored International Working Group. *J Clin Oncol* 1999;17:1244.

14. Cheson BD, Pfistner B, Juweid ME et al. Revised response criteria for malignant lymphoma. *J Clin Oncol* 2007;25:579–586.

15. Bubley GJ, Carducci M, Dahut W et al. Eligibility and response guidelines for phase II clinical trials in androgen-independent prostate cancer: recommendations from the Prostate-Specific Antigen Working Group. *J Clin Oncol* 1999;17:3461–3467.

16. Scher HI, Halabi S, Tannock I et al. Design and end points of clinical trials for patients with progressive prostate cancer and castrate levels of testosterone: Recommendations of the Prostate Cancer Clinical Trials Working Group. *J Clin Oncol* 2008;26:1148–1159.

17. Macdonald DR, Cascino TL, Schold SC, Cairncross JG. Response criteria for phase II studies of supratentorial malignant glioma. *J Clin Oncol* 1990;8:1277–1280.

18. Benjamin RS, Choi H, Macapinlac HA et al. We should desist using RECIST, at least in GIST. *J Clin Oncol* 2007;25:1760–1764.

19. Choi H, Charnsangavej C, Faria SC et al. Correlation of computed tomography and positron emission tomography in patients with metastatic gastrointestinal stromal tumor treated at a single institution with imatinib mesylate: proposal of new computed tomography response criteria. *J Clin Oncol* 2007;25:1753–1759.

20. Oxnard GR, Armato SG, Kindler HL. Modeling of mesothelioma growth demonstrates weaknesses of current response criteria. *Lung Cancer* 2006;52:141–148.

21. Plathow C, Klopp M, Thieke C et al. Therapy response in malignant pleural mesothelioma-role of MRI using RECIST, modified RECIST and volumetric approaches in comparison with CT. *Eur Radiol* 2008;18:1635–1643.

22. Wahl RL, Jacene H, Kasamon Y, Lodge MA. From RECIST to PERCIST: Evolving Considerations for PET response criteria in solid tumors. *J Nucl Med* 2009;50 Suppl 1:122S–150S.

23. Sargent DJ, Rubinstein L, Schwartz L et al. Validation of novel imaging methodologies for use as cancer clinical trial end-points. *Eur J Cancer* 2009;45:290–299.

24. Ford R, Schwartz L, Dancey J et al. Lessons learned from independent central review. *Eur J Cancer* 2009;45:268–274.

25. United States Food and Drug Administration Oncology Drugs Advisory Committee briefing document for NDA 21–649 (oblimersen sodium).

26. United States Food and Drug Administration Center for Drug Evaluation and Research. Approval package for application number NDA 22–059 (lapatinib ditosylate). 2007.

27. United States Food and Drug Administration. FDA Briefing Document Oncology Drugs Advisory Committee meeting NDA 21–801 (satraplatin) 2007.

28. Barker AD, Sigman CC, Kelloff GJ, Hylton NM, Berry DA, and Esserman LJ. I-SPY 2: An adaptive breast cancer trial design in the setting of neoadjuvant chemotherapy. *Clin Pharmacol Ther* 2009;86:97–100.

29. Bogaerts J, Ford R, Sargent D et al. Individual patient data analysis to assess modifications to the RECIST criteria. *Eur J Cancer* 2008;45:248–260.

30. Moskowitz CS, Jia X, Schwartz LH, Gönen M. A simulation study to evaluate the impact of the number of lesions measured on response assessment. *Eur J Cancer* 2009;45:300–310.

31. Rubin DL, Mongkolwat P, Kleper V, Supekar K, Channin D. Medical imaging on the Semantic Web: Annotation and image markup. *AAAI Spring Symposium Series, Semantic Scientific Knowledge Integration* 2008.

32. Rubin DL, Rodriguez C, Shah P, Beaulieu C. iPad: Semantic Annotation and Markup of Radiological Images. *AMIA Annu Symp Proc* 2008:626–630.

33. Langlotz CP. RadLex: A New Method for Indexing Online Educational Materials1. *Radiographics* 2006;26:1595.

34. Levy MA, Rubin DL. Tool support to enable evaluation of the clinical response to treatment. *AMIA Annu Symp Proc* 2008:399–403.

35. Levy MA, O'Connor MJ, Rubin DL. Semantic reasoning with image annotations for tumor assessment. *AMIA Annu Symp Proc* 2009;2009:359–363.

36. Eisenhauer EA, Therasse P, Bogaerts J et al. New response evaluation criteria in solid tumours: Revised RECIST guideline (version 1.1). *Eur J Cancer* 2008;45:228–247.

37. Choi H. Response evaluation of gastrointestinal stromal tumors. *Oncologist* 2008;13 Suppl 2:4–7.

17

Image Segmentation

Shaun S. Gleason
Oak Ridge National Laboratory

Vincent C. Paquit
Oak Ridge National Laboratory

Deniz Aykac
Oak Ridge National Laboratory

Image segmentation in computer vision applications can generally be described as the process of differentiating an object of interest in an image or a volume from its background. Computer vision researchers generally agree that segmentation is one of the most challenging and important tasks in the field of computer vision. Historically, many computer vision problems have been roughly broken down into a three-step process: (1) segmentation, (2) feature extraction and analysis, and (3) recognition (or classification). Segmentation, being the first step in the process in this serialized approach, has direct impact on all tasks that follow. If an object is not accurately segmented from its background, the subsequent tasks of feature extraction and classification may become meaningless or, even worse, inaccurate. Hence, the segmentation task is critical to the success of the entire analysis. To compound matters further, it is often the most difficult task of the three.

The difficulty of the segmentation task depends on many factors, including the predictability of the object of interest. If the object can be described uniquely within an image using features like intensity, texture, and shape, then the likelihood of successful segmentation is generally high. Two main problems can dramatically increase the difficulty of segmentation, and hence decrease the potential for success: (1) inconsistencies in the appearance of the object from image to image, and (2) background clutter and noise in the image. An object whose appearance varies wildly will confound a technique that makes narrowly defined assumptions about object appearance. In addition, background clutter and noise in an image may both

obscure the object of interest and may confuse the segmentation technique by "distracting" the algorithm away from the object of interest. All segmentation algorithms are constructed so as to incorporate some assumptions about the appearance or structure of an object. These assumptions may be very simple (e.g., the object has detectable boundaries or a particular internal intensity/texture) or they may be quite complex such as in the case of highly parameterized, model-based segmentation algorithms. For any given segmentation technique, these underlying assumptions drive the algorithm to a solution. If these assumptions are invalid for a given image, the algorithm is doomed to fail. As an example, thresholding, one of the simplest forms of segmentation, makes the assumption that the object of interest will have a unique range of intensity values within the image. If this assumption is violated, thresholding the image will result in poor segmentation. Some of the most challenging segmentation tasks involve objects that have significant but predictable variations in appearance. These tasks require a segmentation algorithm that makes specific assumptions about the object of interest but at the same time has the flexibility to accommodate all of the acceptable object variations.

Medical imaging is one field where image segmentation methods have been thoroughly researched over the last couple of decades. Anatomical structures may be somewhat predictable in terms of appearance (e.g., organ shape and location), but variations are always encountered from subject to subject and within the same subject over time. Also, there are specific anatomical areas (e.g., the abdominal cavity) where the appearance, location, and background of structures are quite unpredictable, making the segmentation task even more difficult. One advantage of studying medical image segmentation is that the anatomy of the subject being studied is typically very well known, so there is an abundance of a priori information available in the form of one patient's own historical records as well as imagery from a potentially large population of other patients that could be used to improve segmentation algorithm performance.

Of all modalities employed for diagnostic medical imaging, magnetic resonance imaging (MRI) provides superior soft-tissue contrast between different soft-tissue types. This makes it a favorite modality to image, for example, neurological, musculoskeletal, cardiovascular, and oncological tissues. This soft-tissue contrast assists clinicians in delineating the boundaries between soft-tissue structures that are adjacent to one another. MRI is an extremely complex, powerful, and flexible imaging technique that allows different tissues to be highlighted differently via different pulse sequences (T_1 and T_2, as described in Chapter 5), other MRI variations (spin labeling and diffusion weighting, as presented in Chapters 6 and 7), and application of contrast agents (see Chapters 9 and 10). Since this book is dedicated to applications of MRI in cancer, it is useful to note that tumors can have a different contrast from surrounding healthy tissue in an MRI scan because, in a varying electromagnetic field, the protons within diseased and healthy tissues return to their equilibrium state at different rates, which affects the emitted radiofrequency signals that are used in the image formation process. This is one of several reasons that MRI is one of the most effective means of imaging a variety of tumor types and is heavily used in cancer diagnosis and in monitoring the effectiveness of cancer therapy.

This chapter will first provide a qualitative introduction to segmentation techniques that are typically used in MRI studies of cancer. A categorization of segmentation methods will be presented along with their applicability to cancer-related problems so that the reader will be able to make an educated decision as to what category of segmentation algorithm would be most appropriate for a given application. Secondly, this chapter will present the mathematical theory behind three of the most commonly employed and researched segmentation methods for cancer applications in MRI data: pixel-based clustering, level sets, and shape/appearance models. Enough detail will be provided to allow algorithm designers to understand the basic principles of the methodology, and didactic references will be provided to allow them to pursue implementation. Additional reference material will be presented that highlights recent research efforts to improve the performance of a given methodology. The chapter concludes by describing how these segmentation algorithms have been applied in the fields of (1) preclinical applications in cancer, (2) clinical applications in cancer diagnosis, and (3) clinical applications in cancer therapy. Given an overview of a variety of example applications, the readers will have a better foundation on which to build a segmentation strategy for their own specific cancer application.

17.1 Qualitative Introduction to Segmentation Techniques for MRI

One useful way to categorize the variety of different scene analysis techniques is by considering the main driving force behind the algorithm that leads to the final result. One class of algorithms relies more on the data in the image to influence the solution, while another class of algorithms makes up-front assumptions about the analysis problem to build a model that strongly influences the final form of the solution. Finally, a third class of algorithms uses a compromise between the data and the model. These three classes of algorithms are sometimes referred to as (1) bottom-up, (2) top-down, and (3) hybrid analysis techniques, respectively, as illustrated in Figure 17.1 (Gleason 2001). This section will provide a qualitative introduction to each in terms of the type of application in which each class is most applicable.

17.1.1 Data-Driven Techniques (Bottom-Up)

At one end of the spectrum, bottom-up segmentation algorithms are completely data driven in that they rely entirely on the local data patterns within the target image to come to a conclusion. Local data patterns include, for example, pixel intensity/color information, and neighborhood measurements such as texture, roughness, and homogeneity. In purely data-driven strategies, no external sources of information are used as guides and/or constraints to arrive at a solution. Bottom-up approaches have the advantages of being flexible (e.g., they are applicable to a variety of object shapes and sizes) and requiring little or no a priori information to perform their task. If, however, the data are corrupted by noise, clutter, occlusions, etc., then bottom-up strategies will not perform well because they have no other source of information to rely upon. Examples of data-driven analysis strategies are pixel-based clustering (Castellani et al. 2008; Chao et al. 2009; Chen et al. 2007; Lee, Yeung et al. 2005; Madabhushi et al. 2005; Mazzara et al. 2004; Solomon, Butman, and Sood 2006), adaptive thresholding (Baboi et al. 2010; Brekke et al. 2007; Pallotta et al. 2006), and region growing (Behrens

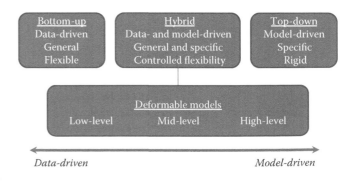

FIGURE 17.1 Illustration of top-down versus bottom-up scene analysis techniques. Note that deformable models fall under the "hybrid" category of techniques that rely on a combination of data-driven and model-driven strategies.

et al. 2007; Dastidar et al. 2000; Pallotta et al. 2006). Generally, bottom-up approaches are simpler in theory and implementation than their counterparts in the top-down analysis category. Bottom-up techniques generally cannot be relied upon for object recognition (i.e., identification of the segmented object) because postprocessing of the result is necessary to identify the object.

Later in this chapter, we will see that bottom-up approaches, in particular, pixel-level clustering, have been commonly applied to segment brain tissue into multiple categories: gray matter, white matter, cerebral spinal fluid, and, in some cases, tumor tissue (Castellani et al. 2008; Chao et al. 2008; Clark et al. 2006; Ferreira da Silva 2007; Weldeselassie and Hamarneh 2007). The superior soft-tissue contrast provided by MRI scans of the brain makes data-driven techniques attractive, but as will be discussed later, noise in the MRI data that corrupts the brain tissue pixel intensities creates challenges in performing accurate segmentation using data-driven strategies.

17.1.2 Model-Driven Techniques (Top-Down)

At the other end of the spectrum are top-down analysis algorithms that start with very specific assumptions about the characteristics of the target image, the object of interest, and, hence, the final solution. These assumptions are based upon a priori information about the particular analysis problem at hand and might include expected object characteristics (e.g., shape, color, texture) as well as background characteristics (e.g., consistency, homogeneity, intensity). These assumptions are used in many cases to preconstruct an image model with characteristics and constraints that correspond to the up-front assumptions. In top-down strategies, the data are examined to determine how to perturb the model in an attempt to fit it to the object(s) of interest. Later in this chapter, a class of segmentation strategies will be introduced that fit well into this category called *atlas-based segmentation* (Artaechevarria, Munoz-Barrutia, and Oritz-de-Solorzano 2009; Bondiau et al. 2005; Heckemann et al. 2006; Luts et al. 2009; Sims et al. 2009), where an a priori atlas of prelabeled structures is created that represents the patient's area of interest (e.g., the brain). The image of a newly scanned patient can then be deformed to fit the atlas through an optimization process, resulting in a fully segmented and labeled image. Template-based approaches that use a correlation strategy to locate and segment structures also fit into the model-based category. Active shape models (Cootes et al. 1995) and active appearance models (Cootes, Edwards, and Taylor 1998), which incorporate the typical variations of shapes and pixel intensities across a population, may also be considered model-driven approaches.

As has been discussed, the use of a template or an atlas comes with the requirements that examples of the structure of interest exist a priori and that the structure does not vary significantly from subject to subject in terms of its shape and appearance. It is this requirement that renders top-down techniques that rely on shape information relatively useless for tumor segmentation because of the unpredictability of the tumor shape. Note, however, that these techniques are very useful in cancer applications in at least two situations. First, if the a priori model does not fit the patient's image very well, it may be an indication that there is a shape or intensity variation that may be an indirect indicator that an abnormality such as a tumor has distorted the normal shape of the anatomy. Also, model-based approaches like atlas- or template-driven algorithms have been demonstrated as useful for mapping out the location and size of critical organs/structures around a tumor that must be avoided when developing a radiation therapy treatment plan.

17.1.3 Hybrid Techniques (Low, Mid, and High Level)

Very few scene analysis algorithms are purely bottom-up or top-down. Most fall on a continuum of hybrid approaches between these two ends of the spectrum. *Deformable models* belong to a class of algorithms that typically use a boundary representation (two-dimensional [2D] deformable contour) or a surface representation (three-dimensional [3D] deformable surface) of an object for segmentation purposes. The deformable contour is typically initialized by placing a contour on the image and then iteratively adjusting its position and shape to best fit the object's boundary. The morphological characteristics of the contour (e.g., local roughness, overall shape) depend on a set of parameters that is adjusted and tuned via an optimization process to best fit the information about the object as represented by the image data. These parameters that control the general form of the resultant contour are typically referred to as *internal constraints*. The image data generate forces, referred to as *external forces*, that attract or repel the model. In their foundational work on active contours, or snakes, Kass, Witkin, and Terzopoulos (1988) formulated the internal constraints and external forces as internal and external energy terms in an objective function that, when minimized, corresponds to the best fit of the model to the image data. The theory behind this approach will be presented in the next section.

The use of deformable models in segmentation of MRI data as well as other medical imaging modalities (e.g., X-ray, ultrasound) is ubiquitous. Level sets, which will be described in detail in the next section, started out inherently as a bottom-up, data-driven approach, but there have been several researchers that have incorporated boundary shape constraints into the level-set algorithm (Taheri, Ong, and Chong 2009; Tsai et al. 2003; Wang and Vermuri 2005; Xie et al. 2005; Zhukov et al. 2003). These changes move level sets into what can be described as a "low-level" deformable model, in that they are primarily data driven but do rely on shape information to drive the boundary to a proper solution. This makes the algorithm applicable to segmentation of both tumors and normal anatomical structures such as brain, prostate, and heart. Active contours, or snakes, are considered a "mid-level" deformable model. They have at their theoretical foundation an "internal" energy component that influences the final shape of the contour, so they also are broadly applicable to both the unpredictable shapes of tumors and the more predictable shapes of organs.

17.2 Quantitative Introduction to Segmentation Techniques for MRI

Given this introduction to the categorization of segmentation methods shown in Figure 17.1, some theoretical foundation will be provided for some of the most commonly researched and employed methods, starting with bottom-up techniques and finishing with top-down, model-based strategies.

17.2.1 Bottom-Up Techniques

MRI techniques provide high-contrast images of the internal anatomy where identifiable regions relate to physical, structural, and functional characteristics of those tissues. From an image processing point of view, performing tissue segmentation in those images consists of identifying regions of interest (ROIs) that are grouping pixels based on their intensity values and their neighborhood connectivity. On a related note, one should remember that MRI data can be multispectral, multiscan, and multisequence, and for that reason, intensity and connectivity should be considered as multidimensional parameters. Therefore, pixel-based segmentation methods, as compared with edge-based methods detailed in the next section, can range from a simple threshold function to a complex multidimensional statistical model.

As part of the pattern recognition field, pixel-based segmentation is a classification problem where one must identify a pixel or group of pixels by defining characteristic descriptors that gather intraclass pixels and scatter interclass pixels within a description space. The classification process includes a learning phase in order to identify discriminatory features, and a decision function is formed to classify or label each pixel and also provide a classification confidence score. One can categorize these bottom-up approaches into four categories as follows: (1) *Bayesian decision theory*: each pixel is defined by a feature vector in the feature space. By defining a statistical partitioning of the feature space, one can label unclassified pixels based on their likelihood probability. (2) *Supervised classification*: each feature vector is given a class label by an expert, and then the system is trained by learning from the intraclass similarities and the interclass differences. (3) *Unsupervised classification*: this is a self-classification method where the final number of classes and the characteristic class parameters are unknown. Using only the classes grouping rules known a priori, this method clusters similar samples and creates new classes as needed. (4) *Classification with knowledge of the distribution model*: sometimes in a statistical classification approach, we have total knowledge of the initial probability laws. Then, parametric discrimination consists of adapting general rules to the realities on the samples. These probability distributions are most often assumed to be Gaussian in nature. However, when probability laws that govern the classes are completely unknown, a probability estimator based on a nonparametric approach can be derived.

17.2.1.1 Bayesian Decision Theory

In Bayesian decision theory, there are multiple ways to formulate the classification problem. Here, three formulations are introduced: (1) error minimization, (2) cost minimization, and (3) likelihood maximization.

17.2.1.1.1 Error Minimization

This method assumes that the pixel classification problem can be represented by a probabilistic model that encapsulates the total knowledge of the probability laws governing the observations. A pixel is assigned to the class for which the error is minimized. Given the domain R^n, a pixel x defined by the feature vector $\{x_1, x_2, ..., x_n\}$, where, for example, x_i can be spectral values, specific statistical features, pixel neighborhood statistics, or a combination of those, and a partition $\Omega = \{w_1, w_2, ..., w_m\}$ of all possible classes, w_i. Given $p(x|w_i)$ as the known n-dimensional probability law of x in class w_i, $P(w_i)$ the known probability of each class w_i with $\sum P(w_i) = 1$, and $d(x)$ the decision function linking a pixel x to one of the m classes, one defines the decision boundary, $d_m(x)$, that minimizes the classification error as

$$d_m(x) = w \Rightarrow \forall w_i \in \Omega, p(w|x) \geq p(w_i|x)$$

verifying

$$\forall x \in R^d, p(d_m(x)|x) \geq p(d(x)|x) \quad (17.1)$$

In order for the global error probability to be minimal, one needs to associate x with the class with the highest probability value of $p(w_i|x)$ such as

$$p(w_i|x) = p(x|w_i)P(w_i)/p(x) \quad (17.2)$$

where

$$p(x) = \sum p(x|w_i)P(w_i) \quad (17.3)$$

If one applies the decision d_m, the global classification error is given by

$$Err(d) = 1 - \int_{R^d} p(d_m(x)|x)p(x).dx \quad (17.4)$$

A sample joint probability distribution over two variables is shown in Figure 17.2.

17.2.1.1.2 Cost Minimization

Pixel classification techniques in MRI for cancer applications must observe that nondetection of an abnormality has more potentially devastating consequences as compared with a false positive. To take this into account, one associates a decision cost, ρ_{ij}, to weight the probability law of each class with regard to the misclassification impact between class i and class j.

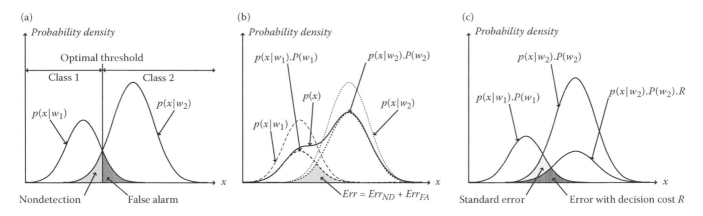

FIGURE 17.2 Bayesian decision: Example for a one-dimensional problem with two classes. (a) Assuming that both probability density functions are known, the optimal threshold between both classes is represented as the orthogonal line to the data value axis and passing through the probability density function intersection point. In this situation, one sees that two regions of the class distributions are a part of the decision threshold and therefore in the wrong classes. In this example, class 1 corresponds to normal objects and class 2 to abnormal objects; therefore, the elements of class 1 detected as abnormal correspond to "false alarms" and the elements of class 2 detected as normal correspond to "nondetections." (b) Knowing the probability density functions $p(x|w_i)$, one can then retrieve the probabilities $p(x|w_i)P(w_i)$ from the a priori probabilities $P(w_i)$ and then apply Bayes' rule. (c) Cost-based decision: by applying a decision cost R to one of the probability functions, it is possible to introduce a quantification factor regarding the importance of a nondetection against a false alarm.

The optimal decision function can be written as

$$C_i(x) = \sum_j \rho_{ij} \cdot p(w_j|x) \tag{17.5}$$

Considering a two-class problem (e.g., tumor versus nontumor pixels), the decision cost functions can be written as

$$C_1(x) = \rho_{11} \cdot p(w_1|x) - \rho_{12} \cdot p(w_2|x) \tag{17.6}$$

and

$$C_2(x) = \rho_{21} \cdot p(w_1|x) - \rho_{22} \cdot p(w_2|x) \tag{17.7}$$

To minimize the cost, one associates x to the w_1 class if $C_1(x) \leq C_2(x)$, which can be rewritten using Bayes' rule as

$$(\rho_{21} - \rho_{11}) \cdot p(x|w_1) \cdot P(w_1) \leq (\rho_{12} - \rho_{22}) \cdot p(x|w_2) \cdot P(w_2) \tag{17.8}$$

17.2.1.1.3 Likelihood Maximization

The optimal decision rules presented previously involve initial knowledge of the probability laws and of the a priori probabilities. In general, this information is not available. Therefore, during the initial training phase, it is necessary to estimate $p(x|w_i)$ and $P(w_i)$ from a subsample $X_i = \{x_1, x_2, ..., x_n\}$ from class w_i. If the number of samples per class is significant, one can easily estimate the a priori probabilities as the ratio between the number of elements per class and the total number of samples. Assuming $p(x|w_i)$ is known, estimating the probability density requires evaluation of the θ_i parameters characterizing $p(x,\theta)$, and estimating the maximum likelihood consists of finding the θ_i parameters that maximize the probability to draw the X_i samples

from w_i. Since the probability to draw X_i from n-independent samples is defined by

$$p(X_i,\theta_i) = \prod p(x_k|\theta_i) \tag{17.9}$$

maximizing $p(X_i,\theta_i)$ is accomplished by maximizing $\sum \log (p(x_k|\theta_i))$, and therefore the parameters must be found that solve the partial derivatives:

$$\delta/\delta\theta_i \sum \log(p(x_k|\theta_i)) = 0 \tag{17.10}$$

17.2.1.2 Unsupervised Classification

An unsupervised classification method attempts to automatically identify every class present in a population when the number of classes and the number of parameters that characterize them are a priori unknown. A class is defined as a group of individual pixels that are considered similar for given predefined criteria. If an individual pixel does not meet the similarity requirements, it becomes automatically part of a new class having its own characteristic features. At a pixel level, unsupervised classification is used to create an ROI patchwork based on similarity metrics. The k-means and fuzzy c-means (FCM) clustering algorithms are introduced as examples of unsupervised clustering techniques.

17.2.1.2.1 Clustering Algorithm: k-Means

Introduced by Macqueen (1967), the popular k-means method consists of subdividing a data set into k homogeneous classes by iteratively moving the center of each class in the definition domain until their positions become stable, indicating optimal intraclass similarity and optimal interclass separability (see Figure 17.3). During this process, each cluster may be divided

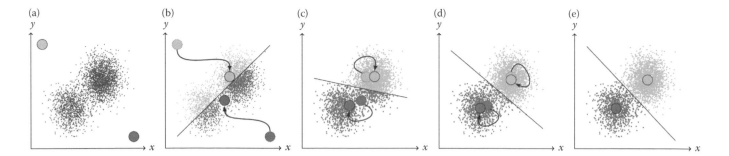

FIGURE 17.3 Illustration of the standard k-means algorithm on a randomly generated dataset. (a) The dark gray dots represent the dataset in a 2D Euclidian space. In this example, two centers have been arbitrarily positioned (light gray and dark gray circles) corresponding to the two clusters desired. (b) In an initial step, each data point is assigned to the cluster of the nearest center, and their color is changed accordingly. The black line represents the limit between the two intermediary clusters, i.e., the equidistant line to the centers. Then each center is recomputed as the mean of the points assigned to each cluster, and then repositioned in the Euclidian space. (c)–(d) Two intermediary iterations that lead to the final result (e). This iterative algorithm stops when the center position is varying under a given threshold.

or merged based on intraclass variance requirements. Easy to implement, k-means–based methods require a limited initialization consisting of (1) defining the number of classes to find and (2) an initial selection of the central positions of the classes. Methods exist to preliminarily evaluate the optimal number of clusters (Hamerly and Elkan 2003), but initial class positioning cannot be assessed. An alternative is using the iterative self-organizing data analysis technique (ISODATA) algorithm (Ball and Hall 1965), which is completely autonomous but tends to oversegment the image. Barcke et al. (2009), Lee et al. (2007), Singh et al. (1996), and Soltanian-Zadeh et al. (2007) present interesting applications of this clustering approach.

17.2.1.2.2 Clustering Algorithm: FCM

FCM clustering is an extension of k-means clustering. This method considers that each sample has a different probability to belong to one of the c classes instead of being part of only a single class. The mathematical formulation introduces a weighting coefficient for each cluster based on the likelihood. Chen et al. (2007) use this method to segment lesions in 3D contrast-enhanced MRI of breast tissues. Zhou et al. (2003) present a modified version of FCM, called *semisupervised FCM* (SFCM), that includes a training set initialization in order to help the regular FCM method in segmenting brain tumors in 3D MRI. Vaidyanathan et al. (1997) and Clark et al. (1994) present multispectral segmentation results using SFCM applied to brain tissue analysis. Klifa et al. (2010) use FCM to segment adipose and fibroglandular areas in breast MRI.

17.2.1.3 Supervised Classification

A supervised classification method attempts to associate each individual pixel to predefined reference classes of known characteristics. There are many approaches based either on distance or probability rules. A selected number of reference classes are initially isolated by a subject matter expert during a training phase where a set of specific features define the similarity requirements for each class. The supervised classification process is divided into five phases: (1) define the number of classes based on the number

of objects to isolate, (2) select representative samples that describe each class, (3) extract characteristic features, (4) select the classification algorithm and the decision laws based on distance metrics or probability criteria, and (5) classify and evaluate.

17.2.1.3.1 Distance-Based Methods: k-Nearest Neighbors

k-Nearest neighbors (k-NN) is an intuitive technique to achieve a simple predictive pixel classification. The basis of this technique is that an individual pixel in the feature space has a high probability of being located in the vicinity of pixels belonging to the same class. For each new point, one searches for its k-nearest neighbors among the training points. Using a class voting scheme, the most represented class in its nearest neighbors is then chosen as the class for this new item. Figure 17.4 shows an illustration of the k-NN algorithm.

17.2.1.3.2 Support Vector Machine

Introduced by Vapnik (1996), the support vector machine (SVM) method searches the feature space in order to find the geometrical separation between classes by defining the optimal representative subset of training samples that delineate the class boundaries and that maximize the distance between the closest points and the boundaries. Representative samples of this subset are then called *support vectors*. Once trained, the SVM classification method maps each new sample in the definition space and predicts in which class it belongs. The SVM classification quality and tolerances are measured using two misclassification error metrics: (1) the expected risk and (2) the empirical risk, which corresponds to the average measurement error of misclassification. In general, SVM methods are used for nonlinear surface decision calculation, thus solving pixel classification problems where tissue classes are not linearly separable in the definition space. The idea is to project the data into a new dimension space that is larger than the initial space where data become linearly separable. The use of linear SVM in this new space then allows the determination of nonlinear surfaces in the original space. A sample illustration of a two-class decision function found using SVM is shown in Figure 17.5.

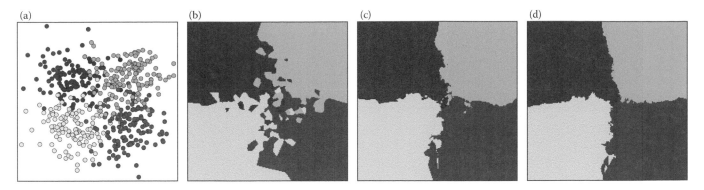

FIGURE 17.4 k-NN classification in a four-class problem. (a) The varied gray dots have been generated randomly and belong to one of four clusters. In this example, each point is defined by a 2D feature vector. (b)–(d) Illustrations of cluster boundary delineation for (from left to right) $k = 1$, $k = 9$, and $k = 17$.

One can summarize the class separation approach presented as follows: considering a learning base of n samples $\{x_i, y_i\}$ with $x_i \in R^m$, $y_i = \pm 1$, where the sample i is defined by an m-dimensional feature vector x_i and a label y_i, the SVM classification method finds the optimal hyperplane (represented by the dark central gray line and defined by $wx + b = 0$) separating both classes such that +1 and –1 labels are separated and present in majority on each side of the hyperplane, and the distance between the closest vectors from each class (light gray and dark gray dots) and the hyperplane is maximized. The closest vectors found are called *support vectors,* since they support optimal margins of each class. Those margins are represented with light gray and dark gray lines and are respectively defined by $wx + b = +1$ and $wx + b = -1$, assuming that +1 is the label of the dark gray class and –1 is the label of the light gray class. Here, w is a vector normal to the hyperplane; b is the minimal distance between the origin of the referential and the hyperplane. The classification phase consists then to evaluate $sign(x_j w + b)$ for a test sample j defined by a feature vector x_j. Classification success is determined by evaluating $\hat{i}_i$ in $wx_i + b \geq \pm 1 - \xi_i$ (correct if $\xi_i \geq 0$, misclassified if $\xi_i \geq 1$). Interested readers can refer to Vapnik (1996) for algorithmic details.

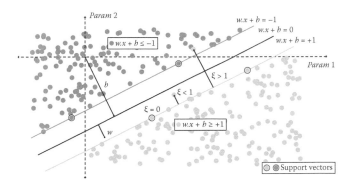

FIGURE 17.5 SVM classification for a two-class problem: in this example, two classes are represented by light gray and dark gray dots in the projection plane (*Param 1, Param 2*), where *Param 1* and *Param 2* are elements of the feature vectors.

17.2.2 Hybrid Techniques

Now that some of the most relevant bottom-up strategies for segmentation in MRI cancer applications have been presented, some hybrid approaches will be introduced that combine the use of data-driven techniques and some a priori information within the segmentation strategy.

17.2.2.1 Active Contours

Although deformable model research dates back to 1973 (Widrow 1973a, 1973b) the original research on piecewise splines by Kass, Witkin, and Terzopoulos (1988) is considered by most to be the foundational work in the field of deformable models. Better known as "snakes," or active contours, these deformable models are considered low level because they are constrained only by local smoothness parameters and have no overall shape constraints. These researchers coined the internal and external energy concepts that are still widely used in current algorithm development strategies for deformable models. Prior to this, Gritton and Parrish (1983) introduced a form of snake called a *bead chain*, but the optimization of this chain was not as sophisticated as that by Kass et al. Their energy function is modeled by internal and external energy forces as follows:

$$E_{snake} = \int_0^1 E_{snake}(v(s))\mathrm{d}s = \int_0^1 E_{int}(v(s)) + E_{ext}(v(s))\mathrm{d}s \quad (17.11)$$

where

$$E_{ext}(v(s)) = E_{image}(v(s)) + E_{con}(v(s)) \quad (17.12)$$

$v(s)$ is the parameterized contour, E_{int} represents the internal energy of the spline due to bending, E_{image} represents forces imposed by the target image, and E_{con} represents external constraint forces that the user may impose. For snakes, the internal energy is related to the smoothness of the active contour, and the external energy is based on edge information in the target image. Because snakes are more of a bottom-up, data-driven concept, they are very useful and flexible in images with clear and consistent intensity differences along the boundary between neighboring objects. They were designed to be interactive and require accurate initial placement of the boundary by the operator to converge to the correct solution. The accuracy of this initial placement depends on many factors, but generally the initial boundary placement must be closer to the target boundary of interest than any other competing boundary in the image. Images with larger number of objects (and, therefore, many competing boundaries) in the scene require better initial guesses to be successful. Snakes tend to become trapped in local minima, especially when the image is cluttered with spurious edges. If the global shape of the object is known a priori, then there is not a convenient way to include this knowledge in the snake deformation process.

The internal and external energy concepts for snakes are important to almost every other deformable model approach. The internal energy term can be represented by

$$E_{int} = \frac{\left[\alpha(s) |v_s(s)|^2 + \beta(s) |v_{ss}(s)|^2 \right]}{2} \quad (17.13)$$

where $v_s(s)$ and $v_{ss}(s)$ are the first- and second-order derivatives of v. From this energy equation, one can see that the values of α and β control the first- and second-order smoothness of the spline $v(s)$. If we assume that the external energy force is composed only of image forces with no additional user-imposed constraints ($E_{con} = 0$), then the only contribution is by E_{image}. In fact, in most snake implementations, the image is the only contributor to the external energy terms. In practice, a variety of image-derived forces are used, but the most common form of image energy is given by edge content

$$E_{ext}(x,y) = E_{edge}(x,y) = -|\nabla I(x,y)|^2 \quad (17.14)$$

where $I(x,y)$ is the image of interest. Snakes and their derivative works have been successfully applied to a wide variety of segmentation problems in MRI-based cancer applications. Some examples are given in Atkins and Mackiewich (1998); Colliot, Camara, and Bloch (2006); Huyskens et al. (2009); Pasquier et al. (2007); Wang, Cheng, and Basu (2009); and Wu et al. (2008).

The classical active contours are topologically limited in that they cannot split, merge, fold over on themselves, etc., so work has been done by McInerney and Terzopoulos (1997) and by DeCarlo and Metaxas (1998) to develop a topologically adaptable snake. In this approach, a decomposition grid is iteratively reparameterized to allow topological changes of the contours. O'Donnell, Dubuisson-Jolly, and Gupta (1998) have adapted snakes for application to branching cylindrical structures. Theoretically, the development of snakes is not conducive to topological variations (e.g., objects with overlapping edges, objects that consist of multiple, discrete pieces, or objects that surround/encompass another object, and, hence, require multiple separate boundaries to segment), and published approaches are typically "add-on" adaptations that attempt to overcome this limitation.

The deformable models known as geodesic snakes lend themselves better to handling topological changes. This category of low-level approaches to boundary-based image segmentation (Casselles et al. 1993; Casselles, Kimmel, and Sapiro 1995; Casalles 1997; Malladi, Sethian, and Vemuri 1995; Sapiro 1996) evolved out of the field of differential geometry, where curve evolution and geometric flow theory are well characterized mathematically, but had not been adapted and applied to the field of image segmentation. As described by Whitaker (1998), these 2D geodesic snakes can be characterized as a level set (contour) of an image (or an isosurface of a volume) and therefore can take on any shape allowed by the discrete pixel (voxel) grid in the image. A geodesic snake can split and merge to allow detection of multiple objects simultaneously. Also, these level sets can be extended directly to 3D (Kelemen, Szekely, and Gerig 1998) because the math supports this without any additional theoretical development.

Geodesic snakes are very flexible in terms of overall shape and are good at capturing the details of complex objects but, like other more bottom-up approaches, are less accurate in the presence of noise and image clutter. Also, incorporation of a priori information (e.g., global shape constraints) to limit variability has not been actively researched and/or reported. Most implementations are data driven, with some simple smoothness constraints. Because geodesic snakes are nonparameterized, they have the flexibility to handle the topological variations described in the snake discussion above, but the results are also more difficult to interpret. Although one is likely to know the general form of the objects delineated by geodesic snakes based on the current application, their nonparameterized form requires additional postsegmentation analysis to complete the task of object recognition. A detailed description of geodesic snakes based on level sets is presented in the next section.

17.2.2.2 Level Sets (Geodesic Snakes)

As introduced in the earlier section, level sets can be categorized as a "low-level" deformable model, which puts them in the hybrid category of segmentation approaches. The level-set formalism (Osher and Sethian 1988) is, in simple terms, a deformable model approach where a closed curve representing the structure of interest iteratively propagates toward the desired boundary through the evolution of an implicit function. Level sets yield a representation of regions by employing variational methods and numerical analysis for minimizing the energy functional. In the literature, there are many examples of such region-based level-set approaches. Level sets can describe topological changes in the segmentation; that is, regions may split and/or merge. Overall, in traditional level-set approaches, the curve evolves based on the image gradient. Hence, only objects with well-defined edges can be found.

Many region-based geometric active contours have been inspired by the piecewise smooth Mumford–Shah model (Mumford and Shah 1989). Like Wang and Vermuri (2005), many researchers started from the region-based active contour model and found different ways to improve the segmentation of the regions in images. Wang and Vermuri (2005) also focused on the level-set formalism to handle matrix-valued images such as diffusion tensor imaging (DTI; see Chapter 7). Their work was based on Chan and Vese's model on active contours without edges and the curve evolution implementation of Tsai et al. (2003). Tsai et al. (2003) incorporated the level-set formalism into a parametric shape model to represent shape variability by using the boundaries of aligned shapes of object of interest. The collections of these aligned shapes were utilized as the zero level set of signed distance functions to distinguish the inside and outside of the ROI. In this iterative scheme during the curve evolution, the statistics about the region are recalculated and updated. The emphasis is on the 2D application of cardiac MRI segmentation. Xie et al. (2005) used a level-set segmentation method incorporating region and boundary information simultaneously in which region information drives the curve propagation and boundary information is used as the stopping criteria. Their semiautomated technique was utilized as a clinical image analysis tool for the diagnosis of brain tumor and edema in MR images and also to set the treatment planning based on the volume measurements. Uberti, Boska, and Liu (2009) presented a semiautomatic mouse brain extraction technique based on level-set formalism. User-defined constraints from orthogonal contours on sagittal and axial MRI slices were incorporated into the level-set method to avoid curve evolution beyond the constraint points and also to initialize the zero level-set function. The only expert time needed on the analysis was to manually draw orthogonal contours and selection of constraint points. Zhukov et al. (2003) focused on the application of DTI data of the human head. The two-volume data sets, one that was the trace of the tensor matrix for each voxel and the other one that represented a measure of the magnitude of the anisotropy within the volume, were used for their implementation. The level-set method was utilized to extract smoothed models from those derived volume data sets. They took advantage of level-set segmentation approach for initialization and selection of different feature extracting terms in the level-set equation to produce a desirable surface deformation. Taheri, Ong, and Chong (2009) proposed a control mechanism for the curve evolution by applying adaptive thresholding scheme for tumor segmentation. They argued that their method can segment a variety of tumors if tumor and nontumor regions differ sufficiently in intensity.

Level-set methods represent object boundaries as the zero level set of an implicit function defined in a higher dimension, and evolution of the level-set function (curve evolution) occurs according to a partial differential equation (PDE). This can be defined as an energy minimization problem in which the energy is minimized when the evolving curve reaches the desired boundary. The level-set method can be explained in simple terms as in the Chan–Vese model in which the assumption is that the image, I_o, is formed by two approximately homogeneous regions, one corresponding to the object of interest and the other to the background. The average intensities inside and outside the object are c_1 and c_2, respectively, and the evolving curve C is defined as the boundary of the object. λ_1 and λ_2 are two weighting terms for the inside and outside region terms. For a smooth boundary, a regularizing term

based on the length of the curve C is also added. The net energy term to be minimized can then be written as

$$E(c_1,c_2,C)=\mu \cdot \text{Length}(C)+\lambda_1 \int\limits_{inside(C)} \left| I_o(x,y)-c_1 \right|^2 dx\,dy$$

$$+ \lambda_2 \int\limits_{outside(C)} \left| I_o(x,y)-c_2 \right|^2 dx\,dy$$

$$(17.15)$$

The balance between the homogeneity (last two terms) and smoothness (first term) is controlled by the single parameter μ, when $\lambda_1 = \lambda_2 = 1.0$. The above energy term is minimized using a level-set approach where the curve C, $C \subset \Omega$, is represented as the zero level set of a Lipschitz function (see Figure 17.6) $\phi : \Omega \to \mathfrak{R}$

$$C=\partial\omega=\left\{(x,y)\in\Omega:\phi(x,y)=0\right\}$$

$$\text{inside}(C)=\omega=\left\{(x,y)\in\Omega:\phi(x,y)>0\right\} \quad (17.16)$$

$$\text{outside}(C)=\Omega\setminus\overline{\omega}=\left\{(x,y)\in\Omega:\phi(x,y)<0\right\}$$

The energy functional is minimized using an Euler–Lagrange approach with time t (corresponding to iterations) resulting in the following PDE:

$$\frac{\partial\phi}{\partial t}=\delta_\varepsilon\left(\phi\right)\left[\mu\cdot\text{div}\left(\frac{\nabla\phi}{|\nabla\phi|}\right)-(I_o-c_1)^2+(I_o-c_2)^2\right] \quad (17.17)$$

where $\dfrac{\partial\phi}{\partial t}=0$ for the optimal ϕ and c_1 and c_2 are defined—using the regularized Heaviside, $H_\varepsilon(\phi^n)$, and Dirac functions, $\delta_\varepsilon(\phi^n)$, as

$$c_1=\frac{\sum\limits_{i,j}\phi^n H_\varepsilon(\phi^n)}{\sum\limits_{i,j}H_\varepsilon(\phi^n)}, c_2=\frac{\sum\limits_{i,j}\phi^n(1-H_\varepsilon(\phi^n))}{\sum\limits_{i,j}(1-H_\varepsilon(\phi^n))} \quad (17.18)$$

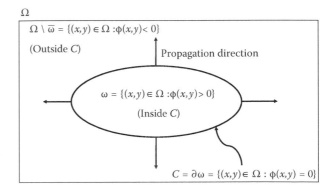

FIGURE 17.6 Illustration of the propagating boundary, C, within the image space, Ω.

$$\delta_\varepsilon(z) = H'_\varepsilon(z) = \frac{1}{\pi} \frac{\varepsilon}{\varepsilon^2 + z^2}$$

$$H_\varepsilon(z) = \frac{1}{2}\left(1 + \left(\frac{2}{\pi}\right)\arctan\left(\frac{z}{\varepsilon}\right)\right)$$

Numerically the iteration of Equation 17.17 can be approximated using the same finite difference scheme as described in Chan and Vese (2001) and based upon earlier work by Aubert and Vese (1997):

$$
\phi_{i,j}^{n+1} = \phi_{i,j}^n + \Delta t \cdot \delta_h\left(\phi_{i,j}^n\right) \cdot \left[\frac{\mu}{h^2}\Delta_-^x\left(\frac{\Delta_+^x\phi_{i,j}^n}{\sqrt{\left(\Delta_+^x\phi_{i,j}^n\right)^2/h^2 + \left(\phi_{i,j+1}^n - \phi_{i,j-1}^n\right)^2/\left(2h\right)^2}}\right) \right.
$$
$$
+ \frac{\mu}{h^2}\Delta_-^y\left(\frac{\Delta_+^y\phi_{i,j}^n}{\sqrt{\left(\Delta_+^y\phi_{i,j}^n\right)^2/h^2 + \left(\phi_{i+1,j}^n - \phi_{i-1,j}^n\right)^2/\left(2h\right)^2}}\right)
$$
$$
\left. - \left(I_{o,i,j} - c_1\left(\phi_{i,j}^n\right)\right)^2 + \left(I_{o,i,j} - c_2\left(\phi_{i,j}^n\right)\right)^2 \right]
$$

$$(17.19)$$

where h is the space step ($h = 1$), Δt is the time step, $\phi_{i,j}^{n+1} = \phi\left(n\Delta t, x_i, y_j\right)$, and the finite differences are given by

$$\Delta_-^x\phi_{i,j} = \phi_{i,j} - \phi_{i-1,j}; \;\; \Delta_+^x\phi_{i,j} = \phi_{i+1,j} - \phi_{i,j}$$
$$\Delta_-^y\phi_{i,j} = \phi_{i,j} - \phi_{i,j-1}; \;\; \Delta_+^y\phi_{i,j} = \phi_{i,j+1} - \phi_{i,j}$$

$$(17.20)$$

The initial $\phi_{i,j}^0$ is set to be the signed distance function to the initial curve, where the points inside the zero level set are positive and those outside are negative. At the end of the above iterative process, determined by essentially no change between successive iterations, the zero level set of $\phi_{i,j}^n$ defines the border of the object of interest. In most cases the process begins with a single seed point near the center of the ROI that has been selected either manually by an expert reviewing the images or by an automated selection mechanism. The initial boundary is usually set to a circle around this seed point and it is then iterated to find the true segmentation boundary.

17.2.3 Model-Based Techniques

This section will conclude by introducing the theory behind a couple of the routinely applied model-based strategies for segmentation of cancer in MRI data. First, atlas-based segmentation strategies are commonly used for segmentation of brain structures. An a priori atlas of prelabeled structures is created

that represents the patient's area of interest (e.g., the brain). The image of a newly scanned patient can then be deformed to fit the atlas through an optimization process, resulting in a fully segmented and labeled image. Although a thorough treatment of the theory behind atlas-based approaches is beyond the scope of this chapter, there are many published works applying atlas-based segmentation methods to brain segmentation within MRI (e.g., Aljabar et al. 2009; Ciofolo and Barillot 2009; Rodionov et al. 2009; Lötjönen et al. 2010).

17.2.3.1 Active Shape Models

It was not until the work of Cootes et al. (1995) that a deformable model approach was introduced that made extensive use of training data in the form of user-defined landmark points (LPs) that delineate significant shape features. These authors coined the term *active shape model* (ASM), or "smart snakes," to describe these contours because of the a priori knowledge that is built into them via a training procedure. ASM is shape-constrained by the measured shape variations within the training set and, at the same time, is attracted to edges in the target image. An optimization strategy is used to balance the shape constraints imposed by the model and the attraction force generated by object edges within the target image.

A simplified view of the ASM training and optimization process is shown in Figures 17.7 and 17.8, respectively, to illustrate the main components of the algorithm. The training process consists of two separate tasks: (1) training the shape model (SM) and (2) training the gray-level model (GLM). Together, SM and GLM constitute the complete ASM. Both of these training processes require a set of training images with manually labeled LPs that define the boundary of the object(s) to be segmented. SM constrains the final shape of the resulting ASM boundaries to look like those represented by the training set. Comparing this technique to the original work in active contours by Kass, Witkin, and Terzopoulos (1988), we see that SM represents the internal forces. Optimizing the fit of SM to a test image is comparable to minimizing the internal energy term in Equation 17.11. The GLM is used to update the position of the ASM boundary under the influence of the data within the image. Optimizing the

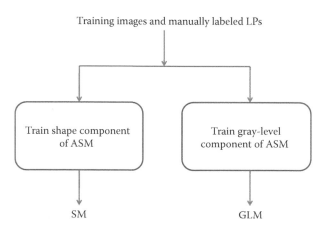

FIGURE 17.7   Training process for ASM.

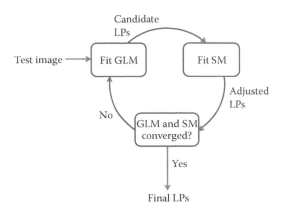

FIGURE 17.8 ASM optimization process.

GLM can be compared to minimizing the external energy term of Equation 17.12. Note that fitting the ASM to a new image is an iterative process in which the GLM and SM are independently applied to the current set of LPs. This independence of GLM and SM is important to note because it leads to some problems with the ASM technique that can be solved by improving the level of communication between these two models. This limitation was overcome in the work of Gleason et al. (1999), where a unified probabilistic framework was developed for the ASM. Finally, Cootes, Edwards, and Taylor (1998) extended the ASM method to include internal region-based intensity information into the model and called this extension the active appearance model.

17.3 Preclinical Applications in Cancer

In the late 1990s, there was a tremendous increase in development of medical imaging modalities and applications for small animal, or preclinical, research. This research boom was in large part due to a focused funding effort by the National Institutes of Health in the preclinical imaging research field. Grants were funded for (1) researchers to develop new small animal imaging modalities and (2) institutions to purchase multiple modalities of small animal imaging equipment to create imaging centers for preclinical research. Focused efforts led to development and implementation of a variety of new and/or improved preclinical imaging modalities including micro X-ray computed tomography (micro-CT), micro positron emission tomography (micro-PET), micro single photon emission CT (micro-SPECT), functional micro optical imaging, micro ultrasound, and, of course, micro-MRI.

Historically, MRI has not been the most accessible preclinical imaging modality, due to the high cost of the equipment and its operation. However, it quickly became the modality of choice for preclinical applications in cancer diagnosis and therapy due to its combination of high spatial resolution (50 μm or better) and superior soft-tissue contrast resolution as compared to other anatomic imaging modalities such as X-ray, CT, and ultrasound. Preclinical researchers can use micro-MRI to image small tumors in longitudinal studies of the same animal and can also image tumors across a statistically large population of animals. Both of these types of preclinical studies generate tremendous amounts of image data (up to gigabytes per animal) that simply cannot be manually analyzed in a consistent, quantifiable manner. As a result, the research and development of automated segmentation tools for preclinical cancer studies has become essential. A useful overview of techniques and challenges associated with high-throughput MRI research in the preclinical setting, including the data analysis challenges, is presented in Nieman et al. (2007). The next section describes some specific applications of segmentation methods for preclinical cancer research.

17.3.1 Tumor Assessment Using Bottom-Up Methods

As has been discussed, preclinical MRI is an excellent modality for tumor imaging because of its flexibility in terms of generating high soft-tissue contrast. Researchers commonly seek to quantify the progression of tumor burden over time and with the application of various therapeutic treatments. It is essential to have (semi) automated segmentation tools to measure tumor burden within large volumes of image data with quantifiable accuracy. Brekke et al. (2007) performed longitudinal studies of gliomas in rats using a gadolinium-based contrast agent and T_1-, T_2-, and T_2^*-weighted MRI imaging protocols. Automated segmentation of tumors was performed by applying a simple pixel intensity-based thresholding technique (an example of a purely data-driven approach). This approach worked due to the ability to deliver contrast to the tumors, which caused them to appear brighter within the MRI scans. Using automatic segmentation, they were able to quantitatively demonstrate that tumors treated with neural stem cell therapy grew at a slower rate as compared to untreated tumors within the corpus callosum. A set of images from this study is shown in Figure 17.9. Pagel et al. (2006) applied a similar simple threshold approach to segment and measure the volume of bone tumors in mouse legs that resulted from metastases of breast cancer. They were able to quantitatively demonstrate the effectiveness of antimetastatic drugs on inhibition of tumor growth, and segmentation results were validated by histology, which is the most effective means of measuring the accuracy of tumor volume quantification via segmentation algorithms. Finally, Barcke et al. (2009) applied a three-stage k-means clustering algorithm followed by some morphological operators and size thresholds to automatically classify breast tumor tissue in mice with an accuracy of 88.9%.

Simple pixel intensity thresholding is rarely an effective bottom-up segmentation approach when there are a large number of different subjects and/or slight variations in the imaging protocols. Adaptive thresholding is a superior approach that can better manage intensity variations across multiple studies. For adaptive threshold approaches, it is commonly helpful to place reference materials in the field of view that can be used to calibrate the measured signal within each image. Baboi et al. (2010) have applied a more sophisticated threshold technique for quantifying the tumor burden within longitudinal MRI studies

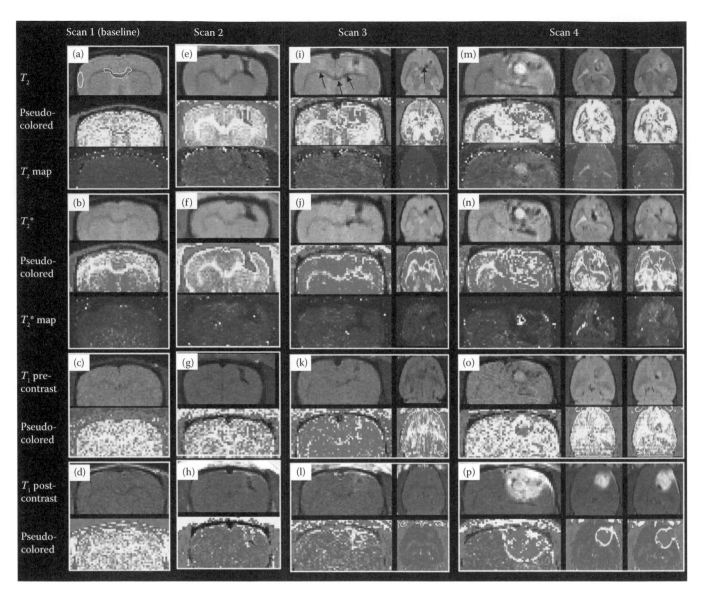

FIGURE 17.9 **(See color insert.)** MR images of a tumor/GRID-labeled neuronal stem cell (NSC) cell implanted animal acquired at baseline (scan 1, a–d), at 6 days (scan 2, e–h), at 10 days (scan 3, i–l), and at 17 days (scan 4, m–p). Axial orientations were obtained from multiplanar reconstructions using the nICE software. To provide a better visualization of the different signal intensities, pseudocolor-coded images were generated. Cold and warm colors indicate low and high signal intensities, respectively. Arrows point to regions with a signal attenuation in the corpus callosum caused by migrating GRID-labeled NSCs. At the last time point (scan 4, m–p), the tumor appeared very heterogeneous on all image sequences. (Reprinted from Brekke, C. et al., *NeuroImage*, 37 (3), 769–782, 2007. With permission.)

of the liver that were the result of metastases of gastroenteropancreatic tumors in mice. An adaptive dual-reference threshold method was employed where the threshold is calculated on a per-image basis. The calculation is performed by first measuring the average reference intensities of (1) water in a reference capillary (I_w) and (2) the image background signal (I_{bg}). Then the segmentation threshold, T, was calculated as 30% of the difference between these two average intensities. The result of this work was the development of an effective MRI protocol that allows automatic calculation of the hepatic lesion volume fraction and can be applied to future studies of effectiveness of therapeutic methods.

17.3.2 Organ Segmentation Using Hybrid Methods

Note that the two previous examples describe bottom-up, or data-driven, approaches for tumor segmentation. A significant reason that data-driven approaches are commonly used for tumor segmentation is that tumors are not typically prone to have a specific shape and/or size that can be used as a priori information in an object model. Both hybrid and top-down approaches require some level of a priori information and are therefore not generally applicable for tumor segmentation. The presence of tumors within the subject can in some cases be indirectly found through the automatic identification of volume and or shape variations of an organ or structure of interest. For example, a glioma within the brain may affect the size and shape of the brain, so one can employ a brain segmentation method to extract the brain from the MRI study, and then measure volume and shape characteristics to determine if they vary from a normal/healthy brain. When a priori information on size and shape of the object of interest is known, then hybrid methods such as deformable models are very effective methods for segmentation. As described previously, deformable models include active contours, or snakes (along with their many adaptations and variations), and implementations of level sets that incorporate a priori information on object appearance.

Uberti, Boska, and Liu (2009) have done this by developing a brain segmentation method based on level sets for brain extraction from T_1- and T_2-weighted MRI scans of a mouse. The level-set algorithm presented is constrained in that it incorporates shape prior information to drive the segmentation solution, making it a true hybrid deformable model. They quantified the accuracy of their segmentation results through comparison to manual segmentation by an expert. Note that level sets are typically a semiautomated approach requiring initial seed points placed

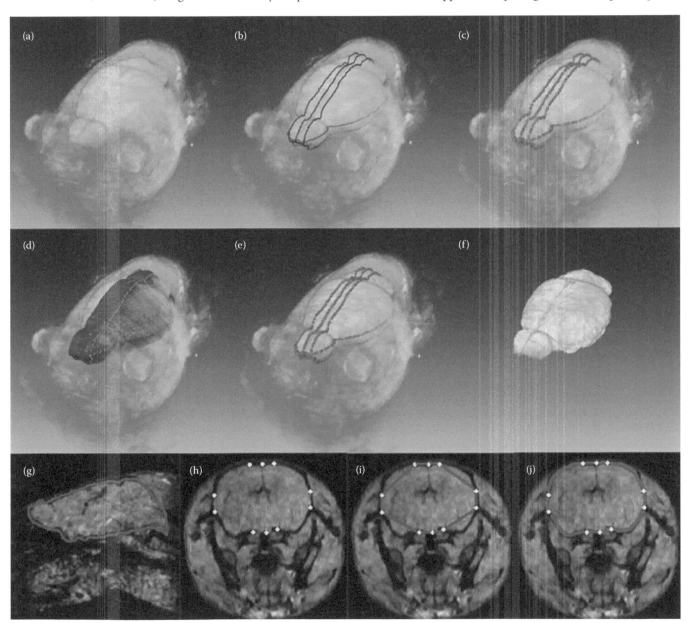

FIGURE 17.10 **(See color insert.)** Results of the brain extraction using 3D constraint level set (CLS) on the T_1 image of a mouse. (a) Mouse head MRI. (b) Brain contours were drawn on sagittal and axial slices. (c) Constraint points were extracted from the contours. (d) Initial zero level surface was constructed using the constraint points as vertices. (e) Final surface after 60 iterations. (f) Segmented brain using CLS method. (g) Contour drawn on the sagittal slice. (h) Constraint points extracted from contours on a coronal slice. (i) Initial surface contour on the coronal slice. (j) Final surface contour on the coronal slice. (Reprinted from Uberti, M. G. et al., *J. Neurosci. Methods*, 179 (2), 338–344, 2009. With permission.)

by the user, and this was the case here. An example of a brain segmentation result using the constrained level-set approach is shown in Figure 17.10.

17.3.3 Organ Segmentation Using Top-Down Methods

In the field of MRI segmentation for preclinical applications, the most commonly referenced top-down techniques for anatomic structure segmentation are atlas-based methods. Recall that atlas-based approaches rely on the generation of a 2D or 3D atlas that represents the "typical" subject. This atlas is developed by imaging a population of "normal" subjects and then merging the resulting image data sets into a single image, or atlas, that best represents the population. New images are then spatially registered to this prelabeled atlas as a means of segmenting the data into the anatomical ROIs. Adopting this approach, Sims et al. (2009) applied an atlas-based segmentation method to the problem of organ segmentation in the head and neck of human patients. The atlas was generated from a set of 45 normal (non-tumor) patients. Target patient scans were then registered to the atlas using a combination of affine and nonlinear registration methods. The brainstem, parotids, and mandible were then segmented so that the size and position of these structures could be measured to determine an effective plan for radiation treatment that would avoid radiation exposure of critical organs.

17.4 Applications in Clinical Cancer Diagnosis

In MRI studies where cancer diagnosis is the goal, clinicians will typically visually review the MRI scan and look for anatomic signs of cancer that are candidates for biopsy. There are classes of segmentation algorithms that are categorized as tumor detection methods that could be considered a diagnostic method in that they are searching for the presence of cancer. Even if those methods are used, they usually indicate the locations of "suspicious" regions, and the final decision is made by the physician. Breast cancer screening has been a fertile area for automatic segmentation research for cancer detection, and several examples of computer aided diagnosis are presented later in this section. Method development for detecting and staging prostate cancer has also been a specific focus of the segmentation research community and we will describe some illustrative examples.

17.4.1 Cancer Detection

This section describes recent examples of segmentation methods for initial detection of cancer. The following text is organized by cancer type, and examples of developments are presented for cancer of the breast, prostate, and brain.

As discussed previously, segmentation methods for breast cancer detection and staging in MRI studies are among the most thoroughly investigated. There are a variety of reasons for this, but the most prominent motivation is that according to

the American Cancer Society, 1 out of every 8 women (12.5%) will be diagnosed with breast cancer in their lifetime (see http://www.cancer.org/Cancer/BreastCancer/OverviewGuide/breast-cancer-overview-key-statistics). This has led to aggressive screening programs that include X-ray mammography as the primary screening technique with breast MRI as a follow-up screening tool. Radiologists are in need of computer-aided diagnostic tools to help them accurately and efficiently read both the mammogram and breast MRI screening studies. Although the majority of segmentation method development has been for mammograms, the more recent developments are targeting breast MRI studies.

A few recent developments include breast lesion detection via neural networks in breast MRI (Ertas et al. 2008). This study included 19 women and 39 marked lesions, which are used as ground truth to evaluate algorithm performance. By adjusting the algorithm's parameters to be sure that no tumors went undetected, the authors report a sensitivity of 100%. As usual, however, tuning an algorithm to be this sensitive comes with a corresponding price of a relatively low specificity. The authors report a 10% false detection rate per slice, which is problematic in that it can potentially lead to a large number of unnecessary biopsies. Woods et al. (2007) also report on the use of MRI breast lesion detection using a neural network classifier based on texture features. The authors report that their results compared with human segmentation and reported a tumor segmentation confidence of 95%. Texture is a commonly used differentiator between healthy and tumor breast tissue, and another example of that is the use of gray-level co-occurrence matrices to calculate breast tissue texture followed by an FCM pixel clustering approach to segment 3D breast tumors in contrast enhanced MRI (Chen et al. 2007).

Prostate cancer is also a popular target for computer-aided diagnosis due to the fact published by the American Cancer Society that about 1 in every 6 men will be diagnosed with prostate cancer in their lifetime (see http://www.cancer.org/Cancer/ProstateCancer/OverviewGuide/prostate-cancer-overview-key-statistics). A few specific and recent examples of this will be presented here, but a more thorough historical review of computer-aided diagnostic tools for prostate cancer, including automated segmentation technologies, can be found in Zhu, Williams, and Zwiggelaar (2006). Automated detection of prostatic adenocarcinoma in MRI scans using voxel-based Bayesian classification based on 3D texture features is presented in Madabhushi et al. (2005). Their results showed equal or better performance on cancer detection as compared to experts on a case study that included 33 patients. An example of another approach is present in Liu et al. (2009), where prostate cancer segmentation is performed using an unsupervised classification strategy based on fuzzy Markov random fields. The results report 89% specificity and 87% sensitivity achieved on full multispectral MRI feature set on a small 11-patient data set.

As will be shown in the following section, there is a large research effort to develop tools for brain tumor staging and monitoring of tumor volume during therapy. One interesting example of brain tumor detection in MRI is presented in Luts et al. (2009), where a supervised pixel classification technique is used to generate a nosologic image (pseudocolored image showing healthy versus tumor tissue) that aids the physician in the decision-making process.

17.4.2 Cancer Staging

This section presents recent examples of segmentation method development for the staging of cancer in three specific areas: breast, nasopharyngeal, and brain. Because of the high sensitivity of multiple organs in the head/neck area, precise location and quantification of tumor volume is required for nasopharyngeal cases during treatment. Finally, accurate knowledge of brain tumor size and volume is critical as it is used for surgical and nonsurgical (e.g., chemotherapy, radiation) treatment planning.

For breast cancer staging, a 3D region growing approach followed by morphological operators is employed to separate breast tumors from surrounding structures in dynamic contrast-enhanced MRI (Behrens et al. 2007). One requirement of a region growing approach is the placement of a seed point somewhere in the tumor, and this is usually done manually by an expert. The requirement to place a seed point within the tumor renders region growing an ineffective segmentation strategy for cancer detection. For tumor staging, however, this is not typically a disadvantage because the location of the tumor is already known and a seed point can be easily placed by the operator.

For staging of nasopharyngeal carcinoma (NPC), a technique for volume estimation of tumors is presented in Zhou et al. (2003). A substantial challenge presented by these types of tumors is the infiltrative growth patterns and irregular shape of tumors, so they do not lend themselves well to top-down or model-based approaches. Consequently, most research requires that data-driven methods be employed. The presented study consisted of 10 patients that had MRI scans with and without a gadolinium-based contrast. An FCM clustering approach was

implemented followed by refinement of the tumor filtering by using prior knowledge of tumor asymmetry and morphological refinement (i.e., hole-filling via dilation). The authors report an 82%–94% match based on comparison to an expert radiologist. Another segmentation technique is presented in Zhang et al. (2004), where a one-class support vector machine learns the nonlinear distribution of NPC tumor pixel data without the use of any prior knowledge. A range of 86% to 94% match was reported by comparing the automated tumor segmentation results to manual results captured by an expert radiologist.

There has been a large investigative effort of segmentation techniques for brain cancer staging. Table 17.1 shows a variety of brain tumor segmentation strategies, which has been republished here from Corso et al. (2008). This paper also presents a segmentation algorithm that focuses on glioblastoma multiforme (GBM) tumors. It is based on a multiscale Bayesian approach to perform tissue classification (tumor versus nontumor). The method's performance is also included in the comparison table. The paper acknowledges that although much progress has been made in the last decade on automated tools for tumor staging, very few are clinic ready.

Several algorithms based on active contours and level sets are being developed. An algorithm for brain tumor segmentation is presented in Wang, Cheng, and Basu (2009) that reports a new active contour approach that improves capture range (i.e., less sensitivity to initial contour placement) and convergence properties, which have been noted as two of the main challenges associated with the automated use of active contours. The paper presents a study using image data from the Internet brain segmentation repository (http://www.cma.mgh.harvard.edu/ibsr/), which can be a useful data set for testing other segmentation strategies as well.

Segmentation strategies based on level sets are also commonly used for brain tumors. As noted earlier in the chapter, level sets require initialization, and to converge, they need a proper speed

TABLE 17.1 Summary of Related Methods in Automatic Brain Tumor Segmentation

Authors	Description	Type	No. of Cases	Accuracy	Time (min)
(Liu et al. 2005)	Fuzzy clustering (semiautomatic)	GBM	5	99%	16
(Phillips et al. 1995)	Fuzzy clustering	GBM	1	N/A	N/A
(Clark et al. 1994)	Knowledge-based fuzzy clustering	GBM	7	70%	N/A
(Fletcher-Heath et al. 2001)	Knowledge-based fuzzy clustering	NE	6	53%–90%	N/A
(Karayiannis and Pai 1999)	Fuzzy clustering (VQ)	MG	1	N/A	N/A
(Prastawa et al. 2004)	Knowledge-based/outlier detection	GBM	4	68%–80%	90
(Prastawa et al. 2003)	Statistical classification via EM	GBM	5	49%–71%	100
(Kaus et al. 2001)	Statistical classification with atlas prior	LGG, MG	20	99%	10
(Vinitski et al. 1999)	k-NN	N/A	9	N/A	2
(Ho, Bullitt, and Gerig 2002)	3D level sets	GBM	3	85%–93%	N/A
(Lee, Schmidt et al. 2005)	Discriminative random fields and SVM	GBM, AST	7	40%–89%	N/A
(Peck et al. 1996)	Eigenimage analysis	N/A	10	N/A	N/A
(Zhu and Yan 1997)	Hopfield neural network and active contours	N/A	2	N/A	N/A
(Zhang et al. 2004)	SVM	N/A	9	60%–87%	N/A
(Corso et al. 2008)	Multilevel Bayesian segmentation	GBM	20	27%–88%	7

Source: Corso, J. et al. *IEEE Trans. Med. Imaging*, 27(5), 629–640, 2008. With permission.

Note: Abbreviations are glioblastoma multiforme (GBM), astrocytoma (AST), nonenhancing (NE), low-grade glioma (LGG), meningioma (MG), not applicable (N/A), vector quantization (VQ), expectation maximization (EM).

function, which defines the direction and the distance that the boundary will move at each iteration. To overcome some of the challenges with setting up the speed function, a method that includes a dynamically updated speed function is presented in Taheri, Ong, and Chong (2009). Typically, level sets rely on boundary information for convergence, but an approach that includes both region and boundary information simultaneously (i.e., "hybrid level set") is reported in Xie et al. (2005). The study consists of 10 patients with 246 brain tumors and reports a percent match (PM) range of 79%–93% based on comparison to a radiologist's manual segmentation. The PM is defined as PM = TPs/GT*100, where TPs is the number of true positive pixels found by the algorithm and GT is the total number of ground truth pixels as defined by the radiologist.

An advantage of MRI is the clinician's ability to perform a variety of scans to generate images with various contrast (e.g., T_1, T_2, and fluid-attenuated inversion recovery) between tissue types. The multiple scans considered simultaneously can be viewed as a multispectral MRI study. A typical MRI session will include both high and low spatial resolution images. This difference in resolution presents challenges for algorithms that wish to make use of multispectral MRI studies for tumor segmentation. An illustrative example of an algorithm that presents a strategy for making use of multiple scans with varying spatial resolution is presented in Nie et al. (2009), where hidden Markov random fields are used that include spatial accuracy weighting terms for each of the different resolution MRI scans. The algorithm was applied to malignant glioma brain tumor segmentation and results were comparable to manually segmented tumor results by multiple subject matter experts.

This section will conclude by considering the problem of performance evaluation. All of the approaches presented thus far evaluate the algorithm's performance by comparing the automated segmentation result to the manual segmentation result provided by an expert. There are two significant problems with this. First, manual segmentation is prone to inconsistency, even when performed by experts. The second and more significant problem is that accurate "ground truth" data are often very hard to come by for many specific applications of interest. In an attempt to address this problem, a physical and statistical modeling system was developed to generate synthetic multimodal 3D brain MRI with both tumor and edema in Prastawa, Bullitt, and Gerig (2009). The system also generates the underlying anatomical ground truth through the use of existing MRI brain scans so that realistic cases can be generated for algorithm evaluation. Several examples of images from this system are shown in Figure 17.11.

17.5 Applications in Clinical Cancer Therapy

MRI is currently the only noninvasive technique for providing 2D and/or 3D anatomical and physiological tissue assessments. MRI has become a strong asset in cancer therapy due to the increased efficiency and accuracy in performing quantitative

and qualitative assessment of normal and abnormal tissues for cancer treatment planning and therapy monitoring.

Automated image processing detection techniques aim at finding features characteristic to a given problem and do not intend to replace trained personnel in locating cancerous lesions in MRI. However, by selecting algorithms suitable to a specific problem, one can provide efficient complementary tools to assist clinicians in cumbersome activities such as locating and delineating the tissue regions present in the images. In the following, a variety of examples are provided to illustrate how region-, model-, or contour-based image processing algorithms can contribute in cancer treatment planning.

17.5.1 Data-Driven Techniques

MRI of the brain that displays distinct regions of white and gray matter will have unique pixel intensities. In the case of brain tumors, treatment planning requires a thorough quantitative and qualitative evaluation of MRI slices. One needs to (1) obtain the cranial and brain volume measurements, (2) localize gray and white and encephalic liquid as well, and (3) identify and localize potential lesions. To perform this quantitative analysis, Ferreira da Silva (2007) uses a statistical approach for pixel classification of brain MRI. Usually performed using standard parametric methods and Gaussian mixture models, the author identifies a limitation and lack of performance of those approaches in the case of an increasing number of parts including presence of abnormal tissues. Moving toward a nonparametric model, the author presents a Bayesian approach based on the Dirichlet mixture model. An important feature of this model is that inference on the number of components in the mixture is incorporated, a Markov random field provides some noise immunity as well. Refer to Figure 17.12 for an example result.

A different approach presented in Mazzara et al. (2004) applies supervised approaches such as k-NN (see Section 17.2.1.3.1) and automatic knowledge-guided segmentation methods for brain MRI analysis. As expected, these methods performed well in segmenting abnormal tissues characterized by significantly different texture characteristics compared to the surroundings. However, the segmentation accuracy was only reported to be 75% at best (as compared to expert manual results), so it appears difficult to delineate precisely the entire tumor with these techniques. In this context, their method can at least trigger an alarm regarding the presence of a suspicious lesion, but cannot be considered as an accurate quantitative approach for the reported cases. An extension to this work, Beyer et al. (2006) investigate the combination of both techniques with regard to treatment selection as compared to the physician's decision. Castellani et al. (2008) introduce a software package that performs data clustering using unsupervised classification method called *MRI-mean shift*. This software package aims at finding tumors in a variety of soft tissues. In this research the concept of pixel is seen as a vector of features. Thus, clustering depends on multidimensional and multispectral characteristics. Vaidyanathan et al. (1997) present a 3D SFCM clustering method for monitoring

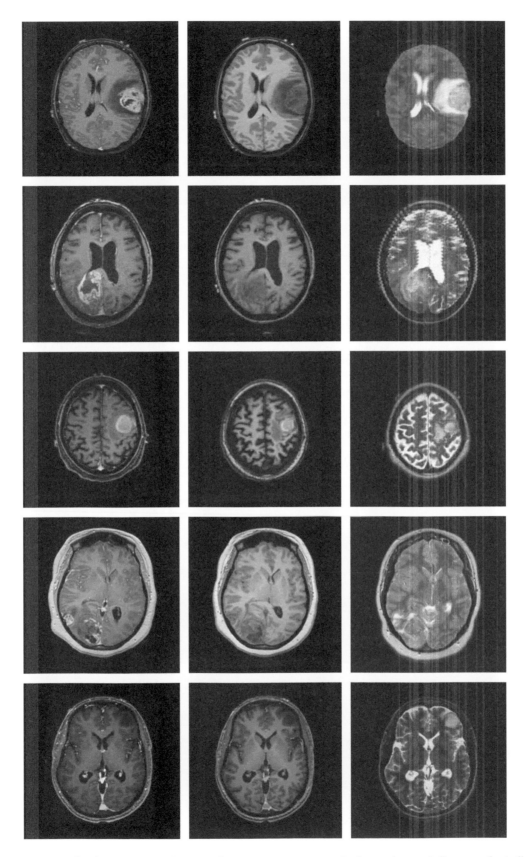

FIGURE 17.11 Axial views of real MR images with varying brain tumor appearances. Each row shows a different synthetically created tumor. From left to right: contrast-enhanced T_1-, T_1-, and T_2-weighted images. (Reprinted from Prastawa, M. et al., *Med. Image Anal.*, 8 (3), 275–283, 2009. With permission.)

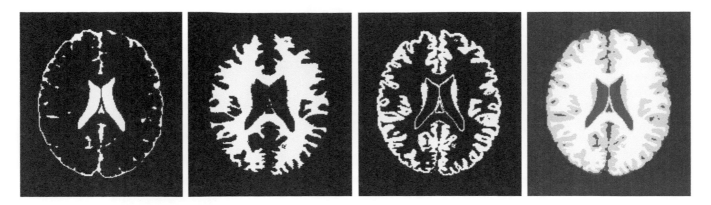

FIGURE 17.12 Clustering results obtained with (from left to right) cerebrospinal fluid, white matter, and gray matter, reconstructed image with the estimated tissue types. (Reprinted from Ferreira da Silva, A. R., *Med. Image Anal.*, 11 (2), 169–182, 2007. With permission.)

brain tumor volume changes during the course of routine clinical radiation-therapeutic and chemotherapeutic regimens.

17.5.2 Hybrid Techniques

Several example cancer therapy applications have been developed using hybrid segmentation approaches. Pasquier et al. (2007) evaluate a 3D-deformable model and seed-based region growing approaches to segment prostate and organs-at-risk in the pelvic region. The authors investigate the performance/time ratio of these automated techniques against manual delineation. They note performance results in terms of a volume overlap

measurement, V_o, which is defined as the ratio of the intersection to the union of the automatic and manual segmentation volume results (optimal value is 1.0 where automatic segmentation is exactly the same as manual). For the rectum and bladder, the V_o values were 0.78 and 0.88, respectively. In addition, Duncan et al. (2004) investigate the segmentation of cortical and subcortical structures and 3D matching algorithms in order to quantify the variations in the human brain as a sign of neuropsychiatric disorders. Simultaneous segmentation of both cortical surfaces is performed using a coupled level-set approach. An illustration of a result using this approach is shown in Figure 17.13. Tsai et al. (2003) applied level sets to prostate segmentation. This 3D

(a)   (b)   (c)

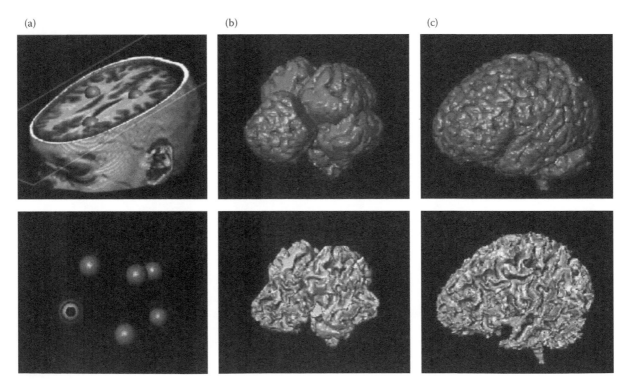

FIGURE 17.13 Results of cortical gray-matter segmentation using coupled level sets. (a) Initialization of pairs of concentric spheres in 3D MR brain images, (b) intermediate step, (c) final result of the outer (top) and inner (bottom) cortical surfaces of the frontal lobe. (Reprinted from Duncan, J. S. et al., *NeuroImage*, 23 (Suppl 1), S34–S45, 2004. With permission.)

approach uses a priori knowledge of the search pattern to initialize a level-set algorithm. During a learning phase, a set of user-segmented ROIs are aligned and used to retrieve shape features. Then a statistical shape model approach segments the position of the ROI in the image without user input.

Huyskens et al. (2009) present a qualitative study on tissue delineation performed by one commercial software for automated segmentation of CT scans for prostate cancer treatment planning and ground truth data provided by medical professionals. Segmentation results were scored using classical similarity metric algorithms, showing (1) that the evaluated algorithm was performing inconsistently along the CT scan and (2) that the expert opinions are subjective on evaluating the same case. For example, for the automatic segmentation of the rectum on 44 total patients, 54% of the patient results were considered to be "acceptable," "good," or "excellent," while 45% were considered "not acceptable."

Pallotta et al. (2006) evaluate three segmentation algorithms (thresholding, region growing, and isocontour) and one registration technique (surface matching) for prostate cancer treatment planning. The authors apply these algorithms on a phantom and a patient data set, and use metrics based on a comparison of the total number of pixels within the segmented ROI (rectum, bladder, prostate, etc.) found by the expert versus the ROI found by each of the three algorithms. Pixel differences between the two range from 5.2% down to 0.2%, showing good agreement between the two techniques. Lee, Yeung et al. (2005) present a segmentation algorithm combining morphological operators, Bayesian classification, and histogram analysis applied to NPC. Their approach is designed to be flexible so that segmentation results (details of which are left to the paper) can be easily refined by tweaking the input parameters. Wu et al. (2008) present a modified version of the traditional active contour model for the specific problem of cervical tumor localization. Quantitative assessment of the delineated area based on the Jaccard similarity metric (J_s) (Jaccard 1901) provides information of the treatment progress. Häme (2008) details a method for semiautomated liver tumor segmentation. At first, the image is thresholded to extract the position of the liver, and then morphological operators are applied to extract the tumor candidates. FCM clustering is used to refine the tumor region. Finally, a deformable model is used to extract the tumor border.

17.5.3 Model-Based Approaches for Cancer Therapy

Bondiau et al. (2005) present an atlas-based method consisting of a learning set of cranial slices where the brain structures are identified and stored in an associated file. By mapping new MRI slices to the learning set and applying deformable constraints to the different labeled regions, it is then possible to perform a block matching of both data sets and then to segment and to label simultaneously multiple areas of the input data. The performance evaluation against ground truth expertise indicates acceptable segmentation results that could be applied to a variety of organs, but points out also that a lack of abnormal tissue

data in the learning set implies difficulties in segmenting tumor. Solomon, Butman, and Sood (2006) present the combination of a spatiotemporal model and a Markov model to perform four-dimensional brain tumor segmentation. This approach analyzes consecutive 3D segmentation results from an MRI series to identify the brain tumor position, and then uses temporal information to refine the segmentation.

17.6 Conclusion

This chapter has provided an overview of image segmentation technology as applied to cancer within images acquired via MRI. The superior soft-tissue contrast and high resolution provided by MRI makes it one of the most effective diagnostic techniques for detecting, staging, and therapy-monitoring for a variety of cancer types. Segmentation strategies range from data-driven (bottom-up) methods to model-driven (top-down) methods, and also include a host of techniques that are combinations of both (hybrid techniques). For cancer-specific applications, the most commonly researched and employed techniques include various types of pixel-based classification (e.g., Bayesian decision theory, clustering) and deformable models (e.g., active contours, level sets). Other approaches gaining popularity for brain segmentation are model-based strategies such as atlas-based segmentation. There is a wealth of research being published in each of these categories, and the purpose of this chapter was to introduce the reader to the most promising techniques so they can (1) choose a type of segmentation strategy most applicable to their cancer imaging research and (2) use the information as a starting point for implementing their own segmentation techniques that build upon the wealth of research being performed.

References

Aljabar, P., R. A. Heckemann, A. Hammers, J. V. Hajnal, and D. Rueckert. 2009. Multi-atlas based segmentation of brain images: Atlas selection and its effect on accuracy. *NeuroImage* 46 (3):726–738.

Artaechevarria, X., A. Munoz-Barrutia, and C. Oritz-de-Solorzano. 2009. Combination strategies in multi-atlas image segmentation: Application to brain MR Data. *IEEE Transactions on Medical Imaging* 28 (8):1266–1277.

Atkins, M. S., and B. T. Mackiewich. 1998. Fully automatic segmentation of the brain in MRI. *IEEE Transactions on Medical Imaging* 17 (1):98–107.

Aubert, G., and L. Vese. 1997. A variational method in image recovery. *SIAM Journal on Numerical Analysis* 34 (5):1948–1979.

Baboi, L., F. Pilleul, L. Milot, C. Lartizien, G. Poncet, C. Roche, J.-Y. Scoazec, and O. Beuf. 2010. Magnetic resonance imaging follow-up of liver growth of neuroendocrine tumors in an experimental mouse model. *Magnetic Resonance Imaging* 28 (2):264–272.

Ball, G., and D. Hall. 1965. ISODATA: A novel method of data analysis and pattern classification. *Transactions on Pattern Analysis and Machine Intelligence* 13:583–598.

Barcke, K., B. Willis, J. Ross, D. French, E. Filvaroff, and R. Carano. 2009. Viable tumor tissue detection in murine metastatic breast cancer by whole-body MRI and multispectral analysis. *Magnetic Resonance in Medicine* 62 (6):1423–1430.

Behrens, S., H. Laue, M. Althaus, T. Boehler, B. Kuemmerlen, H. K. Hahn, and H.-O. Peitgen. 2007. Computer assistance for MR based diagnosis of breast cancer: Present and future challenges. *Computerized Medical Imaging and Graphics* 31 (4–5):236–247.

Beyer, G., R. Velthuizen, F. Murtagh, and J. Pearlman. 2006. Technical aspects and evaluation methodology for the application of two automated brain MRI tumor segmentation methods in radiation therapy planning. *Magnetic Resonance Imaging* 24 (9):1167–1178.

Bondiau, P.-Y., G. Malandain, S. Chanalet, P.-Y. Marcy, J.-L. Habrand, F. Fauchon, P. Paquis, A. Courdi, O. Commowick, I. Rutten, and N. Ayache. 2005. Atlas-based automatic segmentation of MR images: Validation study on the brainstem in radiotherapy context. *International Journal of Radiation Oncology, Biology, Physics* 61 (1):289–298.

Brekke, C., S. C. Williams, J. Price, F. Thorsen, and M. Modo. 2007. Cellular multiparametric MRI of neural stem cell therapy in a rat glioma model. *NeuroImage* 37 (3):769–782.

Casselles, V., F. Catte, T. Coll, and F. Dibos. 1993. A geometric model for active contours in image processing. *Numerische Mathematique* 66:7–37.

Casselles, V., R. Kimmel, and G. Sapiro. 1995. Geodesic active contours. Paper read at IEEE International Conference on Computer Vision.

Casselles, V. 1997. Geodesic active contours. *International Journal of Computer Vision* 22 (1):61–79.

Castellani, U., M. Cristani, C. Combi, V. Murino, A. Sbarbati, and P. Marzola. 2008. Visual MRI: Merging information visualization and non-parametric clustering techniques for MRI dataset analysis. *Artificial Intelligence in Medicine* 44 (3):183–199.

Chan, T., and A. Vese. 2001. Active contours without edges. Paper read at IEEE Transactions on Image Processing.

Chao, W.-H., Y.-Y. Chen, C.-W. Cho, S.-H. Lin, Y.-Y. I. Shih, and S. Tsang. 2008. Improving segmentation accuracy for magnetic resonance imaging using a boosted decision tree. *Journal of Neuroscience Methods* 175 (2):206–217.

Chao, W.-H., Y.-Y. Chen, S.-H. Lin, Y.-Y. I. Shih, and S. Tsang. 2009. Automatic segmentation of magnetic resonance images using a decision tree with spatial information. *Computerized Medical Imaging and Graphics* 33 (2):111–121.

Chen, W., M. Giger, H. Li, U. Bick, and G. Newstead. 2007. Volumetric texture analysis of breast lesions on contrast-enhanced magnetic resonance images. *Magnetic Resonance in Medicine* 58 (3):562–571.

Ciofolo, C., and C. Barillot. 2009. Atlas-based segmentation of 3D cerebral structures with competitive level sets and fuzzy control. *Medical Image Analysis* 13 (3):456–470.

Clark, K. A., R. P. Woods, D. A. Rottenberg, A. W. Toga, and J. C. Mazziotta. 2006. Impact of acquisition protocols and processing streams on tissue segmentation of T1 weighted MR images. *NeuroImage* 29 (1):185–202.

Clark, M. C., L. O. Hall, D. B. Goldgof, L. P. Clarke, R. P. Velthuizen, and M. S. Silbiger. 1994. MRI segmentation using fuzzy clustering techniques. *Engineering in Medicine and Biology Magazine, IEEE* 13 (5):730–742.

Colliot, O., O. Camara, and I. Bloch. 2006. Integration of fuzzy spatial relations in deformable models—Application to brain MRI segmentation. *Pattern Recognition* 39 (8):1401–1414.

Cootes, T., G. Edwards, and C. Taylor. 1998. Active appearance models. Paper read at Proceedings of European Conference on Computer Vision.

Cootes, T., C. Taylor, D. Cooper, and J. Graham. 1995. Active shape models—Their training and application. *Computer Vision and Image Understanding* 61 (1):38–59.

Corso, J., E. Sharon, S. Dube, S. El-Saden, U. Sinha, and A. Yuille. 2008. Efficient multilevel brain tumor segmentation with integrated Bayesian model classification. *IEEE Transactions on Medical Imaging* 27 (5):629–640.

Dastidar, P., J. Mäenpää, T. Heinonen, T. Kuoppala, M. Van Meer, R. Punnonen, and E. Laasonen. 2000. Magnetic resonance imaging based volume estimation of ovarian tumours: Use of a segmentation and 3D reformation software. *Computers in Biology and Medicine* 30 (6):329–340.

DeCarlo, D., and D. Metaxas. 1998. Shape evolution with structural and topological changes using blending. *IEEE Transactions on Pattern Analysis and Machine Intelligence* 20 (11):1186–1205.

Duncan, J. S., X. Papademetris, J. Yang, M. Jackowski, X. Zeng, and L. H. Staib. 2004. Geometric strategies for neuroanatomic analysis from MRI. *NeuroImage* 23 (Suppl 1):S34–S45.

Ertas, G., H. O. Gulcur, O. Osman, O. Ucan, M. Tunac, and M. Dursun. 2008. Breast MR segmentation and lesion detection with cellular neural networks and 3D template matching. *Comput. Biol. Med.* 38 (1):116–126.

Ferreira da Silva, A. R. 2007. A Dirichlet process mixture model for brain MRI tissue classification. *Medical Image Analysis* 11 (2):169–182.

Fletcher-Heath, L., L. Hall, D. Goldgof, and F. Murtagh. 2001. Automatic segmentation of nonenhancing brain tumors in magnetic resonance images. *Artificial Intelligence in Medicine* 21:43–63.

Gleason, S., H. Sari-Sarraf, M. Paulus, D. Johnson, and M. Abidi. 1999. Deformable model-based X-ray CT image segmentation for automatic phenotype identification in laboratory mice. In *The Second Conference on Biomedical Engineering*. Vanderbilt University, Nashville, TN.

Gleason, S. 2001. Development of a unified probabilistic framework for segmentation and recognition of semi-rigid objects in complex backgrounds via deformable shape models. PhD Thesis.

Gritton, C., and E. Parrish. 1983. Boundary location from an initial plan: The bead chain algorithm. Paper read at IEEE Transactions on Pattern Analysis and Machine Intelligence.

Häme, Y. 2008. Liver tumor segmentation using implicit surface evolution. *The MIDAS Journal*. http://www.insight-journal .org/browse/publication/603.

Hamerly, G., and C. Elkan. 2003. Learning the *k* in *k*-means. In *Neural Information Processing Systems*. MIT Press, Cambridge, MA.

Heckemann, R. A., J. V. Hajnal, P. Aljabar, D. Rueckert, and A. Hammers. 2006. Automatic anatomical brain MRI segmentation combining label propagation and decision fusion. *NeuroImage* 33 (1):115–126.

Ho, S., E. Bullitt, and G. Gerig. 2002. Level set evolution with region competition: Automatic 3-D segmentation of brain. Paper read at International Conference on Pattern Recognition.

Huyskens, D. P., P. Maingon, L. Vanuytsel, V. Remouchamps, T. Roques, B. Dubray, B. Haas, P. Kunz, T. Coradi, R. Bühlman, R. Reddick, A. Van Esch, and E. Salamon. 2009. A qualitative and a quantitative analysis of an auto-segmentation module for prostate cancer. *Radiotherapy and Oncology* 90 (3):337–345.

Jaccard, P. 1901. Étude Comparative De La Distribution Florale Dans Une Portion Des Alpes Et Des Jura (Comparative study of the floral distribution in a portion of the Alpes and Jura regions). *Bulletin de la Société Vaudoise des Sciences Naturelles* 37:547–79.

Karayiannis, N., and P.-I. Pai. 1999. Segmentation of magnetic resonance images using fuzzy algorithms from learning vector quantization. *IEEE Transactions on Medical Imaging* 18 (2):172–180.

Kass, M., A. Witkin, and D. Terzopoulos. 1988. Snakes: Active contour models. *International Journal of Computer Vision* 1:321–331.

Kaus, M., S. Warfield, A. Nabavi, P. Black, F. Jolesz, and R. Kikinis. 2001. Automated segmentation of MRI of brain tumors. *Radiology* 218:586–591.

Kelemen, A., G. Szekely, and G. Gerig. 1998. Three-dimensional model-based segmentation. Paper read at SPIE Workshop on Biomedical Image Analysis.

Klifa, C., J. Carballido-Gamio, L. Wilmes, A. Laprie, J. Shepherd, J. Gibbs, B. Fan, S. Noworolski, and N. Hylton. 2010. Magnetic resonance imaging for secondary assessment of breast density in a high-risk cohort. *Magnetic Resonance Imaging* 28 (1):8–15.

Lee, F., M. Schmidt, A. Murtha, A. Bistritz, J. Sander, and R. Greiner. 2005. Segmenting brain tumor with conditional random fields and support vector machines. Paper read at Proceedings of Workshop on Computer Vision for Biomedical Image Application at International Conference on Computer Vision.

Lee, F., D. Yeung, A. King, S. Leung, and A. Ahuja. 2005. Segmentation of nasopharyngeal carcinoma (NPC) lesions in MR images. *International Journal of Radiation Oncology, Biology, Physics* 61 (2):608–620.

Lee, S. H., J. H. Kim, K. G. Kim, J. S. Park, S. J. Park, and W. K. Moon. 2007. Optimal Clustering of kinetic patterns on malignant breast lesions: Comparison between *k*-means clustering and three-time-points method in dynamic contrast-enhanced MRI. Paper read at 29th Annual International Conference of the IEEE Engineering in Medicine and Biology Society at Lyon, France.

Liu, J., J. Udupa, D. Odhner, D. Hackney, and G. Moonis. 2005. A system for brain tumor volume estimation via MR imaging and fuzzy connectedness. *Computerized Medical Imaging and Graphics* 29 (1):21–34.

Liu, X., D. Langer, M. Haider, Y. Yang, M. Wernick, and I. Yetik. 2009. Prostate cancer segmentation with simultaneous estimation of Markov random field parameters and class. *IEEE Transactions on Medical Imaging* 28 (6):906–915.

Lötjönen, J. M. P., R. Wolz, J. R. Koikkalainen, L. Thurfjell, G. Waldemar, H. Soininen, and D. Rueckert. 2010. Fast and robust multi-atlas segmentation of brain magnetic resonance images. *NeuroImage* 49 (3):2352–2365.

Luts, J., T. Laudadio, A. Idema, A. Simonetti, A. Heerschap, D. Vandermeulen, J. Suykens, and S. Huffel. 2009. Nosologic imaging of the brain: Segmentation and classification using MRI and MRSI. *NMR in Biomedicine* 22 (4):374–390.

Macqueen, J. B. 1967. Some methods of classification and analysis of multivariate observations. Paper read at Proceedings of the Fifth Berkeley Symposium on Mathematical Statistics and Probability.

Madabhushi, A., M. Feldman, D. Metaxas, J. Tomasezweski, and D. Chute. 2005. Automated detection of prostatic adenocarcinoma from high resolution ex vivo MRI. *IEEE Transactions on Medical Imaging* 24 (12):1611–1625.

Malladi, R., J. Sethian, and B. Vemuri. 1995. Shape modeling with front propagation: A level set approach. Paper read at IEEE Pattern Analysis and Machine Intelligence.

Mazzara, G., R. Velthuizen, J. Pearlman, H. Greenberg, and H. Wagner. 2004. Brain tumor target volume determination for radiation treatment planning through automated MRI segmentation. *International Journal of Radiation Oncology, Biology, Physics* 59 (1):300–312.

McInerney, T., and D. Terzopoulos. 1997. Medical image segmentation using topologically adaptable surfaces. Paper read at Proceedings of CVRMed, Grenoble, France.

Mumford, D., and J. Shah. 1989. Optimal approximations by piecewise smooth functions and associated variational problems. *Communications on Pure and Applied Mathematics* 42 (5):577–685.

Nie, J., Z. Xue, T. Liu, G. S. Young, K. Setayesh, L. Guo, and S. T. C. Wong. 2009. Automated brain tumor segmentation using spatial accuracy-weighted hidden Markov random field. *Computerized Medical Imaging and Graphics* 33 (6):431–441.

Nieman, B., J. Bishop, J. Dazai, N. Bock, J. Lerch, A. Feintuch, X. Chen, J. Sled, and R. Henkelman. 2007. MR technology for biological studies in mice. *NMR in Biomedicine* 20 (3):291–303.

O'Donnell, T., M. Dubuisson-Jolly, and A. Gupta. 1998. Cooperative framework for segmentation using 2D active contours and 3D hybrid models as applied to branching cylindrical structures. Paper read at International Conference on Computer Vision.

Osher, S., and J. A. Sethian. 1988. Fronts propagating with curvature-dependent speed: Algorithms based on Hamilton–Jacobi formulations. *Journal of Computational Physics* 79 (1):12–49.

Pagel, M. D., S. J. Baldwin, R. K. Rader, and J. J. Kotyk. 2006. Assessment of anti-metastatic drug efficacy via localization and quantification of ex vivo murine bone tumor load using high-throughput MRI T1 parametric analysis. *NMR in Biomedicine* 19 (1):1–9.

Pallotta, S., M. Bucciolini, S. Russo, T. Cinzia, and G. Biti. 2006. Accuracy evaluation of image registration and segmentation tools used in conformal treatment planning of prostate cancer. *Computerized Medical Imaging and Graphics* 30 (1):1–7.

Pasquier, D., T. Lacornerie, M. Vermandel, J. Rousseau, E. Lartigau, and N. Betrouni. 2007. Automatic segmentation of pelvic structures from magnetic resonance images for prostate cancer radiotherapy. *International Journal of Radiation Oncology, Biology, Physics* 68 (2):592–600.

Peck, D., J. Windham, L. Emery, H. Soltanian-Zadeh, D. Hearshen, and T. Mikkelsen. 1996. Cerebral tumor volume calculations using planimetric and eigenimage analysis. *Medical Physics* 23 (12):2035–2042.

Phillips, W., R. Velthuizen, L. Phupanich, L. Hall, L. Clarke, and M. Silbiger. 1995. Application of fuzzy c-means segmentation technique for tissue differentiation in MR images of a hemorrhagic glioblastoma multiforme. *Magnetic Resonance Imaging* 13 (2):277–290.

Prastawa, M. E. Bullitt, N. Bullitt, K. Leemput, and G. Gerig. 2003. Automatic brain tumor segmentation by subject specific modification of atlas priors. *Academic Radiology* 10:1341–1348.

Prastawa, M., E. Bullitt, S. Ho, and G. Gerig. 2004. A brain tumor segmentation framework based on outlier detection. *Medical Image Analysis* 8 (3):275–283.

Prastawa, M., E. Bullitt, and G. Gerig. 2009. Simulation of brain tumors in MR images for evaluation of segmentation efficacy. *Medical Image Analysis* 13 (2):297–311.

Rodionov, R., M. Chupin, E. Williams, A. Hammers, C. Kesavadas, and L. Lemieux. 2009. Evaluation of atlas-based segmentation of hippocampi in healthy humans. *Magnetic Resonance Imaging* 27 (8):1104–1109.

Sapiro, G. 1996. Vector-valued active contours. Paper read at IEEE Computer Vision and Pattern Recognition.

Sims, R., A. Isambert, V. Grégoire, F. Bidault, L. Fresco, J. Sage, J. Mills, J. Bourhis, D. Lefkopoulos, O. Commowick, M. Benkebil, and G. Malandain. 2009. A pre-clinical assessment of an atlas-based automatic segmentation tool for the head and neck. *Radiotherapy and Oncology* 93 (3):474–478.

Singh, M., P. Patel, D. Khosla, and T. Kim. 1996. Segmentation of functional MRI by *k*-means clustering. Paper read at IEEE Transactions on Nuclear Science.

Solomon, J., J. A. Butman, and A. Sood. 2006. Segmentation of brain tumors in 4D MR images using the hidden Markov model. *Computer Methods and Programs in Biomedicine* 84 (2–3):76–85.

Soltanian-Zadeh, H., H. Bagher-Ebadian, J. Ewing, P. Mitsias, A. Kapke, M. Lu, Q. Jiang, S. Patel, and M. Chopp. 2007. Multiparametric iterative self-organizing data analysis of ischemic lesions using pre- or post-Gd T1 MRI. *Cerebrovascular Diseases* 23 (2–3):91–102.

Taheri, S., S. H. Ong, and V. F. H. Chong. 2009. Level-set segmentation of brain tumors using a threshold-based speed function. *Image and Vision Computing* 28 (1):26–37.

Tsai, A., A. Yezzi, W. Wells, C. Tempany, D. Tucker, A. Fan, E. Grimson, and A. Willsky. 2003. A shape-based approach to the segmentation of medical imagery using level sets. *IEEE Transactions on Medical Imaging* 22 (2):137–154.

Uberti, M. G., M. D. Boska, and Y. Liu. 2009. A semi-automatic image segmentation method for extraction of brain volume from in vivo mouse head magnetic resonance imaging using constraint level sets. *Journal of Neuroscience Methods* 179 (2):338–344.

Vaidyanathan, M., L. P. Clarke, C. Heidtman, R. P. Velthuizen, and L. O. Hall. 1997. Normal brain volume measurements using multispectral MRI segmentation. *Magnetic Resonance Imaging* 15 (1):87–97.

Vapnik, V. 1996. The nature of statistical learning theory. In *Computational Learning and Probabilistic Reasoning*, edited by A. Gammerman. John Wiley and Sons.

Vinitski, S., C. Gonzalez, R. Knobler, D. Andrews, T. Iwanaga, and M. Curtis. 1999. Fast tissue segmentation based on a 4D feature map in characterization of intracranial lesions. *Journal of Magnetic Resonance Imaging* 9 (6):768–776.

Wang, T., I. Cheng, and A. Basu. 2009. Fluid vector flow and applications in brain tumor segmentation. *IEEE Transactions on Biomedical Engineering* 56 (3):781–789.

Wang, Z., and B. Vemuri. 2005. DTI segmentation using an information theoretic tensor dissimilarity measure. *IEEE Transactions on Medical Imaging* 24 (10): 1267–1277.

Weldeselassie, Y., and G. Hamarneh. 2007. DT-MRI segmentation using graph cuts. *Progress in Biomedical Optics and Imaging* 8 (31):65121K.1–65121K.9.

Whitaker, R. 1998. A level-set approach to 3D reconstruction from range data. *International Journal of Computer Vision* 29 (3):203–231.

Widrow, B. 1973a. The "rubber-mask" technique—I. Pattern measurement and analysis. *Pattern Recognition* 5:175–197.

Widrow, B. 1973b. The "rubber-mask" technique—II. Pattern storage and recognition. *Pattern Recognition* 5:199–211.

Woods, B., B. Clymer, T. Kurc, J. Heverhagen, R. Stevens, A. Orsdemir, O. Bulan, and M. Knopp. 2007. Malignant-lesion segmentation using 4D co-occurrence texture analysis applied to dynamic contrast-enhanced magnetic resonance breast image data. *Journal of Magnetic Resonance Imaging* 25 (3):495–501.

Wu, D. H., A. D. Shaffer, D. M. Thompson, Z. Yang, V. A. Magnotta, R. Alam, J. Suri, W. T. C. Yuh, and N. A. Mayr.

2008. Iterative active deformational methodology for tumor delineation: Evaluation across radiation treatment stage and volume. *Journal of Magnetic Resonance Imaging* 28 (5): 1188–1194.

Xie, K., J. Yang, Z. G. Zhang, and Y. M. Zhu. 2005. Semi-automated brain tumor and edema segmentation using MRI. *European Journal of Radiology* 56 (1):12–19.

Zhang, J., K.-K. Ma, M. Er, and V. Chong. 2004. Tumor segmentation from magnetic resonance imaging by learning via one-class support vector machine. Paper read at Proceedings of the 2004 International Conference of Intellectual Mechatronics Automation at Chengdu, China.

Zhou, J., T.-K. Lim, V. Chong, and J. Huang. 2003. Segmentation and visualization of nasopharyngeal carcinoma using MRI. *Computers in Biology and Medicine* 33 (5):407–424.

Zhu, Y., and H. Yan. 1997. Computerized tumor boundary detection using a Hopfield neural network. *IEEE Transactions on Medical Imaging* 16 (1):55–67.

Zhu, Y., S. Williams, and R. Zwiggelaar. 2006. Computer technology in detection and staging of prostate carcinoma: A review. *Medical Image Analysis* 10 (2):178–199.

Zhukov, L., K. Museth, D. Breen, R. Whitaker, and A. Barr. 2003. Level set modeling and segmentation of DT-MRI brain data. *Journal of Electronic Imaging* 12 (1):125–133.

18

Spatial and Temporal Image Registration

Xia Li
Vanderbilt University

Image registration is the process of finding a transformation that aligns two images. It is widely used in many fields, including video compression, motion tracking, and medical imaging. Medical image registration is a central component in medical image analysis and has been applied widely to follow disease progression, to assess response to therapy, or to compare populations. In this chapter, the basic idea of medical image registration is introduced, and the mathematical aspects of commonly used registration techniques are explained. Applications in preclinical and clinical cancer research are also reviewed and discussed.

18.1 Qualitative Introduction to Image Registration

One application of medical image registration is the integration of information from multiple imaging modalities. Various anatomical and functional imaging modalities have been developed to support cancer-related research: magnetic resonance (MR) imaging (MRI), computed tomography (CT), positron emission tomography (PET), single photon emission CT, or ultrasound (US). Each modality possesses its own characteristics, strengths, and disadvantages. The combination of images from different modalities can provide complementary and more comprehensive information on disease state. For example, dynamic contrast-enhanced MRI (DCE-MRI, Chapter 12) reports on physiological parameters such as blood vessel perfusion and permeability, whereas PET can provide information of glucose metabolism [1]. The ability to combine such data is of great interest and is currently an area of active investigation, as the synthesis of two such data sets promises a more comprehensive characterization of tumor status. For instance, Moy et al. [2] fused MRI and PET

three-dimensional (3D) breast cancer data sets and examined the effect of multimodal registration between MRI and PET on breast cancer characterization. (Please see more details in Section 18.4.)

Medical image registration also allows for the longitudinal tracking of tumors over time, thereby permitting the assessment of tumor response to therapy. For instance, Chittineni et al. [3] registered serial breast MR cancer images obtained at different time points throughout treatment to monitor the tumor changes. Li et al. [4] scanned breast MR images before neoadjuvant chemotherapy, within 1 week of therapy, and after all cycles of therapy. By registering the longitudinal images, the parametric maps obtained from DCE-MRI were aligned. The spatial information contained in the images was thus preserved, and response to treatment could be assessed on a voxel-by-voxel basis rather than over an entire region of interest.

Registration techniques can also be used to compare multiple patients across time. For instance, Thompson et al. [5] found accelerated gray matter loss in early schizophrenia through registering and comparing two different groups over time. MR images were obtained from 12 healthy subjects and 12 patients with schizophrenia over a 5-year time interval. For each individual, MR images at different time points were aligned, and a deformation vector field, which captures the displacement necessary at each voxel to align two images, was computed. The shape changes encoded in the deformation fields were investigated for the two groups, and accelerated gray matter loss was found in the schizophrenia patients. The findings were consistent with previous studies and provided important information on schizophrenia progression.

Medical image registration can also be used to develop atlases. An atlas can be viewed as an anatomical model in which different tissues/organs are identified and classified so that it is considered as a standard reference. An atlas can also be viewed as a virtual model for a particular population. It allows investigators to estimate the variation in anatomical structures from a set of deformation fields, which encode the morphological differences between individuals or between different groups. Furthermore, atlases are useful for the purpose of automatic segmentation and recognition. Individual subjects are coregistered to an atlas in which each image voxel is classified to a specific tissue. After registration, labels assigned to voxels in the atlas can be projected onto the individual image volumes. Recently, Zacharaki et al. [6] have presented a registration method that performs the registration of MR image volumes with tumors to a common reference or atlas, therefore allowing, for group analysis and comparison, automatic segmentation of anatomical structures in the atlas, as well as estimation of brain tissue loss and tumor growth.

Medical image registration can be classified using different criteria (see references [7–13] for very good overviews). The broadest criterion is based on the imaging modalities involved; that is, registration can be divided into monomodal and multimodal registration. A second major criterion in classifying registration techniques involves whether or not multiple subjects are involved (intersubject registration) or if it involves only a single subject (intrasubject registration).

Beyond the above coarse divisions, registration algorithms are typically classified according to the geometrical transformations generated during the registration. Rigid, affine, and nonrigid are transformations commonly used in medical image registration. Only rotation and translation are involved with a rigid body registration, whereas an affine registration also includes scaling and shearing and, therefore, does not preserve the size and angles of objects, but only parallelism (parallel lines remain parallel after affine transformation). Nonrigid transformations are used when there are more significant deformations between images that need to be registered. These deformations can be due, for example, to properties of different tissues, tumor compression or expansion, interventions during surgery, or differences between individuals or groups. There are two main categories of nonrigid registration techniques: parametric and nonparametric. In general, the transformation is modeled by various basis functions (e.g., splines or polynomial functions) in parametric registration. A spline is a piecewise polynomial curve. Hence, this kind of transformation is also called a curved transformation [10]. Parametric registration algorithms generate the transformation through computing the optimal coefficients for the basis functions to match two images. In contrast, nonparametric nonrigid registration algorithms compute each voxel's displacement directly, and there are no underlying basis functions that model the transformation. Most nonparametric algorithms require the determination of material properties to drive biomechanical models [14–16], which are outside the scope of this chapter.

In the next sections, we will systematically discuss the mathematical aspects of the techniques introduced above and then provide a review of several applications in preclinical and clinical cancer research.

18.2 Quantitative Introduction to Image Registration

18.2.1 Types of Transformations

18.2.1.1 Rigid Transformation

Recall that the goal of registration is to generate a transformation to map each point in the source image to the corresponding point in the target image. As mentioned above, there are only rotation and translation involved in a rigid body transformation, which is given explicitly by the following:

$$x' = Rx + t \qquad (18.1)$$

where x and x' are the original and new coordinate in $\mathbb{R}^d$, with d being the dimensionality of the images, and t is the translation vector (t_x, t_y, t_z). For example, if x is the point (x, y, z), then x' is the transformed point of (x', y', z') in three dimensions. R is the rotation matrix for 3D images:

$$R = \begin{bmatrix} \cos\theta_y \cos\theta_z & -\cos\theta_x \sin\theta_z & \sin\theta_x \sin\theta_z \\ & +\sin\theta_x \sin\theta_y \cos\theta_z & +\cos\theta_x \sin\theta_y \cos\theta_z \\ \cos\theta_y \sin\theta_z & \cos\theta_x \cos\theta_z & -\sin\theta_x \cos\theta_z \\ & +\sin\theta_x \sin\theta_y \sin\theta_z & +\cos\theta_x \sin\theta_y \sin\theta_z \\ -\sin\theta_y & \sin\theta_x \cos\theta_y & \cos\theta_x \cos\theta_y \end{bmatrix}$$

$$(18.2)$$

where θ_x, θ_y, and θ_z are three angles of rotation in x, y, and z directions, respectively. Hence, R should be an orthogonal matrix.

The affine transformation is similar to the rigid body transformation:

$$x' = Ax + t \qquad (18.3)$$

where A is the affine matrix. The only difference between the rigid and affine transformation is that the affine matrix is a linear transformation, thereby allowing not only rotation but also scaling and shear mapping. The rigid body transformation reorients objects and keeps their shape and size unchanged, whereas the affine transformation only preserves parallelism. An example of human brain image is shown in Figure 18.1a, and the transformed images after rigid and affine transformations are displayed in Figures 18.1b and 18.1c, respectively. The corresponding transformations are also applied to a regular grid (Figure 18.1e) to illustrate how rigid and affine transformations result in different images (Figures 18.1f and 18.1g). The affine transformation applied here includes scaling and shearing, as well as rotation, leading to a change of size and angles of the image while maintaining the parallelism.

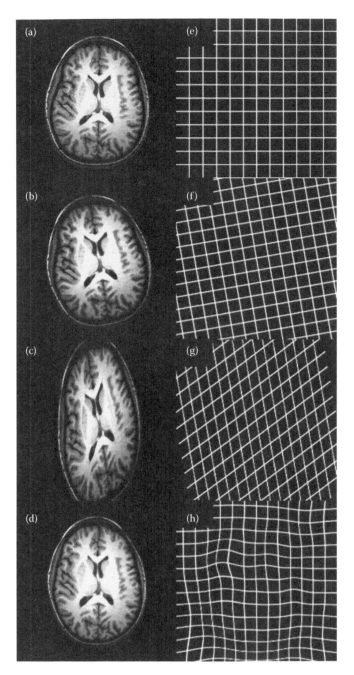

FIGURE 18.1 One example human brain image is shown in panel a. The transformed images are shown in panels b through d after the rigid, affine, and nonrigid transformations, respectively. The corresponding transformations are applied to a regular grid (panel e) to illustrate how different transformations result in different images (panels f through h).

18.2.1.2 Nonrigid Transformation

Different from the transformations described above, a curved transformation cannot be represented by matrices. The sampling coordinates in a nonrigid transformation can be represented as

$$x' = x + v(x) \tag{18.4}$$

where v is known as the deformation field. In general, it is calculated as a linear combination of basis functions:

$$v(x) = \sum_{i=1}^{N} c_i \Phi(x - x_i) \tag{18.5}$$

where

$$c_i = (c_i^x, \quad c_i^y, \quad c_i^z) \tag{18.6}$$

is the vector of coefficients for the ith basis function, N is the total number of functions, and x_i is the location of the ith basis function. The basis function Φ can have various expressions, including Gaussian functions [17] or polynomial functions [18,19] in image registration applications. The thin-plate spline (TPS) [20–22] is another basis function, which has been heavily used in medical image registration. A TPS transformation is defined as

$$x' = Ax + t + \sum_{i=1}^{N} c_i \Phi(r) \tag{18.7}$$

where

$$\Phi(r) = r^2 \log(r^2) \tag{18.8}$$

in two dimensions, and

$$\Phi(r) = r \tag{18.9}$$

in three dimensions, with $r = \|x - x_i\|$.

Another popular basis function used to model the deformation field in many automatic nonrigid registration problems is the cubic B-spline [23,24]. Using such a formulation, the deformation field becomes

$$v(x,y,z) = \sum_{l,m,n=0}^{3} c_{l+k_x-1,m+k_y-1,n+k_z-1} B_l\left(\frac{x}{n_x} - k_x\right)$$
$$\times B_m\left(\frac{y}{n_y} - k_y\right) B_n\left(\frac{z}{n_z} - k_z\right) \tag{18.10}$$

where $k_x = \lfloor x/n_x \rfloor$, $k_y = \lfloor y/n_y \rfloor$, and $k_z = \lfloor z/n_z \rfloor$, and n_x, n_y, and n_z are the number of control points. Typically, control points are the locations at which splines are placed in the images. Control points, splines, and coefficients determine the transformation. The $B_i(r)$'s are segments of the cubic B-spline:

$$B_0(r) = (1-r)^3/6$$
$$B_1(r) = (3r^3 - 6r^2 + 4)/6 \tag{18.11}$$
$$B_2(r) = (-3r^3 + 3r^2 + 3r + 1)/6$$
$$B_3(r) = r^3/6$$

Rohde et al. [25] used another basis function with compact support for intensity-based nonrigid registration task. They used one of Wu's [26] radial basis functions to model the deformation field

$$\Phi_i(x) = R\left(\frac{\|x - x_i\|}{a_i}\right) \tag{18.12}$$

$$R(r) = (1-r)_+^4 \left(3r^3 + 12r^2 + 16r + 4\right), \text{ for } r \geq 0 \tag{18.13}$$

where $(1 - r)_+ = \max(1 - r, 0)$ and a_i is the support of the *i*th basis function. In parametric nonrigid registration tasks, although the deformation field may be modeled by different basis functions, the fundamental goal is the same: to search the optimal coefficients c_i in Equations 18.5 and 18.6 to match the source image and target image maximally. Various schemes to achieve this will be introduced in the following sections. An example of nonrigid transformation is also shown in Figures 18.1d and 18.1h.

18.2.2 Point-Based Registration Methods

18.2.2.1 Point-Based Rigid Body Registration Method

To align two images using point-based registration algorithms, fiducial points must be identified and localized in both source and target images. The accurate localization of fiducial points is important in order to make the fiducial localization error (FLE) as small as possible. Fiducial points can be intrinsic markers placed in the imaging scene intentionally to facilitate identification. For instance, these can be steel pellets embedded in a frame-based stereotactic system [27] or chemical materials such as $CuSO_4$-filled tubes in MR-based systems [28]. They can also be salient points from anatomical structures, for example, the anterior commissure and posterior commissure, the nasion, or the most anterior part of the frontozygomatic suture.

The goal of point-based rigid body registration is to find the optimal rotation matrix R and translation vector t to minimize the distance between the set of fiducial points in the source image and the one in the target image after the registration. Let $X = \{x_i, i = 1,2,\ldots N\}$ and $Y = \{y_i, i = 1,2,\ldots N\}$ be two sets of fiducial points localized in the source and target images, respectively, where N is the number of fiducial points. The distance between these two sets of points after rigid body transformation is

$$D = (1/N)\sum_{i=1}^{N} \|Rx_i + t - y_i\|^2 \tag{18.14}$$

R and t are calculated to minimize the distance D. A number of solutions have been proposed to solve this problem (see reference [10] for an excellent historical coverage of these solutions). A commonly used method can be found in Arun et al. [29] and

is reproduced here (this solution does not take the FLE into consideration). First of all, the center points for the sets X and Y are computed:

$$\bar{x} = \frac{\sum_{i=1}^{N} x_i}{N} \text{ and } \bar{y} = \frac{\sum_{i=1}^{N} y_i}{N} \tag{18.15}$$

Let H be the fiducial covariance matrix:

$$H = \sum_{i=1}^{N} (x_i - \bar{x})(y_i - \bar{y})^t \tag{18.16}$$

Through performing singular value decomposition of H, we have

$$H = U\Lambda V^t \tag{18.17}$$

where U and V are orthogonal matrices and Λ is the diagonal matrix. The rotation matrix can then be calculated as

$$R = V \text{ diag}(1,1,\det(VU))U^t \tag{18.18}$$

and

$$t = \bar{y} - R\bar{x} \tag{18.19}$$

Solutions, which take into account the FLE, can be found in the work of Fitzpatrick et al. [10].

18.2.2.2 Point-Based Nonrigid Registration Method

Nonrigid transformations based on polynomials, cubic B-splines, or TPSs also can be used to align two sets of points. Here we use the TPS to introduce the approach, but it can be used with other basis functions.

Suppose that P_A and P_B, two sets of fiducial points, are selected in the two-dimensional (2D; the method can be extended easily to 3D) source image A and target image B, respectively, with $P_A = \left\{ \left(x_1^A, y_1^A\right), \left(x_2^A, y_2^A\right), \ldots \left(x_i^A, y_i^A\right), \ldots \left(x_N^A, y_N^A\right) \right\}$ and $P_B = \left\{ \left(x_1^B, y_1^B\right), \left(x_2^B, y_2^B\right), \ldots \left(x_i^B, y_i^B\right), \ldots \left(x_N^B, y_N^B\right) \right\}$.

For any point $\left(x_i^A, y_i^A\right)$ in image A, we use the following TPS transformation to compute the coordinates $\left(x_i^{A'}, y_i^{A'}\right)$:

$$x_i^{A'} = a_0 + a_1 x_i^A + a_2 y_i^A + \sum_{j=1}^{N} f_i r_{ij}^2 \ln r_{ij}^2$$

$$y_i^{A'} = b_0 + b_1 x_i^A + b_2 y_i^A + \sum_{j=1}^{N} q_i r_{ij}^2 \ln r_{ij}^2 \tag{18.20}$$

In addition, there are three boundary conditions in x and y directions, respectively, which ensure that the elastic transformation at an infinite point is zero:

$$\sum_{i=1}^{N} f_i = \sum_{i=1}^{N} x_i^A f_i = \sum_{i=1}^{N} y_i^A f_i = 0$$

$$\sum_{i=1}^{N} q_i = \sum_{i=1}^{N} x_i^A q_i = \sum_{i=1}^{N} y_i^A q_i = 0$$

(18.21)

Here, in practice and using the x component as an example, the coefficients $(a_0, a_1, a_2, f_1, ..., f_N)$ and $(b_0, b_1, b_2, q_1, ..., q_N)$ are found by first combining Equations 18.20 and 18.21 and creating the following matrix:

$$\begin{bmatrix} x_1^B \\ x_2^B \\ \vdots \\ x_N^B \\ 0 \\ 0 \\ 0 \end{bmatrix} = \begin{bmatrix} 1 & x_1^A & y_1^A & r_{11}^2 \ln r_{11}^2 & r_{21}^2 \ln r_{21}^2 & \cdots & r_{N1}^2 \ln r_{N1}^2 \\ 1 & x_2^A & y_2^A & r_{12}^2 \ln r_{12}^2 & r_{22}^2 \ln r_{22}^2 & \cdots & r_{N2}^2 \ln r_{N2}^2 \\ \vdots & \vdots & \vdots & \vdots & \vdots & \vdots & \vdots \\ 1 & x_N^A & y_N^A & r_{1N}^2 \ln r_{1N}^2 & r_{2N}^2 \ln r_{2N}^2 & \cdots & r_{NN}^2 \ln r_{NN}^2 \\ 0 & 0 & 0 & 1 & 1 & \cdots & 1 \\ 0 & 0 & 0 & x_1^A & x_2^A & \cdots & x_N^A \\ 0 & 0 & 0 & y_1^A & y_2^A & \cdots & y_N^A \end{bmatrix} \begin{bmatrix} a_0 \\ a_1 \\ a_2 \\ f_1 \\ f_2 \\ \vdots \\ f_N \end{bmatrix}$$

(18.22)

If we define the left side of Equation 18.22 as $\boldsymbol{B}_x$, the matrix on the right side as A, and $[a_0, a_1, a_2, f_1, ..., f_N]^T$ as $\boldsymbol{c}$, the vector of coefficients $\boldsymbol{c}$ can be computed using the Moore–Penrose pseudo-inverse matrix [30]:

$$\boldsymbol{c} = \left(\boldsymbol{B}_x^T \boldsymbol{B}_x \right)^{-1} \cdot \boldsymbol{B}_x^T \cdot \boldsymbol{A}.$$

(18.23)

The coefficients $(b_0, b_1, b_2, q_1, ..., q_N)$ for the y component can be determined similarly.

18.2.3 Surface-Based Registration Methods

18.2.3.1 Surface-Based Rigid Body Registration Method

For either rigid or nonrigid registration tasks, surface-based registration methods require the surface representation of the objects to register. Following the taxonomy presented by Audette et al. [33], surfaces can be represented by feature-, point-, and model-based methods. Feature-based representation includes point features (e.g., salient points or vertices), curve features (e.g., boundaries of anatomical structures), or region features (e.g., regions with homogeneous intensities). Point-based representation consists of a set of points obtained by sampling the surfaces. Model-based methods include physical-based models and surface evolution models [34–36]. Based on this classification, point-based (fiducial point–based) registration methods described in Section 2.2 can be classified as a simple case of feature-based surface registration.

Surface-based rigid body registration also requires determining a disparity or cost function. As is the case for point-based rigid body registration, the rotation and translation parameters are computed through minimizing the disparity function. A few of the common surface-based registration schemes are described in the following paragraphs.

The "head and hat" method is the first surface-based rigid body registration algorithm proposed for medical images. It was developed by Pelizzari et al. [31] to align MR and PET head images. The "hat" is defined as a set of points from the source image, and the "head" is a model of the surface segmented from the target image. The purpose of the head and hat algorithm is to find a transformation that fits the hat to the head surface.

The iterative closest point (ICP) algorithm [32] is a technique that has been applied widely to surface-based registration tasks for medical images. This algorithm calculates the transformation and distance between a point set and a surface iteratively. Once the metric converges to a minimum, the algorithm is terminated. Suppose that $X = \{x_i, i = 1,2,...N\}$ is a point set in the source image and Y is the surface in the target image. The algorithm then proceeds as follows (please also refer to reference [10]):

1. $\forall \boldsymbol{x}_i \in \boldsymbol{X}$, find the closest point y_i to the surface Y.
2. Compute the mean square disparity function D:

$$D = \sqrt{(1/N) \sum_{i}^{N} \left\| \boldsymbol{R}\boldsymbol{x}_i + \boldsymbol{t} - \boldsymbol{y}_i \right\|^2}$$

(18.24)

3. Compute the rotation R and translation t using the algorithm introduced in Section 18.2.2.1 to obtain the transformation T.
4. Apply the rigid body transformation T to the point set X to obtain the new $X'=T(X)$.
5. Compute the new distance D' between X' and Y. If the absolute difference between D and D' is less than a user-selected threshold, terminate the procedure. Otherwise, let $X = X'$, and repeat the procedure from step 1.

18.2.3.2 Surface-Based Nonrigid Registration Method

Audette et al. [33] provide a good summary of both rigid and nonrigid surface-based registration methods, and the reader is referred to this article for a more in-depth treatment of the topic than what can be provided in this chapter. Here we present a registration algorithm proposed after the work of Audette et al., i.e., the robust point matching (RPM) algorithm [37], which has recently gained popularity for medical image analysis applications. The input to the RPM algorithm is two sets of points extracted from the source and target images. The cardinality of these two sets of points does not need to be the same. The RPM algorithm computes the correspondence and the transformation between these points iteratively. The algorithm is also able to label some source points, which have no corresponding points in the target images, as outliers and ignore them while iterating.

Suppose that $X = \{x_i, i = 1, 2, \ldots K\}$ and $Y = \{y_j, j = 1, 2, \ldots N\}$ are two sets of points extracted from the source and target images, respectively, and K and N can have different values. Instead of finding a one-to-one correspondence (e.g., finding the closest point as is done in the ICP algorithm), the RPM algorithm builds the correspondence through a fuzzy correspondence matrix [39], the elements of which are as follows:

$$m_{ij} = \frac{1}{T} \exp\left(-\frac{(y_j - f(x_i))^T (y_j - f(x_i))}{2T} \right) \qquad (18.25)$$

for $i = 1, 2, \ldots, K$ and $j = 1, 2, \ldots, N$. The $K + 1$th column and $N + 1$th row are introduced in the matrix to reject the points that have no counterparts and label them as outliers. For the outlier entries $i = K + 1$ and $j = 1, 2, \ldots, N$, and the outlier entries $i = 1, 2, \ldots, K$ and $j = N + 1$, the elements are

$$m_{K+1,j} = \frac{1}{T_0} \exp\left(-\frac{(y_j - x_{K+1})^T (y_j - x_{K+1})}{2T_0} \right) \qquad (18.26)$$

and

$$m_{i,N+1} = \frac{1}{T_0} \exp\left(-\frac{(y_{N+1} - f(x_i))^T (y_{N+1} - f(x_i))}{2T_0} \right) \qquad (18.27)$$

Here x_{K+1} and y_{N+1} are the outlier clusters, which are calculated as the center of each set of points. f is the nonrigid transformation or mapping function, which is used to transform the source image. T is called the temperature parameter because it is used to control the fuzziness of the correspondence matrix in an optimization process similar to physical annealing [38]. In their original work [37], the authors suggest that the algorithm with $T_0 = 0.5$ be initialized. The row and column elements should be normalized as follows:

$$m_{ij} = \frac{m_{ij}}{\sum\limits_{a=1}^{K+1} m_{aj}}, j = 1, 2, \ldots, N \qquad (18.28)$$

and

$$m_{ij} = \frac{m_{ij}}{\sum\limits_{b=1}^{N+1} m_{ib}}, i = 1, 2, \ldots, K \qquad (18.29)$$

The strength of the association of each point x_i in the source image to the point y_j in the target image is measured by m_{ij}. The larger the distance between x_i and y_j and the higher the temperature T, the smaller the weight m_{ij}. As the algorithm progresses, the temperature T decreases and the fuzziness of the assignment decreases, thus approaching a closest point correspondence.

At each iteration, once the fuzzy correspondence matrix is built, a transformation f is computed, which minimizes the following least-squares function:

$$\min_f E(f) = \min_f \sum_{i=1}^{K} \|\bar{y}_i - f(x_i)\|^2 + \lambda \|Lf\|^2 \qquad (18.30)$$

where $\bar{y}_i = \sum\limits_{j=1}^{N} m_{ij} y_j$ is a virtual correspondence for the point x_i.

All the points in the target image have an influence on the calculation of the correspondence of each point x_i in the source image, using the fuzzy correspondence matrix. L is an operator used to estimate the smoothness of the mapping function f, and λ is a parameter that permits the weighting of the contribution of the smoothness term. In the original work [37], f is a TPS transformation and L is the integral of the square of the second-order derivative of f in 2D:

$$L(f) = \iint \left[\left(\frac{\partial^2 f}{\partial x^2} \right)^2 + 2 \left(\frac{\partial^2 f}{\partial x \partial y} \right)^2 + \left(\frac{\partial^2 f}{\partial y^2} \right)^2 \right] dx\, dy \qquad (18.31)$$

18.2.4 Intensity-Based Registration Methods

As their name suggests, intensity-based registration methods use the image intensity values to drive the registration process. Typically, these methods are applied to the images without preprocessing steps, e.g., selection of points or extraction of surfaces. As a result, they are easier to automate completely.

In intensity-based registration algorithms, the similarity/disparity functions and the type of transformation used to register the images must be selected. Transformations are parameterized, and the set of parameters that maximize the similarity function is then found using one of several optimization algorithms. Hence, the similarity function, the transformation, and the optimization algorithm characterize intensity-based registration algorithms.

Intensity-based registration methods have a long history in the area of medical imaging, and various similarity measures have been proposed. The simplest similarity function is the sum of squared differences, in which the intensity difference between two images is computed and minimized. This measure is suitable for monomodal registration because images acquired with the same modality frequently possess the same intensity level and range. A similar criterion is the mean squared difference. The correlation coefficient (CC) is another similarity function commonly used in intensity-based registration:

$$CC = \frac{\sum\limits_i \left((A(x_i) - \bar{A})(B(x_i) - \bar{B}) \right)}{\sqrt{\sum\limits_i (A(x_i) - \bar{A})^2 \sum\limits_i (B(x_i) - \bar{B})^2}} \qquad (18.32)$$

where x_i is the ith pixel/voxel in the images, $A(x_i)$ and $B(x_i)$ are the intensity of x_i in the source and target images, respectively, and $\bar{A}$ and $\bar{B}$ are the mean intensity values of two images. Other similarity measures include the correlation ratio [40], ratio–image uniformity [41], partitioned intensity uniformity [42], joint entropy, mutual information (MI) [43, 44], and normalized MI (NMI) [45]. Among these measures, MI and NMI are the most successful and frequently used approaches in medical image registration, since a large intersite validation study showed that these led to the most accurate results on a set of images for which the ground truth was known [46].

MI, proposed independently by Collingnon et al. [43] and Wells et al. [44] as a similarity measure for registration, is an information theory concept. Broadly speaking (a detailed treatment of the topic and its application to medical images can be found in reference [47]), it measures the mutual dependency of two random variables. In the context of image registration, MI is defined as

$$MI(A,B) = H(A) + H(B) - H(A,B) \qquad (18.33)$$

where $H(A)$ and $H(B)$ are the marginal entropy of images A and B, respectively, and $H(A,B)$ is the joint entropy:

$$H(A) = -\sum_a p_A(a) \log p_A(a)$$

$$H(B) = -\sum_b p_B(b) \log p_B(b) \qquad (18.34)$$

$$H(A,B) = -\sum_{a,b} p_{AB}(a,b) \log p_{AB}(a,b)$$

where $p_A(a)$ and $p_B(b)$ are the marginal intensity probability distributions, $p_{AB}(a,b)$ is the joint probability distribution, and a and b are intensity values in images A and B, respectively. Entropy is a measure of the amount of information contained in one image. As stated in the work of Viola and Wells [48], maximizing MI will tend to find a transformation that both retains as much as possible the information in each of the images (maximizing the first two terms of Equation 18.33) and is such that the images explain each other well (minimizing the last term of Equation 18.33).

Although it has been shown that MI is a reliable similarity measure [49–51], Studholme et al. [45] showed that MI can be sensitive to the overlap size between two images; in particular, if the overlap size is relatively small, compared with the whole field of view, MI-based registration algorithm tends to perform poorly. To address this issue, the authors have proposed an alternative measure, i.e., NMI:

$$NMI = \frac{H(A) + H(B)}{H(A,B)} \qquad (18.35)$$

Both MI and NMI have been used in a large number of studies. Pluim et al. [12] present an overview of MI-based medical image registration algorithms. They show that MI or NMI is used in a wide range of algorithms, including rigid and nonrigid registration methods, monomodal and multimodal registration methods, and intrasubject and intersubject methods. These MI-based algorithms have also been used for a number of organs, i.e., images of the brain, lungs, spine, heart, breast, abdomen, etc.

Once the similarity measure is selected, an algorithm needs to be chosen to find the optimal transformation parameters. Popular optimization techniques used in medical image registration include Powell's conjugate direction method, Brent's optimization algorithm, Levenberg–Marquardt optimization, the downhill simplex method, the gradient descent technique, Newton–Raphson iteration, etc. Theoretical and implementation details, including computer code, for many of these methods can be found in the work of Press et al. [55]. A detailed description of the various MI-based registration methods that have been developed is beyond the scope of this chapter, but three representative techniques are discussed to illustrate how this metric is used in practice. The first one, proposed by Maes et al. [51], is a rigid body algorithm. The other two, by Rueckert et al. [54] and Rohde et al. [25], are nonrigid algorithms.

18.2.4.1 Intensity-Based Rigid Registration Method

The rigid registration algorithm proposed by Maes et al. [51] estimates three rotation angles (ϕ_x, ϕ_y, ϕ_z) and three translation distances (t_x, t_y, t_z) in x, y, and z directions, by maximizing MI using Powell's multidimensional direction set method. The accuracy of this implementation has been validated by comparing the computed transformations for MRI, CT, and PET registration tasks with a known ground truth [52,53]. This study showed that the MI criterion leads to subvoxel accuracy in multimodal image registration. The authors also reported that the optimization order used to estimate the six parameters has a strong influence on the algorithm performance. The experimental results showed that better results can be obtained when the parameters t_x, t_y, and ϕ_z are optimized before the other three parameters.

The authors also evaluated the influence of different interpolation schemes on the robustness of the MI similarity measure. Interpolation of the source image is necessary because, in general, the new voxel locations of the reoriented source image will not coincide with regular grid points. Three interpolation methods are compared [51]: nearest neighbor (NN), trilinear (TRI), and partial volume distribution (PV) interpolation. The simplest approach is NN, which uses the intensity of the nearest voxel. TRI interpolation assigns different weights to the eight neighbor voxels (four pixels in 2D images) based on the distances between the new location and the neighbor voxels. However, this method may introduce new intensity values, which may affect the MI criterion unpredictably. The recommended interpolation method is PV, which updates the joint histogram distribution directly based on the same weights as those computed in TRI. Experiments performed in [51] suggested that PV interpolation results in a similarity measure with fewer local minima.

The robustness of the algorithm also depends on subsampling, partial overlap, and image degradation [51]. Subsampling can improve the algorithm performance. It has been recommended [51] to start the registration algorithm from a coarse image and increase the resolution as the program proceeds. Partial overlap happens when the field of view of one image covers only a portion of the other image during the registration. Image degradation includes noise, intensity inhomogeneity, and geometric distortion. A number of experiments performed in reference [51] showed that MI is a similarity measure that is robust to these artifacts.

18.2.4.2 Intensity-Based Nonrigid Registration Method

One of the most commonly used intensity-based nonrigid registration algorithms has been proposed by Rueckert et al. [54]. In their work, a combined transformation T is employed to control the deformation of images:

$$T = T_{\text{global}} + T_{\text{local}} \qquad (18.36)$$

where T_{global} is a global transformation, which can be selected to be either rigid or affine. It is used to correct for global orientation or position differences between images. T_{local} is a local transformation that permits deformation. In the work of Rueckert et al. [54], the deformation is modeled with B-splines (refer to Equations 18.10 and 18.11). The parameters $c_{l+k_x-1,m+k_y-1,n+k_z-1}$ in Equation 18.10 are referred to as the control points. The number of control points defines the number of splines and thus the number of degrees of freedom in the transformation. This, in turn, governs both the locality of the transformation and the numerical complexity of the algorithm. In practice, a hierarchical approach is used. First, the image is downsampled, and a limited number of splines are used. This leads to a transformation that takes into account large differences between images. The image resolution and the number of splines are iteratively increased until the desired resolution is achieved. To constrain the spline-based transformation to be smooth, a term based on the bending energy of the transformation is also often utilized.

$$E_{\text{smooth}} = \frac{1}{V} \iiint ((\partial^2 T/\partial x^2)^2 + (\partial^2 T/\partial y^2)^2 + (\partial^2 T/\partial z^2)^2$$
$$+ 2(\partial^2 T/\partial x \partial y)^2 + 2(\partial^2 T/\partial x \partial z)^2 + 2(\partial^2 T/\partial y \partial z)^2)dxdydz \qquad (18.37)$$

where V is the image volume. This term is added to the cost function, leading to the following expression:

$$E_{\text{cost}} = -E_{\text{similarity}} + \lambda E_{\text{smooth}} \qquad (18.38)$$

where $E_{\text{similarity}}$ and E_{smooth} are the NMI and the bending energy, respectively, and the parameter λ is used to control the weight of the smoothness term.

When the algorithm is used, the optimal rigid or affine transformation parameters are first estimated by maximizing the NMI. Next, the parameters $c_{l+k_x-1,m+k_y-1,n+k_z-1}$ are optimized through minimizing the cost function (Equation 18.38) using a gradient descent technique [55]. The program is terminated automatically when the gradient of the cost function with respect to the B-spline parameters is less than a user-selected threshold. Originally, the algorithm was applied successfully to breast MR images [54]. This original publication showed the ability of this algorithm to correct for breast motion between acquisitions. Since then, it has been used successfully on a variety of images such as brain, heart, prostate, lungs, etc.

Another popular MI-based nonrigid registration algorithm called the adaptive bases algorithm (ABA) has been proposed by Rohde et al. [25]. The algorithm also uses NMI or MI as the similarity measure in the cost function. Here, a linear combination of one of Wu's [26] radial basis functions is employed to model the deformation field (refer to Equations 18.2 and 18.13). As is done in Rueckert's algorithm, the ABA computes the final deformation field iteratively through gradually increasing the image resolutions and the density of basis functions. The algorithm is initialized on a low-resolution image with a few basis functions. It then progresses to higher resolutions with more basis functions. One difference between ABA and Rueckert's algorithm is the fact that, in ABA, the basis functions do not have to be placed on a regular grid. Thanks to this, it is possible to focus the algorithm on regions that are misregistered and to ignore those that are not. This is implemented as follows. First, basis functions are placed on a regular grid, and then the derivative of the similarity with respect to the coefficients of the basis functions is computed. If this derivative is low, it indicates that the related basis function is either in an area where the similarity measure is in a local extremum or in an area where the similarity measure is not very sensitive to displacement. In either case, adjusting the coefficient of basis functions in these areas is not productive. The algorithm thus ignores these regions and operates only over regions where the gradient is high. A regular grid is replaced over the images each time the algorithm moves from one image resolution to the other or when the density of basis functions is increased.

Rather than constraining the transformation to be smoothed using its overall bending energy, the smoothness of the transformation is controlled by constraining the difference between the values of the coefficients of adjacent basis functions, which is called ε. A large value of ε allows transformations to vary rapidly and capture large deformations, whereas a small ε value leads to relatively smooth and stiff transformations. The threshold ε can also be changed spatially, which permits adapting the transformation properties. This strategy has been used by Duay et al. [56] to generate transformations that could expand the ventricles more than brain tissues to register brain image volumes with large space-occupying lesions to an atlas without pathology. Li et al. [57] also used this technique to control the transformations of bones and soft tissues during the registration of the whole-body CT mice images. A larger value of ε was assigned to the soft tissue regions than to the bony regions to produce more rigid transformation for bony structures than for soft tissues.

18.3 Preclinical Applications in Cancer

The ability to combine multiple imaging modalities when reporting on tumor status is highly desirable. Since different imaging modalities offer different strengths and weaknesses and offer complimentary biological information, there is currently much ongoing effort to register tumor images obtained from multiple techniques. The last decade has seen an explosion in the number of quantitative imaging techniques that can be used to interrogate tumor physiology. For instance, DCE-MRI is a promising technique that can offer information on tumor vessel perfusion and microvascular vessel wall permeability, extracellular volume fraction, and blood volume (please refer to Chapter 12). Fluorodeoxyglucose PET can report on the metabolic status of a cancer. Thus, the ability to register and synthesize composite images from these two techniques offers the promise of simultaneously characterizing tumor vascular function and metabolism.

Cho et al. [61] recently performed a study of tumor hypoxia using registered DCE-MRI and fluoromisonidazole PET (FMISO-PET) data in a preclinical rat tumor model of prostate cancer. A special coil-marker system [59,60] was used to immobilize the mice, thereby making the image registration accurate and convenient. To acquire MRI and PET images, the animal was placed on a Plexiglas holder, with its tumor-bearing leg stretched into the MR coil with marker assembly. An animal-position mold [62] was then made to fix the animal position. After animal immobilization, the mold with the MR coil-marker system was placed inside the magnet, with the contrast agent gadopentetate dimeglumine (Gd-DTPA) filling markers. The DCE-MRI scans were then performed, and the markers were replaced with FMISO in preparation for PET scanning. Dynamic microPET was then performed with FMISO. After the PET scans, the animal was sacrificed and removed from the holder for tumor excision. The tumor was sectioned and histological stains were obtained.

The MRI and PET images were coregistered using a rigid body registration. In particular, the optimal rotation and translation of the PET marker images were obtained through maximizing the CC between MRI and PET images. The generated rigid transformation was applied to not only MRI and PET images but also the *ex vivo* histological stains that were used to validate the imaging measures. One of the analyses performed in this study was choosing three representative voxels from perfused, hypoxic, and necrotic regions of a tumor, respectively. Both the DCE dynamic curves and PET curves based on time demonstrated three different characteristics for three different regions. Other analyses included the evaluation of the spatial relationship between perfusion and hypoxia. The registration allowed the voxel-by-voxel scatterplot between these two parameters, and the results showed that there was a significant negative correlation between them. Thus, the registration played a crucial role to

allow the comparison between DCE-MRI and PET parameters voxel by voxel.

US and MRI data were coregistered in the study of Loveless et al. [64] to examine the correlation of the physiological and molecular biomarkers. In their study, breast cancer cells were injected into nine mice and tumors were scanned with a 4.7-T MRI and US after 6 and 9 days' growth. Mice were placed in a custom animal holder during the scanning to avoid movement between MRI and US scans. To register tumors in MRI and US images, two operators segmented tumors manually from each modality. The segmented surfaces were aligned manually before they were input to the surface-based registration algorithm, ICP (please refer to Section 18.2.3.1), because it prevented the local convergence of ICP efficiently. ICP computed the rigid transformation by searching closest point pairs iteratively. The obtained transformation was applied to physiological parametric maps derived from MRI so that the tumor information from MRI and US could be combined and analyzed. This study showed that the registration between US and MRI allowed for quantitative comparison of two potential biomarkers.

18.4 Clinical Applications in Cancer

Multimodal registration is also applied widely in clinical applications. Moy et al. [2] studied the effect of the fused MRI and PET breast cancer images on the breast cancer diagnosis. Twenty-three patients with suspected breast cancer were involved in the study. Each individual patient was examined with PET and MRI scans. For each pair of MRI and PET data, 20 to 40 landmarks were selected manually from nipples and skin boundaries for each patient. The landmarks were used to perform a point-based registration [65]. This registration technique calculated the cross-correlation coefficient between each landmark in the MRI data and the corresponding landmark and the neighboring points in the PET data. The pair of points with the strongest correlation was used to compute the transformation, which was modeled by the polynomials. Once the PET and MRI data were fused, the total of 45 lesions was assessed. Similar assessment was also performed on the MRI data alone. The comparison of these two kinds of assessment showed that the registration between PET and MRI increased the readers' confidence in the assessment.

In addition to multimodal registration, registration algorithms can be used to compare imaging data acquired at different time points to assess tumor status and response to treatment. For example, Galbán et al. [66] recently used a novel approach to detect hemodynamic alterations in high-grade glioma through applying the longitudinal registration method. The investigators studied 44 individuals with grades III and IV gliomas. Dynamic susceptibility-weighted contrast MRI (DSC-MRI, please see Chapter 13) scans were performed 1 week before and 1 and 3 weeks after the start of radiotherapy. Two perfusion parameters, the relative cerebral blood volume (rCBV) and flow (rCBF) maps, were generated from DSC images [67]. An MI-based rigid body registration [68] (please also refer to Section 18.2.4.1) was used to register rCBV and rCBF maps before and after treatment to T_1-weighted images before treatment. After coregistration, brain tumors were segmented manually by a neuroradiologist. Since tumors may shrink or expand during treatment, only the overlapped tumor voxels in both before- and after-treatment images were analyzed. Hence, the alignment of parametric maps enabled the following analysis to be performed.

The analysis of hemodynamic alterations was performed based on the parametric response map (PRM). PRM_{rCBV} was determined through calculating the different $\Delta rCBV$ between within-treatment rCBV and pretreatment rCBV for each voxel. When $\Delta rCBV$ was greater than a predetermined threshold set, it suggested a substantial increase in the microvascular density or increased blood volume, and the corresponding voxels were coded as red. Voxels with a decrease in rCBV were coded as blue. Green was used for the unchanged case. PRM_{rCBF} was determined similarly. The study compared the PRM approach with the standard approach in which the mean values of perfusion parameters were calculated. Results of rCBV from one nonresponsive and one responsive subject were analyzed. The mean values of rCBV within the tumor before and after 1 week of therapy for the nonresponsive patient were unchanged. However, substantial changes in PRM_{rCBV} were observed. Similar results were obtained for rCBF. For the responsive subject, the mean rCBV values did not change significantly, but PRM_{rCBV} found negligible changes. Statistical analysis for all 44 patients showed that the standard approach did not provide a significantly early treatment response prognosis; however, the PRM approach can quantify the alterations in rCBV and rCBF accurately. Both increases and decreases in perfusion parameters desensitized the measure of the average change, resulting in a relative insensitivity to treatment outcome using the standard approach. However, the coregistration preserved the spatial information, which was included within individual tumor voxels and allowed a sensitive method for quantification of perfusion values.

Longitudinal coregistration plays a similarly important role in the work of Li et al. [4]. The motivation of their work is to find a more sensitive method to monitor tumor response to therapy. The response evaluation criteria in solid tumors (RECIST; please also see Chapter 16) is currently the most widely used method of measuring tumor response. Although RECIST provides a simple and practical method to extract the salient features of imaging data, it discards abundant underlying information. Hence, Li et al. aligned longitudinal breast DCE-MRI data sets to a common image space in order to retain spatial information in the data. In their work, one patient with locally advanced breast cancer was selected, and DCE-MRI was performed on a Philips 3.0T Achieva MR scanner. Following the dynamic scan, a 3D T_1-weighted high-resolution isotropic volume examination (THRIVE) scan was acquired. This patient was scanned prior to neoadjuvant chemotherapy (t_1), after the first cycle of therapy (t_2), and after completion of all cycles of therapy (t_3). DCE-MRI data were analyzed, and two parametric maps were generated: the contrast agent extravasation rate constant and the extravascular extracellular volume fraction.

The challenge in this work is that most of the intensity-based nonrigid registration algorithms will deform both normal tissues and tumors. The compression or expansion of tumors will lead to misleading results in regard to tumor response. Hence, in their work, an MI-based rigid body registration algorithm [51] (please refer to Section 18.2.4.1) was applied to the THRIVE volumes at three time points to obtain the global transformation first. The ABA [25] was then extended through adding a volume-preserving constraint term [69] to the MI-based cost function. The extended algorithm was applied to the MR volumes to register the normal tissues maximally and preserve the tumor volume simultaneously.

The generated deformation fields were then applied to the parametric maps. Hence, the parameters acquired at t_1 and t_2 were transformed and coregistered to the corresponding parameters at t_3. Therefore, the registration provides a novel approach to allow not only the voxel-by-voxel analysis of tumor-related parameters longitudinally and quantitatively, but also the comparison of different parameters in regard to tumor response to therapy.

18.5 Summary

With the current development of medical imaging, image registration has been widely used in medical image analysis to provide important information for diagnosis, monitoring disease and treatment response, or guiding and planning treatment. This chapter introduces the mathematical aspects of medical image registration as well as a few of the commonly used registration techniques.

The purpose of medical image registration is to search an optimized transformation to match two medical images maximally. Rigid transformation is calculated during the registration when only the rotation and translation occur between objects; nonrigid transformation is required when more complex and nonlinear deformation is involved between objects, e.g., intersubject registration or pretreatment and posttreatment registration. Nonrigid transformation can be represented by the combination of splines, and hence the calculation of transformation can be translated to the calculation of the coefficients of splines.

Point- and surface-based registration techniques require landmark selection or surface identification. The cost function could be the distance between two sets of points or surfaces or with specific constraints, such as the smoothness of deformation. Different pairs of landmarks can also be assigned different weights during registration based on the confidence of the accuracy of landmark selections. Intensity-based registration is performed based on the intensity values of images. MI or NMI is the most investigated measure because of its high robustness and accuracy. Similarly, the cost function in intensity-based registration can be composed of a similarity term and other application-based terms, such as the smoothness term or volume-preservation term.

In summary, medical image registration is an indispensable tool in preclinical and clinical applications. The development and validation of registration algorithms are still an active research area in medical image analysis in the long term.

Acknowledgments

The author wishes to acknowledge Dr. Benoit M. Dawant, Ph.D., and Dr. Thomas E. Yankeelov, Ph.D., for their support, encouragement, and stimulating discussion during this work. Also, the author wishes to thank the National Institutes of Health for funding through NCI 5R01CA129961 and NCI 1U01CA142565.

References

1. Phelps ME. PET: *Molecular Imaging and Its Biological Applications.* Springer, New York 2004.

2. Moy L, Ponzo F, Noz ME, Maguire GQ Jr, Murphy-Walcott AD, Deans AE, Kitazono MT, Travascio L, Kramer EL. Improving specificity of breast MRI using prone PET and fused MRI and PET 3D volume datasets. *J Nucl Med* 2007;48(4):528–537.

3. Chittineni R, Su MY, Nalcioglu O. Breast MR registration for evaluation of neoadjuvant chemotherapy response. *Proc Intl Soc Magn Reson Med* 2008;16:3095.

4. Li X, Dawant BM, Welch EB, Chakravarthy AB, Freehardt D, Mayer I, Kelley M, Meszoely I, Gore JC, Yankeelov TE. A nonrigid registration algorithm for longitudinal breast MR images and the analysis of breast tumor response. *MRI* 2009;27(9):1258–1270.

5. Thompson PM, Vidal CN, Giedd JN, Gochman P, Blumenthal J, Nicolson R, Toga AW, Rapoport JL. Mapping adolescent brain change reveals dynamic wave of accelerated gray matter loss in very early-onset schizophrenia. *Proc Natl Acad Sci U S A* 2001;98(20):11650–11655.

6. Zacharaki EI, Hogea CS, Shen D, Biros G, Davatzikos C. Non-diffeomorphic registration of brain tumor images by simulating tissue loss and tumor growth. *Neuroimage* 2009;46(3):762–774.

7. Zitová B, Flusser J. Image registration methods: a survey. *Image Vis Comput* 2003;21(11): 977–1000.

8. Maintz JBA, Viergever MA. A survey of medical image registration. *Med Image Anal.* 1998;2:1–36.

9. Maurer C, Fitzpatrick JM. A review of medical image registration. In R. Maciunas (Ed.), *Interactive Image-Guided Neurosurgery,* American Association of Neurological Surgeons, Park Ridge, IL; 1993, 17–44.

10. Fitzpatrick JM, Hill DLG, Maurer CR. Image Registration. In M Sonka, JM Fitapatrick (Eds.), *Handbook of Medical Imaging, Volume 2. Medical Image Processing and Analysis.* SPIE International Society for Optical Engineering, Bellingham, Wash; 2000, 447–513.

11. Brown LG. A survey of image registration techniques. *ACM Comput Surv* 1992;24(4):325–376.

12. Pluim JPW, Maintz JBA, Viergever MA. Mutual-information-based registration of medical images: a survey. *IEEE Trans Med Imag* 2003;22:986–1004.

13. Crum WR, Hartkens T, Hill DLG. Non-rigid image registration: theory and practice. *Br J Radiol* 2004;77:140–153.

14. Thirion JP. Image matching as a diffusion process: an analogy with Maxwell's demons. *Med Image Anal* 1998;2(3):243–260.

15. Bajcsy R, Kovacic S. Multiresolution elastic matching. *Comp Vision Graph Image Process* 1989;46:1–21.

16. Chirstensen GE, Rabbitt RD, Miller MI. Deformable templates using large deformation kinematics. *IEEE Trans Med Imag* 1996;5(10):1435–1447.

17. Gaens T, Maes F, Dirk V, Suetens T. Non-rigid multimodal image registration using mutual information. In W. M. Wells, A. Colchester, S. Delp (Eds.), *Medical Image Computing and Computer-Assisted Intervention—MICCAI'98.* Springer-Verlag, Berlin, Germany; 1998; 1496, 1099–1106.

18. Goshtasby A. Piecewise linear mapping functions for image registration. *Pattern Recogn* 1986;19:459–466.

19. Maguire GQ Jr, Noz ME, Rusinek H, Jaeger J, Kramer EL, Sanger JJ, Smith G. Graphics applied to medical image registration. *IEEE Comput Graph Appl* 1991;11:20–29.

20. Bookstein FL. Principal warps: thin-plate splines and the decomposition of deformations. *IEEE Trans Pattern Anal Machine Intell* 1989;11:567–585.

21. Goshtasby A. Registration of images with geometric distortions. *IEEE Trans Geosci Remote Sens* 1988;26:60–64.

22. Harder RL, Desmarais RN. Interpolation using surface splines. *J Aircraft* 1972;9:189–191.

23. Lee S, Wolberg G, Chwa K-Y, Shin SY. Image metamorphosis with scattered feature constraints. *IEEE Trans Vis Comput Graph* 1996;2:337–354.

24. Lee S, Wolberg G, Shin SY. Scattered data interpolation with multilevel B-splines. *IEEE Trans Vis Comput Graph* 1997;3:228–244.

25. Rohde GK, Aldroubi A, Dawant BM. The adaptive bases algorithm for intensity-based nonrigid image registration. *IEEE Trans Med Imag* 2003;22:1470–1479.

26. Wu Z. Multivariate compactly supported positive definite radial functions. *Adv Comput Math* 1995;4:283–292.

27. Peters T, Davey B, Munger P, Comeau R, Evans A, Olivier A. Three-dimensional multimodal image guidance for neurosurgery. *IEEE Trans Med Imag* 1996;15:121–128.

28. Evans AC, Beil C, Marrett S, Thompson CJ, Hakjm A. Anatomical–functional correlation using an adjustable MRI based region of interest atlas with positron emission tomography. *J Cereb Blood Flow Metab* 1988;8:513–530.

29. Arun KS, Huang TS, Blostein SD. Least-squares fitting of two 3-D point sets. *IEEE Trans Pattern Anal Mach Intell* 1987;9:698–700.

30. Ben-Israel A, Greville TNE. *Generalized Inverses: Theory and Applications.* Wiley, New York 1997.

31. Pelizzari CA, Chen GTY, Spelbring DR, Weichselbaum RR, Chen CT. Accurate three-dimensional registration of CT, PET, and/or MR images of the brain. *J Comput Assist Tomogr* 1989;13:20–26.

32. Besl PJ, McKay ND. A method for registration of 3-D shapes. *IEEE Trans Pattern Anal Mach Intell* 1992;14:239–256.

33. Audette MA, Ferrie FP, Peters TM. An algorithmic overview of surface registration techniques for medical imaging. *Med Image Anal* 2000;4:201–217.

34. McInerney T, Terzopoulos D. Deformable models in medical image analysis: a survey. *Med Image Anal* 1996;1(2):91–108.

35. Sethian JA. A review of the theory, algorithms, and applications of level set methods for propagating interfaces. *Acta Numerica* 1996;5:309–395.

36. Kimmel R, Kiryati N, Bruckstein AM. Analyzing and synthesizing images by evolving curves with the Osher–Sethian method. *Int J Comp Vision* 1997;24(1):37–56.

37. Chui H, Rangarajan A. A new point matching algorithm for nonrigid registration. *Comput Vis Image Underst* 2003; 89:114–141.

38. Yuille AL. Generalized deformable models, statistical physics, and matching problems. *Neural Comput* 1990;2(1):1–24.

39. Gold S, Rangarajan A, Lu CP, Pappu S, Mjolsness E. New algorithms for 2D and 3D point matching: pose estimation and correspondence. *Pattern Recogn* 1998;31:1019–1031.

40. Roche A, Malandain G, Pennec X, Ayache N. The correlation ratio as a new similarity measure for multimodal image registration. *Proceedings of the First International Conference on Medical Image Computing and Computer-Assisted Intervention (MICCAI'98), Lecture Notes in Computer Science,* Springer-Verlag, Berlin, Germany; 1998, 1496, 1115–1124.

41. Woods RP, Cherry SR, Mazziotta JC. Rapid automated algorithm for aligning and reslicing PET images. *J Comput Assist Tomogr* 1992;16:620–633.

42. Woods RP, Mazziotta JC, Cherry SR. MRI-PET registration with automated algorithm. *J Comput Assist Tomogr* 1993;17:536–546.

43. Collingnon A, Maes F, Delaere D, Vandermeulen D, Suetens P, Marcha G. Automated multimodality medical image registration using information theory. *Proc. XIV'th Int. Conf. Information Processing in Medical Imaging, Computational Imaging and Vision 3,* Boston: Kluwer; 1995, 263–274.

44. Wells WM III, Viola P, Atsumi H, Nakajima S, Kikinis R. Multi-modal volume registration by maximization of mutual information. *Med Image Anal* 1996;1(1):35–51.

45. Studholme C, Hill DLG, Hawkes DJ. An overlap invariant entropy measure of 3D medical image alignment. *Pattern Recogn* 1999;32(1):71–86.

46. West J, Fitzpatrick JM, Wang MY et al. Comparison and evaluation of retrospective intermodality brain image registration techniques. *JCAT,* Hawkes DJ 1997;21(4):554–566.

47. Maes F, Vandermeulen D, Suetens P. Medical image registration using mutual information. *Proc IEEE* 2003;91(10):1699–1722.

48. Viola P, Wells III WM. Alignment by maximization of mutual information. *Proc 5th Int Conf Comp Vis* 1995:16–23.

49. West JB, Fitzpatrick JM, Wang MY, Dawant BM, Maurer CR Jr., Kessler RM, Maciunas RJ, Barillot C, Lemoine D, Collignon A, Maes F, Suetens P, Vandermeulen D, van den Elsen PA, Napel S, Sumanaweera TS, Harkness B, Hemler PF, Hill DLG, Hawkes

DJ, Studholme C, Maintz JBA, Viergever MA, Malandain G, Pennec X, Noz ME, Maguire GQ Jr., Pollack M, Pelizzari CA, Robb RA, Hanson D, and Woods RP. Comparison and evaluation of retrospective intermodality image registration techniques. *J Comput Assist Tomogr* 1997;21:554–566.

50. Wells WM III, Viola P, Kikinis R 1995. Multi-modal volume registration by maximization of mutual information. In *Medical Robotics and Computer Assisted Surgery.* Wiley-Liss, New York 1995; 55–62.

51. Maes F, Collignon A, Vandermeulen D, Marchal G, Suetens P. Multimodality image registration by maximization of mutual information. *IEEE Trans Med Imag* 1997;16:187–198.

52. Van den Elsen PA, Pol EJD, Sumanaweera TS, Hemler PF, Napel S, Adler J. Grey value correlation techniques used for automatic matching of CT and MR brain and spine images. *Proc Visual Biomed Comput* 1994;2359:227–237.

53. Fitzpatrick JM. Evaluation of retrospective image registration. Vanderbilt Univ., Nashville, TN, National Institutes of Health, Project Number 1 R01 NS33926-01, 1994.

54. Rueckert D, Sonoda LI, Hayes C, Hill DLJ, Leach MO, Hawkes DJ. Non-rigid registration using free-form deformations: application to breast MR images. *IEEE Trans Med Imag* 1999;18:712–721.

55. Press W, Teukolsky S, Vetterling W, Flannery B. *Numerical Recipes in C. The Art of Scientific Computing, 2nd ed.* Cambridge University Press, Cambridge 1994.

56. Duay V, D'Haese P, Li R, Dawant BM. Non-rigid registration algorithm with spatially varying stiffness properties. *Proceedings of the IEEE International Symposium on Biomedical Imaging (ISBI): From Nano to Macro,* Institute of Electrical and Electronics Engineers (IEEE), Arlington: VA; 2004, 408–411.

57. Li X, Yankeelov TE, Peterson TE, Gore JC, Dawant BM. Automatic nonrigid registration of whole body CT mice images. *Med Phys* 2008;35(4):1507–1520.

58. Shen D, Davatzikos C. HAMMER: Hierarchical attribute matching mechanism for elastic registration. *IEEE Trans Med Imag* 2002;21(11):1421–1439.

59. Humm JL, Ballon D, Hu YC, Ruan S, Chui C, Tulipano PK, Erdi A, Koutcher J, Zakian K, Urano M, Zanzonico P, Mattis C, Dyke J, Chen Y, Harrington P, O'Donoghue JA, Ling CC. A stereotactic method for the three-dimensional registration of multi-modality biologic images in animals: NMR, PET, histology, and autoradiography. *Med Phys* 2003;30:2303–2314.

60. Zhang M, Huang M, Le C, Zanzonico PB, Claus F, Kolbert KS, Martin K, Ling CC, Koutcher JA, Humm JL. Accuracy and reproducibility of tumor positioning during prolonged and multi-modality animal imaging studies. *Phys Med Biol* 2008;53:5867–5882.

61. Cho H, Ackerstaff E, Carlin S, Lupu ME, Wang Y, Rizwan A, O'Donoghue J, Ling CC, Humm JL, Zanzonico PB, Koutcher JA. Noninvasive multimodality imaging of the tumor microenvironment: registered dynamic magnetic resonance imaging and positron emission tomography

studies of a preclinical tumor model of tumor hypoxia. *Neoplasia* 2009;11(3):247–259.

62. Zanzonico P, Campa J, Polycarpe-Holman D, Forster G, Finn R, Larson S, Humm J, Ling C. Animal-specific positioning molds for registration of repeat imaging studies: comparative microPET imaging of F18-labeled fluoro-deoxyglucose and fluoro-misonidazole in rodent tumors. *Nucl Med Biol* 2006;33:65–70.

63. Thorwarth D, Eschmann SM, Paulsen F, Alber M. A kinetic model for dynamic [18F]-FMISO PET data to analyse tumour hypoxia. *Phys Med Biol* 2005;50:2209–2224.

64. Loveless ME, Whisenant JG, Wilson K, Lyshchik A, Sinha TK, Gore JC, Yankeelov TE. Coregistration of ultrasonography and magnetic resonance imaging with a preliminary investigation of the spatial colocalization of VEGFR2 expression and tumor perfusion in a murine tumor model. *Mol Imaging* 2009;8:187–198.

65. Maguire GQ, Jr., Noz ME, Rusinek H, Jaeger J, Kramer EL, Sanger JJ, Smith G. Graphics applied to image registration. *IEEE Comput Graph Appl* 1991;11:20–29.

66. Galbán CJ, Chenevert TL, Meyer CR, Tsien C, Lawrence TS, Hamstra DA, Junck L, Sundgren PC, Johnson TD, Ross DJ, Rehemtulla A, Ross BD. The parametric response map is an imaging biomarker for early cancer treatment outcome. *Nat Med* 2009;15(5):572–576.

67. Cao Y, Tsien CI, Nagesh V, Junck L, Haken RT, Ross BD, Chenevert TL, Lawrence TS. Clinical investigation survival prediction in high-grade gliomas by MRI perfusion before and during early stage of RT. *Int J Radiat Oncol Biol Phys* 2006;64:876–885.

68. Meyer CR, Boes JL, Kim B, Bland PH, Zasadnyb KR, Kison PV, Koral K, Frey KA, Wahla RL. Demonstration of accuracy and clinical versatility of mutual information for automatic multimodality image fusion using affine and thin-plate spline warped geometric deformations. *Med Image Anal* 1997;1:195–206.

69. Rohlfing T, Maurer CR, Bluemke JDA, Jacobs MA. Volume-preserving nonrigid registration of MR breast images using free-form deformation with an incompressibility constraint. *IEEE Trans Med Imag* 2003;22:730–741.

Synthesis of Multiparametric Data

Kathryn M. McMillan
GE Healthcare

C. Chad Quarles
Vanderbilt University

Ronald R. Price
Vanderbilt University

19.1 Introduction

Multiparametric image analysis has become increasingly relevant within oncology research. Despite the exquisite spatial detail offered by the traditional T_1- and T_2-weighted images and even with the use of contrast agents, it is well known that these images are not uniquely specific for any malignancy. However, as evidenced by the prior chapters, magnetic resonance (MR) imaging (MRI) now offers a multitude of new parameters (Table 19.1), with each offering a portion of the physiological and anatomical picture. The goal of this chapter is to describe three of the more common approaches and techniques used to synthesize multiparameter data sets, identify areas of uncertainty, and present examples where these methods have been applied to cancer, and to discuss future directions for multiparameter analyses.

At present, multiparameter analysis packages that are available on current clinical instruments are typically limited to analyzing only a few parameters. Specifically, the predominant commercial strategy has been limited to "drag and drop" image fusion that allows a user to select one parameter map and place it as an overlay on another. This limitation is illustrated by the images shown in Figure 19.1. In this example, we have four MRI parameter maps to correlate: (1) a T_1-weighted gray-scale image, (2) a functional MRI (fMRI) statistical activation map, (3) a diffusion fractional anisotropy (FA) map, and (4) a magnetization transfer ratio (MTR) map. Figure 19.1b illustrates the superposition of the fMRI activation map (yellow and red) and the T_1-weighted image. This is currently the most common approach for presenting fMRI results. For this application, this approach provides an easily interpreted result. However, if we go one step further and use the same approach to also include the green FA map, we have the result shown in Figure 19.1c. Since we have control over transparency, it is possible

to improve it somewhat by allowing all three to be visualized. In Figure 19.1d, we further confuse and complicate the result by adding the blue MTR map. This simple illustration points out how simple image fusion has not been found to provide an adequate approach to multiparameter analyses. The goal of the following sections is to review some of the more common approaches to synthesizing multiparameter data sets. We have chosen to specifically illustrate three algorithms: composite mapping, a statistically based method [iteratively self-organizing of k-means (ISODATA)], and a random field model–based method (Markov random fields). These were chosen in part because each of these algorithms is available as public-domain freeware software packages.

19.2 Methods for Multiparameter Mapping

19.2.1 Composite Mapping

The composite mapping method is similar to the overlay method described in Section 19.1, except that, with the composite mapping method, a threshold function is applied prior to the overlay. The choice of the threshold value for a specific parameter is typically chosen to select only those voxels with a high likelihood of predicting the condition or state under consideration. For example, if quantitative values of the choline/N-acetyl-aspartate ratio (Ch/NAA; see Chapter 9 for a detailed description of MR spectroscopy) above 2.2 were known to be highly indicative of recurrent tumors, the appropriate threshold map would be derived from the original Ch/NAA parametric image map and would only show only those voxels in which the predetermined threshold has been exceeded.

To illustrate how the composite mapping method is used to answer a specific question, assume that we have three different

TABLE 19.1 MRI Parameter Maps That Could Be Applied to Tumor Mapping

Technique	Map	Reference
T_1, T_2, proton density	Structure	Chapter 5
Arterial spin labeling	Non–contrast-enhanced cerebral blood flow	Chapter 6
Diffusion weighted imaging	ADC	Chapter 7
	FA	
	Fiber tracks	
Magnetization transfer imaging	MTR	Chapter 8
DCE perfusion	K^{trans}, v_e	Chapter 12
DSC perfusion	Cerebral blood volume, cerebral blood flow	Chapter 13
MR angiography	MRA vasculature maps	Chapter 14
Spectroscopy	Metabolite maps, including choline, creatine, NAA, lactate, citrate, lipids	Chapter 9
	pH	
BOLD contrast (T_2^*)	Task-based function (fMRI), hypoxia (carbogen)	Chapter 15 [35,51]
Susceptibility weighted	Vessel maps	Lupo et al. 2009; Sehgal et al. 2006
MR elastography	Stiffness maps	Glaser et al. 2006; Sinkus et al. 2007; Venkatesh et al. 2008
Hyperpolarized C13	Metabolism	Chapter 22

MR parameter maps shown in Figure 19.2, each of which has its own individual threshold value. From each of the parametric images, a binary image is generated identifying only those voxels that exceed the threshold. Specifically, voxels exceeding the threshold are assigned a value of 1, whereas voxels below the threshold value are assigned 0.

The result is a series of three binary maps (Figures 19.2.2a through 19.2.2c), each containing a parameter of physiologic or anatomical information relevant to predicting the likelihood of a specific tumor. In order to retain the knowledge of the origin of various sources of information, each binary threshold map is then assigned a unique number. The circle voxels are assigned a value of 1, the rectangle voxels are assigned a value of 2, and the irregular map is assigned a value of 4. The resulting maps are then superimposed and the contents summed. With this scheme, the resulting summed image can be decoded numerically and interpreted using the following key, or alternatively, one could use a 7-color-coded map to be interpreted visually:

0 = No relevant techniques contributed to this voxel
1 = Suprathreshold voxels from Circle only
2 = Suprathreshold voxels from Rectangle only
3 = Suprathreshold voxels from Circle and Rectangle
4 = Suprathreshold voxels from Irregular only
5 = Suprathreshold voxels from Circle and Irregular
6 = Suprathreshold voxels from Rectangle and Irregular
7 = Suprathreshold voxels from Circle, Rectangle, and Irregular

If a color-coded map is used as the output, a three-parameter map would require seven colors, a four-parameter map would require 15 colors, and a five-parameter map would require 31 colors.

A concern with the composite mapping technique is the degree of subjectivity that may be introduced by the uncertainty in the choice of threshold values. False-positive areas will in general be avoided due to the redundancy in the overlay calculations; however, assessing the true extent of the diseased region may require multiple maps with different thresholds.

The accuracy of the method is clearly dependent upon the initial feature extraction step. Choosing a threshold utilizing more quantitative techniques such as edge detection [1,2] or region growing algorithms [3] instead of a global threshold may also improve the sensitivity and specificity of the method. However, the composite mapping method is very simple to implement and provides an easily obtained map of a small number of relevant tumor features.

The composite mapping algorithm is available for free download from the National Institutes of Health Web site:

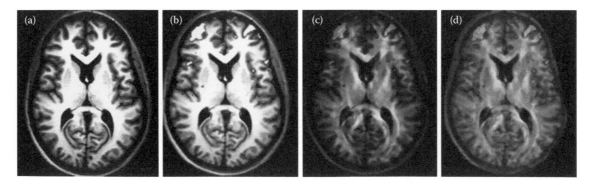

FIGURE 19.1 **(See color insert.)** Simple overlay of four MRI parameter maps: T_1-weighted (gray), fMRI activation map (yellow/red), FA map (green), and MTR map (blue).

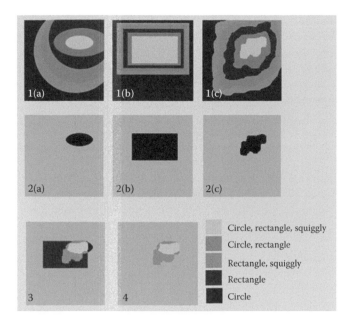

FIGURE 19.2 Composite Mapping. Step 1: Input parameter maps Circle (1a), Rectangle (1b), and Irregular (1c) show a process of interest. Step 2: Binary threshold maps of suprathreshold voxels in Circle (2a), Rectangle (2b) and Irregular (2c). Step 3: Although illustrate here in shades of gray, in practice, color-code maps produced in step 2 are overlayed. Step 4: Simplify the map (and decrease potential for false positives) by showing only regions that contain suprathreshold voxels from two or more maps. Step 5 (not shown): Overlay of composite maps onto structural image for reference.

http://afni.nimh.nih.gov/afni. In order to use the algorithm, parameter images must first be converted to analysis of functional neuro images (AFNI) format (via the to3d command). Once the images have been converted, the calculation can be performed by altering the following command line:

```
3dcalc -prefix Composite -a Circle+orig. -b
Rectangle+orig. -c Irregular+orig. -expr
'1*step(a-Circle_threshold)+2*step(b-
Rectangle_threshold)+4*step(c-Irregular_threshold)'
```

Note that the images must be of the same size, and thus, resampling may be necessary. This can be accomplished by using the command 3dresample.

19.2.2 Statistically Based Methods

Segmentation is the general term used to describe the classification of voxels belonging to the same group or cluster (e.g., disease, anatomy, or function). Statistical segmentation strategies offer some distinct advantages over composite mapping methods and are available as several algorithms [4,5]. The goal of all such methods is essentially the same: to group individual voxels with others of the same type that demonstrate the same characteristics [6,7]. Statistically based methods may be either

supervised with the user interacting with the segmentation process or unsupervised. The unsupervised techniques limit user input to the selection of input parameters and then work automatically to separate the image set into tissue classes based on size, shape, contrast, or connectivity [8,9]. Two of the more common statistics-based multiparameter segmentation algorithms are ISODATA and the hidden Markov random field (HMRF) segmentation algorithm. Below we illustrate the basics of each of these using the same three hypothetical MR parameter maps used for illustrating the composite mapping method as input.

19.2.2.1 ISODATA

With ISODATA, the approach is to define an n-dimensional space of tumor parameters; that is, each MR parameter defines an axis in the n-dimensional space. Different tissue types must be characterized by one of the many available MR image parameters (Table 19.1). These parameter values become the basis for the classification of each tissue type to a particular cluster in the space and specifically define a signature vector, S, for each pixel in a data set. With the same three-parameter data set introduced previously, the S vector may be written as

$$S = [S_{Circle}, S_{Rectangle}, S_{Squiggly}]^T \qquad (19.1)$$

where T denotes that the matrix is transposed and three parameter maps are the input. The process is initialized with a user-defined number of cluster centers. The illustration shown in Figure 19.3 begins with the initial iteration consisting of four clusters (Figure 19.2.2a). Then the intra-Euclidean distances to the cluster center are calculated pixel by pixel. Pixels with large distances will be placed in other clusters. Distances between cluster centers are also determined. Cluster centers located closely together will be merged. The distances are calculated, and further merging and splitting continue until convergence or until the maximum number of iterations is reached. This process is illustrated graphically in Figure 19.3. The result from this analysis is a single-theme map (Figure 19.3.2d), where different clusters are shown by different shades of gray or more commonly by different colors [10–12]. In the final map, the algorithm has concluded that the best solution was that there were five tissue clusters rather than the initial guess of four.

A variation on the method was implemented by Shen et al. [12] in which a Mahalanobis distance rather than Euclidean distance was used to address the coordinate axis scaling, to correct for correlation between different features, and to allow for curved decision boundaries. If the covariance matrix is the identity matrix, the Mahalanobis distance calculation reduces to the standard Euclidean distance calculation. Yovel and Assaf [13] have presented encouraging results for clustering regions within the thalamus using a technique they call "virtual-dot-com imaging," which applies k-means clustering within a multistep algorithm.

Another variation, referred to as "fuzzy c-means," is a generalization of the k-means approach [5]. Fuzzy methods are derived from a fuzzy set theory that deals with reasoning that is

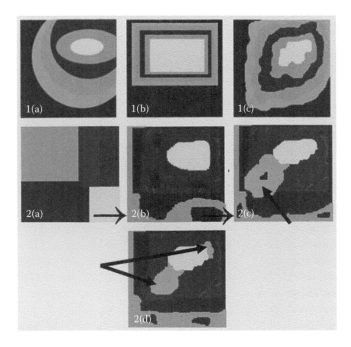

FIGURE 19.3 ISODATA. 1. Iteratively self-organizing variation of the *k*-means algorithm takes Circular region (1a), Rectangular region (1b), and Irregular region (1c) as input parameter images. A new image space is created where each voxel is defined by a combination of these three parameters. 2a. The algorithm begins with the initial iteration assuming four clusters and is then set to allow up to 100 iterations to reach an optimal solution. Four clusters are randomly assigned (2a) as the initial iteration. 2b. Without user interaction, Euclidian distances are calculated and clusters are merged and split so that like elements are appropriately contained. 2c. When it is recognized that an additional cluster is necessary, a fifth class is created (arrows). 2d. After convergence is reached, a final map is output and concludes that five classes best represent the underlying parameter distribution.

approximate rather than precisely accurate. When fuzzy methods are applied, labels are descriptive of how well a pixel fits a given cluster [14,15]. In some cases, this approach allows for better classification when pixels contain a mixture of tissue types. This added dimension allows the classification to more accurately reflect the underlying physiology considering the fact that voxels are very large (> 1 mm³) relative to cellular dimensions. For example, the idea that a voxel can include recurrent disease as well as edema is clearly a much more realistic model. Other investigators have added algorithms to account for image noise, spatial blur during acquisition, and indistinct borders between tissues [16]. ISODATA is a part of the EigenTool package: http://radiologyresearch.org/eigentool.htm.

19.2.2.2 HMRF

Methods that utilize Markov random fields are also statistically based multiparameter classification methods that depend upon certain assumptions or models of the probability distribution function of the image intensities and an associated classification, both of which can be considered random variables. We will assume that *X* and *Y* are two random variables for the class label

and the pixel intensity, respectively. The class-conditional probability density function is $p(y|x)$. Statistical approaches attempt to solve the problem of estimating the associated class label *x*, given only the intensity *y* for each pixel. Such an estimation problem is necessarily formulated from established criteria such as the maximum a posteriori or maximum likelihood (ML) principles. The concept of an HMRF model assumes stochastic processes whose state cannot be observed directly, but only through a sequence of observations. Each observation is assumed to be a stochastic function of the state sequence. A number of investigators have demonstrated that the image segmentation problem can be described by a statistical model and that estimating the parameters which characterize the model can be solved using the estimation maximization (EM) algorithm. The EM algorithm attempts to find the ML of parameters in statistical models where the model depends upon latent or unobserved variables. EM is an iterative method which alternates between calculating an expectation (*E*) step, using the current estimate for the latent variables, and a maximization (*M*) step, which computes parameters maximizing the expected log-likelihood found on the *E* step. These parameter estimates are then used to determine the distribution of the unknown and unobserved "latent" variables in the next *E* step. More traditional Bayesian model classifications have also been used [17–19]. In statistical (nonparametric) methods, the probability density function relies entirely on the data itself, and no assumption is made about the functional form of the distribution. However, a large number of correctly labeled training points are required. A widely used nonparametric method utilizes the mutual influence of nearest neighbors to determine the relationship of the neighborhood.

Image segmentation software utilizing HMRF processing is available from the University of Oxford as part of the Oxford Centre for Functional Magnetic Resonance Imaging of the Brain (FMRIB) software library (FSL) (http://www.fmrib.ox.ac.uk/fsl/). Markov random fields can systematically include constraints about known characteristics of the image as well as implement reasonable algorithms to approximate optimal solutions [20,21]. The details of the HMRF model are explained by Zhang et al. [22], as they pertain to the processing performed using FSL's segmentation tool.

Let us assume that *X* is the true classification of a pixel with an associated state space denoted by *L* and that *Y* is the observed characteristic of the pixel (e.g., the pixel intensity) that depends on the true classification and has a state space denoted by *D*. A configuration of *X* and *Y* can be given, respectively, as

$$X = \{x = (x_1, \cdots, x_N) | x_i \in L, i \in S\} \qquad (19.2)$$

$$Y = \{y = (y_1, \cdots, y_N) | y_i \in D, i \in S\} \qquad (19.3)$$

Since $X_i = l$ (realization of the random field by way of configuration), Y_i has the conditional probability such that

$$p(y_i|l) = f(y_i; \theta_l), \forall l \in L \qquad (19.4)$$

where θ_l is a set of parameters. A value from the set L is taken to assign a class label to each pixel that has an image intensity value y_i from set D.

$$\hat{x} = \arg\max_{x \in X}\{P(y|x)P(x)\} \qquad (19.5)$$

The prior probability of x can be given by way of the finite mixture model,

$$P(x) = \frac{1}{Z^{-U(x)}} \qquad (19.6)$$

Given the class label $x = l$, one can assume that the intensity, y_i, follows a Gaussian distribution:

$$p(y_i|x_i) = g(y_i;\theta_l) = \frac{1}{\sqrt{2\pi\sigma_l^2}}e^{-\frac{(y_i-\mu_l)^2}{2\sigma_l^2}} \qquad (19.7)$$

The joint likelihood probability can then be expressed as

$$P(y-x) = \frac{1}{Z'^{-U(y|x)}} \qquad (19.8)$$

where the likelihood energy is given by

$$U(y-x) = \sum_{i \in S}\left[\frac{(y_i-\mu_{xi})^2}{2\sigma_{xi}^2} + \log(\sigma_{xi})\right] \qquad (19.9)$$

with a constant normalization:

$$Z' = (2\pi)^{\frac{N}{2}} \qquad (19.10)$$

$$\log P(x|y) \propto -U(x|y) \qquad (19.11)$$

$$U(x|y) = U(y|x) + U(x) + c \qquad (19.12)$$

where $U(x|y)$ is the posterior energy, or the probability that a pixel with the intensity y will correspond to the classification state x. Minimizing this function is achieved through iteratively solving $\hat{x} = \arg\min_{x \in X}\{U(y|x)+U(x)\}$ by using y (the signal intensities of the image) with their assigned labels $x_{S-\{i\}}^k$ by updating $x_i^{(k)}$ to $x_i^{(k+1)}$ and minimizing the conditional posterior probability, $U(x_i|y, x_{S-\{i\}})$. As noted earlier, model fitting in FSL is achieved using the EM algorithm. By fitting a series of Gaussians to the data, EM can cluster a given data set. EM optimization schemes have failed on data sets with a large number of input parameters due to problems with numerical precision [8].

Figure 19.4 illustrates the HMRF segmentation method using the same three-parameter input data sets used in the prior illustrations. The resulting image (Figure 19.4.2d) also identified five distinct classes; however, the spatial distribution is significantly different from the results of either composite mapping or ISODATA. These results illustrate how essential high-quality

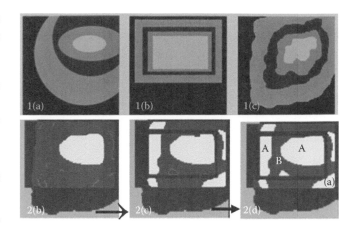

FIGURE 19.4 HMRF segmentation input parameter images (1a, 1b, and 1c). 2b. Initial iteration. 2c. Intermediate result. 2d. Final result showing regions of highest commonality, that is, same classification (A) and (B). Modeling using random fields considers the interactions between neighboring pixels. Since neighboring pixels tend to belong in the same class, the model assumes that a classifiable structure is contained within a single pixel.

training sets are to the accuracy of the classification and clearly establish the need for independent verification of the true tissue classifications. While a multispectral algorithm that operates in three dimensions is potentially of great utility, excessive noise in a region can cause nonideal separation of image regions, yielding a result that fails to realistically map the tumor environment. Additionally, image quality issues due to technical or biological properties can cause nonideal segmentation [22]. Computation times for the multiparameter statistically based methods, especially when high-resolution images are used, will be significantly longer than the image-based composite mapping method. Despite this problem, with faster computers, FSL should be considered a viable analysis option.

19.3 Applications in Cancer

19.3.1 Preclinical Applications in Cancer

As suggested in the previous section, preclinical animal studies are an important approach for validation of multiparameter analyses. In addition to allowing for the use of multiple imaging modalities, sedation, and immobilization for better image coregistration, animal studies have the opportunity to include quantitative histopathological correlation.

To illustrate the methods presented above, we will use the input parameter maps shown in Figure 19.5 derived from a rat model. These maps include (1) a contrast-enhanced T_1 image (Figure 19.5.1a), (2) an apparent diffusion coefficient (ADC) map (Figure 19.5.1b), (3) a cerebral blood volume (CBV) map (Figure 19.5.1c), and (4) a hypoxic fraction map (Figure 19.5.1d) derived from ^{64}Cu-ATSM PET images. Animals were inoculated on day 0 with C6 brain tumor cells. On day 17, rats were imaged with 7.0-T MRI and PET to understand the relationship between perfusion parameters and tumor hypoxia.

Hypoxic fractions were determined using the 24-h-postinjection ^{64}Cu-ATSM images and the Gd-DTPA T_1 weighted images; volume = 135 mm^3, necrotic fraction ≈ 12%, and hypoxic fraction = 88%.

The final composite map is presented in Figure 19.5.4 (with intermediate steps 2 and 3) and enables visualization of the overlapping parameters. Changes in cellular density, estimated by ADC, are also found to correlate with hypoxia, shown by areas in cyan and yellow. Hypoxic areas can be found in regions displaying elevated CBV (red and yellow regions in Figure 19.5.4), possibly because tumor neovasculature is relatively dysfunctional, and the regions it serves may receive inadequate amounts of

oxygenated blood [23]. Additionally, hypoxia promotes angiogenesis. Hypoxia-inducible factor 1 (HIF-1) is upregulated by hypoxic environments. Numerous genes, including glucose transporters, glycolysis enzymes, and vascular endothelial growth factor (VEGF), are then activated to aid the adaptation of malignant cells to the hostile environment [24]. VEGF is the most important of the growth factors, as it promotes angiogenesis and acts as a potent permeability factor [16]. VEGF is specifically required for survival of immature vasculature [8]. While the tumor environment depicted by the composite map is complicated, it indicates a likely aggressive region in the inferior, medial aspect of the contrast-enhancing region. Less labor-intensive but requiring

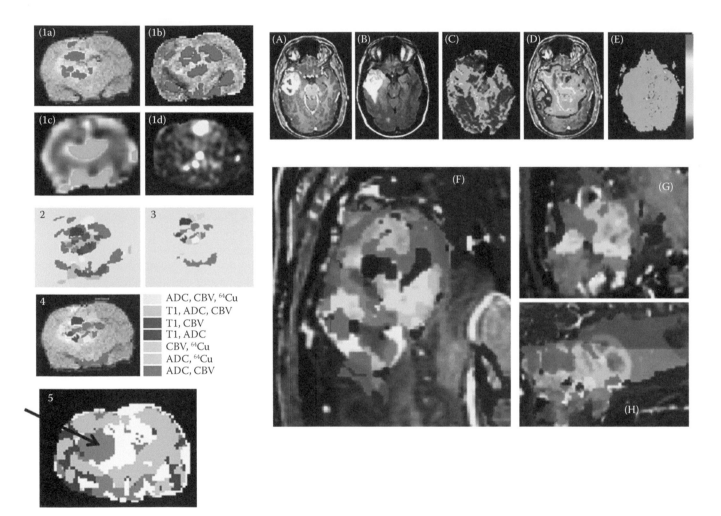

FIGURE 19.5 **(See color insert.)** Composite map analysis of a preclinical animal study (rat). Input images include postcontrast T_1 (1a), ADC (1b), CBV (1c), and PET image with ^{64}Cu-ATSM (1d). Threshold images are shown in panels 2 and 3. Thresholds are applied and overlaid on the postcontrast T_1-weighted image (panel 4). Colors correspond to the legend on the bottom right and show the most concurrent parameters in the lower medial aspect of the contrast-enhancing lesion in yellow and orange. (Panel 5) Using the same input images, analysis was then performed with the HMRF segmentation method. While these computationally intensive, automated methods are more sophisticated, interpretation of the final maps is nontrivial. However, four image classes were identified, and the tumor is clearly seen (arrow) as a separate tissue class. (A–F). Multiparametric imaging performed preoperatively on a glioma patient. (A) Postcontrast T_1. (B) Postcontrast T_2-FLAIR. (C) Cerebral blood volume. (D) Choline/creatine. (E) Apparent diffusion coefficient. The color images use the continuous color bar on the far right. Composite map (F–H), magnified to show tumor area. Axial (F), coronal (G), and sagittal (H) views of the tumor region. The map shows a highly active, heterogeneous environment, but areas in yellow show regions where all four parameters overlap. While surgery to debulk the tumor was done, maps such as these can give treating physicians increased insight into the functional environment in addition to the anatomical structure.

greater computational power, the HMRF segmentation technique applied to the same input parameter images is shown in Figure 19.5.5. The HMRF method resulted in four tissue classes, one of which shows a round region containing the bulk tumor (arrow) with a more irregular region to the medial, inferior edge of the tumor that defines another type of tissue. Careful histopathological analysis could define the utility of maps such as these.

More advanced segmentation strategies have been applied to preclinical data. Early work by Vinitski et al. [25] applied a three-dimensional (3D) feature map to segment tissue in a hamster model of brain tumors. Image analysis using ISODATA clustering has been completed in stroke models using multiparametric MR data [26]. Ali et al. [27] performed 3D segmentation on images acquired at isotropic 90-μm resolution and noted results that are very similar to those achieved on human images. As multiparametric imaging begins to be applied to more interesting therapeutic approaches (e.g., stem cell therapies) [28], better synthesis tools grow increasingly important.

19.3.2 Applications in Cancer Diagnosis

Figures 19.5A through 19.5F illustrate multiparametric imaging performed preoperatively on a glioma patient. The input parameter maps are (A) postcontrast T_1, (B) postcontrast T_2-weighted fluid attenuated inversion recovery (T_2-FLAIR), (C) CBV, (D) choline/creatine, and (E) ADC. The color images use the continuous color bar on the far right.

The postcontrast T_1-weighted images (A) depict a heterogeneous lesion. Given that T_1-weighted images are generally thought to underestimate volumes of high-grade gliomas, it is expected that the abnormality on the T_2-FLAIR image extends farther in the posterior direction (B). Additional data sets can be utilized to better understand the tumor environment and extent.

CBV image (C) appears higher in the anterior portion of the tumor region, suggesting neovascularity in those hyperintense spots. Many quickly growing tumors outgrow their blood supply and form new vasculature in an attempt to deliver nutrients and oxygen necessary for future growth. These vascular networks also offer an avenue for growth into new regions [29]. While CBV maps are nonspecific to angiogenesis given that they also appear bright in normal vascular regions, high blood volume measures could indicate areas of concern.

The large area of elevated choline (D) also highlights the anterior portion of the tumor, suggesting malignant processes in that region [30]. While creatine is thought to remain fairly constant across tissue types, choline has been correlated with tumor grade and outcome [31,32]. Closer examination of the choline/creatine map also shows that the medial portion of the tumor has a metabolic signature indicating aggressive growth.

Increased cellularity can restrict water diffusion by decreasing the extracellular space in a densely packed region of tumor growth. Voxels with decreased intensity (*E*) show reduced ADC and look similar to the T_2-FLAIR enhancement (B). While tumor cellularity is thought to be a major factor when determining the

ADC values of brain tumors, it is not the only variable [33]. For example, changes in ADC over time could show reorganization of the tumor environment [34].

The multiparameter maps in Figures 19.5F through 19.5H offer a great deal of information for this treatment-naive patient. However, as diagnoses are made and treatments are planned, there are two important questions we must answer: Where is the tumor? And are there particularly aggressive areas that should be targeted via surgical intervention and radiation therapies? Composite maps offer a potential way to quantitatively address these questions via imaging.

Using this technique [35,36], we have a series of colors overlaid on the T_1-weighted image. When viewing the axial plane, we see that the posterior tumor extends past the edge of contrast enhancement (combinations of T_2-FLAIR enhancement, ADC abnormalities, and high choline/creatine) and that there is a medial component to the tumor, highlighted by choline/creatine and ADC changes, that requires closer attention. The sagittal view perhaps best demonstrates the value in multiparametric mapping. The diffusion, perfusion, and spectroscopic maps clearly show tumor extending far past the posterior edge of enhancement and define additional regions that may require treatment.

As multiparametric studies become increasingly common in cancer research, handling the large amounts of imaging data becomes increasingly challenging [37,38]. Although a number of papers have demonstrated significantly better definition of tumor volume and extent based on multiparametric data [33,39–41], the synthesis of resulting data remains an open question. Work in brain tumors has been most common and has included approaches covered in this chapter—random field modeling [19] and composite mapping [42]. There are a number of additional techniques that also show promise for cancer diagnosis; among them are self-organizing maps [43].

Texture analysis [44] and an ISODATA approach [45] have been applied to breast cancer. Osteosarcoma has also been covered using a fuzzy c-means approach [46].

Multiparametric imaging applied to prostate cancer is reviewed by Kurhanewicz et al. [47]. Langer et al. [48] completed a logistic regression analysis with perfusion and diffusion parameters. Chen et al. [49] performed imaging in 42 cases of advanced prostate cancer to evaluate the efficacy of multiparametric (T_2, DWI, 3D spectroscopy) in detecting cancer in various regions of the prostate. All 42 cases correlated biopsy specimens with the evaluation of two radiologists who rated four sextants of the prostate images with a five-point confidence scale (1 = definitely cancer, 5 = definitely not cancer). When interpreting ROC curves for the combination of techniques, it was found that the best results compared with biopsy were possible when using all three techniques. It is interesting to note that spectroscopy was almost as powerful.

19.3.3 Applications in Cancer Therapy

Multiparameter mapping can also be of significant benefit for planning cancer therapy. For example, when prescribing radiation therapy, a greater knowledge of the tumor environment could allow advances in radiation delivery systems (e.g., intensity modulated radiotherapy) to deliver more focused, patient-specific treatment [50]. Several groups have attempted multiparametric mapping for planning surgery [51,52]. Treatment planning using spectroscopic images has also been applied [50]. Also, given the challenge introduced by radiation-induced imaging changes, planning treatment and predicting recurrence for gliomas have been studied [35,36,42,53].

Figure 19.6 illustrates a multiparameter analysis of a patient with a metastatic brain tumor. MRIs were performed before (Figures 19.6A through 19.6C) whole-brain radiation therapy and again 6 weeks after (Figures 19.6A$_x$ and 19.6B$_x$) its completion. The scans show a decrease in lesion sizes on the T_1-weighted

image when comparing pretherapy (Figure 19.6A) to posttherapy (Figure 19.6A$_x$). T_2-FLAIR images pretherapy (Figure 19.6B) show more enhancement than their posttherapy (Figure 19.6B$_x$) counterparts. Finally, the ADC map pretherapy (Figure 19.6C) shows more abnormalities than the posttherapy ADC (Figure 19.6 C$_x$).

The results of HMRF segmentation utilizing the combination of T_1, T_2-FLAIR, and ADC maps are shown for pretherapy (Figure 19.6S) and posttherapy (Figure 19.6S$_x$). Note that the HMRF images have been left–right flipped and also note that the large lesion in the left frontal lobe has decreased in size. In addition, the posterior lesion on the left side of the pretherapy lesion appears to have disappeared posttherapy.

Research published by Batchelor et al. [54] utilized a multiparametric approach to understand the normalization of tumor

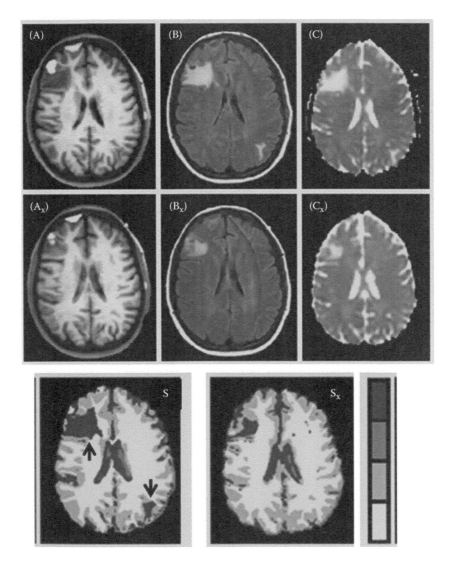

FIGURE 19.6 Metastatic tumor patient MRI before (A–C) whole-brain radiation therapy and again 6 weeks after (A$_x$–B$_x$) its completion. Note the decrease in lesion sizes on T_1-weighted image when comparing pretherapy (A) to posttherapy (A$_x$). T_2-FLAIR images pretherapy (B) show more enhancement than their posttherapy (B$_x$) counterparts. Finally, the ADC map pretherapy (C) shows more abnormalities than the posttherapy ADC (C$_x$). HMRF segmentation defined by a combination of T_1, T_2-FLAIR, and ADC maps pretherapy (S) and posttherapy (S$_x$). Analysis of these images allows for analysis of tumor volume changes. Note that the large lesion in the left frontal lobe has decreased in size. In addition, the posterior lesion on the left size of the pretherapy lesion appears to have disappeared posttherapy.

vasculature following treatment with AZD2171, a pan-VEGF receptor tyrosine kinase inhibitor. Specifically, the parameters used by Batchelor et al. were T_2, T_2-FLAIR, pre- and postcontrast T_1, DCE, DTI, and DSC via MRI along with circulating endothelial cells, progenitor cells, plasma VEGF-A, VEGRFI1 and 2, DIGF, SDF1alpha, and DFGF from blood samples. To analyze these imaging data, regions of interest were placed on a parametric map (manual tracing of volumes on T_2, vessel map size, ADC) and were normalized to unaffected brain tissue, often located distally. Through the analysis of images acquired longitudinally (pre- and post-AZD2171 treatment), the authors were able to better understand the post-anti-angiogenetic effects of treatment for brain tumors. Within their conclusions was the confirmation that the targets were present; the drug did induce a response, and the tumor vessels did rapidly normalize.

19.4 Summary

This chapter has focused on the multiparameter analyses that yield visual output; however, it is also important to recognize parallel efforts of multiparametric image analysis that uses mathematical models [55,56]. Model analysis has been applied to gliomas [57–59] and breast cancer [60].

An increasing number of promising algorithms are freely available and easy to use. Multiparametric imaging is currently in its infancy, with considerable research coming from multiple groups with varying fields of expertise.

References

1. Canny J. A computational approach to edge detection. *IEEE Trans Pattern Anal Machine Intell* 1986;8:679–714.
2. Tang H, Wu E, Ma Q, Gallagher D, Perera G, Zhuang T. MRI brain image segmentation by multi-resolution edge detection and region selection. *Comput Med Imaging Graph* 2000;24:349–357.
3. Brummer ME, Mersereau RM, Eisner RL, Lewine RRJ. Automatic detection of brain contours in MRI data sets. *IEEE Trans Med Imaging* 1993;12(2):153–166.
4. Nattkemper TW. Multivariate image analysis in biomedicine. *J Biomed Inform* 2004;37(5):380–391.
5. Pham D, Xu C, Prince J. A survey of current methods in medical image segmentation. *Ann Rev Biomed Eng* 1998;s:1–27.
6. Clarke L, Velthuizen R, Camacho M, Heine J, Vaidyanathan M, Hall L, Thatcher R, Silbiger M. MRI segmentation: methods and applications. *Magnet Reson Imaging* 1995;13(3):343–368.
7. Xia Y, Feng D, Wang T, Zhao R, Zhang Y. Image segmentation by clustering of spatial patterns. *Pattern Recogn Lett* 2007;28:1548–1555.
8. Benjamin LE, Golijanin D, Itin A, Pode D, Keshet E. Selective ablation of immature blood vessels in established human tumors follows vascular endothelial growth factor withdrawal. *J Clin Invest* 1999;103:159–165.
9. Gibbs P, Buckely D, Blackband S, Horsman A. Tumour volume determination from MR images by morphological segmentation. *Phys Med Biol* 1996;41:2437–2446.
10. Mitsias P, Jacobs M, Hammoud R et al. Multiparametric MRI ISODATA ischemic lesion analysis: correlation with the clinical neurological deficit and single-parameter MRI techniques. *Stroke* 2002;33(12):2839–2844.
11. Omran M, Engelbrecht A. Particle swarm optimization method for image clustering. *Int J Pattern Recogn Image Anal* 2005;19:297–321.
12. Shen Q, Ren H, Fisher M, Bouley J, Duong T. Dynamic tracking of acute ischemic tissue fates using improved unsupervised ISODATA analysis of high-resolution quantitative perfusion and diffusion data. *J Cereb Blood Flow Metab* 2004;24:887–897.
13. Yovel Y, Assaf Y. Virtual definition of neuronal tissue by cluster analysis of multi-parametric imaging (virtual-dot-com imaging). *NeuroImage* 2007;35:58–69.
14. Liu J, Udupa J, Odhner D, Hackney D, Moonis G. A system for brain tumor volume estimation via MR imaging and fuzzy connectedness. *Comput Med Imaging Graph* 2005;29:21–34.
15. Wafa M, Zagrouba E. Tumor extraction from multimodal MRI. *Adv Soft Computing* 2009;57:415–422.
16. Gillies R, Bhujwalla Z, Evelhoch J, Garwood M, Neeman M, Robinson S, Sotak C, van der Saden B. Applications of magnetic resonance in model systems: tumor biology and physiology. *Neoplasia* 2000;2:139–151.
17. Carso JJ, Sharon E, Dube S, El-Saden S, Sinha U, Yuille A. Efficient multilevel brain tumor segmentation with integrated Bayesian model classification. *IEEE Trans Med Imaging* 2008;27(5):629–640.
18. Carso JJ, Sharon E, Yuille A. Multilevel segmentation and integrated Bayesian model classification with an application to brain tumor segmentation. *Lecture Notes in Computer Science* 2006;4191.
19. Lee CH, Wang S, Murtha A, Brown MRG, Greiner R. Segmenting brain tumors of pseudo-conditional random fields. *Lecture Notes in Computer Science* 2008;5241:359–366.
20. Liang Z, Jaszczak R, Coleman R. Parameter estimation of finite mixtures using the EM algorithm and information criteria with application to medical image processing. *IEEE Trans Nucl Sci* 1992;39(4):1126–1133.
21. Marroquin J, Santana E, Botello S. Hidden Markov measure field models for image segmentation. *IEEE Trans Pattern Anal Machine Intell* 2003;25:1380–1387.
22. Zhang Y, Smith S, Brady M. Segmentation of brain MR images through a hidden Markov random field model and the expectation-maximization algorithm. *IEEE Trans Med Imaging* 2001;20:45–57.
23. Collingridge D, Piepmeier J, Rockwell S, Knisely J. Polarographic measurements of oxygen tension in human glioma and surrounding peritumoural brain tissue. *Radiother Oncol* 1999;53:127–131.
24. Kostourou V, Troy H, Murray J, Cullis E, Witley G, Griffiths J, Robinson S. Overexpression of dimethylarginine dimthylaminohydrolase enhances tumor hypoxia: an insight

into the relationship of hypoxia and angiogenesis in vivo. *Neuroplasia* 2004;6:401–411.

25. Vinitski S, Gonzalez C, Andrews D, Knobler R, Curtis M, Mohamed F, Gordon J, Khalili K. In vivo validation of tissue segmentation based on a 3D feature map using both a hamster brain tumor model and stereotactically guided biopsy of brain tumors in man. *J Magn Reson Imaging* 1998;8:814–819.

26. Soltanian-Zadeh H, Pasnoor M, Hammoud R et al. MRI tissue characterization of experimental cerebral ischemia in rat. *J Magn Reson Imaging* 2003;17:398–409.

27. Ali AA, Dale AM, Badea A, Johnson GA. Automated segmentation of neuroanatomical structures in multi-spectral MR microscopy of the mouse brain. *NeuroImage* 2005;27(2):425–435.

28. Brekke C, Williams SC, Price J, Thorsen F, Modo M. Cellular multiparametric MRI of neural stem cell therapy in a rat glioma model. *NeuroImage* 2007;37(3):769–782.

29. Taylor N, Baddeley H, Goodchild K, Powell M, Thoumine M, Hoskin P, Phillips H, Padhani A, Griffiths J. BOLD MRI of human tumor oxygenation during carbogen breathing. *J Magn Reson Imaging* 2001;14:156–163.

30. McKnight TR, Noworolski SM, Vigneron DB, Nelson SJ. An automated technique for the quantitative assessment of 3D-MRSI data for patients with glioma. *JMRI* 2001;13(2);167–117.

31. Shih M, Singh A, Wang A, Patel S. Brain lesions with elevated lactic acid peaks on magnetic resonance spectroscopy. *Curr Probl Diagn Radiol* 2004;33:85–95.

32. Yamagata N, Miller B, McBride D. In vivo proton spectroscopy of intracranial infections and neoplasms. *J Neuroimaging* 1994;4:23–28.

33. Law M, Cha S, Knopp E, Johnson G, Arnett J, Litt A. High-grade gliomas and solitary metastases: differentiation by using perfusion and proton spectroscopic MR imaging. *Radiology* 2002;222(3):715–721.

34. Kono K, Inoue Y, Nakayama K, Shakudo M, Morino M, Ohata K, Wakasa K, Yamada R. The role of diffusion-weighted imaging in patients with brain tumors. *Am J Neuroradiol* 2001;22:1081–1088.

35. McMillan K, Rogers B, Field A, Laird A, Fine J, Meyerand M. Physiologic characterization of glioblastoma multiforme using MRI-based hypoxia mapping, chemical shift imaging, perfusion and diffusion maps. *J Clin Neurosci* 2006;13:811–817.

36. McMillan K, Rogers B, Koay C, Price R, Meyerand M. An objective method for combining multi-parametric MRI datasets to characterize of malignant tumors. *Med Phys* 2007;35(3):1053–1061.

37. Astrakas L, Song YE, Zarifi M, Makedon F, Tzika AA. The clinical perspective of large scale projects: a case study of multiparametric MR imaging of pediatric brain tumors. *Oncol Rep* 2006;15:1065–1069.

38. Costanzo A, Scarabino T, Trojsi F, Giannatempo GM, Popolizio T, Catapano D, Bonavita S, Maggialetti N, Tosetti M, Salvolini U, d'Angelo VA, Tedeschi G. Multiparametric 3T MR approach to the assessment of cerebral gliomas: tumor extent and malignancy. *Neuroradiology* 2006;48(9):622–631.

39. Batra A, Tripathi RP, Singh AK. Perfusion magnetic resonance imaging and magnetic resonance spectroscopy of cerebral gliomas showing imperceptible contrast enhancement on conventional magnetic resonance imaging. *Australasian Radiol* 2004;48:324–332.

40. Henry R, Vigneron D, Fischbein N et al. Comparison of relative cerebral blood volume and proton spectroscopy in patients with treated gliomas. *Am J Neuroradiol* 2000;21:357–366.

41. Tzika A, Zarifi M, Goumnerova L, Astrakas L, Zurakowski D, Young-Poussaint T, Anthony D, Scott R, Black P. Neuroimaging in pediatric brain tumors: Gd-DTPA-enhanced, hemodynamic, and diffusion MR imaging compared with MR spectroscopic imaging. *Am J Neuroradiol* 2002;23(2):322–333.

42. McMillan K, Price R. Multiparametric mapping of gliomas: demonstration of segmentation strategies. *Brain Map Res Dev* 2008:177–196.

43. Vijayakumar C, Damayanti G, Pant R, Sreedhar CM. Segmentation and grading of brain tumors on apparent diffusion coefficient images using self-organizing maps. *Comput Med Imaging Graph* 2007;31(7):473–484.

44. Chen W, Giger ML, Li H, Bick U, Newstead GM. Volumetric texture analysis of breast lesions on contrast-enhanced magnetic resonance images. *Magn Reson Med* 2007;58(3):562–571.

45. Jacobs M, Barker PB, Bluemke DA, Maranto C, Arnold C, Herskovits EH, Bhujwalla Z. Benign and malignant breast lesions: diagnosis with Multiparametric MR Imaging. *Radiology* 2003;229:225–232.

46. Pan J, Li M. Segmentation of MR osteosarcoma images. *Proceedings of the Fifth International Conference on Computational Intelligence and Multimedia Applications* 2003;5:379.

47. Kurhanewicz J, Vigneron DB, Carroll P, Caokley F. Multiparametric magnetic resonance imaging in prostate cancer: present and future. *Curr Opin Urol* 2008;18(1):71–77.

48. Langer D, van der Kwast T, Evans A, Trachtenberg J, Wilson B, Haider M. Prostate cancer detection with multi-parametric MRI: logistic regression analysis of quantitative T2, diffusion-weighted imaging and dynamic contrast-enhanced MRI. *J Magn Reson Imaging* 2009;30(2):327–334.

49. Chen M, Dang H, Wang J, Zhou C, Li S, Wang W, Zhao W, Yang Z, Zhong C, Li G. Prostate cancer detection: comparison of T2-weighted imaging, diffusion-weighted imaging, proton magnetic resonance spectroscopic imaging and the three techniques combined. *Acta Radiol* 2008;49(5):602–610.

50. Chang J, Thakur S, Perera G, Kowalski A et al. Image-fusion of MR spectroscopic images for treatment planning of gliomas. *Med Phys* 2006;33:32–40.

51. Hata N, Muragaki Y, Inomata T, Maruyama T, Iseka H, Hori T, Dohi T. Interoperative tumor segmentation and volume

measurement in MRI-guided glioma surgery for tumor resection rate control. *Acad Radiol* 2005;12:116–122.

52. Sunaert S. Presurgical planning for tumor resectioning. *J Magn Reson Imaging* 2006;23(6):887–905.

52. Verma R, Zacharaki EI, Ou Y, Cai H, Chawla S, Lee SK, Melhelm ER, Wolf R, Davatzikos C. Multiparametric tissue characterization of brain neoplasms and their recurrence using pattern classification of MR images. *Acad Radiol* 2008;15(8):966–977.

54. Batchelor T, Sorensen A, di Tomaso E, Zhang W, Duda D, Cohen K, Kozak K, Cahill D, Chen P, Zhu M, Ancukiewicz M, Mrugala M, Plotkin S, Drappatz J, Louis D, Ivy P, Scadden D, Benner T, Loeffler J, Wen P, Jain R. AZD2171, a pan-VEGF receptor tyrosine kinase inhibitor, normalizes tumor vasculatore and alleviates edema in glioblastoma patients. *Cancer Cell* 2007;11(1):83–95.

55. Anderson AR, Quaranta V. Integrative mathematical oncology. *Nat Rev Cancer* 2008;8:227–234.

56. Sanga S, Frieboes HB, Zheng X, Gatenby R, Bearer EL, Cristini V. Predictive oncology: a review of multidisciplinary, multiscale in silico modeling linking phenotype, morphology and growth. *NeuroImage* 2007;37(Suppl 1):S120–S134.

57. Hogea C, Davatzikos C, Biros G. An image-driven parameter estimation problem for a reaction-diffusion glioma growth model with mass effects. *J Math Biol* 2008;56:793–825.

58. Mandonnet E, Pallud J, Clatz O, Taillandiet L, Konukoglu E, Duffau H, Capelle L. Computational modeling of the WHO grade II glioma dynamics: principles and applications to management paradigm. *Neurosurg Rev* 2008;31:263–269.

59. Roberts T, Liu F, Kassner A, Mori S, Guha A. Fiber density index correlates with reduced fractional anisotropy in white matter of patients with glioblastoma. *Am J Neuroradiol* 2005;26:2183–2186.

60. Weedon-Fekjaer H, Lindqvist BH, Vatten LJ, Aalen OO, Tretli S. Breast cancer tumor growth estimated through mammography screening data. *Breast Cancer Res* 2008;10:R41.

61. Lupo, J., S. Banerjee, et al. (2009). "GRAPPA-based susceptibility-weighted imaging of normal volunteers and patients with brain tumor at 7T." *Magn Reson Imaging* 27(4): 480–488.

62. Sehgal, V., Z. Delproposto, et al. (2006). "Susceptibility-weighted imaging to visualize blood products and improve tumor contrast in the study of brain masses." *J Magn Reson Imaging* 24(1): 41–51.

63. Glaser, K., J. Felmlee, et al. (2006). "Stiffness-weighted magnetic resonance imaging." *Magn Reson Med* 55(1): 59–67.

64. Sinkus, R., K. Siegmann, et al. (2007). "MR elastography of breast lesions: understanding the solid/liquid duality can improve the specificity of contrast-enhanced MR mammography." *Magn Reson Med* 58(6): 1135–1144.

65. Venkatesh, S., M. Yin, et al. (2008). "MR elastography of liver tumors: preliminary results." *Am J Roentgenol* 190(6): 1534–1540.

V

Emerging Trends

MRI in Radiation Therapy Planning

George X. Ding
Vanderbilt University

Eddy S. Yang
Vanderbilt University

Ken J. Niermann
Vanderbilt University

Fen Xia
Vanderbilt University

Anthony Cmelak
Vanderbilt University

20.1 Qualitative Introduction to Technique

Advances in radiation treatment planning dose calculations and radiation delivery are able to tightly conform the radiation to the shape of the intended target while sparing surrounding normal tissues. These have led to the development of treatment machines with integrated planar and volumetric imaging capabilities. Image-guided radiation therapy has dramatically improved the accuracy of radiation delivery and has emerged as the new paradigm for patient positioning and target localization in radiotherapy. It is now possible to deliver the radiation dose to the intended target much more precisely.

Advances in diagnostic imaging provide improved ability to define the tumor volumes accurately. It is increasingly becoming a common practice that magnetic resonance imaging (MRI) is utilized in radiotherapy treatment planning [1] for target volume delineation because commercial radiotherapy treatment planning systems are capable of performing image registration or image fusion between images acquired with different image modalities, such as MRI, computed tomography (CT), positron emission tomography–computed tomography (PET-CT), and so on. These medical images acquired with multiple modalities provide complementary information that is increasingly used in radiotherapy treatment planning. Figure 20.1 shows a cancer metastasis as seen in a radiotherapy treatment planning CT scan (Figure 20.1a) and MRI scan (Figure 20.1b). It is seen that the tumor is hard to visualize in a CT scan even after the target volume is delineated based on MR images. By using MRI in treatment radiotherapy planning, the tumor target can be defined

more accurately. This improved accuracy can lead to better tumor control as well as sparing of normal tissues. Radiation treatment planning is the radiotherapy preparation process in which treatment is defined in terms of planning target volumes (PTVs) and dose distributions.

Treatment planning plays a key role in the advancement of radiotherapy. The accuracy in radiotherapy, especially in conformal therapy treatment planning, is crucial for success in treatment. The accurate radiation dose calculations in patient radiotherapy treatment planning using external megavoltage photon and electron beams depend on patient CT images that provide electron densities. The introduction of tissue inhomogeneity correction in image-based treatment planning has improved the accuracy of radiation dose calculations significantly. The tissue inhomogeneity correction is based on the electron densities, which are related to the CT number or Hounsfield unit. This tissue characterization relationship allows the conversion of CT value in each voxel of the CT images into electron density for use in the dose calculations. The CT images are capable of discriminating structures with different electron densities, while they are incapable of distinguishing between structures with similar electron densities, such as differences in soft-tissue structures, including tumors. The advantage of MRI compared with CT is its superior ability to discriminate soft tissues that pose similar electron density. This is achieved by extensively manipulating MRI parameters of tissue relaxation times (T_1, T_2). The tumors can be visualized, and therefore, radiation target volume can be delineated for treatment planning. The target contours delineated on the MR images have to be accurately mapped to corresponding locations in the three-dimensional (3D) volumetric CT images for radiotherapy treatment planning. In the next section, we

(a)   (b)

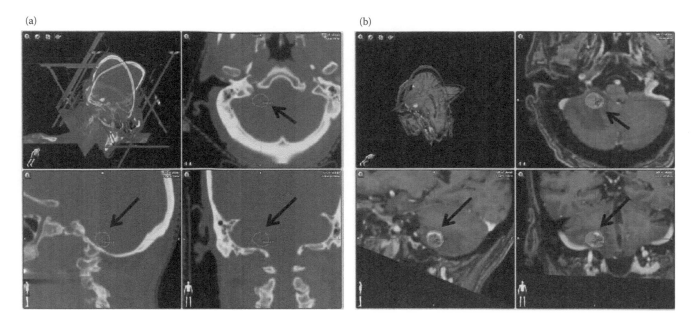

FIGURE 20.1 (a) A cancer metastasis is seen in radiotherapy treatment planning CT images where it is impossible to delineate the target that the contour is mapped from the MR image. (b) The same cancer metastasis is seen in radiotherapy treatment planning MR images where it is easy to identify the target.

will describe a method to ensure that this process is acceptable for patient radiotherapy treatment planning and dose calculations.

20.2 Quantitative Introduction to Techniques of Quality Assurance of Image Registration

Radiation dose calculations have to be performed using patient CT images, which provide electron density that is required for accurate dose calculation for megavoltage radiation beams. Therefore, target volumes delineated on MRI are used to determine the corresponding treatment target on the CT images. This is accomplished by using image registration or fusing between two sets of images. The image registration techniques are based on points, surface, or volume methods. The most common registration algorithm employed in treatment planning systems for automatic image registration is volume based, which uses the optimization of similarity measures calculated from the voxel values between two sets of volumetric images [2]. The optimization process is to relate the image coordinates by minimizing the differences between two sets of images. When two sets of images are registered, contours or a delineated volume on one image set can be seen on the other set. The techniques and algorithms used for image registration can be found in other chapters of this book. In the following sections, we focus on describing a method to evaluate the accuracy and acceptability of CT and MRI fusion software used in a commercial treatment planning system.

Accurate determination of treatment clinical target volume (CTV) and its location is essential for radiotherapy treatment planning. The accuracy of determined target volume on the CT

images is affected by each step in the process, including image registration, image deformation, image acquisition quality, and image resolution. In clinical practice, it is important to estimate the overall uncertainties before a new image modality is

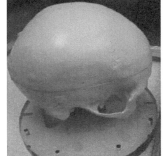

FIGURE 20.2 Photos of the Radionics Brain Phantom. Four geometric objects with known dimensions—a sphere, a cone, a cylinder, and a square column—were mounted on the inside of a manmade skull.

(a)
(b)

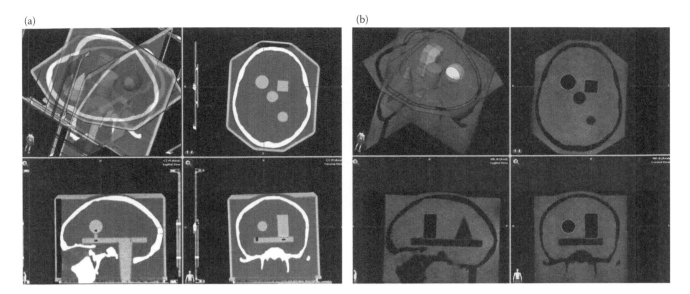

FIGURE 20.3 **(See color insert.)** Comparison of 3D, axial, sagittal, and coronal views of the Radionics Brain Phantom, with (a) CT reconstructed from 2-mm slices and (b) MRI reconstructed from 1.5-mm slices. The four geometric objects were contoured based on CT (a) and MRI (b) images, respectively.

introduced in radiotherapy patient treatment planning. This can be done through a phantom study.

Figure 20.2 shows a Radionics (Radionics Inc., Burlington, MA) head phantom. The phantom consists of a manmade skull placed inside a plastic container that can be filled with water. Four geometric objects with known dimensions—a sphere, a cone, a cylinder, and a square column—are mounted inside the skull. The phantom is filled with water and positioned in the scanners (CT and MRI) using the image acquisition protocols that are typically used for treatment planning purposes.

The scanned digital imaging and communications in medicine (DICOM) images are imported into a radiotherapy treatment planning system, iPlan (BrainLAB Inc, Westchester, IL). Four geometric objects are contoured manually based on CT images and MR images, as shown in Figure 20.3. The manual contouring is based on visual 50% gray level on the CT or MR images. The comparison of object volume size delineated based on CT and MRI, and the actual physical size of each object is listed in Table 20.1. It is seen that the contoured volume based on the visual object edge agrees between CT-based data and MRI-based data but overestimates the size of the volume by a

few percent compared with the actual size of the object. After the MR images are fused (registered) with CT images, each object contour delineated based on the MR images is mapped onto CT images, as shown in Figure 20.4. The good overlaps between each object delineated based on CT and MR images ensure the overall accuracy of both image registration and object contouring. The small discrepancies are results of uncertainties in both object contouring and image registration between CT and MRI. The quality assurance procedure is performed before any

TABLE 20.1 Four Objects Volume Size Delineated Based on CT and MR Images

Contoured Object Volume Size	CT Based (cm³)	MRI Based (cm³)	Actual (cm³)
Sphere	8.9	9.0	8.6
Cone	10.0	9.9	9.4
Cylinder	9.9	9.8	9.4
Square	17.2	17.0	16.0

Note: The actual size is based on the measured physical size of the objects.

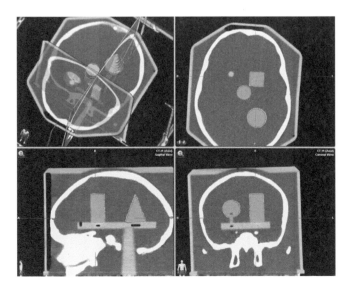

FIGURE 20.4 **(See color insert.)** Contours of four different objects delineated based on CT and MR images mapped onto CT images after image registration are shown in red and blue, respectively, on 3D, axial, sagittal, and coronal views of the Radionics Brain Phantom.

patient image acquired from a new modality is to be used for radiotherapy treatment planning.

20.3 Applications in Cancer Diagnosis

The improvements in MRI technology have advanced the clinical approach to cancer imaging by permitting the use of many techniques, including dynamic contrast-enhanced MRI (DCE-MRI), magnetic resonance spectroscopy, diffusion-weighted imaging (DWI), diffusion-tensor imaging, whole-body MR imaging, blood oxygen level–dependent imaging, and lymphotropic nanoparticle imaging. Each of these is an established clinical tool in the diagnosis of many malignancies, effectively increasing the sensitivity of MRI in detecting various cancers. These are discussed in detail in other chapters.

20.4 Applications in Cancer Therapy

20.4.1 Treatment Planning

The widespread utilization of intensity-modulated radiation therapy (IMRT) in radiotherapy and newer techniques in surgical guidance require sophisticated imaging modalities to accurately direct treatment. MRI confers several advantages over CT in IMRT planning. These include better soft-tissue delineation and improved multiplanar and volumetric imaging. In addition, physiological and biochemical information with MR angiography and spectroscopy can be utilized for treatment via image fusion. Despite these benefits, CT remains the gold standard for radiation planning in most sites because MRI has poor imaging of bone and lacks electron density information required for dosimetry calculations for radiotherapy. Also, MR images have intrinsic systematic and object-induced spatial distortions. Lastly, there has historically been a paucity of widely available computer software to accurately and reliably integrate and manipulate MR images within existing treatment planning systems. New planning systems can utilize angiography, MRI, PET, and single photon emission CT (SPECT) imaging. In addition, due to inherent similarity of these target and delivery systems, manufacturers of radiation systems often also market similar systems for the surgical oncologist to enable accurate tumor resections while minimizing injury to surrounding tissues. Examples of organ systems with tumors where MRI is used extensively for treatment planning and for follow-up are those of the brain, head and neck, lung, gastrointestinal system, and prostate gland.

20.4.2 MRI in the Management of Brain Tumors

MRI is used more than any other imaging modality in the management of tumors of the central nervous system. In radiation treatment planning for both IMRT and stereotactic radiosurgery, and for neurosurgical navigation for biopsy and resection, MRI provides more information on anatomy, metabolic activity, chronological changes, and adjacent critical structures than does CT, PET, or angiography. Its use is imperative when attempting to reduce surgical injury or to reduce irradiation of surrounding normal structures. To illustrate, Chang et al. evaluated 143 patients with previously untreated glioma [3]. Patients were scanned immediately before surgical resection using conventional anatomical MRI. Perfusion-weighted imaging, DW imaging, and proton MR spectroscopic imaging were evaluated to assess tumor burden. Survival was correlated with MR parameters that had historically been found to be predictive of survival in patients with grade IV glioma: tumor grade and histological subtype. Lactate was higher for grade III and grade IV gliomas, whereas lipid was significantly elevated for grade IV gliomas. The authors concluded that metabolic and physiologic imaging characteristics provided by MRI/SPECT provide information about tumor heterogeneity that was important in assisting the surgeon to ensure acquisition of representative histology for diagnosis [3]. This is imperative because treatment algorithms are determined by these characteristics of grade and histology and, to a lesser degree, tumor location and the extent of surgical resection.

Although helpful, perfusion imaging alone as a single parameter cannot distinguish the grade of a tumor. Other characteristics such as elevated lipid distinguish grade IV tumors as well as high blood volume. The presence of lipid in grade IV gliomas is thought to be due to cellular breakdown products after apoptosis and necrosis.

Since Leksell's Gamma Knife was developed in the early 1950s as the first radiosurgical device at the Karolinska Institute, Fakturor, Stockholm, several refinements on the device have been made [4]. Diffusion-tensor tractography, a method which accounts for not only the rate of diffusion (obtained by DW-MRI) but also the direction of the diffusion along white-matter tracts, allows one to produce directional maps of white matter where neoplastic lesions spread. Improved imaging with MR and diffusion-tensor tractography has been shown by Koga et al. to allow safer targeting and minimizing morbidity from treatment [82]. With 36 patients evaluated, diffusion-tensor tractography allowed for a reduction in excess treated volume ($p = .0062$), and Paddick conformity index improved ($p < .001$) [83]. Maximal dose to the white-matter tracts also was reduced significantly ($p < .001$). The authors concluded that for the purpose of coregistration, different imaging techniques carried specific benefits: CT is best applied for obtaining anatomical differential diagnosis, PET for metabolic differential diagnosis, serial MR for clarifying chronological changes, and diffusion-tensor tractography for minimizing morbidities of treatment.

20.4.3 Head and Neck Cancer

The vast majority of cancers of the head and neck region have a squamous histology (head and neck squamous cell carcinoma [HNSCC]). For prognostic factors for HNSCC, primary tumor stage correlates with risk of distant metastases and determines treatment intensity. Nodal metastases also correlate as an adverse prognostic factor in patients with HNSCC and require accurate detection for optimized direction of treatment. Imaging is used in addition to clinical evaluation to improve the detection of nodal metastases. CT and MRI allow simultaneous assessment of primary tumor extent and neck nodal disease and are often used as first-line imaging modalities in the detection of HNSCC. Therefore, conventional MR imaging (without using DW), with morphologic criteria for nodal staging limited and similar to those of CT, has generally yielded results that are similar or slightly inferior to those of CT. However, the assessment of cervical node metastases in HNSCC remains difficult, particularly in smaller lymph nodes (less than 10 mm). Although fluorine 18 fluorodeoxyglucose PET (FDG-PET)/CT has a relatively high sensitivity for the detection of nonpalpable nodal metastases compared with anatomic imaging modalities, its added clinical value in the preoperative assessment of HNSCC has not been proven unambiguously. PET's relatively low spatial resolution can lead to false-negative results in cases of small-volume nodal disease. Owing to the lack of reliable morphologic criteria, anatomic imaging modalities, such as CT and standard MR, rely mainly on node size and internal architecture for determining the presence of nodal metastases [5]. Several criteria based on size for detecting neck nodal disease have been suggested, with each cutoff value constituting a compromise between sensitivity and specificity. The criterion of a short-axis diameter of 10 mm, as proposed by van den Brekel et al. [6], has gained widespread acceptance. In lymph nodes < 10 mm, morphologic features used for diagnosis, such as necrosis and indistinct node borders that indicate extracapsular spread, are relatively uncommon.

Vandecaveye et al. evaluated both CT and diffusion-weighted MR in the detection of tumor in small (<10 mm) lymph nodes preoperatively and correlated findings with pathology from neck dissection [7]. Preliminary studies of MRI showed that DW sequences could depict regions of interest within small nodes representing tumor [8]. The interpretation of DW images in the head and neck, however, is not always straightforward and requires training and expertise. As indicated in this study, the use of DWI could reduce standard MR and CT's limitations, owing to its capability to depict small nodal metastases. Therefore, this examination might be used in conjunction with FDG-PET.

In another study, 22 patients with locally advanced HNSCC underwent contrast-enhanced CT, as well as MRI (with routine and DW sequences) prior to neck dissection. After topographic correlation, lymph nodes were evaluated microscopically with prekeratin immunostaining. Pathology results were correlated with imaging findings, and an RT planning study was performed for these surgically treated patients. One set of target volumes was based on conventional imaging only, and another set was based on the corresponding DW images. A third reference set was contoured based solely on pathology findings. Results for DW-MRI showed a sensitivity of 89% and a specificity of 97% per lymph node. Nodal staging agreement between imaging and pathology was significantly stronger for DW-MRI than for conventional imaging ($p = .019$). Additionally, the absolute differences between RT target volumes determined by CT and DW-MRI and those obtained by pathology were calculated, and the observed difference was significantly larger for CT imaging than for DW-MRI for nodal gross tumor volume (GTV) ($p < 01$), as well as for nodal CTV ($p = .04$).

These results suggest that DW-MRI is superior to conventional CT imaging for preradiotherapy nodal staging of HNSCC and, by decreasing the target volume of cervical nodes, provides a potential impact on organ sparing and tumor control [9,10]. However, it might be premature to speculate on the relevant clinical effect of nodal staging in HNSCC with DWI, since the technique needs to be further refined before it is used in routine clinical practice.

Apparent diffusion coefficient maps had higher accuracy than turbo spin-echo MRI in nodal staging, providing added value in the detection of nodal metastases < 10 mm. Although the combined PET-CT examination may be of value for detecting distant metastases and second primary tumor sites, comparing it with DWI for the detection of small (4–10 mm) lymph nodes has been studied, as shown above.

The success of such highly conformal radiation for sparing healthy tissues and its ability to improve tumor control via dose escalation rely significantly on the accurate delineation of GTVs within any individual patient. This is particularly true for IMRT above the clavicles, where healthy structures and deposits are often in close proximity. To determine the accuracy of DW-MRI in HNSCC patients within nodal basins, Vandecaveye et al. correlated preoperative imaging with pathologic findings at surgery [7,11].

20.4.4 Response to Radiotherapy Treatment in HNSCC

Since its introduction over the last decade, intensity-modulated radiotherapy has been widely applied in HNSCC. This treatment method allows high-dose areas of radiation to be conformed tightly to the target volumes with dose falling off steeply outside these regions. In the head and neck region, where malignancies often lie near critical normal tissues, intensity-modulated radiotherapy has the potential to ensure sufficient target coverage while significantly reducing toxicity, particularly xerostomia. Another promising application, dose escalation, has been attempted to improve tumor control. Because intensity-modulated radiotherapy permits dose conformality to the organs at risk, dose can be minimized to the cartilage, connective tissue, nerves, and bone, lowering late side effects. It has therefore been suggested that dose escalation could be achieved to target areas of increased radio-resistance in the tumor.

Functional imaging (e.g., imaging of tumor burden, proliferation, and hypoxia) is anticipated to be able to determine the biologic target volume for IMRT dose painting. The most commonly used functional imaging modality for radiotherapy planning is undoubtedly FDG-PET. In addition, one study showed that tumor regions (both primary tumor and lymph nodes) where a loco-regional recurrence developed during follow-up had significantly lower apparent diffusion coefficients on DW-MRI during week 4 of radiotherapy ($p = .01$) and at 3 weeks after treatment

($p = .01$) than did lesions that remained controlled. These results confirm the added value of both MR and FDG-PET for radiotherapy planning for HNSCC. They also suggest the potential of DW and dynamic enhanced MRI for dose painting and early response assessment in HNSCC patients.

Despite promising early reports, it remains evident that considerable further research and development is necessary before DW or dynamic enhanced MRI can be routinely used in radiotherapy planning. This is because interpretation of DW or dynamic enhanced MRI scans of the head and neck region is not straightforward, making the use of quantitative or semiquantitative measurements for target volume delineation absolutely necessary. Ultimately, successful introduction of functional imaging other than FDG-PET into clinical routine will likely depend on an improved standardization of imaging technique, interpretation, and registration.

20.4.5 MR Detects Changes in Pharyngeal Constrictors to Radiation

The pharyngeal constrictors are important for normal deglutition. Radiation dose to these muscles has been correlated with incidence of posttreatment dysphagia, aspiration, and percutaneous endoscopic gastrostomy dependency. An imaging modality that could characterize changes correlating with dysphagia would be a useful clinical tool. In one study, the inferior pharyngeal constrictor muscles were compared by MR with the sternocleidomastoid muscles, which receive high doses during HNSCC radiation, but are not related to swallowing function [12]. T_1-weighted signal decreased in both constrictors and sternocleidomastoid muscles receiving > 50 Gy ($p < .03$), but not in muscles receiving lower doses. T_2-weighted signals in the constrictors increased significantly as the dose increased above 50 Gy ($p = .34$, $p = .01$) and were significantly higher than the sternocleidomastoid muscle T_2 changes ($p < .001$). Increased thickness was noted in all constrictor muscles, especially with dose > 50 Gy. Additionally, constrictor muscles gained significantly more thickness than constrictors receiving less dose ($p = .02$), whereas sternocleidomastoid muscle thickness, in contrast, decreased posttherapy ($p = .002$). These investigators concluded that MRI differences between these groups of muscles characterize underlying causes of constrictor dysfunction. Etiologies include inflammation and edema, which are likely a consequence of acute mucositis affecting the submucosa overlying the pharyngeal constrictors. These results corroborate clinical retrospective studies that recommend reducing mean doses to the inferior constrictors to 50 Gy or less, and by other attempts at reducing acute mucositis, in order to improve long-term dysphagia.

20.4.6 Lung Tumors

Historically, radiation treatment planning in non–small-cell lung cancer (NSCLC) patients relied almost exclusively on CT. However, CT imaging does not reliably indicate the actual extent of the tumor and its demarcation from atelectasis, leading to a substantial interobserver variation in tumor delineation and, potentially, overly large radiation portals. With the advance of IMRT and dose painting, the exact identification of GTV and, in turn, PTV could lead to a substantial reduction in radiation dose of organs at risk, especially of the adjacent spinal cord, heart, and esophagus. This is particularly important clinically for pulmonary function, since saving even a small amount of healthy lung tissue might be relevant in a significant amount of NSCLC patients, who often cannot undergo definitive surgery due to limited lung reserve. Several articles have been published about the benefit of integrating FDG-PET information into radiotherapy planning in NSCLC [13]. However, various studies describe both an increase and a decrease in PTV [14,15]. Particularly in patients with atelectasis, the contribution of FDG-PET has been shown to lead to significant changes in irradiation fields [78]. In most studies, such as that reported by Nestle et al., use of FDG-PET translated into a reduction in the irradiated region. This finding may be explained by the ability of PET to discriminate between atelectasis and tumor. However, the use of FDG-PET for target volume definition in the treatment of NSCLC has several limitations [79]. The tumor edge definition is restricted by PET's low spatial resolution of about 5 to 7 mm, and by the fact that there is no threshold value applicable for all patients (standardized uptake value). Furthermore, FDG-PET requires long acquisition times, which may lead to artifacts caused by patient or organ motion and by tumor motion during non–breathhold PET imaging. For image fusion techniques for radiation planning, this can pose a significant problem for the radiation physicist and oncologist, since tumor borders may not be delineated precisely.

Currently, the use of magnetization transfer (MT)–prepared MRI has been utilized within the brain where organ movement is limited and where the typically long acquisition times necessary for MT experiments do not pose as many problems [80,81]. MT-MRI has been proposed as a new alternative modality to gain biologic information within inhomogeneous masses in NSCLC patients. Theoretically, MT-MRI is fast enough for breathhold acquisition. Arnold et al. reported encouraging results that engender further exploration of MT-MRI as an adjunct for radiotherapy planning in NSCLC [84]. Others have reported that DW-MRI is equivalent to PET in distinguishing NSCLC from benign pulmonary nodules and may even have fewer false-positive results compared with FDG-PET. Although there is no significant difference in sensitivity between the two methods, DWI has a significantly higher specificity than PET because of fewer false positives for active inflammatory lesions ($p = .03$).

DW-MRI can be used in place of PET-CT for N staging of NSCLC with fewer false-positive results [16,17]. Furthermore, dynamic MRI is useful for assessing tumor interstitium, vascularity, and vascular endothelial growth factor (VEGF) expression. It has also shown usefulness in predicting survival outcome among patients with peripheral pulmonary carcinoma above and beyond staging. These advantages make dynamic MRI a promising method and a potential biomarker for characterizing tumor response to antiangiogenic treatment as well as for predicting survival outcomes after treatment. Although the clinical role of MRI for evaluating tumor stage, treatment response, and etiology of solitary pulmonary nodules remains limited, considerable experience has been gained with MRI of thoracic diseases. Its utilization will become more prominent in future patients.

20.4.7 Breast Tumors

Breast cancer is the most prevalent cancer and the second leading cause of cancer-related death in women. Early detection and analysis of tumor characteristics are vital in achieving disease control. It is well established that DCE-MRI is highly sensitive in detecting breast cancers and provides detailed information regarding tumor kinetics [18–22]. Additionally, DCE-MRI findings have been correlated with histology, grade, and biomarker status of breast tumors [23–26].

Although the role of MRI is increasing in breast cancer diagnosis, there are currently no adequate radiological studies to detect or predict tumor response to treatment. In locally advanced breast cancer patients, the current standard of care often involves neoadjuvant chemotherapy [27,28]. Oftentimes, invasive techniques such as repeat biopsies are needed to evaluate efficacy of therapy [29–31]. With DCE-MRI techniques, this information may be obtained noninvasively.

The enhancement patterns of tumors following DCE-MRI have been shown to change reliably following treatment and are correlated with the extent and integrity of the tumor vasculature [32,33]. Supporting this notion was that tumor response to neoadjuvant chemotherapy was correlated with a decrease in K^{trans} as measured by DCE-MRI in 11 patients with locally advanced breast cancer [34]. Similarly, DCE-MRI has also been shown in larger cohorts of patients undergoing neoadjuvant chemotherapy to predict tumor vascular changes following only one or two cycles of chemotherapy [35,36]. Interestingly, the potential to predict 5-year survival from DCE-MRI results was also observed [36]. A larger study to validate the use of DCE-MRI parameter kinetics in breast cancer patients is warranted.

Axillary nodal status is a negative prognostic indicator in breast cancer [37]. The presence of disease in four or more axillary lymph nodes prompts the addition of a third radiation field to treat the high axilla and supraclavicular nodes. Recently, axillary nodal status in breast cancer patients has retrospectively been correlated with tumor kinetics measured by DCE-MRI techniques [38,39]. Although these results need to be confirmed in larger longitudinal studies, it is intriguing that DCE-MRI findings could be used to define and plan treatment.

Finally, the use of antiangiogenic agents in treating breast cancer is currently being investigated. AG-013736, a novel VEGF receptor tyrosine kinase inhibitor, has been shown to inhibit the growth of human breast cancer xenografts and decrease vascular permeability in these tumors as measured by DCE-MRI [40]. Similarly, DCE-MRI also effectively assessed the response of a rat model of breast cancer bone metastases to bevacizumab, a VEGF antibody [41]. These studies again emphasize the utility of DCE-MRI techniques in assessing the response of tumors at an early time point following treatment as well as the evaluation of novel antiangiogenic agents in cancer therapy.

20.4.8 Gastrointestinal Tumors

Locally advanced rectal cancer requires extensive surgical resection to remove the tumor with a clear margin. Over the past 15 years, neoadjuvant therapy has been utilized to downstage and downsize the tumor to improve resectability and obtain better local control. Accurate pretreatment and posttreatment imaging is essential to determining which patients are appropriate for sphincter-sparing surgery. CT has historically been the mainstay of imaging in this group of patients. However, other imaging techniques are finding usefulness. The sensitivity and specificity of FDG-PET in identifying response are 100% (CT 54%, MRI 71%) and 60% (CT 80%, MRI 67%), respectively. Positive and negative predictive values were 77% (CT 78%, MRI 83%) and 100% (CT 57%, MRI 50%) (PET p = .002, CT p = .197, MRI p = 0.500), respectively. These results suggest that FDG-PET is superior to CT in predicting response to preoperative multimodal treatment of locally advanced primary rectal cancer [42]. MRI has also been an excellent diagnostic tool for predicting the circumferential resection margin (CRM) as well as mesorectal fascia (MRF) invasion in primary rectal cancer [43]. Others have demonstrated the accuracy of high-resolution MRI for predicting the tumor-free resection margin in rectal cancer surgery [44]. Because over half of rectal cancers are assumed to be locally advanced cancers [45,46], neoadjuvant chemoradiotherapy (CRT) has been accepted as the standard of care for treating locally advanced rectal cancer (T3–T4 or node-positive), since several investigators demonstrated that it decreased the chance of local recurrence and improved the likelihood of avoiding an abdominoperitoneal resection [47–50]. Using preoperative combined modality treatment, pathologic complete response (pCR) rates have increased as high as 30% with newer regimens [51], which downstages tumor in 46%–60% of patients [52,53]. Therefore, it is important for investigators who advocate less extensive surgery to select in advance favorable responders (pCR or near-complete regression) for patients who would benefit from less extensive surgical resection by using a noninvasive imaging modality. MRI appears to be the best imaging modality to accomplish this. In addition, clear sagittal and coronal sectional pelvic images with MRI can give a lot of information about adjacent organ invasion or any invasion of *levator ani* muscle and can be useful for choosing an appropriate extent of lymph node dissection and type of surgery. It also appears to be superior to endorectal ultrasound in determining depth of transmural invasion and comparable in detecting lymph node metastases (N stage) [54].

Kim et al. [55] evaluated 65 consecutive patients with locally advanced rectal cancer (≥T3 or lymph node positive) who underwent neoadjuvant CRT and subsequent surgery. Two blinded radiologists independently reviewed both the pre- and post-CRT rectal MR images and measured the post-CRT CRM; they recorded their confidence level with respect to the tumor response and evidence of MRF invasion. This study demonstrated that the sensitivity of MRI for determining MRF invasion was 75%; specificity, 88%–98%; and overall accuracy, 85%–92%. MRI had a positive predictive value of 66.7%–92.3% and a negative predictive value of 91.5%–92.3%.

MRI provides accurate information regarding the CRM of locally advanced rectal cancer after neoadjuvant chemoradiation; it also shows relatively high accuracy for predicting MRF invasion and moderate accuracy for assessing tumor response. One caveat is that it is difficult to predict disease activity in patients with mucinous histology; measurement errors were caused by the presence of both mucinous tumors and rectal wall fibrosis.

20.4.9 Prostate Cancer

Recent investigations suggest that dose escalation with IMRT potentially increases the local control while greatly reducing rectal and bladder exposure to high radiation doses [56–59]. As dose levels are increased, the use of new imaging methods to more accurately target the prostate and the accuracy of dose delivery become crucial. MRI provides superior image quality for soft-tissue delineation over CT and is widely used for target and organ delineation in radiotherapy for treatment planning [60]. The prostate volume on CT appears much larger than on MRI [61]. Others have shown that if prostate and rectal volume delineation from MRI could be optimized, then coverage of both target volumes as well as rectal sparing can be improved and radiation toxicity reduced [62]. As a result of its improved soft-tissue delineation, using MRI for radiotherapy planning of prostate cancer is desirable.

Patients with localized prostate cancer often have an abnormality within the prostate on endorectal coil MRI that may not be seen on CT. Early dosimetric analyses reported that, in theory, IMRT-based simultaneous integrated boost could be delivered to an MRI-defined dominant intraprostatic lesion(s) and should have acceptable toxicity [63]. Forward or inversely planned segmental multileaf collimator IMRT and sequential tomotherapy can treat multiple dominant intraprostatic lesions of prostate cancer to 90 Gy. Older reports by Pickett et al. showed that an early form of IMRT could be used to deliver 90 Gy to a single MRI-defined intraprostatic lesion while treating the rest of the prostate to 70 Gy [87]. This corroborated the study by van Lin et al., showing the rectum-sparing ability of IMRT with MRI-based planning [85]. This, in theory, could provide tumor dose escalation while simultaneously lowering toxicity. Singh et al. showed that during radiation planning, a nodular lesion (plus 3-mm additional margin) could receive 94.5 Gy in 42 fractions, and the remainder of the prostate (with 7-mm additional margin) could receive 75.6 Gy simultaneously using a dose-painting technique, with acceptable dosimetry and toxicity [86]. Theoretically, this dose escalation can provide improved tumor control.

Results in this small study showed that one patient out of four had no acute toxicity, two experienced acute radiation therapy oncology group (RTOG) grade 2 genitourinary toxicity, and one had grade 1 gastrointestinal toxicity. All symptoms in the

three patients had completely resolved by 3 months. These early results demonstrate the feasibility of using IMRT for simultaneous integrated boost to biopsy-proven intraprostatic nodules visible on MRI. Consistent with earlier dosimetric studies, all three patients in the current study achieved resolution of acute treatment-related gastrointestinal and genitourinary symptoms as described by the RTOG scale. No late toxicities have been observed. In fact, in follow-up, one patient has shown marked improvement from baseline pretreatment symptoms of urinary frequency.

Certainly, these early results using MRI in prostate cancer radiotherapy are encouraging. However, two substantial hurdles remain prior to wide implementation of this approach. First, it remains unclear how well MR scans differentiate regions of prostate cancer from regions of prostate inflammation. The second is the widespread use of image-fusion capabilities in radiation facilities. The latter is quickly becoming a reality with modern treatment planning systems.

20.4.10 Cervical Cancer

Cervical cancer is the fifth most prevalent cancer in women worldwide. Radiation therapy is an integral part of cervical cancer treatment. It is well established that cervical tumors are characteristically hypoxic, which renders them more aggressive and more likely to metastasize [65,66]. Additionally, hypoxia also increases the resistance of tumors to radiation therapy [67], and an increased frequency for local failure and lower overall survival is seen in patients with highly hypoxic tumors [68,69]. It is thus important that a noninvasive technique to assess the extent of hypoxia in cervical tumors is established.

Dynamic MRI has been a well-established technique in delineating cervical tumors [70]. Additionally, an improved local control was observed in patients with tumors exhibiting high signal intensity [71]. It was hypothesized that this high signal intensity is due to greater blood perfusion, which results in increased oxygenation and subsequent radiosensitivity. In support of this notion, pharmacokinetic analysis of DCE-MRI revealed that areas of increased contrast enhancement indeed correlated with tumor and areas of angiogenesis, and improved local control was observed in patients whose tumors were well enhanced [72–75]. An evaluation of tumor hypoxia, however, was not performed in these studies.

A recent report investigates the use of DCE-MRI in assessing the fraction of hypoxic cells in human tumor xenografts [76]. It was found that the fraction of radiobiologically hypoxic cells inversely correlated with K^{trans}. Additionally, semiquantitative and quantitative DCE-MRI measurements were shown to predict radiation responsiveness in 13 cervical cancer patients [77]. Thus, DCE-MRI appears to have the potential for clinical utility in determining the oxygenation status of cervical tumors.

20.5 Conclusion

Clinical studies over the past two decades have provided evidence that MRI is a useful modality for tumor staging and evaluating treatment response in a multitude of human cancers. It also has been proven that tumor microenvironment and hemodynamics play a major role in how neoplasms can be imaged and followed during the course of therapy. Through its use in human clinical trials and within the cancer clinic, MRI provides the clinician with a powerful tool as an imaging end point for determining optimal treatment.

References

1. Khoo VS. MRI—"Magic radiotherapy imaging" for treatment planning? *Br J Radiol* 2000;73:229–233.
2. Wells WM, 3rd, Viola P, Atsumi H, Nakajima S, Kikinis R. Multi-modal volume registration by maximization of mutual information. *Medical Image Anal* 1996;1:35–51.
3. Chang SM, Nelson S, Vandenberg S, Cha S, Prados M, Butowski N, McDermott M, Parsa AT, Aghi M, Clarke J, Berger M. Integration of preoperative anatomic and metabolic physiologic imaging of newly diagnosed glioma. *J Neurooncol* 2009;92:401–415.
4. Leksell L. The stereotaxic method and radiosurgery of the brain. *Acta Chir Scand* 1951;102:316–319.
5. van den Brekel MW. Lymph node metastases: CT and MRI. *Eur J Radiol* 2000;33:230–238.
6. van den Brekel MW, Stel HV, Castelijns JA, Nauta JJ, van der Waal I, Valk J, Meyer CJ, Snow GB. Cervical lymph node metastasis: assessment of radiologic criteria. *Radiology* 1990;177:379–384.
7. Vandecaveye V, De Keyzer F, Verslype C, Op de Beeck K, Komuta M, Topal B, Roebben I, Bielen D, Roskams T, Nevens F, Dymarkowski S. Diffusion-weighted MRI provides additional value to conventional dynamic contrast-enhanced MRI for detection of hepatocellular carcinoma. *Eur Radiol* 2009;19:2456–2466.
8. Takahara T, Imai Y, Yamashita T, Yasuda S, Nasu S, Van Cauteren M. Diffusion weighted whole body imaging with background body signal suppression (DWIBS): technical improvement using free breathing, STIR and high resolution 3D display. *Radiat Med* 2004;22:275–282.
9. Dirix P, Vandecaveye V, De Keyzer F, Op de Beeck K, Poorten VV, Delaere P, Verbeken E, Hermans R, Nuyts S. Diffusion-weighted MRI for nodal staging of head and neck squamous cell carcinoma: impact on radiotherapy planning. *Int J Radiat Oncol Biol Phys* 2009;76:761–766.
10. Dirix P, Vandecaveye V, De Keyzer F, Stroobants S, Hermans R, Nuyts S. Dose painting in radiotherapy for head and neck squamous cell carcinoma: value of repeated functional imaging with (18)F-FDG PET, (18)F-fluoromisonidazole PET, diffusion-weighted MRI, and dynamic contrast-enhanced MRI. *J Nucl Med* 2009;50:1020–1027.
11. Vandecaveye V, De Keyzer F, Vander Poorten V, Dirix P, Verbeken E, Nuyts S, Hermans R. Head and neck squamous cell carcinoma: value of diffusion-weighted MR imaging for nodal staging. *Radiology* 2009;251:134–146.
12. Popovtzer A, Cao Y, Feng FY, Eisbruch A. Anatomical changes in the pharyngeal constrictors after chemo-irradiation of head and neck cancer and their dose-effect relationships: MRI-based study. *Radiother Oncol* 2009;93:510–515.
13. Kiffer JD, Berlangieri SU, Scott AM, Quong G, Feigen M, Schumer W, Clarke CP, Knight SR, Daniel FJ. The contribution of 18F-fluoro-2-deoxy-glucose positron emission tomographic imaging to radiotherapy planning in lung cancer. *Lung Cancer* 1998;19:167–177.
14. van Baardwijk A, Baumert BG, Bosmans G, van Kroonenburgh M, Stroobants S, Gregoire V, Lambin P, De Ruysscher D. The current status of FDG-PET in tumour volume definition in radiotherapy treatment planning. *Cancer Treat Rev* 2006;32:245–260.
15. van Baardwijk A, de Jong J, Arens A, Thimister P, Verseput G, Kremer B, Lambin P. False-positive FDG-PET scan due to brown tumours. *Eur J Nucl Med Mol Imaging* 2006;33:393–394.
16. Nomori H, Mori T, Ikeda K, Kawanaka K, Shiraishi S, Katahira K, Yamashita Y. Diffusion-weighted magnetic resonance imaging can be used in place of positron emission tomography for N staging of non-small cell lung cancer with fewer false-positive results. *J Thorac Cardiovasc Surg* 2008;135:816–822.
17. Nomori H, Shibata H, Uno K, Iyama K, Honda Y, Nakashima R, Sakaguchi K, Goya T, Takanami I, Koizumi K, Suzuki T, Kaji M, Horio H. 11C-Acetate can be used in place of 18F-fluorodeoxyglucose for positron emission tomography imaging of non-small cell lung cancer with higher sensitivity for well-differentiated adenocarcinoma. *J Thorac Oncol* 2008;3:1427–1432.
18. Kuhl C. The current status of breast MR imaging. Part I. Choice of technique, image interpretation, diagnostic accuracy, and transfer to clinical practice. *Radiology* 2007;244:356–378.
19. Lehman CD, Schnall MD. Imaging in breast cancer: magnetic resonance imaging. *Breast Cancer Res* 2005;7:215–219.
20. Warner E, Plewes DB, Hill KA, Causer PA, Zubovits JT, Jong RA, Cutrara MR, DeBoer G, Yaffe MJ, Messner SJ, Meschino WS, Piron CA, Narod SA. Surveillance of BRCA1 and BRCA2 mutation carriers with magnetic resonance imaging, ultrasound, mammography, and clinical breast examination. *JAMA* 2004;292:1317–1325.
21. Kriege M, Brekelmans CT, Boetes C, Besnard PE, Zonderland HM, Obdeijn IM, Manoliu RA, Kok T, Peterse H, Tilanus-Linthorst MM, Muller SH, Meijer S, Oosterwijk JC, Beex LV, Tollenaar RA, de Koning HJ, Rutgers EJ, Klijn JG. Efficacy of MRI and mammography for breast-cancer screening in women with a familial or genetic predisposition. *N Engl J Med* 2004;351:427–437.
22. Lehman CD, Isaacs C, Schnall MD, Pisano ED, Ascher SM, Weatherall PT, Bluemke DA, Bowen DJ, Marcom PK, Armstrong DK, Domchek SM, Tomlinson G, Skates SJ, Gatsonis C. Cancer yield of mammography, MR, and US in

high-risk women: prospective multi-institution breast cancer screening study. *Radiology* 2007;244:381–388.

23. Szabo BK, Aspelin P, Kristoffersen Wiberg M, Tot T, Bone B. Invasive breast cancer: correlation of dynamic MR features with prognostic factors. *Eur Radiol* 2003;13:2425–2435.

24. Tuncbilek N, Karakas HM, Okten OO. Dynamic magnetic resonance imaging in determining histopathological prognostic factors of invasive breast cancers. *Eur J Radiol* 2005;53:199–205.

25. Montemurro F, Martincich L, Sarotto I, Bertotto I, Ponzone R, Cellini L, Redana S, Sismondi P, Aglietta M, Regge D. Relationship between DCE-MRI morphological and functional features and histopathological characteristics of breast cancer. *Eur Radiol* 2007;17:1490–1497.

26. Agrawal G, Chen JH, Baek HM, Hsiang D, Mehta RS, Nalcioglu O, Su MY. MRI features of breast cancer: a correlation study with HER-2 receptor. *Ann Oncol* 2007;18:1903–1904.

27. Rastogi P, Anderson SJ, Bear HD, Geyer CE, Kahlenberg MS, Robidoux A, Margolese RG, Hoehn JL, Vogel VG, Dakhil SR, Tamkus D, King KM, Pajon ER, Wright MJ, Robert J, Paik S, Mamounas EP, Wolmark N. Preoperative chemotherapy: updates of National Surgical Adjuvant Breast and Bowel Project protocols B-18 and B-27. *J Clin Oncol* 2008;26:778–785.

28. Bear HD, Anderson S, Smith RE, Geyer CE, Jr., Mamounas EP, Fisher B, Brown AM, Robidoux A, Margolese R, Kahlenberg MS, Paik S, Soran A, Wickerham DL, Wolmark N. Sequential preoperative or postoperative docetaxel added to preoperative doxorubicin plus cyclophosphamide for operable breast cancer: National Surgical Adjuvant Breast and Bowel Project protocol B-27. *J Clin Oncol* 2006;24:2019–2027.

29. Stearns V, Singh B, Tsangaris T, Crawford JG, Novielli A, Ellis MJ, Isaacs C, Pennanen M, Tibery C, Farhad A, Slack R, Hayes DF. A prospective randomized pilot study to evaluate predictors of response in serial core biopsies to single agent neoadjuvant doxorubicin or paclitaxel for patients with locally advanced breast cancer. *Clin Cancer Res* 2003;9:124–133.

30. Buchholz TA, Hunt KK, Whitman GJ, Sahin AA, Hortobagyi GN. Neoadjuvant chemotherapy for breast carcinoma: multidisciplinary considerations of benefits and risks. *Cancer* 2003;98:1150–1160.

31. Chakravarthy A, Nicholson B, Kelley M, Beauchamp D, Johnson D, Frexes-Steed M, Simpson J, Shyr Y, Pietenpol J. A pilot study of neoadjuvant paclitaxel and radiation with correlative molecular studies in stage II/III breast cancer. *Clin Breast Cancer* 2000;1:68–71.

32. Padhani AR, Ah-See ML, Makris A. MRI in the detection and management of breast cancer. *Expert Rev Anticancer Ther* 2005;5:239–252.

33. Knopp MV, von Tengg-Kobligk H, Choyke PL. Functional magnetic resonance imaging in oncology for diagnosis and therapy monitoring. *Mol Cancer Ther* 2003;2:419–426.

34. Yankeelov TE, Lepage M, Chakravarthy A, Broome EE, Niermann KJ, Kelley MC, Meszoely I, Mayer IA, Herman CR, McManus K, Price RR, Gore JC. Integration of quantitative DCE-MRI and ADC mapping to monitor treatment response in human breast cancer: initial results. *Magn Reson Imaging* 2007;25:1–13.

35. Ah-See ML, Makris A, Taylor NJ, Harrison M, Richman PI, Burcombe RJ, Stirling JJ, d'Arcy JA, Collins DJ, Pittam MR, Ravichandran D, Padhani AR. Early changes in functional dynamic magnetic resonance imaging predict for pathologic response to neoadjuvant chemotherapy in primary breast cancer. *Clin Cancer Res* 2008;14:6580–6589.

36. Johansen R, Jensen LR, Rydland J, Goa PE, Kvistad KA, Bathen TF, Axelson DE, Lundgren S, Gribbestad IS. Predicting survival and early clinical response to primary chemotherapy for patients with locally advanced breast cancer using DCE-MRI. *J Magn Reson Imaging* 2009;29:1300–1307.

37. Fisher B, Bauer M, Wickerham DL, Redmond CK, Fisher ER, Cruz AB, Foster R, Gardner B, Lerner H, Margolese R, Poisson R, Shibata H, Volk H. 1983. Relation of number of positive axillary nodes to the prognosis of patients with primary breast cancer. An NSABP update. *Cancer* 1983;52:1551–1557.

38. Hsiang DJ, Yamamoto M, Mehta RS, Su MY, Baick CH, Lane KT, Butler JA. Predicting nodal status using dynamic contrast-enhanced magnetic resonance imaging in patients with locally advanced breast cancer undergoing neoadjuvant chemotherapy with and without sequential trastuzumab. *Arch Surg* 2007;142:855–861; discussion 860–861.

39. Loiselle CR, Eby PR, Demartini WB, Peacock S, Bittner N, Lehman CD, Kim JN. Dynamic contrast-enhanced MRI kinetics of invasive breast cancer: a potential prognostic marker for radiation therapy. *Int J Radiat Oncol Biol Phys* 2009;13:13.

40. Wilmes LJ, Pallavicini MG, Fleming LM, Gibbs J, Wang D, Li KL, Partridge SC, Henry RG, Shalinsky DR, Hu-Lowe D, Park JW, McShane TM, Lu Y, Brasch RC, Hylton NM. AG-013736, a novel inhibitor of VEGF receptor tyrosine kinases, inhibits breast cancer growth and decreases vascular permeability as detected by dynamic contrast-enhanced magnetic resonance imaging. *Magn Reson Imaging* 2007;25:319–327.

41. Bauerle T, Bartling S, Berger M, Schmitt-Graff A, Hilbig H, Kauczor HU, Delorme S, Kiessling F. Imaging anti-angiogenic treatment response with DCE-VCT, DCE-MRI and DWI in an animal model of breast cancer bone metastasis. *Eur J Radiol* 2008;11:11.

42. Denecke T, Rau B, Hoffmann KT, Hildebrandt B, Ruf J, Gutberlet M, Hunerbein M, Felix R, Wust P, Amthauer H. Comparison of CT, MRI and FDG-PET in response prediction of patients with locally advanced rectal can-

cer after multimodal preoperative therapy: is there a benefit in using functional imaging? *Eur Radiol* 2005;15: 1658–1666.

43. Gollub MJ, Schwartz LH, Akhurst T. Update on colorectal cancer imaging. *Radiolog Clin North Am* 2007;45:85–118.

44. Beets-Tan RG. Uses of error: radiological. *Lancet* 2001;358: 1542.

45. Jessup JM, Loda M, Bleday R. Clinical and molecular prognostic factors in sphincter-preserving surgery for rectal cancer. *Semin Radiat Oncol* 1998;8:54–69.

46. Jessup JM, Stewart AK, Menck HR. The National Cancer Data Base report on patterns of care for adenocarcinoma of the rectum, 1985–95. *Cancer* 1998;83:2408–2418.

47. Bosset JF. Distal rectal cancer: sphincter-sparing is also a challenge for the radiation oncologist. *Radiother Oncol* 2006;80:1–3.

48. Bosset JF, Collette L, Calais G, Mineur L, Maingon P, Radosevic-Jelic L, Daban A, Bardet E, Beny A, Ollier JC. Chemotherapy with preoperative radiotherapy in rectal cancer. *New Engl J Med* 2006;355:1114–1123.

49. Bosset M, Bosset JF, Maingon P. What to do with a rising PSA after complete remission post-prostatectomy? *Cancer Radiother* 2006;10:168–174.

50. Bosset M, Maingon P, Bosset JF. [Radiotherapy for PSA failure after prostatectomy: which volumes?]. *Cancer Radiother* 2006;10:117–123.

51. Habr-Gama A, Perez RO, Nadalin W, Sabbaga J, Ribeiro U, Jr., Silva e Sousa AH, Jr., Campos FG, Kiss DR, Gama-Rodrigues J. Operative versus nonoperative treatment for stage 0 distal rectal cancer following chemoradiation therapy: long-term results. *Ann Surg* 2004;240:711–717; discussion 717–718.

52. Janjan NA, Abbruzzese J, Pazdur R, Khoo VS, Cleary K, Dubrow R, Ajani J, Rich TA, Goswitz MS, Evetts PA, Allen PK, Lynch PM, Skibber JM. Prognostic implications of response to preoperative infusional chemoradiation in locally advanced rectal cancer. *Radiother Oncol* 1999;51:153–160.

53. Janjan NA, Khoo VS, Abbruzzese J, Pazdur R, Dubrow R, Cleary KR, Allen PK, Lynch PM, Glober G, Wolff R, Rich TA, Skibber J. Tumor downstaging and sphincter preservation with preoperative chemoradiation in locally advanced rectal cancer: the M. D. Anderson Cancer Center experience. *Int J Radiat Oncol, Biol, Physics* 1999;44:1027–1038.

54. Halefoglu AM, Yildirim S, Avlanmis O, Sakiz D, Baykan A. Endorectal ultrasonography versus phased-array magnetic resonance imaging for preoperative staging of rectal cancer. *World J Gastroenterol* 2008;14:3504–3510.

55. Kim SH, Lee JM, Park HS, Eun HW, Han JK, Choi BI. Accuracy of MRI for predicting the circumferential resection margin, mesorectal fascia invasion, and tumor response to neoadjuvant chemoradiotherapy for locally advanced rectal cancer. *J Magn Reson Imaging* 2009;29:1093–1101.

56. Hanks GE, Hanlon AL, Schultheiss TE, Pinover WH, Movsas B, Epstein BE, Hunt MA. Dose escalation with 3D conformal treatment: five year outcomes, treatment optimi-

zation, and future directions. *Int J Radiat Oncol Biol Phys* 1998;41:501–510.

57. Zelefsky MJ, Leibel SA, Gaudin PB, Kutcher GJ, Fleshner NE, Venkatramen ES, Reuter VE, Fair WR, Ling CC, Fuks Z. Dose escalation with three-dimensional conformal radiation therapy affects the outcome in prostate cancer. *Int J Radiat Oncol Biol Phys* 1998;41:491–500.

58. Pollack A, Zagars GK, Rosen, II. Prostate cancer treatment with radiotherapy: maturing methods that minimize morbidity. *Semin Oncol* 1999;26:150–161.

59. Yeoh EE, Fraser RJ, McGowan RE, Botten RJ, Di Matteo AC, Roos DE, Penniment MG, Borg MF. Evidence for efficacy without increased toxicity of hypofractionated radiotherapy for prostate carcinoma: early results of a Phase III randomized trial. *Int J Radiat Oncology Biol Phys* 2003;55:943–955.

60. Tanner SF, Finnigan DJ, Khoo VS, Mayles P, Dearnaley DP, Leach MO. Radiotherapy planning of the pelvis using distortion corrected MR images: the removal of system distortions. *Phys Med Biol* 2000;45:2117–2132.

61. Rasch C, Barillot I, Remeijer P, Touw A, van Herk M, Lebesque JV. Definition of the prostate in CT and MRI: a multi-observer study. *Int J Radiat Oncology Biol Phys* 1999;43:57–66.

62. Debois M, Oyen R, Maes F, Verswijvel G, Gatti G, Bosmans H, Feron M, Bellon E, Kutcher G, Van Poppel H, Vanuytsel L. The contribution of magnetic resonance imaging to the three-dimensional treatment planning of localized prostate cancer. *Int J Radiat Oncol Biol Phys* 1999;45:857–865.

63. Pickett B, Vigneault E, Kurhanewicz J, Verhey L, Roach M. Static field intensity modulation to treat a dominant intra-prostatic lesion to 90 Gy compared to seven field 3-dimensional radiotherapy. *Int J Radiat Oncol Biol Phys* 1999;44:921–929.

64. Singh AK, Guion P, Sears-Crouse N, Ullman K, Smith S, Albert PS, Fichtinger G, Choyke PL, Xu S, Kruecker J, Wood BJ, Krieger A, Ning H. Simultaneous integrated boost of biopsy proven, MRI defined dominant intra-prostatic lesions to 95 Gray with IMRT: early results of a phase I NCI study. *Radiat Oncol* 2007;2:36.

65. Sundfor K, Lyng H, Rofstad EK. Tumour hypoxia and vascular density as predictors of metastasis in squamous cell carcinoma of the uterine cervix. *Br J Cancer* 1998;78:822–827.

66. Pitson G, Fyles A, Milosevic M, Wylie J, Pintilie M, Hill R. Tumor size and oxygenation are independent predictors of nodal diseases in patients with cervix cancer. *Int J Radiat Oncol Biol Phys* 2001;51:699–703.

67. Fyles AW, Milosevic M, Wong R, Kavanagh MC, Pintilie M, Sun A, Chapman W, Levin W, Manchul L, Keane TJ, Hill RP. Oxygenation predicts radiation response and survival in patients with cervix cancer. *Radiother Oncol* 1998;48:149–156.

68. Fyles A, Milosevic M, Pintilie M, Syed A, Levin W, Manchul L, Hill RP. Long-term performance of interstitial fluid

pressure and hypoxia as prognostic factors in cervix cancer. *Radiother Oncol* 2006;80:132–137.

69. Nordsmark M, Loncaster J, Aquino-Parsons C, Chou SC, Gebski V, West C, Lindegaard JC, Havsteen H, Davidson SE, Hunter R, Raleigh JA, Overgaard J. The prognostic value of pimonidazole and tumour pO2 in human cervix carcinomas after radiation therapy: a prospective international multi-center study. *Radiother Oncol* 2006;80:123–131.

70. Yamashita Y, Takahashi M, Sawada T, Miyazaki K, Okamura H. Carcinoma of the cervix: dynamic MR imaging. *Radiology* 1992;182:643–648.

71. Mayr NA, Yuh WT, Zheng J, Ehrhardt JC, Magnotta VA, Sorosky JI, Pelsang RE, Oberley LW, Hussey DH. Prediction of tumor control in patients with cervical cancer: analysis of combined volume and dynamic enhancement pattern by MR imaging. *AJR Am J Roentgenol* 1998;170:177–182.

72. Yamashita Y, Baba T, Baba Y, Nishimura R, Ikeda S, Takahashi M, Ohtake H, Okamura H. Dynamic contrast-enhanced MR imaging of uterine cervical cancer: pharmacokinetic analysis with histopathologic correlation and its importance in predicting the outcome of radiation therapy. *Radiology* 2000;216:803–809.

73. Hawighorst H, Weikel W, Knapstein PG, Knopp MV, Zuna I, Schonberg SO, Vaupel P, van Kaick G. Angiogenic activity of cervical carcinoma: assessment by functional magnetic resonance imaging-based parameters and a histomorphological approach in correlation with disease outcome. *Clin Cancer Res* 1998;4:2305–2312.

74. Cooper RA, Carrington BM, Loncaster JA, Todd SM, Davidson SE, Logue JP, Luthra AD, Jones AP, Stratford I, Hunter RD, West CM. Tumour oxygenation levels correlate with dynamic contrast-enhanced magnetic resonance imaging parameters in carcinoma of the cervix. *Radiother Oncol* 2000;57:53–59.

75. Loncaster JA, Carrington BM, Sykes JR, Jones AP, Todd SM, Cooper R, Buckley DL, Davidson SE, Logue JP, Hunter RD, West CM. Prediction of radiotherapy outcome using dynamic contrast enhanced MRI of carcinoma of the cervix. *Int J Radiat Oncol Biol Phys* 2002;54:759–767.

76. Ellingsen C, Egeland TA, Gulliksrud K, Gaustad JV, Mathiesen B, Rofstad EK. Assessment of hypoxia in human cervical carcinoma xenografts by dynamic contrast-enhanced magnetic resonance imaging. *Int J Radiat Oncol Biol Phys* 2009;73:838–845.

77. Zahra MA, Tan LT, Priest AN, Graves MJ, Arends M, Crawford RA, Brenton JD, Lomas DJ, Sala E. Semi-quantitative and quantitative dynamic contrast-enhanced magnetic resonance imaging measurements predict radiation response in cervix cancer. *Int J Radiat Oncol Biol Phys* 2009;74:766–773.

78. Nestle U, Walter K, Schmidt S, Licht N, Nieder C, Motaref B, Hellwig D, Niewald M, Ukena D, Kirsch CM, Sybrecht GW, Schnabel K. 18F-deoxyglucose positron emission tomography (FDG-PET) for the planning of radiotherapy in lung cancer: high impact in patients with atelectasis. *Int J Radiat Oncol Biol Phys* 1999; 44:593–597.

79. Lavrenkov K, Partridge M, Cook G, Brada M. Positron emission tomography for target volume definition in the treatment of non-small cell lung cancer. *Radiother Oncol* 2005;77:1–4.

80. Horsfield MA. Magnetization transfer imaging in multiple sclerosis. *J Neuroimaging* 2005;15:58S-67S.

81. Henkelman RM, Stanisz GJ, Graham SJ. Magnetization transfer in MRI: a review. *NMR Biomed* 2001;14:57–64

82. Koga T, Maruyama K, Igaki H, Tago M, Saito N. The value of image coregistration during stereotactic radiosurgery. *Acta Neurochir (Wein)* 2009;151:465–471.

83. Paddick I. A simple scoring ratio to index the conformity of radiosurgical treatment plans. Technical Note. *J Neurosurg* 2000; 93 Suppl 3: 219–222.

84. Arnold JF, Kotas M, Pyzalsky RW, Pracht ED, Flentje M, Jakob PM. Potential of magnetization transfer MRI for target volume definition in patients with non-small-cell lung cancer. *J Magn Reson Imaging* 2008; 28:1417–1424.

85. van Lin EN, Futterer JJ, Heijmink SW, van der Vight LP, Hoffmann AL, van Kollenburg P, Huisman HJ, Scheenen TW, Witjes JA, Leer JW, Barentsz JO, Visser AG. IMRT boost dose planning on dominant intraprostatic lesions: gold marker-based three-dimensional fusion of CT with dynamic contrast-enhanced and 1H-spectroscopic MRI. *Int J Radiat Oncol Biol Phys* 2006;65:291–303.

86. Singh AK, Guion P, Sears-Crouse N, Ullman K, Smith S, Albert PS, Fichtinger G, Choyke PL, Xu S, Kruecker J, Wood BJ, Krieger A, Ning H. Simultaneous integrated boost of biopsy proven, MRI defined dominant intra-prostatic lesions to 95 Gray with IMRT: early results of a phase I NCI study. *Radiat Oncol* 2007; 2: 36.

87. Pickett B, Vigneault E, Kurhanewicz J, Verhey L, Roach M. Static field intensity modulation to treat a dominant intra-prostatic lesion to 90 Gy compared to seven field 3-dimensional radiotherapy. *Int J Radiat Oncol Biol Phys* 1999;44:921–929.

21

Molecular and Cellular Imaging*

Martin Lepage
Université de Sherbrooke

21.1 Molecular Imaging

Molecular imaging aims at the visualization of molecular targets or molecular processes occurring at the molecular and cellular levels. Exogenous agents that target specific cell receptors or that interact with specific enzymes and proteins *in vivo* will be covered in this section, while the imaging of cells labeled *in vitro* and subsequently injected into a subject will be covered in Section 21.2. This is a subjective classification, and other authors have grouped both of these topics under "molecular imaging," while others have included the targeting of cellular targets by exogenous compounds under "cellular imaging." Quantitation of results in molecular imaging refers to the ability to estimate the concentration of an exogenous agent that has reached a specific location at a specific time and, in special cases, to estimate the rate of a biochemical process, such as enzymatic cleavage. Magnetic resonance imaging (MRI) does not detect the agent itself but, rather, is sensitive to the contrast agent's effects on the magnetic properties of the neighboring water molecules. (See Chapter 9 for more details on contrast mechanisms.) This indirect detection makes quantitative molecular MRI challenging. However, several excellent publications have described methods and types of contrast agents that can be synthesized and imaged quantitatively. Here, the focus will be on the application of these molecules and nanoparticles in cancer imaging.

21.1.1 Design and Testing of MRI Molecular Imaging Agents

There have been an impressive number of publications describing compounds tested *in vitro* with potential for *in vivo* imaging;

---

* In memory of William C. Dow (1955–2009).

an overview of these is available [1]. As there is a substantial leap from *in vitro* to *in vivo* testing, as the experimental conditions change drastically between test tubes and a living organism, we will restrict the discussion to those agents with demonstrated *in vivo* applications.

21.1.1.1 Biology and Tumor Biology at a Glance

While more details can be found in Chapter 1, we briefly provide the salient features of tumor biology that are relevant for the present chapter.

Solid tumors are composed of cancer cells that divide at an accelerated pace and that do not (typically) naturally undergo apoptosis. This is a tremendous advantage over neighboring normal cells, as cancer cells are able to proliferate more efficiently, eventually outgrowing, invading, and even replacing normal tissues. These processes require energy, which is obtained from nutrients in the blood stream. Those nutrients are available to the first few layers of cells around a blood vessel, which would in principle limit the development of tumors. However, cancer cells have the ability to recruit new blood vessels (neovasculature) that allow them to develop further [2]. These neovessels are grown rapidly and imperfectly, resulting in a "leaky" vasculature (see Figure 21.1). The epithelium of normal blood vessels is smooth and relatively impermeable to exogenous contrast agents. However, the tumor neovessels are characterized by a higher permeability. The fenestration of tumor neovessels has diameters in the micrometer range [3], such that compounds present in the vasculature, such as glucose, amino acids, or contrast agents, can readily escape the vasculature and enter the interstitium, where they can interact with tumor cells. The size of the fenestration is an upper limit for molecular imaging agents designed to target cancer cells, which may be located several micrometers away from a blood vessel. Here, diffusion can be a rate-limiting step

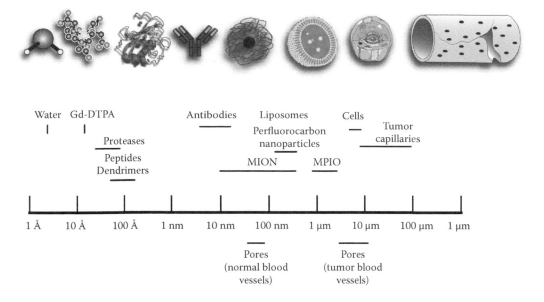

FIGURE 21.1 Logarithmic size scale of features relevant to molecular imaging. Schematic illustrations of a water molecule, a Gd-DTPA molecule, MMP-8 with an inhibitor, an antibody, a dextran-coated MION, a liposome, a cell, and a tumor capillary shown with fenestration and discontinuity. The diffusion coefficient of compounds drops considerably with an increase in molecular weight [94]. The representation of MMP-8 is courtesy of Dr. Oded Kleifeld and Dr. Christopher Overall.

in the delivery of a contrast agent to the desired target, as is the case for the delivery of therapeutic agents [4]. The diffusion coefficient decreases as a function of the molecular weight and size (Figure 21.2) such that smaller agents can diffuse further than larger agents for a given time. For example, the time constant for an antibody to diffuse over 100 μm in a tumor is around 1 h [4]. In the same interval, a low-molecular-weight compound would have diffused over 1 mm. There is, however, a trade-off; small compounds can easily extravasate and diffuse over several hundred micrometers, but they may also be rapidly eliminated by the kidneys. Larger contrast agents typically have a longer blood circulation time, and this

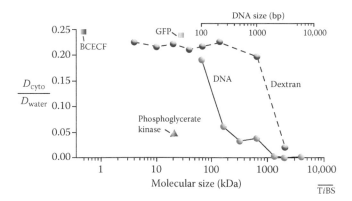

FIGURE 21.2 Diffusion coefficient of compounds in the cytoplasm (D_{cyto}) relative to that in pure water (D_{water}), as a function of molecular weight. The diffusion is already hindered in the cytoplasm, and the effect is exacerbated for large compounds. GFP stands for green fluorescent proteins, BCECF stands for 2,7-bis(carboxyethyl)-5,6-carboxyfluorescein, a small fluorescent probe. (Reproduced from Verkman, A. S., *Trends Biochem. Sci.*, 27, 27–33, 2002. With permission.)

can improve the total quantity of compound delivered to a tumor [3]. However, it may become more difficult to determine whether the accumulation of a contrast agent in a tumor is due to a specific event or due to an unspecific effect such as the "enhanced permeability and retention" for macromolecules and lipids [5]. Elimination of a compound (and its degradation products) from a tissue is normally either due to lymphatic drainage or the return of the compound or its degradation products to the bloodstream. Tumors lack a fully functional lymphatic drainage system, and this has contradictory consequences. First, this contributes to an elevated interstitial pressure, which decreases the transport of molecules from the vasculature to the tumor tissue; second, this prolongs the retention and favors the accumulation of a molecule that did reach the tumor interstitium. While it is certainly desirable to obtain the largest concentration possible of a drug in a tumor, the design of a specific contrast agent may require a different approach in order to optimize the specificity of the agent. Finally, it must be noted that once inside the cytosol, the diffusion of compounds is reduced in relation to their molecular weight [6].

21.1.1.2 Delivery

Molecular imaging agents and therapeutic drugs face similar limitations. In a review article on the delivery of therapeutic agents in tumors, Jain identified three physiological factors responsible for the poor localization of macromolecules in tumors: (i) heterogeneous blood supply, (ii) elevated interstitial pressure, and (iii) large transport distances in the interstitium [4]. The accumulation of contrast agents in tumors depends on the molecular weight, the charge, the resistance to degradation, the hydrophobicity/hydrophilicity of the compound, the state of the neovasculature, and the presence of target-specific ligands.

A contrast agent will normally be injected into the blood circulation, where enzymes and peptidases are present. The agent must thus offer some resistance to degradation by these natural defense mechanisms. Accumulation of the agent at the target site will inevitably be a competition against elimination by the kidneys and/or the liver. Molecular imaging agents (except perhaps fluorescence agents based on the Förster resonance energy transfer principle [7]; see below) face the additional constraint that nonspecific accumulation at a target site will obscure the desired information. It should be noted that the binding reaction of a compound to a target will lower the apparent diffusion rate of the compound [8], such that targets proximal to vessels may efficiently be reached, at the expense of similar targets located farther from a vessel. A clear example of this is found in hypoxic conditions. Let us pretend that oxygen is a contrast agent; it is a very small molecule and it should readily diffuse in a tissue. However, active uptake by the first cell layers surrounding blood vessels will absorb most of the oxygen, resulting in the "starvation" of cells that lie at a larger distance, and will result in tumor hypoxia when the intercapillary distance exceeds ~200 μm in human tumor xenografts [9].

The intracellular targeting of molecular imaging compounds imposes additional requirements to the design, and many strategies have been tested. First, nonspecific internalization by phagocytosis or pinocytosis is possible for large compounds [10]. The agent may be designed to bind to a receptor that will be internalized; this process is specifically referred to as receptor-mediated endocytosis. Finally, the agent may be actively transported into the cytosol by a receptor; this is the case for the glucose transporter-1 protein, which transports the 18-F fluorodeoxyglucose positron emission tomography tracer into cells, where it is phosphorylated and trapped [11].

21.1.1.3 Concentration and Specificity

The concentration of a paramagnetic contrast agent in a tissue can be determined if its relaxivity is known and constant in a tissue. Thus, quantitative concentration maps can reveal the nonuniform distribution of the agent resulting from a nonuniform delivery and the heterogenous tissue properties.

The interpretation of whether the accumulation of a compound is specific (i.e., directly linked to the presence of the target) or nonspecific (i.e., passive accumulation in the tumor interstitium) invariably requires control experiments. In principle, quantitation of a biochemical process would require knowledge of the concentration of the agent in blood, the fractional blood volume, the fractional interstitial volume, and the concentration of the agent either bound to its target or cleaved by an enzyme, with all these as a function of time. The difficulty is exacerbated by the heterogeneity of tumor tissues characterized by nonuniform vascularization, nonuniform vessel permeability, and interstitial pressure [4]. A single experiment with a contrast agent cannot provide all this information simultaneously such that specificity must be assessed by control experiments. Quantitation of biochemical processes has not yet been reported for MRI contrast agents. For example, if one were to test a compound designed to bind to a specific cell receptor via a peptide ligand, then an appropriate set of control experiments should include a control compound that is identical in every point except for the ligand-binding properties. In this particular case, simply removing the peptide may not be adequate, since this will change the molecular weight, molecular size, and possibly the hydrophilicity of the compound. Instead, a peptide of similar weight and charge that does not interact strongly with the receptor could be used. Competition studies where the targeting moiety is injected alone prior to the molecular imaging agent are also informative. It must be emphasized that simple comparisons with commercially available low-molecular-weight contrast agents may not be appropriate. Similarly, tests may also be conducted with a different cell line not expressing the targeted receptor. Caution must also be exercised in this case, since the phenotype of the control cells may be different even though they originated from the same cell line. Since different tumors, or similar tumors with a different grade [12], may have different vascular permeability, vascular volume, extravascular extracellular volume, interstitial pressure, or pH, the accumulation of a given compound may be drastically different between tumor types. (In fact, this is even true for simple nonfunctionalized compounds [13]). Finally, it must also be remembered that different animal species have different heart rates, which may directly affect the circulation time of compounds. Thus, it is preferable to perform the experiments and their control using the same species.

In 2002, Weissleder divided molecular imaging agents, particularly optical imaging probes, into three classes: (i) nonspecific probes that do not interact with a target and that are simply distributed in the vascular and extravascular extracellular space (these provide information on changes in blood volume, perfusion, and blood flow in angiogenesis); (ii) targeted probes, where the contrast agent is functionalized with an antibody, a protein, or a peptide targeted specifically to tumor cells; and (iii) activatable probes that can only be detected once they have interacted with their substrate [14]. This classification remains valid for MRI contrast agents, with a slight modification of the third definition where activatable probes display a change in relaxivity or a change in solubility after interaction with their target. This takes into account the difficulty of reducing the relaxivity of an agent to zero and also includes agents in which the pharmacokinetic, and not the relaxivity, is modified by interaction with a target. A limitation in activatable agents was that their design required knowledge of the local concentration in order to determine whether the signal change is specific or unspecific [15]. There are, of course, several other ways to categorize MRI contrast agents [16].

21.1.2 Types of Molecular Imaging Agents

The different classes of contrast agents for MRI are detailed in Chapters 10 and 11. The contrast agents currently approved for clinical use cannot be classified as "molecular imaging contrast agents" since they are not endowed with a specific biological function (i.e., they are nonspecific). However, they serve as the basis for a large portion of molecular imaging agents reported to date.

21.1.2.1 Paramagnetic Chelates, Nanoparticles, and Liposomes

This chapter covers molecular imaging agents in preclinical applications to cancer. One notable exception will be made to include the work of Meade et al., whose pioneering work led to many innovations in the development of molecular imaging agents for MRI [17,18]. Meade et al. designed a compound that restricted the access of water to a gadolinium atom. Upon cleavage of a section of the molecule, the access of water to the paramagnetic center was increased, thereby resulting in an increase in relaxivity. Note that this effort has some similarity with the well-known Förster resonance energy transfer mechanism exploited in optical imaging [7], whereby the fluorescence of a compound is quenched until a portion of the compound (the absorber) is released. However, it is extremely difficult to reduce the relaxivity of a compound to zero. In the case of agents based on a "relaxivity switch," an increase in T_1-weighted signal may result from both an increase in the local concentration of a compound or from an activation-induced increase in relaxivity of the agent.

An interesting approach was proposed by Heckl et al. in 2003 that enabled the detection of intracellular targets *in vivo* [19]. These investigators developed a gadolinium chelate linked to a c-*myc* mRNA-targeting peptide nucleic acid and a transmembrane carrier peptide (c-*myc* mRNA is a gene product in cancer cells). They showed specific retention of the contrast agent in a rat model of prostate carcinoma when using a c-*myc*-specific compared to a c-*myc*-nonspecific peptide nucleic acid sequence. Newer agents using a similar approach have not yet been tested *in vivo* [20] or require further control experiments.

The $\alpha_v\beta_3$ integrins are strongly expressed on activated endothelial cells and can therefore serve as a marker of cancer-induced angiogenesis. A gadolinium chelate coupled to a peptidomimetic ligand to these integrins was developed and tested in mouse and rat models of atherosclerosis [21]. The careful experimental design enabled the investigators to demonstrate very well the success that agents may have *in vitro* and the difficulty of translating those successes *in vivo*. An increase in the signal-to-noise ratio of 150% was obtained about 40 min after injection in atherosclerotic plaques. The specificity of this accumulation was tested by a competition experiment where a europium derivative without significant r_1 relaxivity was injected first, with the aim of saturating the integrins. The gadolinium derivative was injected subsequently, and an increase in the signal-to-noise ratio of 55% was detected, which was interpreted as nonspecific accumulation of the compound.

A subclass of the matrix metalloproteinases (MMPs) endopeptidase family is involved in tumor progression and in degradation of the extracellular matrix. We designed and tested an agent consisting of a gadolinium chelate attached to a peptide sequence cleavable by selected MMPs. Upon cleavage of the peptide, a solubilizing moiety was detached from the agent, resulting in a solubility decrease and accumulation at the site of cleavage. These so-called solubility-switchable, activatable contrast agents were shown to detect the activity of MMP-7 in a mouse model of colon cancer [22] and of MMP-2 in a mouse model of breast cancer [23,24]. Each mouse was inoculated with tumor cells expressing the targeted MMP and with control tumor cells expressing a much lower level of the enzyme on the other hind limb. Nonspecific accumulation of the compounds was observed in the control tumor. The same conclusion was reached when injecting a control compound that was resistant to cleavage. In this approach, the accumulation as a function of time in the MMP-expressing tumor was different for the solubility-switchable compound as compared with either control tumor or control compound. The sensitivity of MRI would be too low to detect a compound designed to bind to an MMP, such as an MMP inhibitor. Instead of binding, these contrast agents were based on an amplification strategy and were cleaved by an MMP; one MMP can cleave multiple copies of its substrate peptide.

Another strategy to make Gd-based contrast agents sensitive to their environment is by modulating the hydration of the metal complex as a function of pH, which affects the exchange rate of the water protons associated with the paramagnetic center and thereby changes the longitudinal relaxivity of the agent (reviewed in [15]). Raghunand et al. were able to compute pH maps of mouse kidneys using sequentially a pH-sensitive and a pH-insensitive gadolinium chelate [25]. The longitudinal relaxivity of a low-molecular-weight compound, Gd-DOTA-4AmP^{5-}, increased from about 3.2 to 4.5 mM^{-1}s^{-1} (4.7 T) when pH decreased from 8 to 5.7. This is compared with a pH-independent value of 3.0 mM^{-1}s^{-1} of Gd-DOTP^{5-}. This methodology could clearly differentiate between kidneys in control mice and in mice treated with acetazolamide having abnormally high pH values (pH ~7.7).

It has also been demonstrated that pH can modify the rigidity of a polymer structure, which results in an increase of the rotational correlation time of the gadolinium chelate and thus increases the relaxivity of the agent (reviewed in [15]). In principle, this strategy could also be applied to *in vivo* imaging.

Antibodies have high and specific affinity toward selected ligands. Their large size (see Figure 21.1), however, limits their application for imaging receptors in solid tissues. A two-step approach was proposed by Artemov et al. whereby biotinylated anti-HER-2/*neu* antibodies (~150 kDa) were first administered and then allowed to bind for 12 h, at which point unbound antibodies were cleared from the blood circulation. An avidin-Gd-DTPA conjugate (~70 kDa) was then administered, which could avidly bind to biotin [26]. Although a direct comparison with the antibody prelabeled with Gd-DTPA was not shown, the strategy was deemed to be more efficient in comparison with a single, large-molecular-size contrast agent. The compounds were tested on two different mouse models inoculated either with HER-2 positive (NT-5) or HER-2 negative (EMT-6). In the HER-2-positive tumors, pretreatment with the labeled antibody resulted in a larger T_1-weighted signal increase 24 h after injection of the avidin-Gd-DTPA conjugate. This was also the case when comparing with the T_1-weighted signal increase in the control HER-2-negative tumors, although a direct comparison with a nonspecific contrast agent would have been useful to assess possible differences in blood volume, blood perfusion, and blood vessel permeability between these tumors originating from different cell lines.

Polyamidoamine dendrimers derivatized with folate acid ligands and approximately 45 gadolinium chelates were shown to bind to folate cell surface receptors or soluble receptors in the interstitium in a mouse model of ovarian cancer [27]. Competition experiments consisting of the simultaneous injection of folic acid and folate–dendrimer prevented the decrease in T_2-weighted signal intensity that was observed when the folate–dendrimer was injected alone. Comparison was also made with a nonspecific contrast agent (Gd-HP-DO3A) that had a much lower molecular weight (559 Da), and accumulation of that compound was lower than that of the dendrimer (>30,000 Da) 24 h postinjection, as expected. Injection of the dendrimer with 45 gadolinium chelates, but without folate acid ligands, was not reported.

The same folate cell surface receptors were targeted by a compound containing two gadolinium chelates and a folic acid tested in a rat model of ovarian tumors [28]. The investigators could find a significant change in R_1 in weakly enhancing tumors of the same animals 1 h after injection of folate receptor–targeted or folate receptor–untargeted compounds on different days. The results suggest about half of the increase in R_1 is nonspecific, and the dynamic curves were not shown.

21.1.2.2 Liposomes

The purpose of incorporating paramagnetic gadolinium in or on the surface of liposomes is to increase the local concentration of gadolinium (or other imaging agents [29]), which may incidentally increase the relaxivity of each gadolinium atom by slowing down the tumbling rate of the chelate. Functionalization of the liposome surface may then confer a biological activity to the liposome. Sipkins et al. prepared liposomes incorporating a lipid gadolinium chelate [30] conjugated with an antibody targeting the $\alpha_v\beta_3$ integrins [31]. Approximately 50 antibodies were conjugated to each liposome, resulting in a mean diameter between 300 and 350 nm. In a rabbit model of squamous cell carcinoma, these investigators observed a larger MR signal intensity increase (approximately 15%) when the liposome was injected, as compared with control liposomes without antibody or with isotype-matched immunoglobulins (approximately 8%). Anderson et al. successfully tested a perfluorocarbon nanoparticulate emulsion containing gadolinium chelates and targeting the $\alpha_v\beta_3$ neovasculature integrin [32]. The perfluorocarbon emulsions were linked to antibodies specifically targeting the $\alpha_v\beta_3$ receptor, to an isotype-matched antibody, or to no antibody. Accumulation for the specific antibody was observed 90 min after injection; this was also confirmed by a blocking experiment.

In a recent paper, Erdogan et al. prepared a liposome incorporating a polymer that is able to chelate multiple gadolinium atoms [33]. The authors showed in a mouse model of Lewis lung carcinoma that the longitudinal relaxation rate was 0.48 s^{-1} for a liposome with an unspecific antibody to 0.62 s^{-1} for a liposome derivatized with an antibody directed against surface-bound nucleosomes released from apoptotically dying cancer cells 4 h after injection. Additional information on the circulation time of these two liposomes and the relaxivity of the unspecific liposome is, however, not mentioned.

Winter et al. prepared paramagnetic nanoparticles from perfluorocarbon lipids, covalently coupled with $\alpha_v\beta_3$ peptidomimetics, emulsified with paramagnetic lipids, resulting in a particle size around 273 nm [34]. The nanoparticles were composed of over 90,000 gadolinium atoms. Signal enhancement was observed, and comparisons with similar but untargeted nanoparticles in a rabbit tumor model revealed that 50% of the signal enhancement was nonspecific and related to extravasation of the nanoparticles 2 h after injection. The authors commented that their nanoparticles could extravasate but could not diffuse significantly into the interstitium. Similar nanoparticles were used in a model of melanoma tumors in mouse [35].

Mulder et al. modified the design of Sipkins et al. by replacing the antibodies by 700 tripeptides that bind specifically to $\alpha_v\beta_3$ integrins [36]. The resulting liposomes had a size around 150 nm. Signal enhancement in mouse tumors was observed on T_1-weighted images for this liposome, but a similar enhancement was also observed when a control peptide was used. The liposomes were also fluorescent, which enabled the authors to determine the location of the liposomes within the tumors. The number of animals in each group and the size of the tumors for each experimental condition were not specified.

Recently, Freedman et al. developed immunoliposomes with a diameter of 100 nm composed of a cationic liposome encapsulating gadolinium complexes and decorated with an antitransferrin receptor single-chain antibody fragment [37]. Specificity for the detection of lung metastases in mouse was shown by injecting the targeted and the untargeted liposomes (without the antibody fragment) to the same animals. The number of gadolinium atoms in each liposome was not specified.

Liposomes and perfluorocarbon nanoparticles can carry a large number of gadolinium atoms, resulting in compounds with a high relaxivity. The size and molecular weight of these compounds result in longer circulation times than simple gadolinium chelates such that nonspecific presence of the compounds needs to be accounted for. Additionally, these compounds can likely escape tumor neovasculature but can hardly diffuse thereafter. The usefulness of this class of compounds is thus emphasized for endovascular targets or for targets in the close vicinity of blood capillaries.

21.1.2.3 Superparamagnetic Nanoparticles

Several terms and acronyms have been proposed for superparamagnetic iron oxide (SPIO) particles. The most commonly known is *monocrystalline iron oxide nanoparticles* (MIONs) [38,39] or *monodispersed superparamagnetic iron oxide colloid* (also termed MION). As their names indicate, the diameter of these particles ranges between 5 and 1000 nm. The smaller particles (<10 nm) have been termed *ultrasmall particles of iron oxide* (USPIO) [40]. The larger particles are referred to as *micron-sized particles of iron oxide* (MPIO) [41]. Bare iron oxide nanoparticles accumulate in the liver and may have toxicity. To avoid toxicity issues, the nanoparticles are generally covered with an inert and biocompatible material such as dextran. Typically, particles with a 5-nm core diameter are coated with a dextran coating diameter of 25 nm. Some coating preparations are cross-linked for further stability and to enable surface derivatization; these have been named *cross-linked iron oxide nanoparticles* (CLIO) [42].

SPIOs mostly affect the transverse relaxation time of nearby water molecules and, accordingly, are often referred to as T_2 agents. The transverse relaxivity (r_2) of this class of compound is much larger than the longitudinal relaxivity (r_1) of gadolinium-based contrast agents. Thus, many investigators have designed molecular imaging agents based on superparamagnetic particles.

These relatively large particles have a longer circulation time than small gadolinium chelates, and their excretion is mainly by the liver and spleen. The diffusion of compounds is inversely related to their diameter (or weight), so targets for such compounds must be carefully selected.

Dextran-covered MION with a 3-nm core was conjugated with human holotransferrin, a ligand of the transferrin receptor [43]. A mouse gliosarcoma tumor model with cells engineered to express the transferrin receptor (ETR+) on one flank and ETR– cells on the other flank was studied. A significantly lower value of T_2 was measured 24 h after administration in the ETR+ tumor. The size of each group of tumors was not specified, and unconjugated MIONs were not tested to account for possible differences in the vasculature permeability of the different tumor types.

USPIOs were derivatized with octreotide, a short peptide-binding somatostatin receptor [44]. *In vitro* results demonstrated the binding specificity, but *in vivo* mouse imaging was less convincing, which emphasizes the difficulty in demonstrating the usefulness of probes *in vivo*.

Polymeric micelles loaded with SPIOs and derivatized with cyclic RGD peptide were used to detect $\alpha_v\beta_3$ integrins. Experiments were performed with the micelles (75 ± 11 nm) derivatized or not with the peptide. Blocking experiments confirmed the targeting specificity and revealed that approximately half the signal change is nonspecific [45]. An off-resonance saturation scheme allowed for larger signal changes compared with conventional T_2^* imaging. It should be noted that a conclusive study will require the sequential injection of derivatized and nonderivatized micelles in order to ascertain the specificity of the targeting in individual subjects. The long circulation times of the micelles (~4 and ~9 h) will therefore require imaging on different days.

MIONs with a 10-nm core diameter were covered with amphiphilic polymers and functionalized with peptides targeting the urokinase plasminogen activator receptor, a receptor highly expressed in pancreatic cancer [46]. Control MIONs were not functionalized, and both compounds were derivatized with a fluorophore for optical detection. MRI results from a pancreatic orthotopic tumor nude mouse model were substantiated by optical imaging, which clearly revealed a specific accumulation of the targeted MIONs.

21.1.2.4 Paramagnetic Nanoparticles

Superparamagnetic particles cause a loss of MR signal on standard T_2^*-weighted images; this is often referred to as "negative contrast." A recent review described the advances in the design of nanoparticulate T_1 contrast agents and their applications in target-specific MRI of cancers [47]. These particles generate an increase in T_1-weighted signal, and their longitudinal relaxivity values were over 10^4 and more typically of the order of 10^5 mM^{-1}s^{-1} for various compositions and sizes [48]. Preliminary *in vivo* studies are encouraging [48].

21.1.2.5 CEST and PARACEST Agents

The potential of contrast agents based on the chemical exchange saturation transfer (CEST) effect was demonstrated in 2000 [49]. The CEST effect is based on the exchange of protons between a molecule and water. Irradiation with radiofrequency waves at the resonance frequency of the exchangeable proton of a molecule increases the "spin down" population of these protons. It is assumed that the resonance frequency of these protons is different from that of water protons; hence, the saturation radiofrequency field is called "off-resonance." Proton exchange between this molecule and a neighboring water molecule results in a decreased longitudinal magnetization of water protons. Thus, the intensity on a proton density–weighted image would decrease. The main advantage of this class of contrast agents is the possibility to turn the contrast "on" (irradiation at the resonance frequency) or "off" (irradiation at the opposite frequency, or no irradiation). The rate of proton exchange should be as high as possible without reaching the fast exchange regime where the resonance frequency of both proton pools would coalesce. Attaching a paramagnetic ion to a molecule effectively shifts the RF of water protons bound or associated with the agent; these are so-called PARACEST agents. PARACEST allows for larger saturation transfer rates and hence larger CEST effects, as well as a decreased competition from magnetization transfer effects from tissue macromolecules, although a larger radiofrequency (RF) saturation power is required. Some advantages of this class of agents are that they may be sensitive to pH, temperature, and their chemical environment. Quantitation of these effects requires knowledge of the local agent concentration, which would be unknown *in vivo*. However, it is possible to include different proton exchanging groups on the same molecule, for example, one that would be sensitive to pH and one that would not be, such that the latter could serve as a reference to determine the concentration. *In vivo* applications to cancer of this class of agents have not yet been reported [50]. Compounds that detect enzymatic activity have been developed and tested *in vitro* [51].

21.1.2.6 Imaging/Therapeutic Agents

It was realized early that combined imaging/therapeutic agents could be useful. Radiosensitizing consisting of a porphyrin-like gadolinium chelator was tested in mice models of spontaneous mouse mammary and murine mammary sarcoma [52]. Although tumor accumulation was likely nonspecific, this study highlighted the merits of visualizing, and potentially quantifying, the accumulation of a therapeutic agent.

21.1.3 Future Developments and Refinements

The field of MRI molecular imaging is developing rapidly, and discoveries are likely to impact significantly on this field in the future; as of 2009, no MRI contrast agent is commercially available for molecular imaging [53]. There have already been successful examples of multimodality agents (e.g., visualized with MRI and fluorescence imaging), and, at least for preclinical research, these tools are likely to find newer applications. However, the sensitivity of MRI is low compared to nuclear and optical techniques, which restricts the targets that can be in the scope of MRI agents.

New materials are constantly being developed, which could be used as MRI agents. For example, novel paramagnetic nanoparticles [54], ferrite nanoparticles [47], and gold-coated cobalt nanoparticles [55] may yield potential for active targeting.

An original design to target protease activity in tumors has recently been developed for optical imaging agents [56]. In this approach, a cell-penetrating peptide is initially masked, preventing its accumulation into cells. Following cleavage of a peptide linker by the targeted protease, the cell-penetrating peptide is unmasked, which leads to its adsorption and uptake into cells.

Novel pH-sensitive nanogels have recently been synthesized based on the modulation of ^{19}F signal [57]. This signal modulation resulted in the swelling from ~60 to ~90 nm of a PEGylated nanogel at physiologically relevant pH values.

Strategies to couple multiple gadolinium chelates to a molecule already exist. Another strategy is to synthesize gadolinium oxide nanoparticles [48]. It is to be expected that *in vivo* experiments on cancer models will be available in the near future.

21.2 Cellular Imaging

21.2.1 Overview of Techniques

Cellular imaging was first achieved using nuclear techniques in the early 1970s [58–60]. Early experiments were typically performed on red and white blood cells, although cancer cells have also been labeled and detected *in vivo* [61,62]. The technique of accumulating a compound into a cell *ex vivo* and later administering and imaging the distribution of these cells was later adopted by other imaging techniques such as optical imaging and MRI (recently reviewed in [63]). The majority of applications in MRI utilize MPIO, SPIO, or gadolinium chelates internalized by phagocytosis (e.g., macrophages), by

transfection into selected cells or by receptor-mediated endocytosis. Different labeling techniques have been reviewed [64], and various gene constructions have been compared [65]. The main advantage of MRI cellular imaging is its very high spatial resolution [66]. An amount of contrast agent sufficient to cause a detectable signal change should be incorporated into the cells, but this must not perturb the phenotype of the labeled cells. It was shown that SPIO nanoparticles had a relatively low toxicity *in vivo* [67]. Cells labeled with iron oxide nanoparticles rely on the magnetic field inhomogeneity generated by the large magnetic dipole moments of the iron oxide magnetic domain or domains within the particles. The dipolar field extends well beyond the radius of the particle; the field is a function of the orientation relative to the dipolar moment (for a single magnetic domain), and it decreases with the inverse of the distance cubed [68,69]. When diffusing within this temporally and spatially varying dipolar field (Figure 21.2), the magnetization of water molecules is strongly dephased, since the Larmor frequency of the water protons is proportional to the local magnetic field strength. This leads to a marked signal decrease on T_2^*-weighted images and, to a lesser extent, on T_2-weighted images. The volume of the voxels is much larger than that of labeled cells, and the maximum possible amount of iron is loaded in the cells such that it is possible to detect a signal void corresponding to the presence of a single cell *in vivo* with specialized hardware to achieve high spatial resolution [69,70]. If the spin dephasing caused by a single labeled cell is sufficient to bring the signal in a voxel to zero, then adding a second labeled cell cannot cause an additional signal drop in that voxel. Experimentally, different extents of cell labeling have been tested *in vitro* and *in vivo*. It was shown that the apparent diffusion coefficient (ADC), the value of R_2, and the product of ADC × R_2 correlated with the amount of cells (between 7.5×10^5 and 2.5×10^6 cells/ml) in a gelatin phantom [71]. *In vivo* in the rat brain, results highlighted that change in host tissue properties needs to be considered, but values of ΔR_2 correlated with cell concentrations between $\sim 4 \times 10^4$ and 3.5×10^5 cells/ml [71].

21.2.2 Detectability Limits

An overarching question in cellular imaging is "what is the minimum number of cells needed for detection?" This depends on numerous factors, including the field strength, the gradient strength, the pulse sequence used (including voxel volume and signal-to-noise ratio), the nature of the labeling compound, the amount of label in a cell and its compartmentalization [69,72], and the presence of nearby blood vessels that also cause T_2^* effects. An expression was derived which predicted that the minimum mass of SPIO was in the femtomole range, and a demonstration was made showing that a single cell can be detected in a phantom [69]. Experimentally, the *in vivo* detection limit is in the range of 100–500 cells, with 10–60 pg of iron per cell down to a single cell with Bangs particles (100 pg of iron per cell) (reviewed in [66]). The contribution to T_2^* effects from blood vessels can be decreased in small animal models by changing the inhalation gas to 95% O_2 and 5% CO_2, enabling a more reliable identification of labeled cell clusters [73].

21.2.3 Positive Contrast Imaging of Superparamagnetic Cells

Positive contrast of iron oxide nanoparticle–loaded cells has recently been developed, which circumvented the problem associated with the detection of labeled cells in a tissue with inherent low T_2- or T_2^*-weighted signal intensity [74]. This positive contrast is obtained with spectrally selective RF pulses that refocus the dephased magnetization around labeled cells, resulting in a signal only from these water molecules. In this first demonstration, a linear correlation between the estimated number of cells and the signal was observed *in vitro* [74]. Several techniques for obtaining positive contrast from iron oxide nanoparticles have since been developed and compared [75], and new approaches are still being developed [76]. Quantitation of labeled cells remains an elusive goal, and positive contrast images from MIONs are prone to susceptibility artifacts near air–tissue interfaces.

21.2.4 Limitations of Cell Labeling

Recognized limitations of cellular imaging are (i) dilution of the label by cell proliferation, (ii) difficulty of discriminating between live and dead cells, and (iii) uptake of the label by other cells after cell death [77]. Dilution of the particles after cell division progressively decreases the susceptibility effect caused by the labeled cells, leading to a loss of detection after 7 days [78], although cells were detected for at least 18 days in a mouse glioma model [79]. Injection of dead or live labeled cells produced different results in a mouse model [80], but leakage of contrast agent from a cell is still difficult to assess *in vivo*. This is a particular concern for imaging of stem cell therapy, since iron-avid macrophages are likely to be present in areas of tissue remodeling and reconstruction; iron leaking from dead labeled cells could potentially be accumulated by macrophages, leading to a localized signal loss no longer attributable to the labeled cells.

21.2.5 Recent Advances in Cell Labeling

The most recent advances in cellular imaging have resulted from the development of MRI reporter genes that result in MRI-observable signal changes [81–84]. The most promising strategy to date is the transfection of genes that induce the expression of ferritin receptors. These cells internalize and sequester iron atoms (Figure 21.3), either endogenous or exogenous, which leads to an increase in the spin–spin relaxation rate constant. Such MRI reporter genes have the capability to be more quantitative than the well-known luciferase bioluminescence reporter gene [85] and green fluorescent proteins, and MRI-based approaches may overcome the imaging resolution limitation of positron emission tomography [86]. When under the control of a promoter, these systems can provide "on–off" information about the activity of the gene construct, which could be particularly useful in monitoring

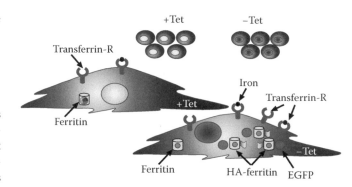

FIGURE 21.3 Conditional expression of the transferrin receptor (Transferrin-R). In the absence of the transactivator TET (–Tet), the cells express the influenza hemagglutinin–tagged ferritin (HA-ferritin), which ultimately leads to accumulation of iron and MR contrast. The mechanism is switched off in the presence of TET (+Tet). This was confirmed with the coexpression of the enhanced green fluorescent protein (EGFP). (Modified from Cohen, B. et al., *Neoplasia*, 7(2), 109–17, 2005. With permission.)

transgene expression *in vivo* and to characterize the duration and location of the gene expression. The main advantages of ferritin-based MRI reporter genes are that the challenges associated with the delivery of exogenous compounds are alleviated and that the daughter cells can also be detected for several weeks [82,83].

While quantitative measurements of the spin–spin relaxation rate constant can be achieved, the relationship between this number and the concentration or number of cells is not trivial. In addition to the concentration of labeled cells, the signal change is dependent upon the magnetic field strength, the voxel volume, and the imaging parameters. In the reporter gene approach, the amount of iron internalized and sequestered is dependent upon the availability of iron in the subject and is thus highly dependent upon the diet and presumably on the time between ingestion of food and imaging.

21.2.6 Alternative Cell Labeling Approaches

In a 2005 letter, Bulte highlighted the advantages of ^{19}F MRI of cells labeled with molecules containing a ^{19}F atom [77]. The main feature is that these cells would appear as a hot spot on an image without background, which could allow quantification of the label [87]. This approach was successfully tested using cells labeled with perfluoropolyether emulsion nanoparticles [88].

Cells can be labeled with paramagnetic liposomes to yield increased T_1-weighted signal intensity *in vitro* [89] and could have potential for *in vivo* imaging.

21.2.7 Cellular Imaging of Cancer

21.2.7.1 Cancer Cell Metastasis

Early efforts into the visualization of the cell invasion process were directed towards multicellular tumor spheroids of brain glioma cells [41]. In this study, small discrete signal loss areas were observed around the original spheroid within 2 days after they were embedded in a collagen gel [41]. This work has recently been extended to an *in vivo* mouse model of brain glioma [79]. In this study, the authors actually show that a trade-off exists between the requirement for detection sensitivity and that for spatial accuracy [79]. Prostate cancer cells were labeled with MPIO, injected into the heart of mice, and their dissemination across the blood–brain barrier and subsequent colonization of the brain were monitored [90].

21.2.7.2 Gene Expression and Metastasis

C6 glioma tumor cells were engineered to express ferritin, which ultimately leads to intracellular accumulation of iron and MR contrast (Figure 21.3) [83]. In this approach, the injection of an exogenous contrast agent is not required such that it will be potentially possible to identify metastases from a primary tumor. Subsequently, a mouse model was developed that expressed the ferritin receptor and the ferritin complex, yielding high R_2^* values [91].

Tumor cells (9L rat glioma cells) were engineered to produce lysine-rich protein that contains exchangeable protons that can be detected by the CEST effect [92]. The authors were able to discriminate between control and engineered cells implanted in the rat brain. These proteins are produced by the engineered cells such that their concentration may be expected to be maintained even after cell divisions.

The expression of Mag A, a bacterial iron transporter from the magnetotactic bacteria family, was induced in a human cell line under doxycycline regulation [84]. When injected into a mouse brain, these cells were shown to accumulate iron *in vivo*.

For these systems, the amount and the organization of iron within the cells are important factors that can influence the R_2^* relaxivity and thus the signal loss [91] such that the relationship between iron content and signal loss is not straightforward.

21.2.7.3 Neovasculature

Mouse bone marrow–derived precursor cells were shown to incorporate into neovasculature in a severe combined immunodeficiency (SCID) mouse brain tumor model [93]. In this case, dextran-coated SPIOs were complexed with poly-L-lysine for incorporation into the precursor cells. Histology confirmed that these cells differentiated into endothelial-like cells. The cells were injected intravenously into a tail vein and were observed to localize in the tumor periphery.

21.3 Summary

The field of molecular imaging with MRI has literally exploded in the past decades, where unforeseen increases in sensitivity made possible the identification of molecular targets and processes *in vivo* in animal models. The design of a contrast agent is dictated by several crucial factors, such as the location of the target (i.e., intravascular, extravascular, intracellular). Diffusion of the contrast agent into a tissue is modulated by its size, such that larger particles and liposomes are mostly confined to the intravascular

space or to targets in close vicinity of blood capillaries. In all cases, the design of a contrast agent must be accompanied by the design of a suitable control agent to enable assessment of the specificity of the compound. Ideally, the control agent must be as similar to the contrast agent as possible to avoid confounds linked to widely different pharmacokinetic properties, diffusion properties, and so on.

Cellular imaging with MRI has also witnessed a phenomenal ascension in the past decade, and single cells may now be detected and monitored over time in animal models. Genetic modifications to cancer cells that make them overexpress ferritin receptors are particularly appealing to the study of cancer metastasis. In fact, issues related to label dilution or to delivery of exogenous contrast agents to specific cells are mostly alleviated since those cells are able to generate their own iron monocrystals from available iron in their microenvironment.

Acknowledgments

Martin Lepage is the Canada Research Chair in Magnetic Resonance Imaging. He is a member of the FRSQ-funded Centre de recherche clinique Étienne-Le Bel.

References

1. Yoo B, Pagel MD. An overview of responsive MRI contrast agents for molecular imaging. *Front Biosci* 2008;13:1733–1752.
2. Hanahan D, Folkman J. Patterns and emerging mechanisms of the angiogenic switch during tumorigenesis. *Cell* 1996;86:353–364.
3. Greish K, Fang J, Inutsuka T, Nagamitsu A, Macda H. Macromolecular therapeutics—Advantages and prospects with special emphasis on solid tumour targeting. *Clin Pharmacokinet* 2003;42(13):1089–1105.
4. Jain RK. Vascular and interstitial barriers to delivery of therapeutic agents in tumors. *Cancer Metastasis Rev* 1990;9(3):253–266.
5. Maeda H, Wu J, Sawa T, Matsumura Y, Hori K. Tumor vascular permeability and the EPR effect in macromolecular therapeutics: A review. *J Control Release* 2000;65(1–2):271–284.
6. Verkman AS. Solute and macromolecule diffusion in cellular aqueous compartments. *Trends Biochem Sci* 2002;27:27–33.
7. Förster T. Zwischenmolekulare energiewanderung und fluoreszenz. *Ann Physik* 1948;47:55–75.
8. Fujimoni K, Covell DG, Fletcher JE, Weinstein JN. Modeling analysis of the global and microscopic distribution of immunoglobulin G, F(ab')2, and Fab in tumors. *Cancer Res* 1989;49:5656–5663.
9. Kallinowski F, Schlenger KH, Runkel S, Kloes M, Stohrer M, Okunieff P, Vaupel P. Blood flow, metabolism, cellular environment, and growth rate of human tumor xenografts. *Cancer Res* 1989;49:3759–3764.
10. Marsh, M. *Endocytosis (Frontiers in Molecular Biology)*. Oxford: Oxford University Press; 2001.

11. Gallagher BM, Fowler JS, Gutterson NI, MacGregor RR, Wan C-N, Wolf AP. Metabolic trapping as a principle of radiopharmaceutical design: Some factors responsible for the biodistribution of [^{18}F] 2-deoxy-2-fluoro-D-glucose. *J Nucl Med* 1978;19:1154–1161.

12. Daldrup H, Shames DM, Wendland M, Okuhata Y, Link TM, Rosenau W, Lu Y, Brasch RC. Correlation of dynamic contrast-enhanced MR imaging with histologic tumor grade: Comparison of macromolecular and small-molecular contrast media. *Am J Roentgenol* 1999;171:941–949.

13. Wintersperger BJ, Runge VM, Tweedle MF, Jackson CB, Reiser MF. Brain tumor enhancement in magnetic resonance imaging dependency on the level of protein binding of applied contrast agents. *Invest Radiol* 2009;44(2):89–94.

14. Weissleder R. Scaling down imaging: Molecular mapping of cancer in mice. *Nat Rev Cancer* 2002;2:1–8.

15. Aime S, Crich SG, Gianolio E, Giovenzana GB, Tei L, Terreno E. High sensitivity lanthanide(III) based probes for MR-medical imaging. *Coord Chem Rev* 2006;250(11–12):1562–1579.

16. Geraldes CFGC, Laurent S. Classification and basic properties of contrast agents for magnetic resonance imaging. *Contrast Media Mol Imaging* 2009;4(1):1–23.

17. Louie AY, Hüber MM, Ahrens ET, Rothbächer U, Moats R, Jacobs RE, Fraser SE, Meade TJ. In vivo visualization of gene expression using magnetic resonance imaging. *Nat Biotechnol* 2000;18:321–325.

18. Moats RA, Fraser SE, Meade TJ. A "smart" magnetic resonance imaging agent that reports on specific enzymatic activity. *Angew Chem Int Edit Engl* 1997;36(7):726–728.

19. Heckl S, Pipkorn R, Waldeck W, Spring H, Jenne J, von der Lieth CW, Corban-Wilhelm H, Debus J, Braun K. Intracellular visualization of prostate cancer using magnetic resonance imaging. *Cancer Res* 2003;63:4766–4772.

20. Su W, Mishra R, Pfeuffer J, Wiesmuller K-H, Ugurbil K, Engelmann J. Synthesis and cellular uptake of a MR contrast agent coupled to an antisense peptide nucleic acid–cell-penetrating peptide conjugate. *Contrast Media Mol Imaging* 2007;2(1):42–49.

21. Burtea C, Laurent S, Murariu O, Rattat D, Toubeau G, Verbruggen A, Vansthertem D, Elst LV, Muller RN. Molecular imaging of alpha(v)beta(3) integrin expression in atherosclerotic plaques with a mimetic of RGD peptide grafted to Gd-DTPA. *Cardiovasc Res* 2008;78(1):148–157.

22. Lepage M, Dow WC, Melchior M, You Y, Fingleton B, Quarles CC, Pépin C, Gore JC, Matrisian LM, McIntyre JO. Non-invasive detection of matrix metalloproteinase activity in vivo using a novel MRI contrast agent with a solubility switch. *Mol Imaging* 2007;6(6):393–403.

23. Lebel R, Jastrzębska B, Therriault H, Cournoyer M-M, McIntyre JO, Neugebauer W, Escher E, Paquette B, Lepage M. Novel solubility-switchable MRI agent allows the non-invasive detection of matrix metalloproteinase-2 activity in vivo in a mouse model. *Magn Reson Med* 2008;60:1056–1065.

24. Jastrzebska B, Lebel R, Therriault H, McIntyre JO, Escher E, Guérin B, Paquette B, Neugebauer WA, Lepage M. New enzyme-activated solubility-switchable contrast agent for magnetic resonance imaging: From synthesis to in vivo imaging. *J Med Chem* 2009;52:1576–1581.

25. Raghunand N, Howison C, Sherry AD, Zhang S, Gillies RJ. Renal and systemic pH imaging by contrast-enhanced MRI. *Magn Reson Med* 2003;49:249–257.

26. Artemov D, Mori N, Ravi R, Buhjwalla ZM. Magnetic resonance molecular imaging of the HER-2/neu receptor. *Cancer Res* 2003;63:2723–2727.

27. Konda SD, Aref M, Wang S, Brechbiel M, Wiener EC. Specific targeting of folate–dendrimer MRI contrast agents to the high affinity folate receptor expressed in ovarian tumor xenografts. *Magn Reson Mat Phys Biol Med* 2001;12(2–3):104–113.

28. Wang ZJ, Boddington S, Wendland M, Meier R, Corot C, Daldrup-Link H. MR imaging of ovarian tumors using folate-receptor-targeted contrast agents. *Pediatr Radiol* 2008;38(5):529–537.

29. Phillips, WT, Goins, B. Targeted delivery of imaging agents by liposomes. In: Torchilin, VP, editor. *Handbook of Targeted Delivery of Imaging Agents*. Boca Raton: CRC Press; 1995. p 149–173.

30. Storrs RW, Tropper FD, Li HY, Song CK, Kuniyoshi JK, Sipkins DA, Li KCP, Bednarski MD. Paramagnetic polymerized liposomes: Synthesis, characterization, and applications for magnetic resonance imaging. *J Am Chem Soc* 1995;117:7301–7306.

31. Sipkins DA, Cheresh DA, Kazemi MR, Nevin LM, Bednarski MD, Li KCP. Detection of tumor angiogenesis in vivo by $\alpha_v\beta_3$-targeted magnetic resonance imaging. *Nat Med* 1998;4(5):623–626.

32. Anderson SA, Rader RK, Westlin WF, Null C, Jackson D, Lanza GM, Wickline SA, Kotyk JJ. Magnetic resonance contrast enhancement of neovasculature with alpha(v)beta(3)-targeted nanoparticles. *Magn Reson Med* 2000;44(3):433–439.

33. Erdogan S, Medarova ZO, Roby A, Moore A, Torchilin VP. Enhanced tumor MR imaging with gadolinium-loaded polychelating polymer-containing tumor-targeted liposomes. *J Magn Reson Imaging* 2008;27(3):574–580.

34. Winter PM, Caruthers SD, Kassner A, Harris TD, Chinen LK, Allen JS, Lacy EK, Zhang H, Robertson JD, Wickline SA, Lanza GM. Molecular imaging of angiogenesis in nascent vx-2 rabbit tumors using a novel $\alpha_v\beta_3$-targeted nanoparticle and 1.5 tesla magnetic resonance imaging. *Cancer Res* 2003;63:5838–5843.

35. Schmieder AH, Winter PM, Caruthers SD, Harris TD, Williams TA, Allen JS, Lacy EK, Zhang H, Scott MJ, Hu G, Robertson JD, Wickline SA, Lanza GM. Molecular MR imaging of melanoma angiogenesis with $\alpha_v\beta_3$-targeted paramagnetic nanoparticles. *Magn Reson Med* 2005;53:621–627.

36. Mulder WJM, Strijkers GJ, Habets JW, Bleeker EJW, van der Schaft DWJ, Storm G, Koning GA, Griffioen AW, Nicolay K.

MR molecular imaging and fluorescence microscopy for identification of activated tumor endothelium using a bimodal lipidic nanoparticle. *FASEB J* 2005;19(14):2008–2010.

37. Freedman M, Chang EH, Zhou Q, Pirollo KF. Nanodelivery of MRI contrast agent enhances sensitivity of detection of lung cancer metastases. *Acad Radiol* 2009;16(5):627–637.

38. Shen T, Weissleder R, Papisov M, Bogdanov Jr. A, Brady TJ. Monocrystalline iron oxide nanocompounds (MION): Physicochemical properties. *Magn Reson Med* 1993;29:599–604.

39. Schaffer BK, Linker C, Papisov M, Tsai E, Nossiff N, Shibata T, Bogdanov A, Brady TJ, Weissleder R. MION-ASF: Biokinetics of an MR receptor agent. *Magn Reson Imaging* 1993;11(3):411–417.

40. Weissleder R, Elizondo G, Wittenberg J, Rabito CA, Bengele HH, Josephson L. Ultrasmall superparamagnetic iron oxide: Characterization of a new class of contrast agents for MR imaging. *Radiology* 1990;175:489–493.

41. Bernas LM, Foster PJ, Rutt BK. Magnetic resonance imaging of in vitro glioma cell invasion. *J Neurosurg* 2007;106(2):306–313.

42. Josephson L, Tung C, Moore A, Weissleder R. High-efficiency intracellular magnetic labeling with novel super-paramagnetic-Tat peptide conjugates. *Bioconjugate Chem* 1999;10:186–191.

43. Weissleder R, Moore A, Mahmood U, Bhorade R, Benveniste H, Chiocca EA, Basilion JP. In vivo magnetic resonance imaging of transgene expression. *Nat Med* 2000;6(3):351–354.

44. Li X, Du X, Huo T, Liu X, Zhang S, Yuan F. Specific targeting of breast tumor by octreotide-conjugated ultra-small superparamagnetic iron oxide particles using a clinical 3.0-tesla magnetic resonance scanner. *Acta Radiol* 2009;50(6):583–594.

45. Khemtong C, Kessinger CW, Ren J, Bey EA, Yang S-G, Guthi JS, Boothman DA, Sherry AD, Gao J. In vivo off-resonance saturation magnetic resonance imaging of $\alpha_v\beta_3$-targeted superparamagnetic nanoparticles. *Cancer Res* 2009;69(4):1651–1658.

46. Yang L, Mao H, Cao Z, Wang YA, Peng X, Wang X, Sajja HK, Wang L, Duan H, Ni C, Staley CA, Wood WC, Gao X, Nie S. Molecular imaging of pancreatic cancer in an animal model using targeted multifunctional nanoparticles. *Gastroenterology* 2009;136(5):1514–1525.

47. Lin W, Hyeon T, Lanza GM, Zhang M, Meade TJ. Magnetic nanoparticles for early detection of cancer by magnetic resonance imaging. *MRS Bull* 2009;34(6):441–448.

48. Bridot J-L, Faure A-C, Laurent S, Riviere C, Billotey C, Hiba B, Janier M, Josserand V, Coll J-L, Vander Elst L, Muller R, Roux S, Perriat P, Tillement O. Hybrid gadolinium oxide nanoparticles: Multimodal contrast agents for in vivo imaging. *J Am Chem Soc* 2007;129(16):5076–5084.

49. Ward KM, Aletras AH, Balaban RS. A new class of contrast agents for MRI based on proton chemical exchange dependent saturation transfer (CEST). *J Magn Reson* 2000;143(1):79–87.

50. Sherry AD, Woods M. Chemical exchange saturation transfer contrast agents for magnetic resonance imaging. *Annu Rev Biomed Eng* 2008;10:391–411.

51. Chauvin T, Durand P, Bernier M, Meudal H, Doan B-T, Noury F, Badet B, Beloeil J-C, Tóth É. Detection of enzymatic activity by PARACEST MRI: A general approach to target a large variety of enzymes. *Angew Chem Int Ed* 2008;47:4370–4372.

52. Young SW, Qing F, Harriman A, Sessler JL, Dow WC, Mody TD, Hemmi GW, Hao YP, Miller RA. Gadolinium(III) texaphyrin: A tumor selective radiation sensitizer that is detectable by MRI. *Proc Natl Acad Sci U S A* 1996; 93(13): 6610–6615.

53. Wong FC, Kim EE. A review of molecular imaging studies reaching the clinical stage. *Eur J Radiol* 2009;70(2):205–211.

54. Turner JL, Pan D, Plummer R, Chen Z, Whittaker AK, Wooley KL. Synthesis of gadolinium-labeled shell-cross-linked nanoparticles for magnetic resonance imaging applications. *Adv Funct Mater* 2005;15:1248–1254.

55. Bouchard L-S, Anwar MS, Liu GL, Hann B, Xie ZH, Gray JW, Wang X, Pines A, Chen FF. Picomolar sensitivity MRI and photoacoustic imaging of cobalt nanoparticles. *Proc Natl Acad Sci U S A* 2009;106(11):4085–4089.

56. Olson ES, Aguilera TA, Jiang T, Ellies LG, Nguyen QT, Wong EH, Gross LA, Tsien RY. In vivo characterization of activatable cell penetrating peptides for targeting protease activity in cancer. *Integr Biol* 2009;1(5–6):382–393.

57. Oishi M, Sumitani S, Nagasaki Y. On-off regulation of ^{19}F magnetic resonance signals based on pH-sensitive PEGylated nanogels for potential tumor-specific smart ^{19}F MRI probes. *Bioconjugate Chem* 2007;18:1379–1382.

58. Hamilton RG, Alderson PO, Harwig JF, Siegel BA. Splenic imaging with 99mTc-labeled erythrocytes: A comparative study of cell-damaging methods. *J Nucl Med* 1976;17:1038–1043.

59. Wagstaff J, Gibson C, Thatcher N, Ford WL, Sharma H, Crowther D. Human lymphocyte traffic assessed by indium-111 oxine labelling: Clinical observations. *Clin Exp Immunol* 1981;43(3):443–449.

60. Becker W, Schomann E, Fischbach W, Borner W, Gruner KR. Comparison of 99Tcm-HMPAO and 111In-oxine labelled granulocytes in man: First clinical results. *Nucl Med Commun* 1988;9:435–447.

61. Oku N, Koike C, Sugawara M, Tsukada H, Irimura T, Okada S. Positron emission tomography analysis of metastatic tumor cell trafficking. *Cancer Res* 1994;54:2573–2576.

62. Koike C, Oku N, Watanabe M, Tsukada H, Kakiuchi T, Irimura T, Okada S. Real-time PET analysis of metastatic tumor cell trafficking in vivo and its relation to adhesion properties. *Biochim Biophys Acta* 1995;1238:99–106.

63. Akins EJ, Dubey P. Noninvasive imaging of cell-mediated therapy for treatment of cancer. *J Nucl Med* 2008;49: 180S–195S.

64. Modo M, Hoehn M, Bulte JW. Cellular MR imaging. *Mol Imaging* 2005;4(3):143–164.

65. Gilad AA, Ziv K, McMahon MT, van Zijl PCM, Neeman M, Bulte JWM. MRI reporter genes. *J Nucl Med* 2008; 49(12):1905–1908.

66. Muja N, Bulte JWM. Magnetic resonance imaging of cells in experimental disease models. *Prog Nucl Magn Reson Spectrosc* 2009;55(1):61–77.

67. Weissleder R, Stark DD, Engelstad BL, Bacon BR, Compton CC, White DL, Jacobs P, Lewis J. Superparamagnetic iron oxide: Pharmacokinetics and toxicity. *Am J Roentgenol* 1989;152(1):167–173.

68. Haacke, EM, Brown, RB, Thompson, MR, Venkatesan, R. *Magnetic Resonance Imaging: Physical Principles and Sequence Design*. New York: John Wiley & Sons; 1999.

69. Heyn C, Bowen CV, Rutt BK, Foster PJ. Detection threshold of single SPIO-labeled cells with FIESTA. *Magn Reson Med* 2005;53(2):312–320.

70. Shapiro EM, Skrtic S, Sharer K, Hill JM, Dunbar CE, Koretsky AP. MRI detection of single particles for cellular imaging. *Proc Natl Acad Sci U S A* 2004;101(30):10901–10906.

71. Athiraman H, Jiang Q, Guang LD, Zhang L, Zheng GZ, Wang L, Arbab AS, Li Q, Panda S, Ledbetter K, Rad AM, Chopp M. Investigation of relationships between transverse relaxation rate, diffusion coefficient, and labeled cell concentration in ischemic rat brain using MRI. *Magn Reson Med* 2009;61(3):587–594.

72. Bowen CV, Zhang X, Saab G, Gareau PJ, Rutt BK. Application of the static dephasing regime theory to superparamagnetic iron-oxide loaded cells. *Magn Reson Med* 2002;48:52–61.

73. Himmelreich U, Weber R, Ramos-Cabrer P, Wegener S, Kandal K, Shapiro EM, Koretsky AP, Hoehn M. Improved stem cell MR detectability in animal models by modification of the inhalation gas. *Mol Imaging* 2005;4(2):104–109.

74. Cunningham CH, Arai T, Yang PC, McConnell MV, Pauly JM, Conolly SM. Positive contrast magnetic resonance imaging of cells labeled with magnetic nanoparticles. *Magn Reson Med* 2005;53(5):999–1005.

75. Liu W, Dahnke H, Jordan EK, Schaeffter T, Frank JA. In vivo MRI using positive-contrast techniques in detection of cells labeled with superparamagnetic iron oxide nanoparticles. *NMR Biomed* 2008;21(3):242–250.

76. Çukur T, Yamada M, Overall WR, Yang P, Nishimura DG. Positive contrast with alternating repetition time SSFP (PARTS): A fast imaging technique for SPIO-labeled cells. *Magn Reson Med* 2010;63:427–437.

77. Bulte JWM. Hot spot MRI emerges from the background. *Nat Biotechnol* 2005;23(8):945–946.

78. Daldrup-Link HE, Rudelius M, Oostendorp RAJ, Settles M, Piontek G, Metz S, Rosenbrock H, Keller U, Heinzmann U, Rummeny EJ, Schlegel J, Link TM. Targeting of hematopoietic progenitor cells with MR contrast agents. *Radiology* 2003;228(3):760–767.

79. Bernas LM, Foster PJ, Rutt BK. Imaging iron-loaded mouse glioma tumors with bSSFP at 3 T. *Magn Reson Med* 2010;64:23–31.

80. Shapiro EM, Sharer K, Skrtic S, Koretsky AP. In vivo detection of single cells by MRI. *Magn Reson Med* 2006;55(2):242–249.

81. Alfke H, Stoppler H, Nocken F, Heverhagen JT, Kleb B, Czubayko F, Klose KJ. In vitro MR imaging of regulated gene expression. *Radiology* 2003;228(2):488–492.

82. Genove G, DeMarco U, Xu HY, Goins WF, Ahrens ET. A new transgene reporter for in vivo magnetic resonance imaging. *Nat Med* 2005;11(4):450–454.

83. Cohen B, Dafni H, Meir G, Harmelin A, Neeman M. Ferritin as an endogenous MRI reporter for noninvasive imaging of gene expression in C6 glioma tumors. *Neoplasia* 2005;7(2):109–117.

84. Zurkiya O, Chan AW, Hu X. MagA is sufficient for producing magnetic nanoparticles in mammalian cells, making it an MRI reporter. *Magn Reson Med* 2008;59(6):1225–1231.

85. Contag CH, Bachmann MH. Advances in vivo bioluminescence imaging of gene expression. *Annu Rev Biomed Eng* 2002;4:235–260.

86. Gambhir SS, Barrio JR, Phelps ME, Iyer M, Namavari M, Satyamurthy N, Wu L, Green LA, Bauer E, MacLaren DC, Nguyen K, Berk AJ, Cherry SR, Herschman HR. Imaging adenoviral-directed reporter gene expression in living animals with positron emission tomography. *Proc Natl Acad Sci U S A* 1999;96(5):2333–2338.

87. Holland GN, Bottomley PA, Hinshaw WS. [19]F magnetic resonance imaging. *J Magn Reson* 1977;28:133–136.

88. Ahrens ET, Flores R, Xu H, Morel PA. In vivo imaging platform for tracking immunotherapeutic cells. *Nat Biotechnol* 2005;23:983–987.

89. Oliver M, Ahmad A, Kamaly N, Perouzel E, Caussin A, Keller M, Herlihy A, Bell J, Miller AD, Jorgensen MR. MAGfect: A novel liposome formulation for MRI labelling and visualization of cells. *Org Biomol Chem* 2006;4(18):3489–3497.

90. JuanYin J, Tracy K, Zhang L, Munasinghe J, Shapiro E, Koretsky A, Kelly K. Noninvasive imaging of the functional effects of anti-VEGF therapy on tumor cell extravasation and regional blood volume in an experimental brain metastasis model. *Clin Exp Metastasis* 2009;26(5):403–414.

91. Deans AE, Wadghiri YZ, Bernas LM, Yu X, Rutt BK, Turnbull DH. Cellular MRI contrast via coexpression of transferrin receptor and ferritin. *Magn Reson Med* 2006;56(1):51–59.

92. Gilad AA, McMahon MT, Walczak P, Winnard PT, Jr., Raman V, van Laarhoven HWM, Skoglund CM, Bulte JWM, van Zijl PCM. Artificial reporter gene providing MRI contrast based on proton exchange. *Nat Biotechnol* 2007;25(2):217–219.

93. Anderson SA, Glod J, Arbab AS, Noel M, Ashari P, Fine HA, Frank JA. Noninvasive MR imaging of magnetically labeled stem cells to directly identify neovasculature in a glioma model. *Blood* 2005;105(1):420–425.

94. Jain RK. Transport of molecules in the tumor interstitium: A review. *Cancer Res* 1987;47:3039–3051.

Hyperpolarized MR of Cancer

Kevin W. Waddell
Vanderbilt University

Eduard Y. Chekmenev
Vanderbilt University

22.1 Qualitative Introduction to Hyperpolarized Imaging Techniques

22.1.1 Introduction to Thermal Polarization of Conventional MR and Hyperpolarization Techniques of MR Signal Enhancement

Few experimental techniques can claim an evolutionary period as long as magnetic resonance (MR). Discovered over 50 years ago [1,2], MR is still finding new applications. This period has been characterized by a steady trend of increasing MR sensitivity. The signal in MR depends on the product of the number of spins and the fractional nuclear spin polarization, which is only 10^{-6}–10^{-5} for typical high-field (~1.5 T) equilibria. This severely limits the use of MR in biomedicine, especially for nuclei with low gyromagnetic ratios (γ) such as carbon-13 (^{13}C) and nitrogen-15 (^{15}N). Nuclear spin polarization scales linearly with magnetic field, and this relationship between magnetic field strength and sensitivity has in turn perpetuated the development and widespread availability of human and preclinical MR systems up to 7 and 12 T, respectively. From standard clinical 1.5-T human instruments to the currently available 7-T systems, nuclear spin polarization has increased by 4.7-fold. While other factors (radiofrequency (RF) coil sensitivity, T_1/T_2 relaxation, and pulse sequences) contribute to the overall MR sensitivity, the lone impact of the nuclear spin polarization theoretically decreases acquisition time by 4.7^2 or 21.8-fold to achieve the same signal-to-noise ratio (SNR) at 7 T compared with that at 1.5 T. These advances have enabled the penetration of new layers of diagnosis and have driven the emergence of new applications.

Notwithstanding the inherent difficulties in achieving yet higher fields, there are limits to the sensitivity gains that are achievable by increasing field strength. For example, a standard estimate from the thermal Boltzmann population shows that detectable spins can be increased from about 1 to 28 parts per million (scaled linearly) when going from 0.25 T to the highest field currently approved for human use, 7 T. There also are numerous disadvantages (B_1 and B_0 homogeneity, shimming, gradient power requirements) of performing high-field MR measurements *in vivo*, and dramatic technological advances will be required to further increase sensitivity by even a modest factor of 2 using this approach. While current fields are adequate for obtaining information from protons in the millimolar concentration range (e.g., lactate, choline), information on metabolism of ^{13}C- or ^{15}N-labeled substrates, which have advantages of greater chemical shift dispersion and negligible background signal, requires extraordinary scan durations (on the order of hours with direct detection) and large amounts of expensive isotopically labeled precursors.

However, emerging hyperpolarization techniques have the potential to increase sensitivity to the highest levels predicted from theory, well beyond achievable sensitivity available from traditional high-field MR (Figure 22.1). The conceptual framework behind these techniques, parahydrogen and synthesis allow dramatically enhanced nuclear alignment (PASADENA) [3–5], also known as parahydrogen-induced polarization (PHIP) [6], and dynamic nuclear polarization (DNP) [7], has been known for many years, and these are collectively referred to as hyperpolarized MR. Recently, signal amplification by reversible exchange (SABRE) [8] using parahydrogen gas was demonstrated as well. With this dramatic increase in sensitivity, these methods

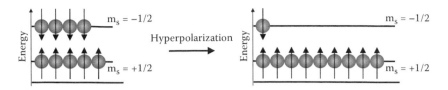

FIGURE 22.1 Hyperpolarization of nuclear spin alignment from several parts per million or <0.01% to theoretical maximum approaching polarization of unity or 100%. The m_s refer to the Zeeman energy levels of a spin 1/2 nucleus.

(Figure 22.2) should be capable of directly imaging metabolic conversion *in vivo* at micromolar concentrations.

This chapter provides background about hyperpolarized tracer and hyperpolarized MR, describes the first emerging preclinical studies, and outlines the broad perspective of potentially useful tracers for cancer imaging on a cellular level in humans.

22.1.2 Sensitivity Enhancement by Hyperpolarization

MR sensitivity enhancements up to 10,000 have been reported using DNP [9], which is now a commercially available technology from Oxford Instruments (UK) and GE Healthcare. To emphasize the impact of this sensitivity improvement, it would take several years of continuous acquisition by conventional MR to equal the sensitivity attained from just one hyperpolarized experiment. Given this potential, it is perhaps not surprising that useful results have already been reported with this technique. For example, the *in vivo* conversion of pyruvate has been mapped, and it was shown that the MR spectral signatures of tumors were readily distinguished from surrounding muscle [10]. This study provides clear evidence that hyperpolarized pyruvate and its metabolites are preserved long enough *in vivo* to be imaged. It also demonstrates that differential conversion of pyruvate between tumor and normal tissue can be imaged from a standard carotid injection. In

this study, ^{13}C molecules were imaged at approximately 30 s from preparation of the hyperpolarized agent. In summary, the agents only last a few minutes in favorable cases, but useful information from tumor metabolism can be obtained in as little as 10 s.

Theoretical sensitivity and the temporal window available for metabolic imaging are actually increased in PASADENA when compared with DNP. PASADENA requires more specialized precursors than DNP, though, and this has likely contributed to the absence of a commercially available apparatus. In PASADENA, para-H_2(g) is the conduit for increased polarization and is added across double bonds to form hyperpolarized molecules [11]. Thus, unsaturated compounds, which can be hydrogenated to yield the metabolic agent of interest, are a prerequisite for PASADENA. Small molecules such as pyruvate and lactate are therefore not accessible by current PASADENA technology. However, there are many metabolites with suitable PASADENA precursors, which can be readily synthesized from established techniques. Glutamine (Gln), glutamate (Glu), and succinate (Suc) are just a few of the metabolically relevant molecules with PASADENA precursors, with Suc already commercially available. To summarize, molecules larger than glycine are amenable to PASADENA preparation, but options are limited for small molecules like pyruvate and lactate.

Access to polarization reservoirs at room temperature is a significant advantage of PASADENA versus DNP, where polarized molecules must be warmed rapidly from approximately 5 K to room temperature. In addition to being technically difficult, this causes a significant reduction in polarization due to T_1 relaxation during warming. Furthermore, while it is inexpensive to create pure parahydrogen (cooling is sufficient), the analogous process in DNP requires more exotic materials (free radicals at <3 K). In addition, the elaborate process required to produce hyperpolarized substrates by DNP cannot quickly be repeated. Hence, optimizing experimental parameters is expensive and makes less efficient use of scanner sessions than PASADENA. The sole (albeit significant) advantage for DNP is the nominal capability to hyperpolarize arbitrary compounds and the commercial availability of the apparatus. For these reasons, PASADENA could prove to be more useful than DNP within its (more restricted) range of applications.

In summary, PASADENA and DNP both offer the potential to dramatically increase sensitivity and to enable fast imaging of *in vivo* metabolism. Pyruvate metabolism already has been imaged *in vivo* with DNP; this key result is convincing evidence that other hyperpolarized substrates with similar magnetic

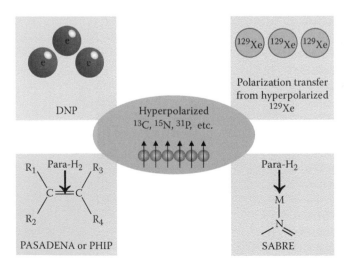

FIGURE 22.2 MR hyperpolarization methods.

relaxation properties can also be imaged. In addition, recent work [12,13] has demonstrated the feasibility of preparing and imaging hyperpolarized Suc and Gln *in vivo* and *in vitro*, respectively.

22.1.3 Fundamentals of DNP

Hyperpolarization of nuclear spins by DNP is based on polarizing electron spins and efficiently transferring this enhanced polarization to nuclear spins [14–16]. For MRI applications, this usually involves the ex situ polarization of electrons associated with an organic free radical, followed by cooling to <3 K in the presence of metabolic agents to achieve a matrix of uniformly distributed radicals in proximity to the target molecules. Microwave radiation is then used to saturate the electron spin states of the radical in the solid state, facilitating the transfer of polarization from electrons to target nuclei. Frozen hyperpolarized target compounds are then released into the solution after rapid (~1 s) dissolution. Rapid dissolution is critical to preserving nuclear spin polarization and making the hyperpolarized tracers injectable for *in vivo* use. This fundamental breakthrough in technology allowed for efficient operation of first-generation DNP polarizers [9].

A typical procedure for preparing hyperpolarized pyruvate by these methods consists of mixing a solution containing the target metabolic agent with a trityl radical/glycerol matrix that allows for efficient polarization transfer from radical free electron to target nucleus such as ^{13}C or ^{15}N. Cooling to 1.2 K produces glass, which is then irradiated for 100 min with microwaves in a magnetic field (94.118 GHz at 3.35 T). The sample is then rapidly thawed by mixing with hot water, and the resulting solution is then available for injection into an organism.

The advantage of DNP is that arbitrary molecules can be hyperpolarized, whereas with PASADENA, a suitable unsaturated precursor is necessary. For this reason, the set of DNP applications that has been reported has more breadth and volume than competing technologies for hyperpolarizing metabolic substrates. DNP has been used to hyperpolarize key molecules such as pyruvate [10,17–19], lactate [20], and bicarbonate [21] to differentiate tumor metabolism cellular pH, respectively, whereas the only metabolic substrate studied with PASADENA to date is succinate [5,13,22,23]. Counterbalancing this strength of DNP are inherent incompatibilities with experimental repetition, substrate phase and temperature, and toxicity.

The procedure for obtaining hyperpolarization by DNP frames these issues. Even though DNP has proven itself to be a capable and flexible methodology for hyperpolarizing metabolic substrates, several challenges need to be addressed for the technology to reach its full potential. DNP utilizes toxic reagents containing free radicals and requires very specialized and expensive equipment employing high magnetic fields and cryogenic systems, to name a few. Therefore, it is most likely best suited for single experiments due to the long preparation time, with the final hyperpolarized product being in the solid state and at cryogenic temperatures. These challenges appear tractable, when

prior art (radical filtration) and economies of scale (equipment) are considered, in addition to recent advances in delivery systems (melting and injection).

22.1.4 Fundamentals of PASADENA and PHIP

The second approach to prepare hyperpolarized tracers for cancer imaging is to utilize spin order of the singlet state of the parahydrogen molecule. Parahydrogen can be routinely produced by passing high-purity hydrogen (25% para and 75% ortho at room temperature) gas through a catalyst-filled column at cryogenic temperatures in the range of 5–20 K, allowing fast conversion of the ortho-/para- mixture of hydrogen gas to parahydrogen. Production of parahydrogen gas can therefore be scaled up considerably to accommodate foreseeable clinical demands [24]. The percentage of parahydrogen enrichment can be readily characterized by ^{1}H spectroscopy by monitoring the residual signal of incompletely converted orthohydrogen, as the singlet state that is characteristic of the parahydrogen molecule is nuclear MR (NMR) invisible [25]. In addition to easy preparation, parahydrogen undergoes back-conversion to orthohydrogen very slowly, on the time scale of weeks, and can also be stored for periods of weeks to months at room temperature.

The discovery of parahydrogen as a source of nuclear spin polarization in NMR experiments occurred in 1986 in the laboratory of Prof. Daniel Weitekamp [3,4], and shortly thereafter by Prof. Joachim Bargon in 1987 [6]. PASADENA or PHIP is unique in its ability to achieve hyperpolarization in the liquid state in seconds, employing *cis* addition of parahydrogen across alkene or alkyne bonds followed by the spin order transfer from nascent protons to biomolecular nuclei such as ^{13}C or ^{15}N, with a theoretical polarization limit of 100% (Figure 22.3). PASADENA is also inexpensive, portable, and easy to maintain, as the chemical reaction and spin order can be transferred in a low-field magnet of only a few milliteslas. Historically, this low-field experimental setup for PASADENA was first demonstrated by the Malmö group, where hyperpolarized ^{13}C molecules were used as angiographic agents [26,27]. As of 2010, this hyperpolarization technology is pursued in basic and translational research by several groups [5,13,28–30].

The molecular hydrogenation of the unsaturated bond and spin order transfer from the added parahydrogen spins is carried out on the time scale of several seconds with fast rhodium-based molecular catalysts tailored for PASADENA. As a result, polarization losses associated with the spin lattice relaxation time (T_1) of X nuclei and the decay of the singlet state of the attached spins are minimized. Moreover, the water-soluble catalysts allow the entire hyperpolarization procedure to be performed in an aqueous medium, a requirement for *in vivo* applications. The toxicity of the catalyst remains a primary concern for extending PASADENA hyperpolarization to *in vivo* clinical applications, and efforts are under way to filter the homogeneous catalysts or uncover appropriate heterogeneous catalysts, which eliminate the need for subsequent filtration [31].

RF pulses necessary to transfer polarization from proton singlet states to X (^{13}C, ^{15}N, etc.) nuclei that possess advantageous

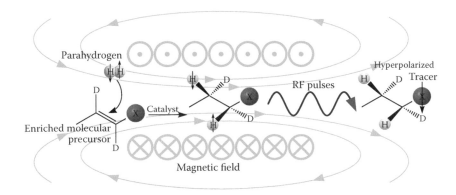

FIGURE 22.3 Molecular addition of parahydrogen to unsaturated molecular precursor followed by RF pulses necessary for spin order transfer from singlet state of parahydrogen spins to X nucleus, where X is ^{13}C, ^{15}N, or other low-gamma nucleus and D is deuterium. This molecular addition requires a catalyst and usually is conducted on the time scale of a few seconds by injection of an aqueous solution containing molecular precursors and the catalyst in the atmosphere of parahydrogen gas (1–12 atm).

chemical shift dispersion and relaxation properties are applied after a suitable delay (typically a few seconds) following the hydrogenation reaction. The efficiency of the polarization transfer [32] depends on scalar couplings of the involved nuclei and requires prior knowledge of the spin–spin couplings between these three nuclei; the relevant couplings can be determined at physiological pH and used without further refinement. To the extent that measurements are available at *in vivo* pH, these couplings can be obtained using standard high-resolution NMR techniques [5]. Moreover, the molecular tracer is typically deuterated in order to simplify the spin system and to increase relaxation times. Polarization of 20%–40% has been achieved with PASADENA [5,22,23,26,30,33], and these ^{13}C agents had T_1's up to 105 s *in vitro* [23]. Preliminary reports indicate that PASADENA ^{15}N tracers can be produced with T_1's in excess of 7 min [34].

22.1.5 SABRE

The source of hyperpolarization in SABRE is also parahydrogen. While PASADENA requires a chemical reaction between unsaturated precursors and parahydrogen, SABRE utilizes reversible binding between metal dihydrides and tracer molecules, as displayed in Figure 22.2. The polarization can be transferred from nascent protons in metal dihydride complexes to the tracer molecule nuclear spins at modest magnetic fields (~20 mT), followed by dissociation of the complex and release of the tracer and hydrogen molecules [8]. As a result, no chemical modification of the tracer is necessary. Moreover, SABRE is significantly less demanding in hardware compared to PASADENA, as the polarization transfer step does not require applied magnetic fields. SABRE has been successfully demonstrated with nicotinamide (vitamin B₃) [8] *in vitro* and, in principle, could be applied to other tracers with aromatic nitrogens.

However, several major challenges have to be addressed before this technique can be successfully applied *in vivo*. As with PASADENA, metal-based hydrogenation catalysts need to be removed to produce nontoxic hyperpolarized tracer agents.

Water-soluble metal hydride complexes are necessary, or toxic organic solvents will need to be evacuated prior to *in vivo* injection (SABRE has been demonstrated only in organic solvents so far). Finally, only modest levels of hyperpolarization on the order of 1% have been achieved, which is at least an order of magnitude less than that routinely produced by PASADENA and DNP.

22.1.6 Xenon ([129]Xe)-Induced Polarization (XIP)

Hyperpolarized [129]Xe gas with polarization in excess of 60% has been routinely obtained by several groups [35,36], and the technology has been commercialized by Xemed LLC (Durham, New Hampshire). Using this commercially available product, >0.5 mol/h (>60 g/h) of hyperpolarized [129]Xe can be produced. In addition, MagniXene™ (Xemed's trade name for hyperpolarized [129]Xe) is an FDA-approved investigational new drug.

While hyperpolarized [129]Xe has found application in lung and brain imaging, it has not yet been efficiently utilized as a potential polarization reservoir for metabolic cancer applications due to the difficulty of efficient polarization transfer to ^{13}C and [15]N [37]. This inherent difficulty arises because the transfer mechanism requires prolonged close contact between the isotopes. This has motivated the scientific community to immobilize hyperpolarized xenon in cages in proximity to modified moieties for functionalization [38], as commonly performed in molecular imaging such as gadolinium-based chelates [39]. In contrast, the alternative approach is to transfer the spin polarization of [129]Xe to nontoxic metabolites prior to injection. Mechanisms for polarization transfer that exploit relatively weak spin–spin (usually referred to as J-) couplings are generally excluded, because covalent linkages with [129]Xe are precluded due to the inert nature of xenon. Dipolar mechanisms utilizing the interactions between nuclear spins in the solid state are the most promising [37,40]; however, even the prolonged interaction of the limited set of solutes, which are soluble in liquid Xe, has been unable to bring the polarization of the target spin to a level close to that of the 20%–60% hyperpolarized [129]Xe. Presumably, this occurs because the intermolecular dipolar cross-relaxation is slow compared with other relaxation mechanisms, which has the effect of driving the polarization to small equilibrium values. A more promising technique mixes hyperpolarized ^{129}Xe intimately with target molecules to maximize static dipolar interactions [41,42] in order to facilitate rapid transfer of polarization.

Hyperpolarized ^{129}Xe has been used to enhance ^{13}C polarization by low-field thermal mixing [16,37,43,44] as well as spin polarization–induced nuclear Overhauser effect [45,46]. The low-field thermal mixing is typically performed within a frozen mixture of hyperpolarized xenon and ^{13}C/^{15}N tracers and requires a uniform distribution of ^{13}C/^{15}N spins in a solid matrix of ^{129}Xe. If the applied magnetic field is comparable to those of local dipolar interactions in solids, nuclear spin polarization will have a tendency to equilibrate during the period of low-field thermal contact. Because a homogeneous phase between ^{129}Xe and ^{13}C/^{15}N tracers is required, the majority of published attempts were limited to tracers with good solubility in liquid xenon. Moreover, many studies have been limited to molecules without protons because ^{1}H nuclei pose additional difficulty due to their significantly greater dipolar couplings, resulting in much higher efficiency of the polarization transfer to ^{1}H sites rather than ^{13}C or ^{15}N sites of interest. However, most cancer imaging tracers have poor xenon solubility and contain protons. A recent study offers an alternative approach where ^{129}Xe- and proton-carrying tracers are mixed in the gas phase, which is followed by condensation and homogeneous matrix formation in liquid N_2 [40]. This approach potentially overcomes the problem of limited xenon solubility and offers a possibility to hyperpolarize proton-containing tracers for cancer imaging.

XIP offers an additional advantage for *in vivo* applications because, unlike parahydrogen- and DNP-based methods, it does not use any toxic chemicals and uses only pure xenon, which is already FDA approved for clinical research, as well as pharmacy-grade tracers already approved for injection in humans. However, the main challenge for XIP methods is improving polarization transfer efficiency, which has been approximately three orders of magnitude less [37,40] than that demonstrated by DNP and PASADENA.

22.2 Hyperpolarized MR Detection

22.2.1 Decay of MR Hyperpolarization

Once the hyperpolarized state is produced, it begins to decay exponentially to its equilibrium state with a time constant corresponding to the spin lattice relaxation time T_1. The half-life $t_{1/2}$ typically used in nuclear medicine relates to T_1 as follows:

$$t_{1/2} = \ln(2) \cdot T_1 \approx 0.693 \cdot T_1 \qquad (22.1)$$

Figure 22.4 provides a direct comparison of $t_{1/2}$ values for some frequently used positron emission tomography (PET) and hyperpolarized MR nuclei [5,9,12,18,19,21,23,26,29,30,33,34, 47,48]. While PET agents have exact $t_{1/2}$ values, the half-lives of hyperpolarized magnetization are influenced by the chemical environment experienced by the hyperpolarized nucleus. A range of half-lives is shown for hyperpolarized ^{15}N, ^{13}C, and ^{1}H nuclei in Figure 22.4. Because of very short $t_{1/2}$, proton sites in molecular tracers are very difficult to use; hyperpolarized states decay to equilibrium rapidly enough so that polarization may be quenched prior to the injection of the agent. Thus, the hyperpolarized MR community is currently focused on development and *in vivo* trials of molecular tracers with hyperpolarized ^{13}C and ^{15}N, which can have substantially longer relaxation times (up to several minutes). This arises in part because dipolar interactions are diminished in nuclei with low gyromagnetic ratios compared with protons ($\gamma_{1H} = 3.98 \cdot \gamma_{13C}$ and $\gamma_{1H} = 9.87 \cdot \gamma_{15N}$). The chemical environment also modulates lifetimes of the hyperpolarized states. For example, the T_1 of ^{15}N-choline analogues is approximately 7 min [34], with >80% molecular deuteration, and that for ^{15}N-choline is only 4.7 min [47], without deuteration. Lifetimes can also be decreased by interaction with blood

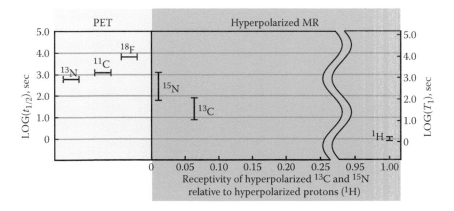

FIGURE 22.4 Decay constants of selected PET and hyperpolarized MR nuclei and relative receptivity of hyperpolarized MR nuclei. The decay constants are provided in units of $t_{1/2}$ and T_1. Note that PET tracers have well-defined values, while hyperpolarized MR nuclei have a wide range of values. The relative receptivity is provided as a percentage receptivity of a single hyperpolarized nucleus with respect to that of a single hyperpolarized proton.

oxygen. For example, the corresponding T_1 of hyperpolarized ^{15}N-choline is only 2 min in blood [47].

^{13}C and ^{15}N hyperpolarized metabolic tracers have much longer lifetimes compared with ^{1}H hyperpolarized tracers. While it would be difficult for MR hyperpolarized agents to approach the lifetime of ^{18}F PET tracers (Figure 22.4), certain ^{15}N MR hyperpolarized tracers have the potential to approach the lifetimes of ^{13}N and ^{11}C PET tracers. For example, ^{15}N-choline [34,47] has a lifetime only several times less than that of ^{11}C-choline (Figure 22.4).

22.2.2 MR Detection of Hyperpolarization

22.2.2.1 Trigonometric Dependence of Magnetization

Hyperpolarized MR detection strategies can be tailored to match the required sensitivity and temporal resolution simply by considering the trigonometric dependence of magnetization in the presence of applied RF fields. MR hyperpolarization at equilibrium aligned along the static field (I_z), needs to be rotated into the orthogonal plane by RF to create detectable magnetization (I_{XY}), in direct analogy to conventional MRI and MR spectroscopy (MRS). As a result of the excitation pulse, the hyperpolarization is given by

$$I_Z = I_{Z0} \cdot \cos(\alpha)$$
$$I_{XY} = I_{Z0} \cdot \sin(\alpha)$$
(22.2)

where I_{Z0} is polarization along the z-axis before rotation, I_Z is the residual polarization along the static field after the RF rotation, I_{XY} is polarization in plane orthogonal to the static field, and α is an excitation pulse angle. Note that I_{XY} decays according to transverse (or spin–spin) relaxation time T_2, with $T_1 \gg T_2$ for hyperpolarized tracers. Moreover, as hyperpolarization decays due to spin lattice relaxation, the consecutive RF excitation pulse i with excitation angle α yields MR signal $I_{XY}(i) < I_{XY}(i-1)$ with respect to the previous ($I-1$) pulse, because a single RF excitation pulse produces

$$I_Z(i) = I_Z(i-1) \cdot \cos(\alpha)$$
$$I_{XY}(i) = I_Z(i-1) \cdot \sin(\alpha)$$
(22.3)

22.2.2.2 Small-Angle Excitation

In cases where excitation angles are approximately 90°, the entire hyperpolarization is rotated to the x–y plane, with the effect that I_Z becomes approximately zero and the maximum MR signal is recorded. While this is beneficial for nonlocalized or single-voxel MRS, most MRI sequences require multiple excitations, and once I_Z becomes zero, this is no longer possible. As a result, hyperpolarized tracers are typically excited with small-angle pulses. For example, a 10° excitation pulse produces signals equivalent to ~17% of those obtained with 90° pulses and retains 98.5% of I_Z. While full sensitivity enhancement benefits cannot be realized in a single acquisition, this approach offers numerous

benefits. Most importantly, images and chemical shift imaging (CSI) [20] can be recorded. Because hyperpolarization is not completely depleted after the MRI or MRS scan is completed, sequential scans can be acquired providing the time course of metabolic flux. In addition, significant acquisition acceleration is achieved, because recovery times (*TR*) are no longer necessary between the consecutive acquisitions and scans. As a result, the individual hyperpolarized CSI or MRI scans can be acquired in several seconds compared to several minutes for conventional CSI and MRI. Because of the trigonometric dependence of the excitation RF pulses, multiple MR images can be recorded without significant loss of hyperpolarization, sampling the temporal regime where spin lattice relaxation T_1 dominates the lifetimes of the hyperpolarized tracer.

22.2.2.3 Hyperpolarized ^{1}H Secondary Polarization Using Nuclear Polarization Storage of Low-γ Sites

^{13}C and ^{15}N metabolic tracers such a 1-^{13}C-pyruvate [18] and ^{15}N-choline [47] benefit from long lifetimes but suffer from low receptivity (directly proportional to MR sensitivity) when compared to protons. Receptivity scales with γ³ in conventional MR, with one γ factor originating from polarization. Because polarization contributions are identical for nuclei that are hyperpolarized to the same fractional polarization, receptivity scales only as γ² for hyperpolarized MR in cases where nuclei are equally polarized. Therefore, hyperpolarized ^{13}C and ^{15}N nuclei are 15.8 and 97.4 times less sensitive than hyperpolarized protons, respectively. This is a substantial penalty for hyperpolarized sensitivity gains in the range of 10,000–100,000. As a result of this sensitivity penalty, hyperpolarized ^{15}N tracers have not been pursued actively *in vivo*.

These sensitivity penalties can be circumvented if polarization from ^{13}C or ^{15}N sites is used only for storage of the hyperpolarization, while MR signals are detected from coupled protons (Figure 22.5). This is achieved by standard methods for polarization transfer (e.g., insensitive nuclei enhanced by polarization transfer (INEPT) [49]) to transfer polarization from low γ nuclei

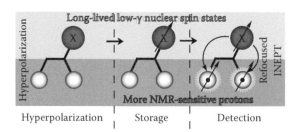

FIGURE 22.5 Hyperpolarization storage using low-γ nucleus and MR signal detection from J-coupled protons [23]. While the long-lived X (^{13}C, ^{15}N, or other low-gamma) nucleus is polarized first and hyperpolarization is stored on this long-lived site during delivery and *in vivo* conversion, the MR signal is detected using the more sensitive proton by employing polarization transfer immediately before the detection event.

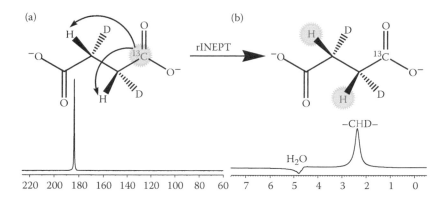

FIGURE 22.6 Hyperpolarization storage on ^{13}C site of succinate followed by polarization transfer to methylene protons and more sensitive proton detection when compared with direct ^{13}C detection. (a) ^{13}C NMR spectrum of hyperpolarized 1-^{13}C-succinate-d$_{2,3}$, ^{13}C polarization of 5.5% after being stored for 70 s, $T_1 = 105$ s. (b) ^{1}H NMR spectrum of the same hyperpolarized 1-^{13}C-succinate-d$_{2,3}$ where net ^{1}H signal enhancement is 1350-fold, with approximately 40% polarization transfer efficiency [23]. rINEPT refers to the refocused INEPT polarization transfer sequence [49]. This demonstrates overall SNR benefit of >6, corresponding to the product of 41% and $(\gamma_{1H}/\gamma_{13C})^2 = 15.8$.

to *J*-coupled protons. This approach was successfully demonstrated for two ^{13}C tracers with up to 50% polarization transfer efficiency and can be generally applied to other ^{13}C and ^{15}N metabolic tracers. The example for hyperpolarized succinate is shown in Figure 22.6 [23].

$$Signal(1H) = Signal(X) \cdot \left(\frac{\gamma_{1H}}{\gamma_X} \right)^2 \cdot n \qquad (22.4)$$

where *n* is the number of protons available for signal detection. Using this strategy with methyl groups of hyperpolarized 1-^{13}C-pyruvate and 1-^{13}C-lactate can give a theoretical signal enhancement of 15.8-fold (T_1 effects not considered). This massive sensitivity gain could be used to (i) enhance the SNR, contrast, or spatial resolution of images, (ii) decrease the quantity of metabolic tracers, or (iii) acquire 282 times more images (SNR scales with the square of number of scans). The above sensitivity gains would be even more dramatic for ^{15}N tracers such as ^{15}N-choline, making it possible to take full advantage of the exceptionally long lifetimes of ^{15}N hyperpolarized metabolic tracers. While this approach requires double resonance (i.e., ^{1}H/^{13}C or ^{1}H/^{15}N) capabilities, most MRI manufacturers already offer double-channel and multinuclear capabilities.

22.2.2.4 Hyperpolarized MR Spectroscopic Imaging (MRSI)

MR hyperpolarized tracers interrogate cancer metabolic profiles by resolving resonances, which arise from reactant and product(s) across the desired pathway. These chemical shift imaging (CSI) [10,12,18,19,21] time-series experiments produce images that are capable of depicting spatial distributions of reactant/product pairs with temporal resolution on the order of seconds. The resulting images therefore reflect the spatial distribution of metabolic flux through the chosen pathway and

have been shown to clearly distinguish tumor, muscle, and blood pools [10]. Hyperpolarized tracers can be rapidly imaged with these methods because recovery periods are minimized with small-angle pulses [17,20].

22.3 Preclinical Applications

22.3.1 Metabolic Contrast Agents

Cancer cells are known to exhibit altered metabolic characteristics, including elevated glucose uptake, increased production of lactate, decreased pH, increased glutaminolysis, and impaired oxidative cycling. The relevant pathways that have been interrogated by hyperpolarized imaging are shown in Figure 22.7. Pyruvate flux to lactate and alanine, shown in Figure 22.8, has been most frequently studied by hyperpolarized MR imaging. Pyruvate occupies a key location in glycolysis and has a long-lived (long T_1) carbonyl carbon that has been used to monitor flux over a period of approximately 1 min after injection [10,13,17–20]. Other molecules that have been successfully hyperpolarized are lactate [10,18], succinate [5,13], glutamine [12], bicarbonate [21], and choline [47].

22.3.2 Pyruvate

Pyruvate occupies a key position in cellular metabolism; it is formed at the last step of glycolysis when its precursor, phosphoenolpyruvate, interacts with pyruvate kinase. A unique isoform of pyruvate kinase is also expressed in tumor cells, with apparent kinetic influences on the rate of glycolysis. Tumors have a disproportionate amount of the dimeric type M2 isoform of pyruvate kinase (PK-M2) [50–52]. Aside from this distinction, enzyme activities are known to vary widely among tumors, and this is potentially related to proliferative capacity. Measuring flux through these pathways directly with hyperpolarized MR

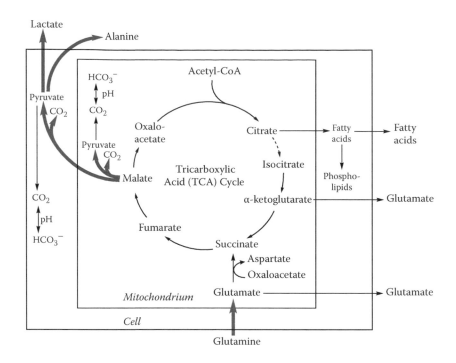

FIGURE 22.7 Biochemical pathways of cellular energetics. Pathways elevated in cancer are in bold gray. Additional details about regulating enzymes are provided in Figures 22.8, 22.10, and 22.11.

could enable more precise staging and provide more specific markers of therapy than those currently available.

Early studies have indeed shown that hyperpolarized pyruvate images can distinguish tissue types [10]. When hyperpolarized $^{13}C_1$-pyruvate is injected, in theory, $^{13}C_1$-lactate, $^{13}C_1$-alanine, $H^{13}CO_3$, and $^{13}CO_2$ will be produced, as outlined in Figure 22.8. In practice, the conversion of pyruvate to alanine and lactate has been observed. The products seen in MR images show clear patterns [10,18]; in P22 tumor tissue, a much larger percentage of pyruvate is converted to lactate relative to alanine production, whereas similar abundances of lactate and alanine are observed for skeletal muscle [10]. Volumes that included the vena cava showed traces of alanine and lactate, but the spectra

were dominated by pyruvate as would be expected for volumes composed primarily of blood. Residual by-products may represent contamination from surrounding active metabolic tissue. Nevertheless, the SNR was sufficiently high, and patterns were sufficiently different to afford facile identification of tumor tissue.

22.3.3 Succinate

Nontoxic hyperpolarized $^{13}C_1$-succinate-$d_{2,3}$ is the most developed PASADENA metabolic agent, with routinely obtained hyperpolarization in excess of 18% [5,25]. It can potentially assess the *in vivo* activity of succinate dehydrogenase (SDH),

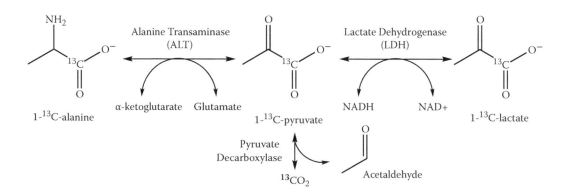

FIGURE 22.8 Metabolic pathways of 1-^{13}C-pyruvate catabolism to 1-^{13}C-lactate, 1-^{13}C-alanine, and $^{13}CO_2$. Conversion of ^{13}C label is shown by bold text. Note that $^{13}CO_2$ could be rapidly exchanged to $H^{13}CO_3^-$.

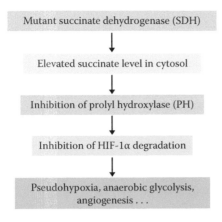

FIGURE 22.9 Effect of SDH mutations in human cells [54]. A series of regulatory steps leads to hypoxia and onset of cancer. SDH therefore plays the role of an oncogene.

the enzyme that was recently tagged as an oncogene due to its crucial role in cell energetics [53,54]. Therefore, it could be possible to detect the deficiency of SDH on the cellular level, which could potentially serve as a more efficient tool for cancer diagnosis and treatment monitoring, especially in early stages and recurrence. These mutations in SDH result in the increase of mitochondrial succinate pool that eventually leaks to the cytosol, where it inhibits prolyl hydroxylase (Figure 22.9). The reduced level of prolyl hydroxylation stabilizes otherwise constantly depleting hypoxia-inducible factor HIF-1α causing pseudohypoxia and hypoxia-inducible factor induction allowing tumor proliferation by means of anaerobic glycolysis. It is anticipated that ultrafast localized and nonlocalized spectroscopy utilizing hyperpolarized ^{13}C-succinate could detect the appearance of key early products of the tricarboxylic acid (TCA) cycle such as fumarate and malate (Figure 22.10), as well as other mitochondrial and cytosolic pathways utilizing TCA cycle intermediates.

Early results [22,25] demonstrate that metabolic profiles of cancer cells are distinct from noncancer cells with $^{13}C_1$-succinate as a contrast agent. Based on the direct link between TCA cycle dysfunction due to impaired SDH transmembrane assembly (as displayed in Figure 22.9), it would be possible to identify metabolic markers for dysfunctional SDH complex and

$$^{13}CO_2 + H_2O \xrightleftharpoons[\text{Anhydrase}]{\text{Carbonic}} {}^{13}CO_3^- + H^+$$

FIGURE 22.11 Bicarbonate and carbon dioxide exchange catalyzed by carbonic anhydrase. Conversion of ^{13}C label is shown by bold text. The ratiometric measurements of hyperpolarized bicarbonate and carbon dioxide allow for noninvasive *in vivo* pH imaging of tumors [21].

provide an opportunity to detect genetic mutations in action. When translated to the clinical environment, these new metabolic markers could also enable detection of the spatial distribution of such genetic mutations by means of fast MRSI, providing an opportunity to image active genetic mutations in humans. Such diagnostic methods have utility beyond diagnosing deficient SDH complexes *in vivo*. They also have potential to guide surgery, radiation therapy, or gene therapy and are particularly well suited towards monitoring treatment response due to the absence of radioactive tracers.

22.3.4 Imaging of Tumor pH

Extracellular and intracellular pH is tightly regulated by carbonic anhydrase (Figure 22.11). The catalytic activity of carbonic anhydrase is very high and usually diffusion limited. Several human pathologies are associated with significant pH changes. For example, many tumors have significantly lower (pH < 7) pH values compared with healthy cells (pH = 7.4). Measuring pH *in vivo* noninvasively by MRI is an attractive approach for cancer diagnosis and probably even more important as a predictor of response to treatment [56,57]. Several translational approaches have been developed for pH imaging of human cancer in animal models. The first group of methods relies on conventional CSI with contrast that arises from pH-sensitive proton chemical shifts [58]. While chemical shifts provide accurate measures of extracellular pH, the main challenge for clinical translation is relatively low sensitivity associated with low agent concentrations and relatively fast renal clearance of these pH-sensing agents.

An attractive alternative recently demonstrated by Brindle et al. is to measure the ratio of bicarbonate and CO$_2$ using injected ^{13}C hyperpolarized bicarbonate as a source of the signal [21]. The ratio of these two metabolites maintains blood and

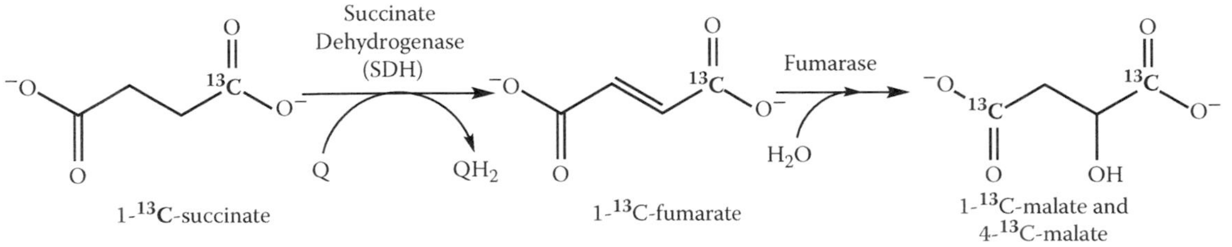

FIGURE 22.10 Metabolic pathway of 1-^{13}C-succinate metabolism in TCA cycle to 1-^{13}C-fumarate, 1-^{13}C-malate, and 4-^{13}C-malate. Conversion of ^{13}C label is shown by bold text.

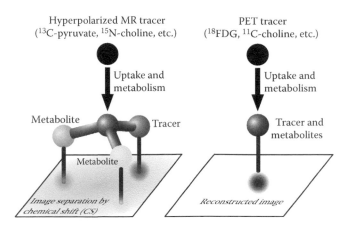

FIGURE 22.12 Metabolic tracers in hyperpolarized MRI and PET. PET provides information about tracer uptake, because the injected tracer and its metabolite(s) cannot be fundamentally distinguished, while hyperpolarized MR offers multiple nuclear properties that can be used to yield images corresponding to the injected tracer and its metabolite(s). Image separation using chemical shift (CS) is provided as an example.

extracellular pH, and pH maps can be reconstructed from MRSI measurements of hyperpolarized bicarbonate and CO_2 using a Henderson–Hasselbalch equation [21]:

$$pH = pK_a + \log_{10}\left(\frac{\left[HCO_3^-\right]}{\left[CO_2\right]}\right) \qquad (22.5)$$

The ratio of the bicarbonate and CO_2 concentrations was obtained as a ratio of their hyperpolarized intensities from CSI voxels. Although the ^{13}C T_1 of fast-exchanging bicarbonate and CO_2 is only 10 s, this was sufficient to acquire hyperpolarized CSI images in mice 10 s after injection of the contrast agents. This approach may potentially become a valuable clinical tool if T_1 hyperpolarization losses will allow imaging with sufficient SNR in humans, where blood circulation is significantly slower compared with small rodents.

22.3.5 ^{15}N-Choline

^{11}C-choline PET augmented by computed tomography (CT) allows for the noninvasive monitoring of response to treatment and is advantageous in some types of cancer. For example, the use of choline-PET over fluorodeoxyglucose (FDG)-PET is advantageous in the prostate where the urinary excretion of radioactive FDG is significant enough to interfere with the imaging of pelvic tumors [59]. Moreover, choline-PET versus FDG-PET serves as a marker of cell proliferation because it measures the elevated uptake of choline necessary for membrane formation in rapidly dividing cancer cells. This is a fundamentally different process from the elevated rate of glycolysis, which is interrogated by FDG-PET. Nevertheless, PET-CT technologies are inherently limited to imaging tissue uptake of radioactive functional contrast agents and cannot directly observe the metabolic events of chemical conversion, as manifest in metabolite flux in cellular energetic and other pathways. This is possible because MR has a capability to multiplex in metabolite and spatial domains, which is missing in PET (Figure 22.12) due to lack of chemical specificity. PET-CT could therefore be complemented by imaging the conversion of hyperpolarized nonradioactive ^{15}N-choline using the parahydrogen-based method of NMR hyperpolarization. The use of ^{15}N hyperpolarized choline potentially allows for noninvasively imaging of choline kinase (ChoK) activity, a measure that is inaccessible from PET-based choline uptake images (Figure 22.13). ChoK is overexpressed in several human cancer cell lines [60]; therefore, ChoK imaging could serve as a metabolic marker useful in diagnosis and treatment monitoring of many other cancers besides prostate cancer, where choline-PET is the most useful. ChoK activity is also enhanced by hypoxia-inducing factor HIF-1α, *ras*, *src*, *mos*, and *raf* oncogenes, shown in Figure 22.14 [61]. As a result of these relatively recent findings, the search for potent ChoK inhibitors is an active and attractive area of cancer chemotherapy. Therefore, ^{15}N hyperpolarized choline [34,47] would be a useful imaging approach in identification of potent therapeutic targets in cellular and rodent preclinical models of human cancer. Moreover, when translated to the clinical environment, ^{15}N hyperpolarized choline would provide (1) images of choline uptake and ChoK activity complimentary to FDG-PET without additional radiation exposure to the patient; (2) information about treatment efficacy of novel therapeutic agents targeting inhibiting ChoK or oncogenes involved in ChoK activity and expression upregulation; and (3) real-time metabolic imaging of ChoK, which may ultimately allow oncologists to adjust the dose of treatment designed to inhibit ChoK and/or related oncogenes and quickly identify nonresponders.

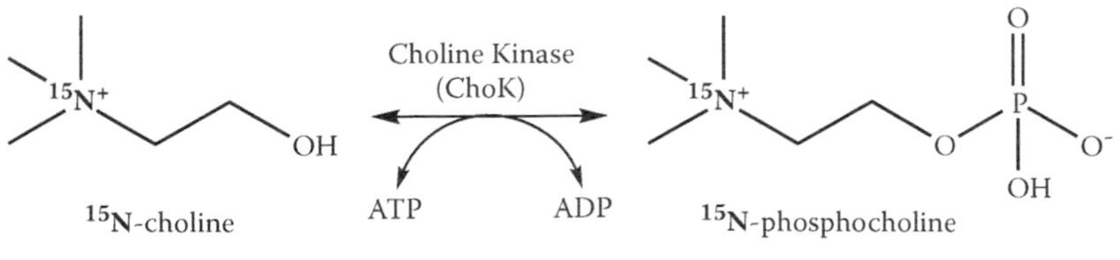

FIGURE 22.13 The first step of ^{15}N-choline metabolism is phosphorylation by ChoK to ^{15}N-phosphocholine.

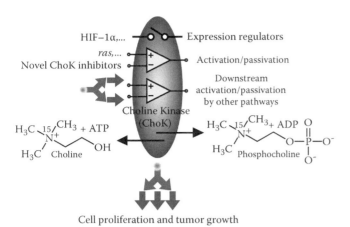

FIGURE 22.14 Regulation of ChoK expression and activity by oncogenes, factors, and inhibitors. The choline to phosphocholine flux is regulated by ChoK. The flux imaging could serve as a measure of cell division and proliferation useful for cancer imaging.

22.4 Applications in Cancer Therapy

22.4.1 Monitor the Efficacy of Chemotherapy, Gene Therapy, and Radiation Therapy

Tumor response to treatment is largely evaluated by anatomical imaging and therefore lags the functional response by several weeks. PET is used most frequently for clinical evaluation of response to treatment, as the measurements of contrast agents' uptake lead changes in disease status. Conventional MRS has also been used to empirically define several key metabolic biomarkers for cancer *in vivo*. For example, peaks associated with lactate and choline increase as tumor status changes and as tumors respond to therapy [62–66]. However, conventional proton imaging reports on steady-state concentrations averaged across several minutes to hours, due to the inherently low sensitivity of traditional MR acquisitions. Hyperpolarized MRI traces the fate of the injected molecular agent by detecting metabolite flux and enzymatic activity on a much shorter time scale, an advantage that is particularly well suited toward confirming the mechanism of action. As a result, after further development and refinement, hyperpolarized metabolic tracers potentially offer a complementary approach or alternative to PET imaging.

So far, only hyperpolarized $^{13}C_1$-pyruvate has been used to monitor the response to treatment in an animal model of human lymphoma [18]. $^{13}C_1$-pyruvate was injected intravenously in these experiments, and the synthesis of $^{13}C_1$-lactate was recorded up to 60 s after injection of the metabolic contrast agent, with a temporal resolution of 2 s. Using DNP, Brindle et al. demonstrated that the metabolic activity of lactate dehydrogenase responsible for pyruvate-to-lactate conversion is reduced 24 h after treatment with etoposide. While etoposide treatment does not target lactate dehydrogenase specifically, the metabolic changes could be seen much earlier before tumor size decreased. It is therefore possible that the effect of other targeted chemotherapeutics could be detected much earlier as well. For example, the effect of ChoK

inhibitors could potentially be imaged upon reaching cancer cells using hyperpolarized choline, serving to report on the metabolic status of ChoK on a time scale of several seconds [47].

22.4.2 Comparison with PET and Clinical Translation Perspectives

The success of PET-CT builds on the use of radioactive molecular agents capable of depicting the spatial distribution of metabolic or biochemical activity in the body. PET-CT imaging is limited to integrated cellular uptake of radioactive cancer biomarkers. The main disadvantages of PET agents include (i) inevitable patient contact with ionizing radiation, (ii) expensive cyclotron equipment required in the production of the radioactive agents, and (iii) high operating costs and lengthy examinations affecting patient stress. While typical PET-CT scans last approximately 1 h, the entire procedure takes 2–3 h when accounting for the 45–60 min that is necessary for FDG uptake prior to scanning. Moreover, fasting is usually required prior to the examination, resulting in an unnecessary metabolic stress in patients, especially for those with cancer cachexia. When translated to clinical settings, hyperpolarized MR imaging potentially (i) provides the same or greater level of metabolic information, (ii) uses no ionizing/radioactive agents, (iii) has significantly reduced costs, and (iv) requires significantly less examination time. Hyperpolarized MRI has been demonstrated *in vivo* in animal models of cancer [18] and has the capability to fulfill all of the above. Hyperpolarized MRI differs from conventional MRI by achieving dramatically enhanced signals (10,000- to 100,000-fold improvement) of exogenous injectable contrast agents, leading to accelerated data acquisition and allowing real-time metabolic imaging. A fundamental advantage of MR is that chemical shifts can discern between metabolic input and product in biochemical pathways.

References

1. Rabi II, Millman S, Kusch P, Zacharias JR. The molecular beam resonance method for measuring nuclear magnetic moments. The magnetic moments of $_3Li^6$, $_3Li^7$ and $_9F^{19}$. *Phys Rev* 1939;55(6):526–535.
2. Bloch F, Siegert A. Magnetic resonance for nonrotating fields. *Phys Rev* 1940;57(6):522–527.
3. Bowers CR, Weitekamp DP. Transformation of symmetrization order to nuclear-spin magnetization by chemical-reaction and nuclear-magnetic-resonance. *Phys Rev Lett* 1986;57(21):2645–2648.
4. Bowers CR, Weitekamp DP. Para-hydrogen and synthesis allow dramatically enhanced nuclear alignment. *J Am Chem Soc* 1987;109(18):5541–5542.
5. Chekmenev EY, Hovener J, Norton VA, Harris K, Batchelder LS, Bhattacharya P, Ross BD, Weitekamp DP. PASADENA hyperpolarization of succinic acid for MRI and NMR spectroscopy. *J Am Chem Soc* 2008;130(13):4212–4213.

6. Eisenschmid TC, Kirss RU, Deutsch PP, Hommeltoft SI, Eisenberg R, Bargon J, Lawler RG, Balch AL. Para hydrogen induced polarization in hydrogenation reactions. *J Am Chem Soc* 1987;109(26):8089–8091.

7. Abragam A. *The Principles of Nuclear Magnetism.* Oxford: Clarendon Press; 1961.

8. Adams RW, Aguilar JA, Atkinson KD, Cowley MJ, Elliott PIP, Duckett SB, Green GGR, Khazal IG, Lopez-Serrano J, Williamson DC. Reversible interactions with para-hydrogen enhance NMR sensitivity by polarization transfer. *Science* 2009;323(5922):1708–1711.

9. Ardenkjaer-Larsen JH, Fridlund B, Gram A, Hansson G, Hansson L, Lerche MH, Servin R, Thaning M, Golman K. Increase in signal-to-noise ratio of > 10,000 times in liquid-state NMR. *Proc Natl Acad Sci U S A* 2003;100(18): 10158–10163.

10. Golman K, in't Zandt R, Lerche M, Pehrson R, Ardenkjaer-Larsen JH. Metabolic imaging by hyperpolarized C-13 magnetic resonance imaging for in vivo tumor diagnosis. *Cancer Res* 2006;66(22):10855–10860.

11. Bowers CR, Weitekamp DP. Transformation of symmetrization order to nuclear-spin magnetization by chemical reaction and nuclear magnetic resonance. *Phys Rev Lett* 1986;57(21):2645–2648.

12. Gallagher FA, Kettunen MI, Day SE, Lerche M, Brindle KM. C-13 MR spectroscopy measurements of glutaminase activity in human hepatocellular carcinoma cells using hyperpolarized C-13-labeled glutamine. *Magn Reson Med* 2008;60(2):253–257.

13. Ross BD, Bhattacharya P, Wagner S, Tran T, Sailasuta N. Hyperpolarized MR imaging: Neurologic applications of hyperpolarized metabolism. *Am J Neuroradiol* 2010;31(1):24–33.

14. Carver TR, Slichter CP. Polarization of nuclear spins in metals. *Phys Rev* 1953;92(1):212–213.

15. Carver TR, Slichter CP. Experimental verification of the Overhauser nuclear polarization effect. *Phys Rev* 1956;102(4):975–980.

16. Abragam A, Goldman M. Principles of dynamic nuclear-polarization. *Rep Prog Phys* 1978;41(3):395–467.

17. Cunningham CH, Chen AP, Albers MJ, Kurhanewicz J, Hurd RE, Yen YF, Pauly JM, Nelson SJ, Vigneron DB. Double spin-echo sequence for rapid spectroscopic imaging of hyperpolarized C-13. *J Magn Reson* 2007;187(2): 357–362.

18. Day SE, Kettunen MI, Gallagher FA, Hu DE, Lerche M, Wolber J, Golman K, Ardenkjaer-Larsen JH, Brindle KM. Detecting tumor response to treatment using hyperpolarized C-13 magnetic resonance imaging and spectroscopy. *Nat Med* 2007;13(11):1382–1387.

19. Golman K, in't Zandt R, Thaning M. Real-time metabolic imaging. *Proc Natl Acad Sci U S A* 2006;103(30):11270–11275.

20. Yen YF, Kohler SJ, Chen AP, Tropp J, Bok R, Wolber J, Albers MJ, Gram KA, Zierhut ML, Park I, Zhang V, Hu S, Nelson SJ, Vigneron DB, Kurhanewicz J, Dirven H, Hurd RE. Imaging considerations for in vivo C-13 metabolic mapping using hyperpolarized C-13-pyruvate. *Magn Reson Med* 2009;62(1):1–10.

21. Gallagher FA, Kettunen MI, Day SE, Hu DE, Ardenkjaer-Larsen JH, in't Zandt R, Jensen PR, Karlsson M, Golman K, Lerche MH, Brindle KM. Magnetic resonance imaging of pH in vivo using hyperpolarized C-13-labelled bicarbonate. *Nature* 2008;453(7197):U940–U973.

22. Bhattacharya P, Chekmenev EY, Perman WH, Harris KC, Lin AP, Norton VA, Tan CT, Ross BD, Weitekamp DP. Towards hyperpolarized 13C-succinate imaging of brain cancer. *J Magn Reson* 2007;186:150–155.

23. Chekmenev EY, Norton VA, Weitekamp DP, Bhattacharya P. Hyperpolarized (1)H NMR employing low gamma nucleus for spin polarization storage. *J Am Chem Soc* 2009;131(9):3164–3165.

24. Hövener J-B, Chekmenev E, Harris K, Perman W, Robertson L, Ross B, Bhattacharya P. PASADENA hyperpolarization of 13C biomolecules: Equipment design and installation. *Magn Reson Mat Phys Biol Med* 2009;22: 111–121.

25. Hövener J-B, Chekmenev E, Harris K, Perman W, Tran T, Ross B, Bhattacharya P. Quality assurance of PASADENA hyperpolarization for 13C biomolecules. *Magn Reson Mat Phys Biol Med* 2009;22:123–134.

26. Goldman M, Johannesson H, Axelsson O, Karlsson M. Hyperpolarization of C-13 through order transfer from parahydrogen: A new contrast agent for MFI. *Magn Reson Imaging* 2005;23(2):153–157.

27. Goldman M, Johannesson H, Axelsson O, Karlsson M. Design and implementation of C-13 hyperpolarization from para-hydrogen, for new MRI contrast agents. *C R Chimie* 2006;9(3–4):357–363.

28. Aime S, Gobetto R, Reineri F, Canet D. Hyperpolarization transfer from parahydrogen to deuterium via carbon-13. *J Chem Phys* 2003;119(17):8890–8896.

29. Golman K, Axelsson O, Johannesson H, Mansson S, Olofsson C, Petersson JS. Parahydrogen-induced polarization in imaging: Subsecond C-13 angiography. *Magn Reson Med* 2001;46(1):1–5.

30. Reineri F, Viale A, Giovenzana G, Santelia D, Dastru W, Gobetto R, Aime S. New hyperpolarized contrast agents for C-13-MRI from para-hydrogenation of oligooxyethylenic alkynes. *J Am Chem Soc* 2008;130(45): 15047–15053.

31. Chan HR, Bhattacharya P, Imam A, Freundlich A, Tran TT, Perman WH, Lin AP, Harris K, Chekmenev EY, Ingram M-L, Ross BD. No clinical toxicity is seen in vivo from hyperpolarized PASADENA MR reagents or catalyst. 17th ISMRM Conference, April 18–24. Berkeley, CA: Honolulu, Hawaii; 2009.

32. Goldman M, Johannesson H. Conversion of a proton pair para order into C-13 polarization by rf irradiation, for use in MRI. *C R Physique* 2005;6(4–5):575–581.

33. Bhattacharya P, Harris K, Lin AP, Mansson M, Norton VA, Perman WH, Weitekamp DP, Ross BD. Ultra-fast three dimensional imaging of hyperpolarized C-13 in vivo. *Magn Reson Mat Phys Biol Med* 2005;18(5):245–256.

34. Bhattacharya P, Wagner SR, Chan HR, Chekmenev EY, Perman WH, Ross BD. Hyperpolarized 15N MR: PASADENA & DNP. 17th ISMRM Conference, April 18–24. Berkeley, CA: Honolulu, Hawaii; 2009.

35. Ruset IC, Ketel S, Hersman FW. Optical pumping system design for large production of hyperpolarized Xe-129. *Phys Rev Lett* 2006;96(5).

36. Zook AL, Adhyaru BB, Bowers CR. High capacity production of > 65% spin polarized xenon-129 for NMR spectroscopy and imaging. *J Magn Reson* 2002;159(2):175–182.

37. Cherubini A, Payne GS, Leach MO, Bifone A. Hyperpolarising C-13 for NMR studies using laser-polarised Xe-129: SPINOE vs thermal mixing. *Chem Phys Lett* 2003; 371(5–6):640–644.

38. Spence MM, Rubin SM, Dimitrov IE, Ruiz EJ, Wemmer DE, Pines A, Yao SQ, Tian F, Schultz PG. Functionalized xenon as a biosensor. *Proc Natl Acad Sci U S A* 2001;98(19):10654–10657.

39. Schroder L, Lowery TJ, Hilty C, Wemmer DE, Pines A. Molecular imaging using a targeted magnetic resonance hyperpolarized biosensor. *Science* 2006;314(5798):446–449.

40. Lisitza N, Muradian I, Frederick E, Patz S, Hatabu H, Chekmenev EY. Toward C-13 hyperpolarized biomarkers produced by thermal mixing with hyperpolarized Xe-129. *J Chem Phys* 2009;131(4):5.

41. Long HW, Gaede HC, Shore J, Reven L, Bowers CR, Kritzenberger J, Pietrass T, Pines A, Tang P, Reimer JA. High-field cross-polarization NMR from laser-polarized xenon to a polymer surface. *J Am Chem Soc* 1993;115(18):8491–8492.

42. Pawsey S, Moudrakovski I, Ripmeester J, Wang LQ, Exarhos GJ, Rowsell JLC, Yaghi OM. Hyperpolarized Xe-129 nuclear magnetic resonance studies of isoreticular metal-organic frameworks. *J Phys Chem C* 2007;111(16):6060–6067.

43. Bowers CR, Long HW, Pietrass T, Gaede HC, Pines A. Cross polarization from laser-polarized solid xenon to (CO2)-C-13 by low-field thermal mixing. *Chem Phys Lett* 1993;205(2–3):168–170.

44. Goldman M. *Spin Temperature and Nuclear Magnetic Resonance in Solids.* Oxford: Oxford University Press; 1970.

45. Fitzgerald RJ, Sauer KL, Happer W. Cross-relaxation in laser-polarized liquid xenon. *Chem Phys Lett* 1998;284(1–2):87–92.

46. Navon G, Song YQ, Room T, Appelt S, Taylor RE, Pines A. Enhancement of solution NMR and MRI with laser-polarized xenon. *Science* 1996;271(5257):1848–1851.

47. Gabellieri C, Reynolds S, Lavie A, Payne GS, Leach MO, Eykyn TR. Therapeutic target metabolism observed using hyperpolarized N-15 choline. *J Am Chem Soc* 2008; 130(14):4598–4599.

48. Pileio G, Carravetta M, Hughes E, Levitt MH. The long-lived nuclear singlet state of 15N-nitrous oxide in solution. *J Am Chem Soc* 2008;130(38):12582–12583.

49. Morris GA, Freeman R. Enhancement of nuclear magnetic-resonance signals by polarization transfer. *J Am Chem Soc* 1979;101(3):760–762.

50. Christofk HR, Vander Heiden MG, Harris MH, Ramanathan A, Gerszten RE, Wei R, Fleming MD, Schreiber SL, Cantley LC. The M2 splice isoform of pyruvate kinase is important for cancer metabolism and tumour growth. *Nature* 2008;452(7184):U230–U274.

51. Christofk HR, Vander Heiden MG, Wu N, Asara JM, Cantley LC. Pyruvate kinase M2 is a phosphotyrosine-binding protein. *Nature* 2008;452(7184):U181–U127.

52. Spoden GA, Rostek U, Lechner S, Mitterberger M, Mazurek S, Zwerschke W. Pyruvate kinase isoenzyme M2 is a glycolytic sensor differentially regulating cell proliferation, cell size and apoptotic cell death dependent on glucose supply. *Exp Cell Res* 2009;315(16):2765–2774.

53. Rustin P, Munnich A, Rotig A. Succinate dehydrogenase and human diseases: New insights into a well-known enzyme. *Eur J Hum Genet* 2002;10(5):289–291.

54. Selak MA, Armour SM, MacKenzie ED, Boulahbel H, Watson DG, Mansfield KD, Pan Y, Simon MC, Thompson CB, Gottlieb E. Succinate links TCA cycle dysfunction to oncogenesis by inhibiting HIF-alpha prolyl hydroxylase. *Cancer Cell* 2005;7(1):77–85.

55. Chekmenev EY, Bhattacharya P, Ross BD. *Development of Hyperpolarized Metabolic Contrast Agents Using PASADENA. Application Note #21.* Andover, MA: Cambridge Isotope Laboratories; 2008.

56. Gillies RJ, Raghunand N, Garcia-Martin ML, Gatenby RA. PH imaging. *IEEE Eng Med Biol Mag* 2004;23(5):57–64.

57. Robey IF, Baggett BK, Kirkpatrick ND, Roe DJ, Dosescu J, Sloane BF, Hashim AI, Morse DL, Raghunand N, Gatenby RA, Gillies RJ. Bicarbonate increases tumor pH and inhibits spontaneous metastases. *Cancer Res* 2009;69(6): 2260–2268.

58. van Sluis R, Bhujwalla ZM, Raghunand N, Ballesteros P, Alvarez J, Cerdan S, Galons JP, Gillies RJ. In vivo imaging of extracellular pH using H-1 MRSI. *Magn Reson Med* 1999;41(4):743–750.

59. Hara T, Kosaka N, Kishi H. PET imaging of prostate cancer using carbon-11-choline. *J Nucl Med* 1998;39(6):990–995.

60. de Molina AR, Rodriguez-Gonzalez A, Gutierrez R, Martinez-Pineiro L, Sanchez JJ, Bonilla F, Rosell R, Lacal JC. Overexpression of choline kinase is a frequent feature in human tumor-derived cell lines and in lung, prostate, and colorectal human cancers. *Biochem Biophys Res Comm* 2002;296(3):580–583.

61. de Molina AR, Gallego-Ortega D, Sarmentero J, Banez-Coronel M, Martin-Cantalejo Y, Lacal JC. Choline kinase is a novel oncogene that potentiates RhoA-induced carcinogenesis. *Cancer Res* 2005;65(13):5647–5653.

62. Eliyahu G, Kreizman T, Degani H. Phosphocholine as a biomarker of breast cancer: Molecular and biochemical studies. *Int J Cancer* 2007;120(8):1721–1730.

63. Furman-Haran E, Margalit R, Grobgeld D, Degani H. Dynamic contrast-enhanced magnetic resonance imaging reveals stress-induced angiogenesis in MCF7 human breast tumors. *Proc Natl Acad Sci U S A* 1996;93(13):6247–6251.

64. Neeman M, Eldar H, Rushkin E, Degani H. Chemotherapy-induced changes in the energetics of human breast-cancer cells — P-31-NMR and C-13-NMR studies. *Biochim Biophys Acta* 1990;1052(2):255–263.

65. Rivenzon-Segal D, Boldin-Adamsky S, Seger D, Seger R, Degani H. Glycolysis and glucose transporter 1 as markers of response to hormonal therapy in breast cancer. *Int J Cancer* 2003;107(2):177–182.

66. Rivenzon-Segal D, Margalit R, Degani H. Glycolysis as a metabolic marker in orthotopic breast cancer, monitored by in vivo C-13 MRS. *Am J Physiol Endocrinol Metab* 2002;283(4):E623–E630.

Index

Page numbers followed by italicized f and t indicate figures and tables, respectively.

T - #0639 - 071024 - C8 - 276/216/20 - PB - 9780367576875 - Gloss Lamination